Tribochemistry

Tribochemistry

by

Prof. Dr. rer. nat. habil. GERHARD HEINICKE †
Berlin

in co-operation with

Dr. rer. nat. Hans-Peter Hennig
Dr. rer. nat. Eberhard Linke
Prof. Dr. rer. nat. habil. Ursula Steinike
Dr. rer. nat. Karsten-Peter Thiessen

and with a preface by

Prof. Dr. phil. habil. Peter-Adolf Thiessen

With 329 Figures and 106 Tables

AKADEMIE-VERLAG · BERLIN

1984

Paragraph 3.2. was written under the leadership of U. STEINIKE, paragraph 3.3. under the leadership of E. LINKE, paragraph 6.6. under the leadership of K.-P. THIESSEN, and paragraph 6.1.2. under the leadership of H.-P. HENNIG and U. STEINIKE. The paragraphs 6.1.3. and 7.2.2.4. were compiled in co-operation with R. PAUDERT and A. HARENZ and the paragraphs 4.7., 6.1.1. and chapter 8 in co-operation with H.-P. HENNIG.

The translation was performed by B. KLEU and revised by P. ROPER.

Erschienen im Akademie-Verlag, DDR - 1086 Berlin, Leipziger Straße 3—4

Lizenznummer: 202 · 100/489/84
Printed in the German Democratic Republic
Gesamtherstellung: VEB Druckhaus „Maxim Gorki", 7400 Altenburg
LSV 1215
Bestellnummer: 763 199 3 (6746)
09800

Table of contents

1. Preface

Tribochemistry deals with the relations between mechanical work and mass transformation. The results of this interaction can be found in various forms throughout nature and in human activity. They are so well-known that they are mostly accepted without further consideration.

A reversar of the question shows the problems. What expenditure is needed to exclude chemical influences occurring during the mechanical interaction of one solid upon another? The experimental difficulties here are very great. It is desirable that highly pure solids mutually influence their structures upon application of mechanical work without causing more than mixed phases in solid or liquid state. At the same time, the demands on the atmosphere, that is to say the environment, would not even be met by inert gases. Desirable would be a vacuum exceeding the emptiness of the universal space.

Seen in terms of the subject and content of the present monograph, such a consideration could seem to be redundant and remote. In fact, it is just as useful through its abstraction as developments of pure mathematics beyond the hitherto existing level. The abstraction can systematically be charged by boundary conditions with the aim of constructing different real cases. Through this, the directed co-ordination of elementary processes leads to defined integrals.

In comparison to this construction, the reality, i.e. the real processes, appear as an extremely complicated and complex accumulation of parameters. This applies both for natural processes as well as for the mode of acting of human practice in the field of tribomechanics. A separation of the complicated integrals into partial ranges has proved its worth, which have as clear an arrangement as possible and which are heuristically usable. The approximation of both modes of procedure will finally lead to explanations, experimental verifications, and ultimately productive application.

Experience shows that the great abundance of phenomena are very irregularly developed in terms of knowledge, the resolving of contradictions, and in application. Some areas have become clear by virtue of the knowledge of molecular mechanics whilst others are partically or completely lacking in this.

A series of least contradiction models is proposed. Their ranges may be far apart from each other. Borderline cases may for instance be processes of a steady-state chain of extremely unsteady-state pulses with a high energy density. The transformation of energy for the sequence of mass transformations can be represented as "Impact chemistry" according to a triboplasma model. In contrast to this, processes influenced by statistically stable, and for a long time constant, vibrations in the molecular structure, are under the influence of elastic deformations by vibrations.

Much progress can be seen in comparison to the 1967 monograph „Grundlagen der Tribochemie". New experiences and knowledge explain complicated relations better than hitherto. Moreover they permit a broader and deeper theoretical treatment, experimental basis and technological application of selective developments.

In line with the current efficiency of capable and proven models and productive demands, the present monograph does not deal uniformly with the different fields of tribochemistry. The dissymmetry does not consist in the nature of the matter but solely in the degree of deepened and broader theoretical understanding. Its application should never cease to conscientiously define the boundaries of the known and to stimulate their broadening. In terms of elasticity and modulation capability, Tribochemistry can be particularly well adapted to the demands of the hierarchy: "Sensor-microprocessor/operator-reactor".

The present monograph should and will serve the further and accelerated opening up of these relations.

P. A. Thiessen

Survey-References

[1] Thiessen, P. A., Meyer, K., Heinicke, G.: „Grundlagen der Tribochemie", Abh. dtsch. Akad. Wiss., Kl. Chem. Geol. u. Biol. Nr. 1 (1966)

[2] Kubo, T.: "Mechanochemistry", Tokio 1971

[3] Barambojm, N. K.: "Mechanochimija vysokomolekuljarnych soedinenij", Moskva 1971

[4] Boldyrev, V. V., Meyer, K.: „Festkörperchemie", p. 497–560, Leipzig 1973

[5] 1. Simpozium „Mechanochimičeskie i mechanoėmissionnye processy v tverdych telach", Voronež 1968, Tezisy, Voronež 1968

[6] 2. Simpozium „Mechanoėmissija i Mechanochimija tverdych tel", Frunze 1969, Materialy, Frunze 1974

[7] 3. Symposium „Triboemission und Tribochemie", Berlin 1971, Thesen, Berlin 1971

[8] 4. Simpozium „Mechanoėmissija i Mechanochimija tverdych tel", Irkutsk 1973, Tezisy, Moskva 1973

[9] 5. Simpozium „Mechanoėmissija i Mechanochimija tverdych tel", Tallin 1975, Materialy, Tallin 1977

[10] 6. Symposium „Mechanoėmission und Mechanochemie", Berlin 1977, Thesen, Berlin 1977

[11] 7. Simpozium „Mechanoėmissija i Mechanochimija" Taškent 1979, Tezisy, Taškent 1979, Doklady Taškent 1981

[12] Avvakumov, E. G.: „Mechaničeskie metody aktivacii chimičeskich processov", Novosibirsk 1979

[13] 8. Simpozium „Mechanoėmissija i Mechanochimija tverdych tel", Tallin 1981, Tezisy, Tallin 1981

[14] 9. Symposium „Mechanoemission und Mechanochemie", Berlin 1983, Kurzreferate, Berlin 1983

2. Introduction

In physical chemistry the individual scientific branches are subdivided according to the kind of energy used for the initiation of chemical reactions. These are inter alia thermochemistry, electrochemistry, magnetochemistry, photochemistry, and radiation chemistry. As early as the beginning of this century W. Ostwald had pointed out that a field of research should also belong to this systematic dealing with the effect of mechanical energy on chemical reactions [1]. At that time this field, which he called "mechanochemistry", was virtually unexplored so that Ostwald was not able to say anything on the independent character and the importance of this field of chemistry.

In the following a short historical review will be provided on the development of this field of science which has only recently been dealt.

It is one of the oldest experiences of mankind that chemical reactions are released by the application of mechanical energy, e.g. by friction or impact. Thus already in prehistoric times, men had used flints to make fire. During the 17th and 18th centuries, lighter-flints played an important role in the firing of weapons. In recent times numerous synthetically produced substances are used for these purposes in the form of initiators susceptible to shocks. Some substances, such as iodine nitrogen, react so sensitively to mechanical influences that they detonate upon the slightest contact with a pigeon's feather [2].

As regards the corrosion of iron, it was already found in the 18th century that the dissolution of metal is greatly accelerated by the action of friction on the surface of the metal [3].

In the field of geology, modification transformations under the influence of geotectonic movements have similarity been long-known for example the micatization of feldspar due to the deformation which took place in geological epochs [4].

By means of the decomposition of halides of silver, gold, platinum and mercury it was shown by Carey-Lea in a first systematic paper at the end of the last century how besides thermal energy, light and electricity, mechanical energy may also initiate chemical reactions [5]. He could easily decompose mercury chloride into its components by trituration in a mortar. Since this substance sublimes undecomposed upon heating the local heating was excluded by Carey-Lea as the cause of the initiation of the reaction. The silver halides also melt undecomposed at red heat, whereas under the influence of a small shear stress a partial decomposition takes place accompanied by the formation of metal [5]. At the beginning of this century Parker conducted double reactions between simple salts influenced by frictional stress. He regarded the heat developing locally under friction as the cause of the release of this reaction [6].

In the 1920s TAMMAN studied the changes in solid reactivity caused by mechanical treatment. He found that not all the mechanical energy is transformed into heat but 5 ··· 15 per cent of the expended energy remains in the metal as potential energy and increases the thermodynamic potential of the solid. As a consequence of the treatment a displacement of the thermodynamic potential and a significant increase of the dissolution velocity occurs (see 7.2.2.2.), [7].

In the subsequent 3 decades, FINK [8] and BOWDEN and TABOR [9] in particular, investigated oxidation reactions of metal and decomposition reactions under conditions of rolling and sliding friction. Thus during the rolling friction of iron rolls on the stressed surface tarnishing layers come into existence within a few minutes, which without mechanical stress would normally require about 10^{17} years for their growth. FINK excluded the influence of temperature for the interpretation of this effect and discussed plastic deformation as a cause of the acceleration of the reaction (see 7.4.4.) [10].

By contrast, linking up with the concept of PARKER, BOWDEN and TABOR developed the "hot-spot-model" for the mechanical initiation of chemical reactions. They found that with friction processes for $10^{-3} \cdots 10^{-4}$ s temperatures of more than 1,000 K can occur, and that these represent an important cause of mechanically initiated reactions (see 4.1. and 5.3.).

At about the same time the effect of mechanical energy on polymers was also investigated. STAUDINGER and HESS in particular, dealt with the mechanical degradation of high polymers, e.g. of polystyrene. They found an increase of the reactivity of polymers, e.g. of the solubility, parallel to the decrease of the molecular weight [11, 12]. It was evident from this that main valence bonds in a carbon chain can be broken by the impact of mechanical energy on polymers. The investigations in this field were continued by GROHN in the fifties [13] and are now being conducted on a major scale by BARAMBOIM (Chapt. 1., [3]), SIMIONESCU [14], and numerous other researchers (see 6.3.).

In the fifties PETERS investigated a whole series of inorganic reactions under the influence of mechanical energy, such as the decomposition and synthesis of carbonates, oxidation and reduction reactions as well as colour reactions for microanalysis. However he disputed the role of temperature in initiating mechanically activated reactions, taking the attitude that only such reactions were to proceed where conditions for reactions are given from the thermodynamic point of view [15].

However, according to investigations by HEINICKE the latter opinion does not hold for a number of mechanically activated reactions. He was able to demonstrate by means of numerous chemical reactions that the thermodynamics of reversible processes is not, or only conditionally applicable to mechanically activated reactions [16]. On the other hand the thermodynamics of irreversible processes is demonstrably for the description of these reactions (see 4.8.2. and 6.6.), even if in turn it is not universally applicable. At the same time he proved that mechanically activated reactions proceed according to kinetics which deviate in comparison with thermally activated reactions (see Chapt. 4.).

In order to clarify the contradictory opinions in the early sixties THIESSEN developed the "magma-plasma model" by starting out from the fact that directly upon mechanical intervention particularly highly excited states occur for a short time. Which inter alia lead to the separation of lattice components and free electrons (Chapt. 1., [1]), (see 3.3. and 5.1.). THIESSEN distinguished between reactions caused by plasmatic states and those proceeding on the surface of the solid in a strongly excited state or immediately after the latter's conclusion. From this resulted the important conclusion for the course of mechanically activated reactions: they do not proceed according to a uniform mechanism. The physical processes occurring with mechanical activation have proved so manifold (see 3.3.) that the number of possible mechanisms via which the initiation of a chemical reaction occurs is not known to a full extent even today.

A multistage pattern of energy dissipation was developed in the form of a "hierarchy" of energy states, which has attained great importance for the mechanical activation of chemical reactions (see Chapt. 4., [1]). According to this model 3 main groups of reactions can be distinguished:

To the 1st group belong stochastic reactions released by short-lived highly excited energy states, which not in a thermal equilibrium, e.g. in the impact plasma.

The 2 nd group of reactions are excited by so-called postplasmatic states which occur with different lifespans in the most varied dissipation processes and which already lead to the local establishment of the MAXWELL-BOLTZMANN distribution of energy. Phonon and dislocation migrations as well as the appearance of "hot spots" are important dissipation processes here. In this phase the chemical processes are directly coupled with the dissipation processes, thus permitting the application of irreversible thermodynamics.

Finally, to the 3 rd group belong those reactions which are initiated or accelerated by the energy stored in the solid in the form of lattice defects due to the dissipation of mechanical energy. If the lattice defects have a long lifespan compared to the velocity of chemical processes then the thermodynamics of reversible processes can be used for their description, using additional functions, for instance of excess free enthalpy (see 4.8.1.).

The reaction model for mechanically activated reactions presented here is based on an entire system of various causes of reaction. The weakness of older reaction models consisted in the fact that for the most part they only took into consideration one cause of reaction for the interpretation isolated from the overall complex of dissipation processes, which inevitably resulted in contradictions in the generalization of the model conceptions.

During the development of the foundations of this field of research, the real definitions also had to be adapted to the actual state of knowledge. The termin "mechanochemistry" was coined by OSTWALD for this field of research. The following definition, generally accepted today, does justice to his concern to classify the individual fields of chemistry according to the kind of acting energy [17–20]:

"*Mechanochemistry is a branch of chemistry dealing with the chemical and physico-chemical changes of substances of all states of aggregation due to the influence of mechanical energy*".

This field has been expanded in the most varied directions during the last decades. It also includes for instance the reactions released by cavitation in liquids, the shock wave reactions in gases, etc.

Mechanical influence plays a special part with regard to chemical reactions where at least one solid component is involved. The chief among these are reactions which occur with milling stress and friction, lubrication and abrasion. The term "tribochemistry" with the following definition has been coined for this special field [18, 21–23]:

"*Tribochemistry is a branch of chemistry dealing with the chemical and physico-chemical changes of solids due to the influence of mechanical energy.*"

During recent years a large number of further terms have been found for the description of sub-fields of science, such as tribocatalysis or triboelectrochemistry, and for the description of individual triboprocesses, such as tribosorption, tribodiffusion, and tribocorrosion, which are based accordingly on the above definition [18].

Parallel to this development in the mid-sixties under the leadership of JOST, tribology was created as a new field of science, and terms such as tribotechnology, tribomechanics, tribometallurgy were introduced [24]. Within the bounds of this field of science tribochemistry, like tribophysics, is a sub-field of tribology (see 7.4.2.).

These terms have proved to be particularly suitable for an international terminology and have contributed to a classification of the most varied processes occuring during mechanical treatment (e.g. Tab. 4.49.).

At the same time, the definition of numerous new terms in this field indicates that a stage of development has been reached which both requires and enables an analytical investigation of the dissipation process down to the individual elementary processes, for further clarification of the correlations.

BOLDYREV has contributed an essential part in the clarification of elementary processes in tribochemical reactions, particularly by emphasizing the role of highly excited vibration states in thermal non-equilibrium for the initiation of tribochemical reactions (see 5.3., [25].

AVVAKUMOV has been able to prove that tribochemical solid reactions run mainly according to diffusionless kinetics (see 5.6.5.), [26]. BUTJAGIN has investigated and generalized in particular the elementary processes leading to the rupture of polymer chains upon the relaxation of elastic energy (see 5.2. and 5.3.). The relations between the electric states of solids caused by treatment and their reactivity have been systematized by the fundamental studies of DERJAGIN and KROTOVA (see 5.4.). Factors releasing reactions in tribochemical explosion reactions have been mainly clarified by FOX [27]. SCHRADER investigated the correlation between structure and reactivity in mechanically pretreated solids, particularly with regard to technological application [28]. In all, such a variety of scientific papers have been published since the first edition of the first monograph in this field, „Grundlagen der Tribochemie", in 1967 that it is impossible now to refer completely to all the publications in this field. Knowledge of the relations in tribochemistry has expanded dramatically and permits already a detailed evaluation of the dissipation model and of the most important causes of reactions, depending on the type and intensity of the acting mechanical energy (see Chapt. 5.).

A great number of studies of the application of tribochemistry in technology have been published during the last 10 ··· 15 years [29—31]. Tribochemical reactions have become important in the chemical, metallurgical, and building industries and have led to the introduction of a number of tribochemical procedures to technology (see 7.2.). The manufacturing process for the high-grade building material "Silicalcit" developed by HINT (see 7.2.3.), [32] and the production of tribomechanically activated phosphor fertilizers TRIBOPHOS (see 7.2.2.2.4.) are examples of this. Today the development of effective activation sets, has mainly been advanced by BERNHARDT and HEEGN [33].

Tribochemical reactions play an important part in ignition processes in the engineering industry (see 7.3.) and now as before in the release of explosion reactions. Tribochemical reactions are also involved in material corrosion, e.g. in vibrational corrosion (see 6.6.). Not least their importance is also increasing in tribology in respect of the reduction of friction and abrasion (see 7.4.). For instance, the effect of the high-temperature lubricant developed by GRAUE has its origin in tribochemical reactions with the base being lubricated (see 7.4.4.1.6.). According to BÜCHNER, KEIL and MEYER tribochemical reactions also play an important part in the anti-abrasion effect of EP-additives (see 7.4.4.1.5.). Reference can also be made to the importance of tribochemistry in geochemical processes as well as in the emergence of life on earth (see Chapt. 8.). In spite of the precipitous development of tribochemistry during the last 20 years the extension of this field is still continuing. May this monograph stimulate this process and lead to a further improvement in the knowledge of the elementary processes of tribochemical reactions. An additional and specific aim of this monograph is to promote the application of scientific basis results in technology, to initiate the development of new processes, and thereby contribute to the fulfilling of high-priority economic tasks.

References

[1] OSTWALD, W.: Handbuch d. allg. Chemie, I, p. 70, Leipzig 1919

[2] HEINICKE, G.: Wissenschaft u. Fortschritt **14** (1964), 163

[3] RINMAN, S.: „Försöck till Järnets Historia med Tillämpning för Slögder och Handtwerk", Stockholm 1782

[4] SANDER, B.: Jahrbuch der geol. Bundesanstalt **73** (1923), 209

[5] CAREY-LEA, M.: Phil. Mag. **34** (1892), 46; **37** (1894), 31, 470

[6] TAMMANN, G.: Z. Elektrochem. **35** (1929), 21

[7] PARKER, L. H.: J. Chem. Soc., London **105** (1914), 1504; **113** (1918), 396

[8] FINK, M.: Trans. Am. Soc. f. Steel Treating **18** (1930), 1026

[9] BOWDEN, F. P. u. TABOR, D.: "The friction and lubrication of solids", Oxford 1958

[10] FINK, M., HOFMANN, U.: Z. anorg. allg. Chem. **210** (1933), 100

[11] STAUDINGER, H., DREHER, E.: Ber. dtsch. chem. Ges. **A 69** (1936), 1091

[12] HESS. K., STEURER, E., FROMM, H.: Kolloid-Z. **98** (1942), 148

[13] GROHN, H. und PAUDERT, R.: Wiss. Z. Techn. Hochschule Chem., Leuna–Merseburg **3** (1961), 203

[14] SIMIONESKU, K., OPREJA, K. V.: „Mechanochimija vysokomolekujarnych soedinenij", Moskva 1970

[15] PETERS, K.: „1. Europäisches Symposium Zerkleinern", Frankfurt 1962, p. 78, Weinheim/Bergstr. 1962

[16] HEINICKE, G.: Habilitationsschrift, Berlin 1965

[17] Chapt. 1., [1], p. 113

[18] HEINICKE, G.: Schmierungstechnik **1** (1970), 65

[19] HEINICKE, G.: Silikattechnik **22** (1971), 18

[20] HEINICKE, G.: „ABC Naturwissenschaften u. Technik", II, p. 652, Leipzig 1969

[21] Chapt. 1, [1], p. 114

[22] Ref. [19], p. 126

[23] HEINICKE, G.: „ABC Naturwissenschaften u. Technik" II, p. 1079, Leipzig 1967

[24] JOST, H. P.: "Lubrication" (Tribology), London 1966

[25] BOLDYREV, V. V.: Kinetika i Kataliz **13** (1972), 1411

[26] Chapt. 1., [12], p. 62

[27] FOX, P. G.: J. Solid State Chem. **2** (1970), 491

[28] SCHRADER, R., HOFFMANN, B.: chapt. 1., [4], p. 522

[29] HEINICKE, G.: Technik **32** (1977), 628, 688

[30] HEINICKE, G.: Chem. prum. **25** (1975), 508

[31] HEINICKE, G.: „Anwendungen der Tribochemie in der Technik", Berlin 1978

[32] HINT, I. A.: chapt. 1., [2], p. 6

[33] HEEGN, H., BERNHARDT, C.: Freiberger Forschungshefte **A 553** (1976), 57

3. Structural and physical changes of solids under mechanical energy influences

3.1. Transmission of mechanical energy to solids

3.1.1. Tribomechanics

Tribomechanics deals with microscopic and submicroscopic processes under the influence of mechanical energy on solids. The outward appearance of these processes is highly varied. Already in early times man was concerned with the working of tools such as hammering, drilling, and grinding, the crushing and pestling of raw materials, and the frictional processes in the transport of objects.

There are a great number of techniques in advanced technology using mechanical energy for the treatment of raw and other materials. In the manufacturing of metallic materials these techniques range from rolling to explosive forming, and in the refinement of material surfaces, cutting, grinding, and polishing processes mainly occur. A great number of techniques for crushing and agglomeration of solid products have been developed in connection with the exploitation and treatment of raw materials. One of the most important properties of materials is their behaviour under the influence of mechanical energy, which is expressed in material parameters, such as strength, and frictional and wear behaviour. Hence, problems of tribomechanics are dealt with according to different technological objectives in various fields of science, such as in materials science, crushing science, tribology etc. They are factors of growing importance in the intensive utilization of raw and other materials and in advanced techniques of engineering, particularly in the automation of technological processes.

All the above-mentioned processes are based on the influence of mechanical energy on solids which is transmitted either by the action of pulses, e.g. by the impingement of solids, by gas and liquid jets, or by the action of tensile and compressive forces onto the solid. The processes thus activated which result in an alteration of the structure and physical properties of solids are extremely complex and depend on the nature and intensity of the mechanical treatment and on the specific properties conditioned by the lattice structure of the solid, especially by the deformation properties.

Independent of the nature of the influence of mechanical energy an elastic deformation occurs in the first stage leading to a change of the bond distances in the solid. Upon exceeding the limit of elasticity, further processes are set in motion, related chiefly to plastic deformation and fracture processes.

In order to investigate these processes the pulsed impact treatment of monocrystals and compact solid surfaces, have proved particularly suitable, since within broad limits the treatment intensity may be varied by the mass and velocity of the moving tool, its form, the size of the contact surface, and the angle of entry by the free choice of direction of the flying particle. The single impact is particularly suitable for the investigation of elementary processes. The tribomechanic process can be studied through repetition and summing-up of impacts in case of endurance stress [1]. The following three processes for the transmission of mechanical energy by impact treatment will be accentuated.

3.1.1.1. Triboplasma

Immediately at the commencement of a grain colliding at high velocity with a solid surface it comes to a quasi-adiabatic energy accumulation and to the formation of an "energy bubble" at the point of action in the sub-microscopic deformation zone. This leads to high excitation states and to strong lattice loosenings and structural disruptions for very short periods which are connected with the detachment of lattice components, photons, and electrons. These aspects are summarized in the phenomenological "magma-plasma model", which is schematically shown in Fig. 3.1., [2]. Further details are discussed in 3.3., 4.1. and 5.1.

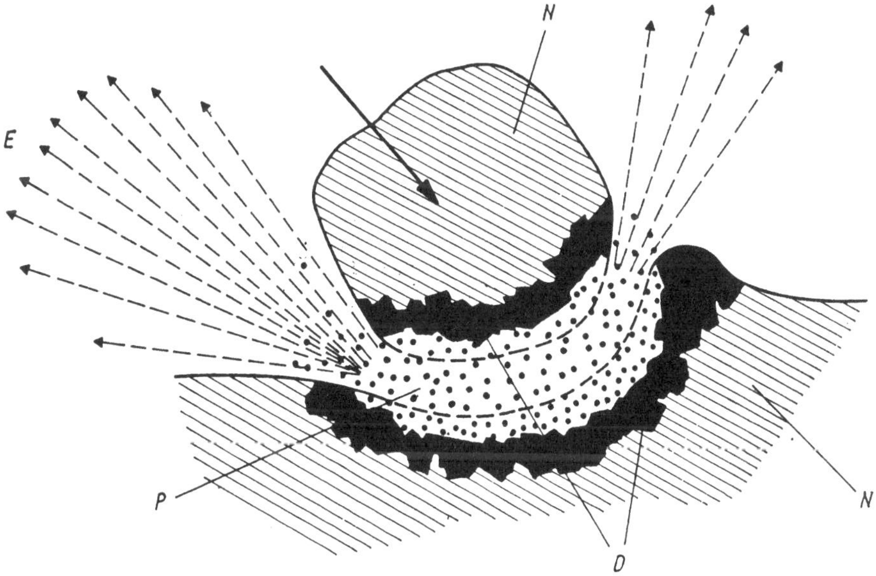

Fig. 3.1. **Magma-plasma model for the impact stress of flying grain (3).**
E — exoemission, N — normal structure, P — plasma, D — disordered structure.

3.1.1.2. Deformation

In the first stage of the impact the elastic stresses occurring with the mechanical action attain for very short periods high values and cause the spread of the energetic disturbance. One of the most important elementary processes by which the propagation of the disturbance occurs is the migration of dislocations (see 3.3.).

It becomes apparent that the microplasticity founded in this process is a general property of solids [4—8]. The occurrence of fracture-free material displacement during the impact treatment can also be observed in both very hard and very brittle solids, such as corundum, diamond, SiC, BaF_2, MgO, etc., if the mechanical action is sufficiently small [9—14]. The occurrence of microplasticity, then, is not dependent on whether

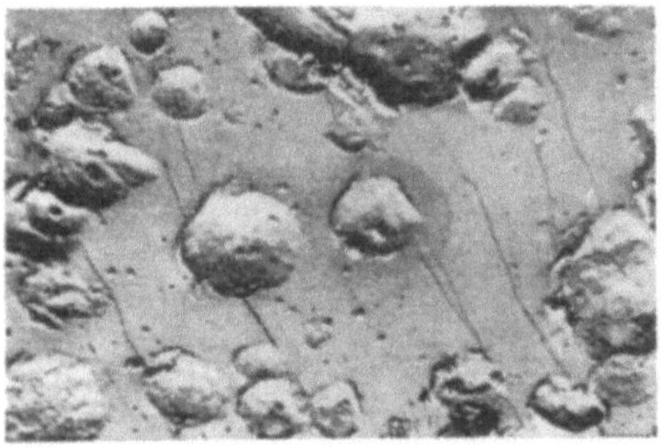

Fig. 3.2. Plastic behaviour of macroscopically brittle galena in submicroscopic ranges during the treatment by irons balls having 10 μm in diameter [15].
|——| – 1 μm.

the stressed solid has macroscopically plastic or brittle properties. The forced material migration during the impact treatment takes place without ascertainable directional influence in respect of the anisotropy of crystal properties [8], (Fig. 3.2.).

Morphological investigations of the impact region have shown a particularly marked disordered structure, with partial destruction, coming into being. Etching experiments

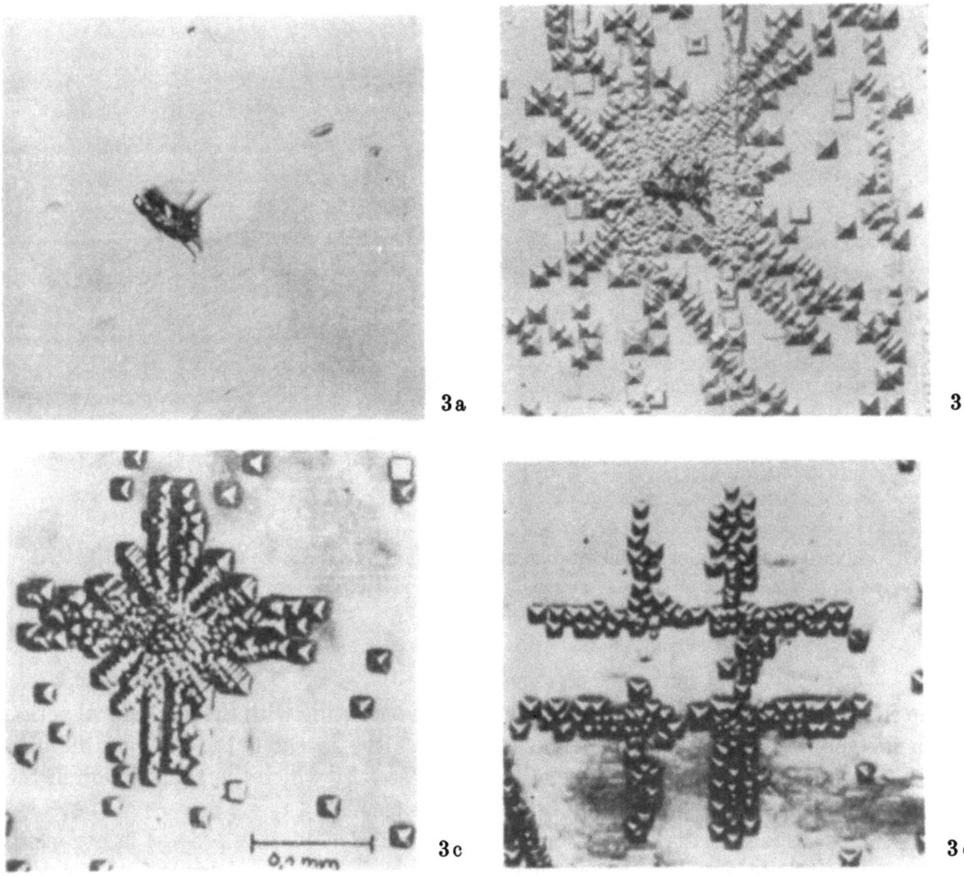

Fig. 3.3. Dislocation structure of a NaCl surface stressed by impact after the edging of surface layers by chemical polishing [16].
a – impact, *b*, *c*, *d* – dislocation structures with increasing duration of edging.

proved that starting from the impact region the dislocations migrated inside the solid. In this way the primary disturbance zone is followed by a less disturbed surround a which is designated as secondary disturbance zone. The primary disturbance zone in particular is characterized by a high dislocation concentration which decreases the further the distance from the centre of the impact region [16], (Fig. 3.3.).

. By repeated impact treatment the zone of marked deformation in solids extends especially in materials having plastic properties. With a large number of impacts, through deformation a flow of the material can be discerned mostly in the middle of the impact area inside the solid and back on its rim to the surface. The texture thus arising depends on the treatment parameters (see 7.4.). By virtue of the material flow the possibility exists that gases or other impurities adsorbed to the surface are transported into the depth of the deformation zone (see tribosorption, 4.7.). Fig. 3.4. presents by way of example the longitudinal grinding of a copper sample stressed by 6,000 impacts. It can be stated that the steel particles put onto the surface after 1,000 impacts were transported inside the stressed solid through further treatment [17].

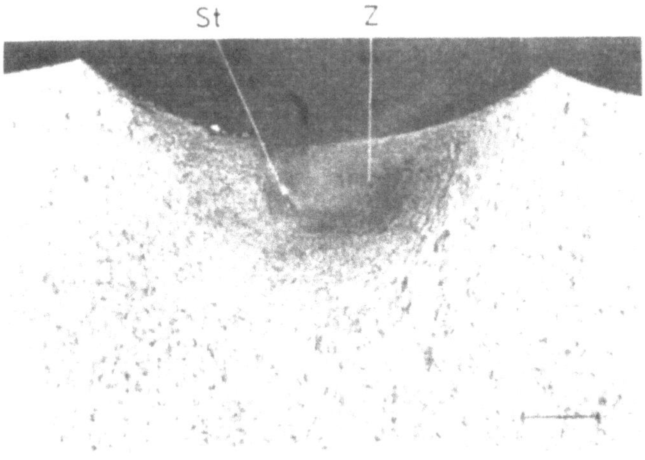

Fig. 3.4. Longitudinal cut through the impact surface of a copper sample after 6,000 impacts [17]. Steel particles were applied to the impact surface after 1,000 impacts.
Z – strongly deformed, ⊢—⊣ – 1 mm.

If there is a statistic stress on the solid surface, e.g. by a frictional process or by a stress of solids in a vibrating grain mixture (apparatus see 4.2.), there then develops a continuous disordered boundary layer on the whole surface which in most cases has a changed material composition in comparison to the backing material through the uptake of components from the enviroment or by material transmission from the tool. It is called the Beilby layer in tribology and it is schematically shown in the case of a solid disturbed by a slipping stress in Fig. 3.5., [18].

There is always a reaction layer on the Beilby layer if the mechanical stress of metals takes place in the air or other reactive media. In the inert medium the reaction layer does not even exist in the milling of oxides, such as SiO_2 and Al_2O_3. The transition from the Beilby layer to to undisturbed metal takes place continuously. It is characterized by an intermediate layer in Fig. 3.5. The solid has the highest disorder degree

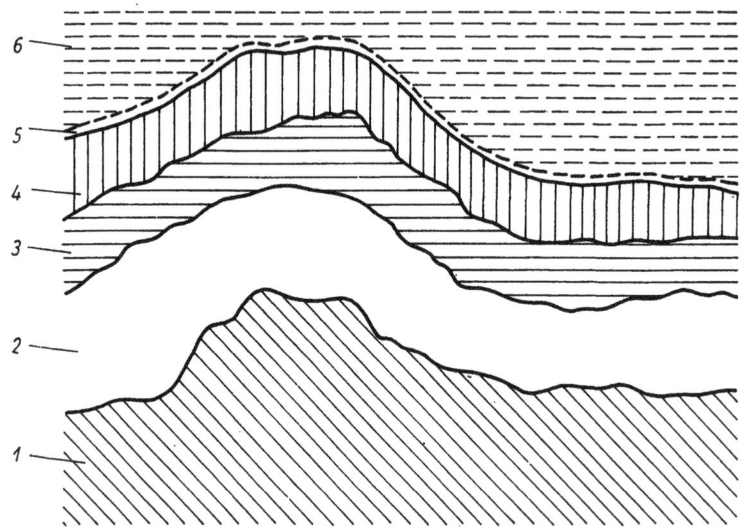

Fig. 3.5. Boundary layer pattern of a mechanically stressed material [18].
1 – undisturbed metal, 2 – transition layer, 3 – lattice-disturbed layer, 4 – reaction layer, 5 – layer formed by adsorption and reaction, 6 – surrounding medium.

immediately on its surface, and it must be added that the disturbance leads to the formation of a quasi-amorphous structure. It is possible that the impurity additions taken up from the environment cause an increase in stability of the heavily meta-stable disorder degree of the boundary layer by the inhibition of recrystallization [19]. The expansion of the separate zones depends on the nature and the intensity of treatment and on the properties of the stressed solid.

3.1.1.3. *Fracture*

If the intensity of the mechanical treatment is increased then the initially plastic deformation changes to a fracture process (see Fig. 3.6.), [20]. Here the binding forces between the atomic components in the crystal are overcome and so-called fresh surfaces are created. If the fracture crosses the entire solid it leads to the crushing of the solid.

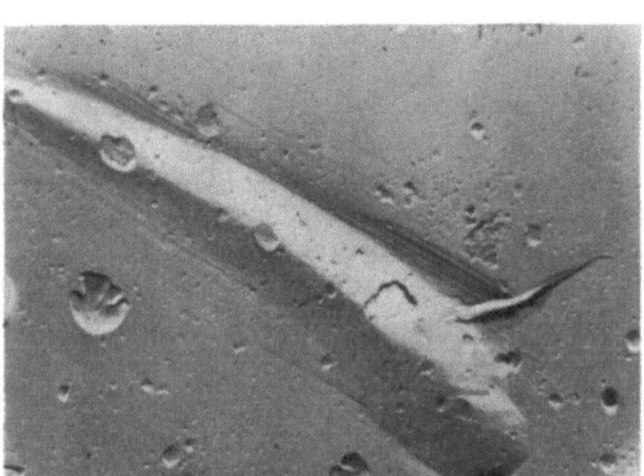

Fig. 3.6. Simultaneous appearance of plastic deformation and fracture during the impact stress of a (100)-LiF-crack area [20].
⊔ – 1 μm.

Two different mechanisms are responsible for the spread of the fracture; firstly the sliding process leading to the complete division of to the slipping fracture of the solid, under the effect of tangential stresses and secondly the rapid "brittle fracture" under the effect of tensile stresses. Whether a slipping fracture or a brittle fracture comes into being under mechanical stress, depends then on the nature of the state of stress, but also on the lattice properties and on the temperature [21]. In crushing science the brittle fracture is of dominant importance. The brittle fracture is bound to the existence of cracks in the solid. Generally a crack propagation occurs when the stored elastic internal potential energy is sufficient for the composition of free interfacial energy of resulting fracture areas (GRIFFITH energy condition) [21]. This is also the reason why the brittle fracture behaviour of solids increases with the growing intensity of mechanical stress. In order to meet the mentioned conditions it is necessary that very high tensions exist at the crack tips. Because of this tension concentration, microplastic deformations are caused at the crack front. The bigger the plastically deformed volume, the more energy is required for the release and acceleration of fracture processes. The fracture area energy may amount to between 10 and 10^6 times the amount of the free interfacial energy.

By the energy conversion in the environment of the crack tip where the stresses are so high that the molecular yield strength and tensile strength are exceeded, a highly active plasma-like state comes into being with a high excess of free energy in the solid [21]. Apparent temperatures of 10^5 K and more dominate for a short period (see Fig. 5.10.). Since the excitation state released by the fracture is restricted to a narrowly limited volume a rapid solidification of non-equilibrium states occurs which are characteristic of the fracture area properties (see 3.3. and 5.3.).

If no cracks exist in the solid as can be the case with whiskers and very thin glass areas, the strength approaches the molecular tensile strength. However, with impact stress of sufficient intensity, crack formation will occur through the blockage of dislocations so that solids originally free of cracks also break far below the molecular tensile strength [22].

If smaller particles are increasingly produced by fracture processes these will have a continuously decreasing concentration of cracks, loose places, and internal tensions. This means that homogenity and strength increase with decreasing particle size so that continuously higher stresses are necessary to release further fracture processes [23].

As shown in 5.5. the primary crystallite size leads with continuous mechanical stress to a boundary value dependent on system and nature of the stress. This value remains constant under further influence of mechanical energy (see Fig. 5.27.). Under these conditions a further stress of the very fine solid particles occurs through microplastic deformations of the surface areas. The influence on fracture processes by external media by reason of the REHBINDER effect [24] or of the chemo-mechanical effect [25] is discussed in 5.5. and 6.6.

Four processes are important for the release or influencing of chemical reactions in fracture processes:

1. The short-term generation of a highly active state at the fracture tip,
2. the separation of chemical bonds occurring during the fracture,
3. the magnification of the solid surface and
4. the alteration of the lattice structure in the layer near the interface formed during the propagation of a fracture through the volume phase.

Since the above-mentioned processes influence the individual chemical reactions differently (see Chapt. 5.) an analysis of the causes of tribochemical reactions in connection with the physical and chemical elementary processes is difficult, the more so as the deformation processes initiated independently of fracture action which are shown in 3.1.1.2. overlap. In most cases the course of chemical reactions is only measureable with repeated stress. On the other hand the latter will complicate the analysis of mechanical elementary processes. Hence it has only been possible in a few cases to follow the chain starting from the mechanical treatment through the physical elementary step occurring at the solid right up to the chemical elementary process. Most of the investigations made until now have been carried out with integral stress, e.g. with the use of mills, where the total effect of a sum of single impacts on the reactivity of the treated solid is counted. As shown in Chapt. 5. it is also possible in such a case to get information on the main causes of the tribochemical activation by indirect means, e.g. by an analysis of the variation of physical and chemical properties of the treated products, and to arrive at a classification of the causes of reactions. In order to deal more extensively with the obtained results however, a comprehensive connection of the elementary processes occurring during the treatment is increasingly required.

3.1.2. Effects of different triboreactors

3.1.2.1. Kinds of stresses in triboreactors

A triboreactor may generally be defined as a reaction apparatus where intended chemical reactions of solids take place under the influence of mechanical energy. There are a great number of possibilities for the design.

A reaction arrangement suitable for the investigation of the kinetics of tribochemical reactions is described in 4.2. The principle of this arrangement is that a reaction material consisting of coarse grains or small plates is subjected to vibrating stress together with hard-material particles of approximately similar size as working tool. In analogy to this a disc-type electrode rotating in a hard-material mixture is used for triboelectrochemical processes (see 6.6.). These kinds of stress represent "mild" stresses, because the impacts occur with relatively small impact energy. They do not exclude the occurrence of high excitation states in the sub-microscopic range.

In order to achieve high material conversions those arrangements are suitable in which a comminution for the formation of large surfaces take place. For this purpose fine mills have particularly proved their worth, such as the ball mill, the vibrating mill, the disintegrator, the jet mill, and the planetary mill. A short description of the operating mode of these mills will be given below.

Generally the different kinds of stress occurring in the mills may be divided according to three possible mechanisms (see Fig. 3.7.), [26, 27]:

I. The particles of the reaction material are stressed individually or collectively between two milling areas. If the influence is vertical a compressive stress takes place and with a tangential relative movement of the milling areas a shearing stress. In practice both the stresses occur together. If the stress between the milling areas occurs by the impingement of one of the milling areas with the velocity v onto the grinding particle then this action is described as double impact stress (see Fig. 3.7., I), [26].

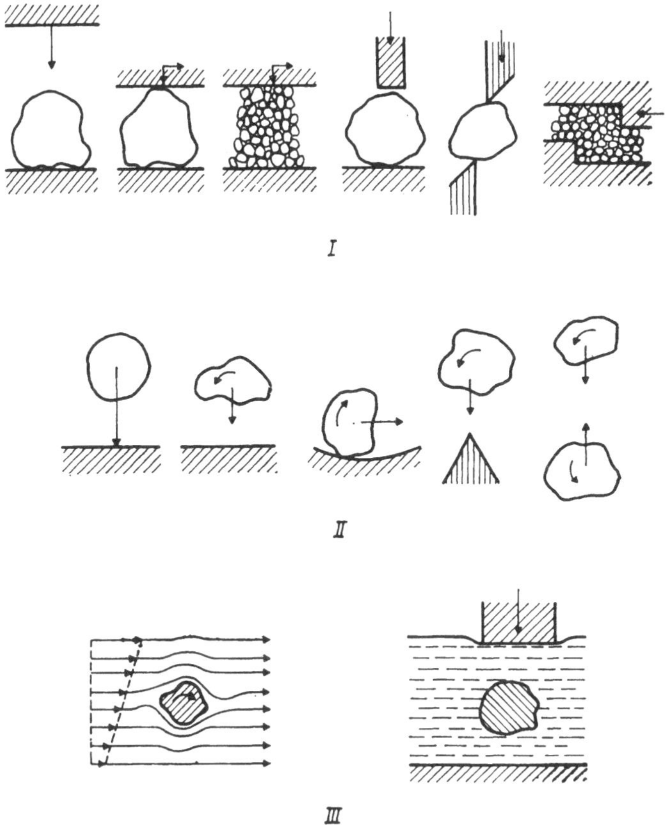

Fig. 3.7. Different stress mechanisms during the influence of mechanical energy on the solid particles [23].

I — stress between two faces, II stress of freely moved parts at a face or to one another, III — stress by the surrounding medium: left: normal forces and tangential forces on the particles surface, right: impact stress, for instance by shock waves.

II. Particles moving freely in a stationary or moved base medium impact against a milling area or against each other. This type of stress is deserted as impact stress. The prototype of the impact stress is the vertical impact of a ball onto an area (see Fig. 3.7., II, pict. 1) or a second ball. Since the particles may be both in a translation movement and in a rotation movement a sliding and rolling stress also takes place in addition to the impact stress (see Fig. 3.7., II, pict. 2—5).

III. Mechanical energy can also be transmitted to solids by surrounding liquid and gaseous media, e.g. by a jet of gases [28] accelerated approximately to sound velocity (see Fig. 3.7., III). For instance, these may bounce against the solid with a high impulse in the form of liquid droplets. The mechanical energy can also act on the solid by shock waves through the liquid medium released for instance by an underwater explosion [29, 30]. Here strong physical [29, 31] and chemical [32] changes of the solid properties occur in the volume phase of the solid, and chemical reactions take place in the presence of a system capable of chemical conversions, such as the formation of Al-[33], Ti-[34], W-carbide [30, 33] from the elements, the formation of Zn_2SiO_4 from ZnO_2 and SiO_2 [35], the polymerization of monomers [36], etc.

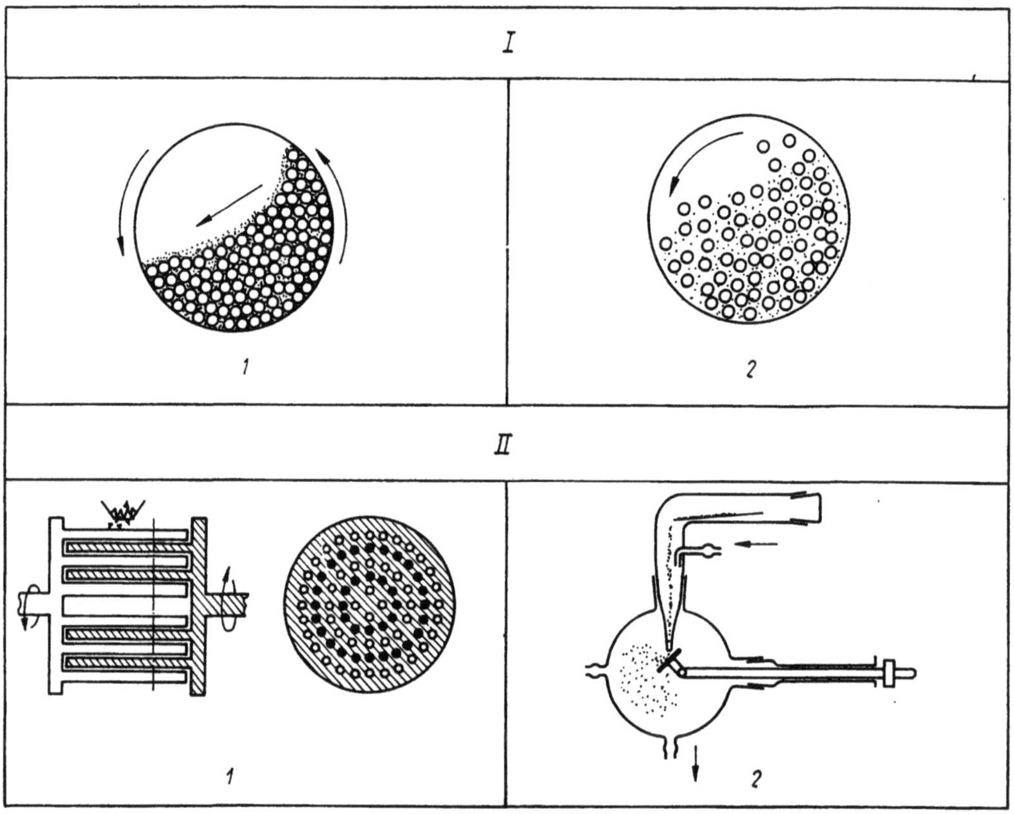

Fig. 3.8. Kinds of stresses for the examination of tribochemical reactions shown as schematic representation [37].

I — stress between two faces, I_1 — ball mill, I_2 — vibration mill: o — milling balls, ⁂ — stressed solids.
II — stress of freely moved particles at a face or to one another, II_1 — pinned disk mill, disintegrator, II_2 — jet mill, ↓ — direction of the gas stream.

In view of their technical significance the most important triboreactors operating according to the stress mechanisms I and II will be represented in the following section. Their principle of operation is represented schematically in Fig. 3.8.

3.1.2.2. *Mode of operation of current triboreactors*

Ball mills, vibrating mills, and planetary mills belong to the devices working according to the stress principle I. Pinned disc mill and jet mill are the chief devices working according to the stress principle II. In the following sections a short description of the most important types of mills is given.

3.1.2.2.1. *Ball mills*

Most ball mills consist of a cylinder which is partially filled with the milling bodies, e.g. balls, and the material to be stressed (see Fig. 3.8., I, pict. 1). The cylinder is rotated with the axle lying horizontally. According to the rotational speed the ball filling at the wall of the cylinder will be raised up with varying strength. With lower rotational speed the top balls roll back onto the slantingly arranged ball layer or complete layers displace one another. With higher rotational speed the top balls detach from the milling wall and fall back in the form of a trajectory parabola to deeper ball layers. The main stress of the grinding material occurs upon the impact of these balls [38].

In industrial practice ball mills are in the forefront in the line grinding of products e.g. cement. However, as discussed in 5.9. they show poorer effects than vibrating and planetary mills when used as triboreactors.

3.1.2.2.2. Vibrating mills

The external design of the reaction bin of a vibrating mill largely corresponds to that of the ball mill. The closed cylindrical tube with the lying axle is partially filled with milling bodies and grinding material. The milling tube is made to swing circularly on a plane perpendicular to its axle. This causes the filling to be accelerated upwards. At a certain point the filling detaches from the wall of the tube and makes a throwing motion while the tube continues swinging on its circular trajectory. After a certain time the filling again comes into contact with the tube wall where an intensive double impact effect occurs. These operations are repeated in each swing period. The grinding material is in a very loose aerosol-like state between the milling bodies and is mainly stressed by impacts (see Fig. 3.8., II), [39]. The vibrating mill has proved its worth as a good triboreactor and belongs to the most used devices for the investigations of tribochemical reactions (see 5.9.).

3.1.2.2.3. Planetary mills

The reaction tube filled with the grinding material and the balls is vertically arranged with its axle. It rotates round a centre lying outside the tube and rotates simultaneously round its own axle in the opposite direction. Through this the balls detach at a certain point from the milling space wall and fall back again with high energy to an other point of the wall [40]. By an appropriate geometric arrangement of the trajectories and by selecting the rotation times accelerations of 60 times the accerelation due to gravity can be achieved. As shown in Fig. 4.14. very high tribochemical effects are obtained in such mills.

3.1.2.2.4. Pinned disk mills

In the pinned disk mill the mechanical stress occurs through impact of the grinding material particles against pins fixed to 2 discs arranged in parallel. Each disc has several rows of pins arranged circularly and concentrically to the rotation axis. The rows of pins of the one disc interlock in the space between the other ones. There are many apparatuses where only one disc rotates at high speed while the second is tightly arranged. Great differences in the relative velocity of the neighbouring rows of pins and hence a higq stress intensity can be obtained if the discs are allowed to rotate at high speed in opposing directions (see Fig. 3.8., II, pict. 1). This principle is applied in the disintegrator, which has proved its value in the activation of solids, particularly building materials (see Fig. 3.9.), [41—45].

Fig. 3.9. Technical mills.
I – ball mill, II – vibration mill, III – disintegrator, IV – oval-tube jet mill.

3.1.2.2.5. Jet mills

The mode of operation of a jet mill consists in the outlet of a gas flow from a nozzle. The gas flow usually has an overpressure of 6 x 10^5 Pa and the powder particles to be stressed are added before or after the nozzle. Under these conditions the gas flow reaches a final velocity of 450 m/s [46]. The mechanical stress mainly occurs through the collision of the individual particles against each another in the nozzle flow or by the impact against a wall (see Fig. 3.8., II, pict. 2). The various types differ in the design of the milling space, which leads to a spiral-like gas flow with a flat cylindrical design (spiral jet mill), to the impact of the jets against each other with a counterflow design, and to an orbital motion of the gas and the solid particles in an oval tube design in an oval-shaped tube system (see Fig. 3.9.), [47]. Due to the high stress intensity to which the grinding material particles are subjected, high degrees of division can be obtained in the jet mill, especially with soft and medium-hard materials. Jet mills have also proved to be efficient as triboreactors in selected chemical reactions [48].

The ball mill, the vibrating mill, the disintegrator, and the jet mill are used in industry (see Fig. 3.9.), the design of planetary mills, however, has only led to the construction of semi-technical aggregates [49]. Altogether the construction of triboreactors specifically optimized for high tribochemical efficiency (see equ. 4.30) is only at the early stages of development, since so far technical problems of the mechanical comminution of solids have to a very large extent been the objective in the construction of mills.

3.1.2.3. *The effect of different triboreactors on changes of physical and chemical properties*

Changes in the solid structure and in the reactivity of the milling products occur during the mechanical stress of solids in the mills described in 3.1.2.2. It is shown now that ball and vibrating mills on the one hand, and disintegrators and jet mills on the other, lead to different changes of the properties. As an example the dissolving velocity of ground quartz in fluorhydric acid is represented in Fig. 3.10. and the change of the relative X-ray intensity of ground quartz (see 3.2.) in dependence on the surface in Fig. 3.11., respectively. In both cases the two mentioned types of mills lead to several dependences.

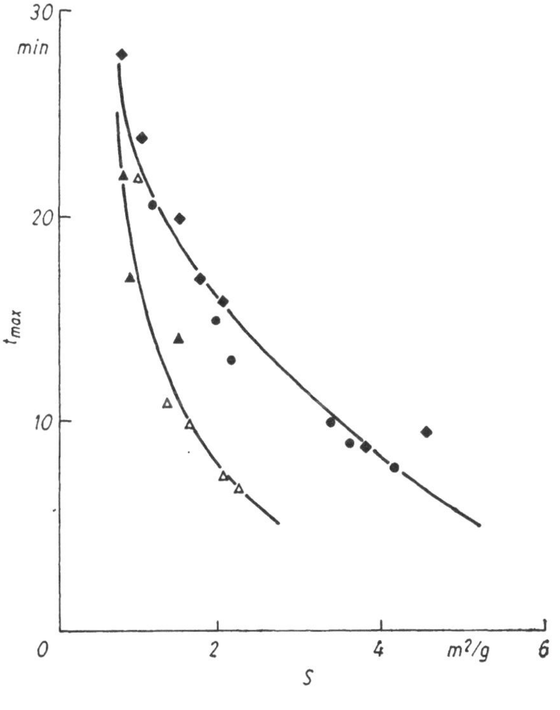

Fig. 3.10. Dissolving velocity of quartz mechanically stressed in different mills in fluorhydric acid in dependence on the specific surface S [50].
t_{max} – maximum time of dissolution, ♦ – ball mill, ● – vibration mill, ▲ – disintegrator, – △ jet mill.

The stronger decrease of the relative X-ray intensity and the increased solubility in fluorhydric acid under impact stress, in comparison with an impact and shearing stress in ball and vibrating mills, can be attributed to the higher working intensity in disintegrators and jet mills. However, as the solubility studies in soda lye show, a high rate of disorder occurs in vibrating milled quartz samples in the peripheral layers of quartz particles. whereas a smaller rate of disorder can be found inside the structure. The latter is a consequence of an impact and shear stress smaller in intenstiy but more frequently affecting the individual particle. Therefore the stress mechanism deter-

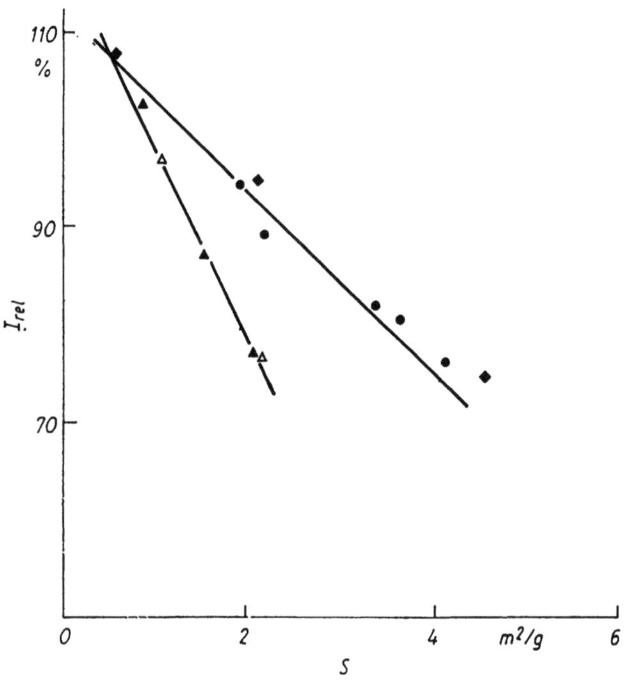

Fig. 3.11. Relative X-ray intensity I_{rel} of quartz mechanically stressed in different mills in dependence on the specific surface S [50].

♦ — ball mill, ● — vibration mill, ▲ — disintegrator, △ — jet mill.

mines the change in the microstructure to a very large extent. The model conceptions shown in Fig. 3.12. were developed for the influence of both mentioned stress types on the microstructure of stressed solid particles [50], (see also 6.12.).

In all the foregoing discussion shows that several mechanisms always come into consideration in the transmission of mechanical energy onto solids by the use of mills.

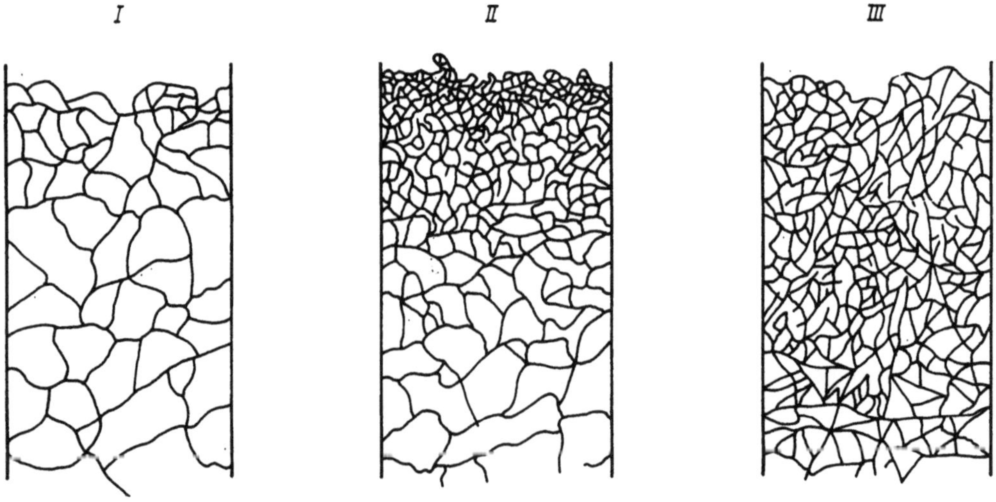

Fig. 3.12. Change of the microstructure of solid particles mechanically stressed by different mechanisms [50].

I — initial material, II — structure after double impact stress, III — structure after impact stress.

The share of these mechanisms in the whole proceedings is only difficult to evaluate. Variations in the design of the mills as well as variations of the milling conditions such as amplitude, frequency, the proportion or degree of filling, and other parameters shown in 4.4. and 5.9. often have an important influence on the efficiency and the nature of the transmission of mechanical energy. Due to the abundance of the effective factors one still depends to a large extent on the phenomenological approach to these problems. By this means it is already possible today to select the mills for a mechanically induced systematic change of the physical and chemical properties of solids and their structure as well as to determine the criteria for the milling conditions.

3.2. Changes of the solid structure by mechanical treatment

The reactivity of a solid is changed by mechanical treatment. Changes in the structure of the solid occur. A deeper understanding of the causes of the changed solid reactivity following the mechanical treatment is primarily possible on the basis of research of the real structure. The real structure is an important influencing factor of solid reactivity. The real structure can be regarded both from the crystallographic point of view and from the thermodynamic point of view. From the crystallographic point of view the real structure has to be understood as a deviation from the strict periodicity of the infinitely extended ideal crystal. The regular periodic arrangement of the atoms is disturbed by the thermal motion and the existence of different statistically distributed defects.

However, the thermal vibrations (phonons) will be to some extent neglegted and reference made to the actual position of the component units. Since in addition the distance between the component units is very small in comparison to the external dimensions, the monocrystal is regarded as infinite. The defects are distributed all over the volume of the crystal structure, but they accumulate in certain areas dividing the crystal in separate spheres.

In the following a summary outline of the different structural disorders of mechanically disturbed solids will be given. The mechanical treatment creater thermodynamically instable states in the solid, with in increased energy content compared to the thermodynamic state of equilibrium. The difference between the free enthalpies of imperfect and perfect structures may be regarded as integral measure for the degree of disorder (see 4.8.).

3.2.1. Principle of imperfection classification

One distinguishes between chemical and structural imperfections. Here in the structural defect will mainly be discussed. The structural defect is the deviation from the strict three-dimensional periodic structure of crystals. There are several synonyms for the term structural defect, such as structural disorder, structural imperfections. Different attempts have been made to classify the possible structural defects. The classification has been made according to geometric aspects [51]. The defects are divided according to the number of the dimensions in which their extent is more than atomic (see Tab. 3.1.). The structural defects seldom occur separately, they influence one another and are capable of reacting with one another. Since a separation into individual defects is difficult one generally speaks of a defect state. Detailed representations can be found in [51—62].

Table 3.1. Classification of imperfections.

1. Zero-dimensional imperfection
   atomic imperfection (vacant sites, interstitial atoms, substituted foreign atoms, crowdions)
   electronic imperfection
   energetic imperfection
2. One-dimensional imperfection
   dislocations (edge and screw dislocations)
3. Two-dimensional imperfection
   surface, grain boundaries, twin boundaries, small-angle grain boundaries, large-angle grain boundaries, stacking fault
4. Three-dimensional imperfection
   amorphous structure regions, pores, cracks, inclusions

3.2.2. *Zero-dimensional imperfection*

The zero-dimensional imperfection has atomic extensions in three dimensions. It differs fundamentally from the one- and two-dimensional imperfections still to be discussed, by the fact that it can occur in a thermodynamic equilibrium. Moreover the zero-dimensional imperfection can in addition result from external influences, for instance by mechanical treatment. In this case deviations from the equilibrium concentration are present.

3.2.2.1. *Atomic lattice defect*

The atomic lattice defect is a structural defect. It leads to deviations from the translational symmetry. If one site is left unoccupied in the perfect structure the resulting configuration is designated as a vacant site. By contrast one speaks of an interstitial atom (intrinsic or extrinsic interstitial atom), if an atom of the host crystal or an atom chemically different from the latter is intercalated into a crystal (chemical defect combined with structural defect) without an occupation of the spatial point being anticipated.

The substitution of extrinsic atoms to existing vacant sites is also possible. In consequence of the changed electron structure of the extrinsic atom, structural variations can occur which go beyond atomic interspaces. The incorporation of an additional atom into a densely occupied grid line leads to a crowdion. The spatially neighbouring occurrence of an interstitial atom with a vacant site results in a FRENKEL disorder. A SCHOTTKY vacancy appears when the atom of the initially occupied lattice site migrates outwards (Fig. 3.13.).

Zero-dimensional point defects interact with each other. Point defect associates may form.

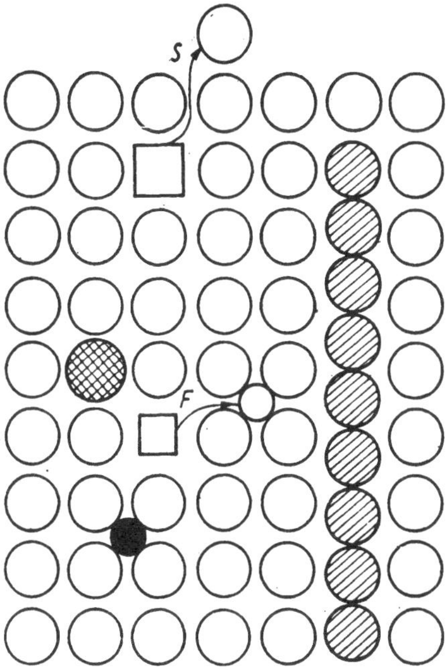

Fig. 3.13. Zero-dimensional (atomic) defect order [59].
□ – vacant site, ⊗ – substituted foreign atom
● – interlattice foreign atom, ○○○ – interlattice atom,
◍ – crowdion, 8 atoms instead of 7 atoms,
F – Frenkel disorder, *S* – Schottky disorder.

3.2.2.2. Electronic imperfection

According to the definition the electronic imperfection is no structural defect. It has no influence on the symmetry, but it belongs to the imperfect structure. The atomic defects are capable of inducing electronic imperfections by disturbing the periodicity and the interaction with the ensemble of electrons of the total crystal. Electronic point defects closely interact with atomic lattice defects.

The atomic lattice defects are able to absorb electrons or defect electrons, and thereby induce another charge state. The *F*-centres, V_k-centres, and exitions are part of the electronic imperfections. The *F*-centres are an associate between an anionic vacant site (e.g. halogen vacant site in the alkali halogenide) and an electron which is captured in this vacant site. The electron is not strictly located in the middle of the vacant site. V_k-centres also primarily occur in the alkali halogenides. The V_k-centre comes into existence through a the loss of an electron from 2 halogen aniona and represents a negative charged halogen molecule. Exitons are atoms or ions, in excited electron states.

3.2.2.3. Energetic defect

An energetic defect appears in all deviations from the translational symmetry. It has a short life span and is also characterized as an exitation state [52], (see 5.3.).

3.2.3. One-dimensional imperfection

The disturbed crystal area extends beyond atomic interspaces in one dimension only. This group of imperfections includes dislocations. A dislocation is characterized by a dislocation line and BURGERS circuit. By a dislocation line is meant a dividing line which separates the undisturbed from the disturbed crystal area. The atomic structure is in a distorted state within the range of this point. The BURGERS vector provides the absolute value and direction of the displacement of atoms in a crystal lattice. It has the same absolute value at every point of the displacement. If the BURGERS vector runs perpendicular to the dislocation line, an edge dislocation is present. With a screw dislocation the BURGERS vector runs parallel to the dislocation line (Fig. 3.14.). A dislocation cannot end inside the crystal. It is either self-contained, of it ends at a surface or in another lattice defect. By dislocation density is meant the number of dislocations which penetrate the unit area of 1 cm². Fig. 3.15. shows dislocation structures in mechanically treated MgO crystals.

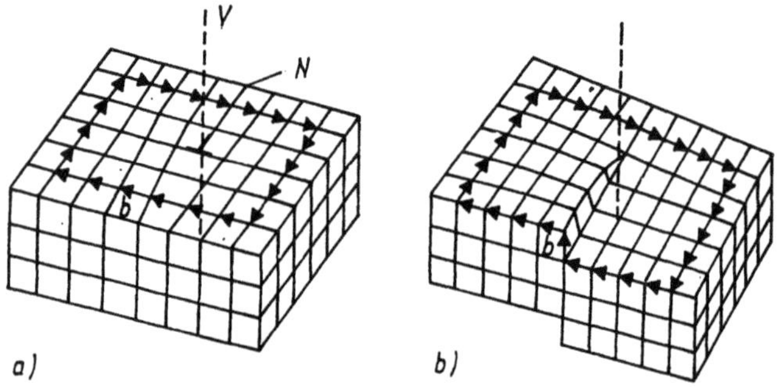

Fig. 3.14. BURGERS circuit in an edge (a) and in a screw dislocation (b) [56].
b — BURGERS vector, V — dislocation line, N — introduced atomic plane.

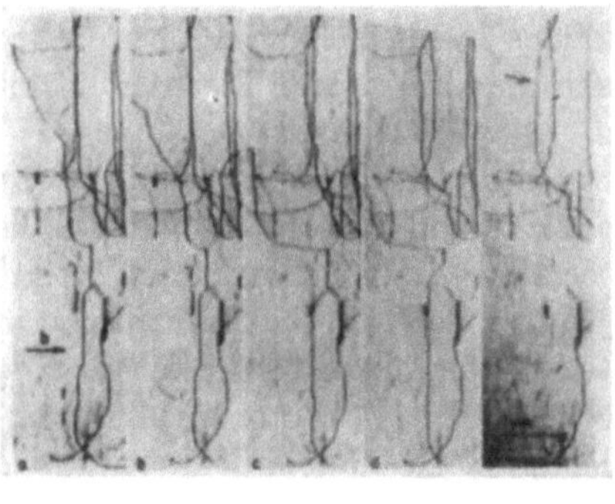

Fig. 3.15. Motion of edge and screw dislocations during the in-situ deformation of a MgO crystal [63].
b — BURGERS vector.

3.2.4. *Two-dimensional imperfection*

The disturbed crystal area extends beyond atomic spaces in two-dimensions. The surface and the grain boundary should be regarded as the most important two-dimensional imperfection, particularly after mechanical treatment. All imperfect crystals are spatially limited. Surfaces and grain boundaries are faces where the periodicity of the structure is interrupted and the chemical bindings are not saturated. But they differ by the fact that the surface is the interface between a disperse or continuously solid phase, and a continuous liquid or gaseous phase, whereas the grain boundary is the interface between two continuous solid phases. In contrast to the so-called primary crystallites (see 3.2.6.) the single grains are often qualified as secondary particles.

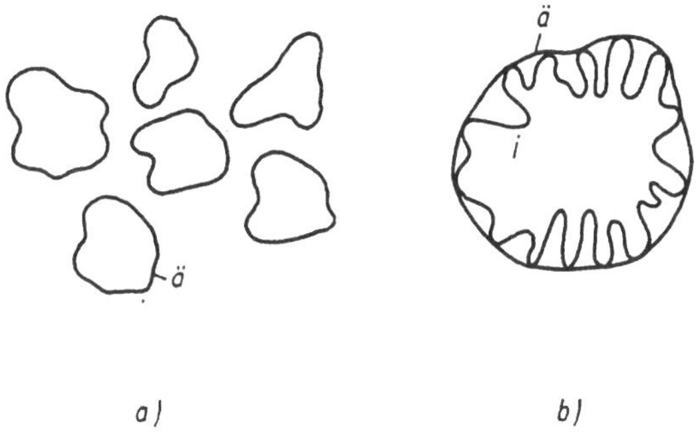

Fig. 3.16. Outside and inside surface.
a) Outside (ä) surface at a polycrystalline powder consisting of single grains (secondary particles.
b) Outside (ä) and inside (i) surface at a magnified single grain.

In respect of the surface a distinction is made between the external and internal surfaces [64] (Fig. 3.16.).

The external surface is the geometrical surface ot the particles without taking into consideration the roughness and the pores. The additional surface of the interface between disperse solid phase and continuous liquid or gaseous phase, the roughness or pores are described as internal surface (interface between continuous solid phase and continuous liquid or gaseous phase). The internal surface may be open or closed. Sporadically one will also find the surface size of the primary crystallites indicated (see 5.5.2.), [65, 66]. This would seem to be problematic and physically of little meaning because of the fact that the latter are not limited by the free surface. But on the other hand, because the primary crystallites are limited by dislocation accumulations and since these influence the reactivity variation (3.2.8.) the determination of their proportion is of interest. It is possible to make a formal evaluation of this proportion.

Single grains are also present in compact polycrystalline samples, such as metals, alloys, sinterings. The single grains collide at the grain boundaries (Fig. 3.17.).

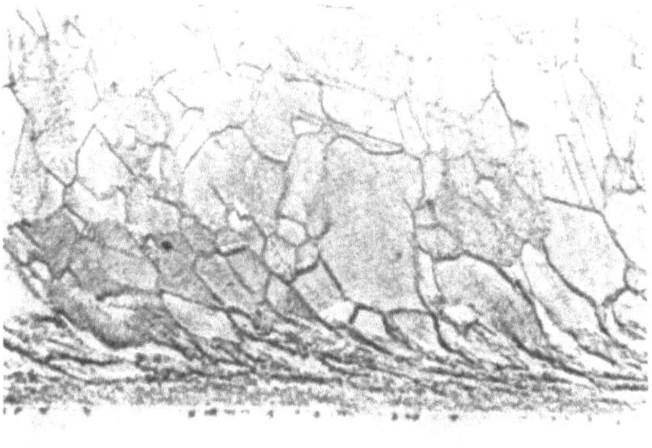

Fig. 3.17. Grain boundaries in compact polycrystalline material mechanically treated [67].

Within a single grain specific "grain boundaries" that is small-angle grain boundaries and large-angle grain boundaries, can arise from an accumulation of dislocations. The name only refers to the angle difference of the same lattice levels in neighbouring crystal particles. Crystal regions having orientation differences are divided by these "grain boundaries". At these "grain boundaries" the structure periodicity is disturbed but not interrupted. Through this the small-angle grain boundaries and largeangle boundaries differ

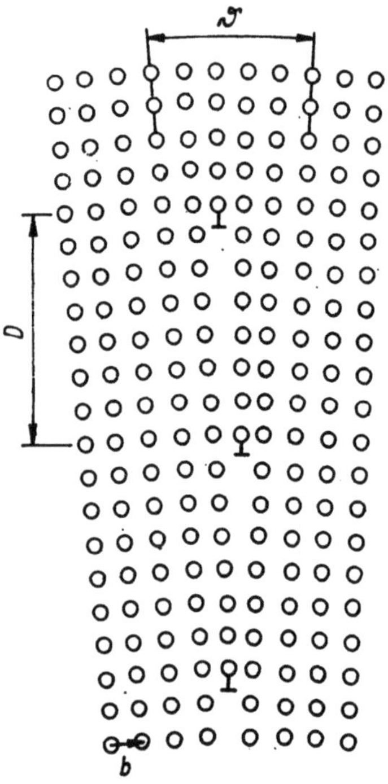

Fig. 3.18. Small-angle grain boundary formed by the edge dislocations [57].
b — Burgers vector, *ϑ* — angle of slope,
D — distance between intersecting points of dislocations, $\vartheta = b/D$.

from the previously described grain boundaries. The orientation difference of the small-angle grain boundary amounts to few angular seconds and is not greater than 4° (see Fig. 3.18.).

If the orientation angle is greater one speaks of a large-angle grain boundary. The large-angle boundaries are extremely manifold and hard to comprehend quantitavely. There are two conceptions of the structure of these large-angle grain boundaries. According to one the transition layer is not defined crystallographically. There is a strongly disturbed layer between two crystal areas, the so-called Beilby layer. According to the second conception the interface contains atoms which can be assigned to both the crystal parts (see Fig. 3.19.).

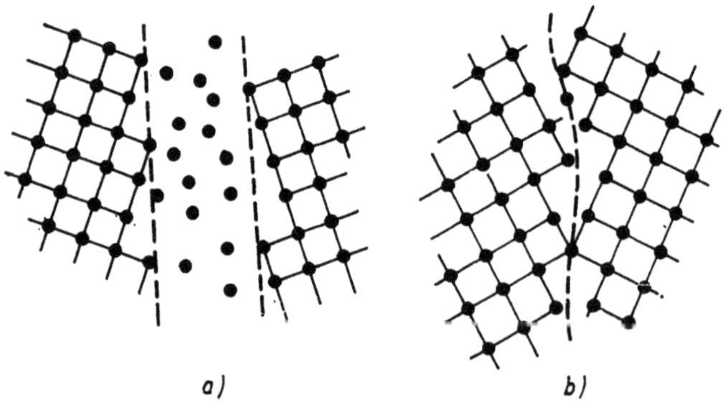

Fig. 3.19. Large-angle grain boundary models [57].
a) Beilby layer, b) probable structure.

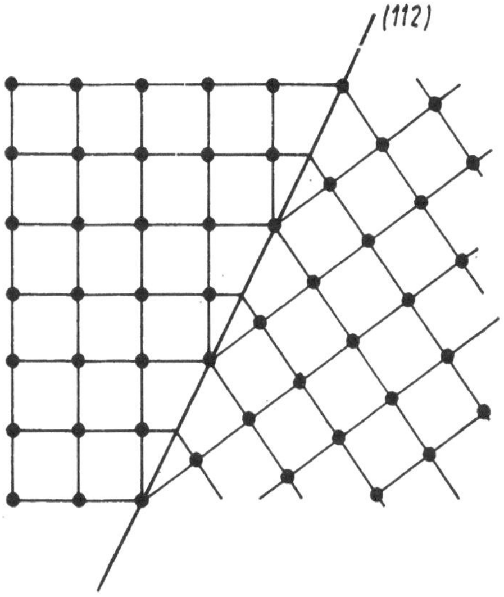

Fig. 3.20. Twin boundaries.
Coherent twin. The right twinned region came into being by reflection on the twin plane [57]

Twin boundaries may arise from mechanical treatment. In such a case two parts of the crystal are identically and coherently connected with each other. They are in twin position. The twin boundary is then a definite reflection plain (see Fig. 3.20.).

In metals and certain laminated structures stacking faults occur representing a disturbance of the normal sequence of layers in the composition of a structure. According to the literature on the subject the stacking faults are listed among the two-dimensional imperfections. By definition they belong to the one-dimensional imperfections because the layer itself is not disordered. The defect occurs only perpendicular to the layer, that means the defect exceeds atomic spaces in one dimension only.

3.2.5. *Three-dimensional defect*

From the atomic point of view the three-dimensional defect can be partially attributed to the other imperfections. It is useful in connection with the problem of mechanical treatment to include the three-dimensional defect, too. As discussed in 5.9.1.1. an amorphization of the solid can frequently be obtained by the mechanical treatment. The structure of amorphous solids is characterized by the absence of any long-range order, the periodic arrangement of the structure units noy longer exists. When the amorphous structure is compared with the perfect and imperfect structures the following differences will result:

perfect structure: spatial periodic arrangement of the structure units,
imperfect structure: disturbed spatial periodic arrangement of the structure units,
amorphous structure: spatial non-periodic arrangement of the structure units.

There is a short-range order in the amorphous structures which only extends to a space of the order of magnitude existing between atomic spaces. The short-range order structure of mechanically treated solids has only been investigated in a few cases till now. However, here a certain particularity becomes apparent [68]. For instance it could be shown that no product analogous to the amorphous silicic acid arises with mechanical activation of low-temperature quartz but a partial-crystalline quartz still containing typical structure components of the low-temperature quartz in the order of space of $r < 0.63$ nm (see 6.1.2.).

The pores are morphologically divided into closed and open pores [69]. The open pores are connected among one another and also with the solid surface. Shape and distribution of the pores play an important part in solid reactivity (see 3.2.8. and 5.5.).

Strong inclusions either have a coherent or an incoherent phase boundary in respect of the matrix, depending on how the several phases deviate from each other in respect to the structure and lattice constants.

3.2.6. Primary crystallite size and lattice distortions

It is common practice to characterize the disturbed state induced by the mechanical treatment among other things by giving the primary crystallite size. Each real crystal (mono-crystal or single grain of a poly-crystal) consists of numerous coherently spread-

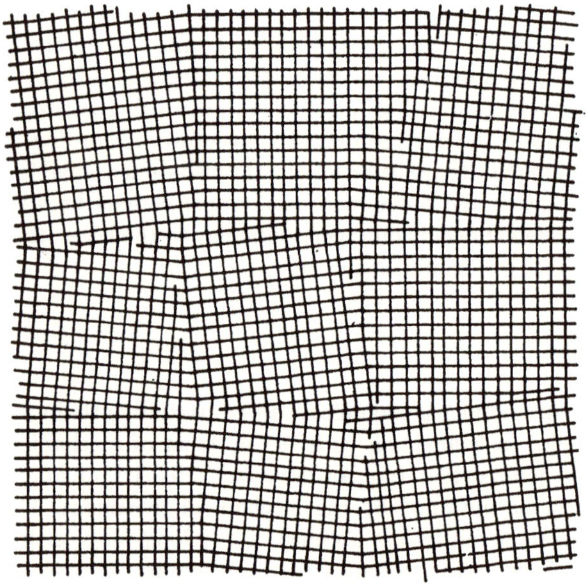

Fig. 3.21. Model of single grain consisting of coherent spreading ranges [55].

ing areas called primary crystallites, crystallites, sub-grains as well as mosaic blocks. Interns of the dislocation theory such a subdivision into primary crystallites is possible. With these crystallites areas of the crystal separated by dislocation walls are involved. The limitations of the areas for instance are small-angle grain boundaries. Fig. 3.21. shows a model of a single grain consisting of coherently spreading areas. The size of these areas corresponds to the space between two dislocation walls, the coherence length [70, 71].

Structural defects produce distortions and strains in their surroundings (Fig. 3.22.). These distortions are the cause of intrinsic strains. The nature, concentration, and

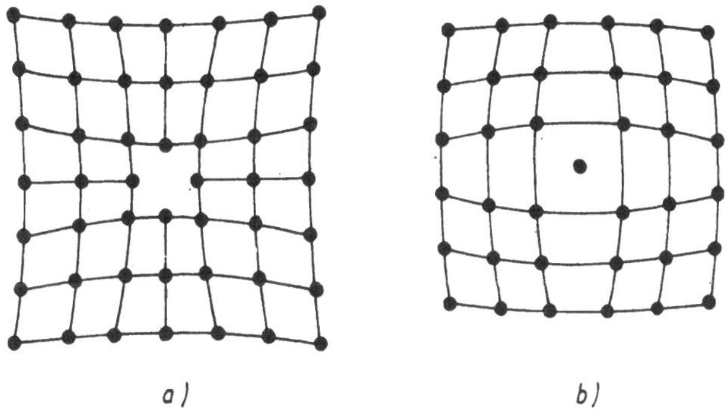

Fig. 3.22. Distortion of the crystal lattice on one plane [54].
a) in the surrounding of a vacant site,
b) in the surrounding of an interstitial atom.

distribution of the defects determine the nature and intensity of the intrinsic strains occurring in solids [72].

A distinction is made between intrinsic strain of first, second, and third order [73, 129], (see Fig. 3.23.).

The first-order intrinsic strains extend almost homogeneously over many coherently spreading areas and cause a homogeneous elastic deformation of macroscopic areas and thus an equar displacement of the atoms from their normal position. This extension becomes apparent in a variation of the lattice constant. The second-order intrinsic strains are only almost homogeneous over one or a few areas. Because of these stresses, respective fluctuations of the lattice constant and of the interlattice plane distance around a mean value, occur. These fluctuations are also frequently described as lattice distortion and serve as a measure of the disturbance occurring in the solid [74—76]. Third-order intrinsic strains are inhomogeneous over smallest areas (several atomic spaces). Impurity atoms fitting into the lattice in accordance with their dimensions, can be incorporated into substitutional or interstitial lattice sites. They are distributed within the coherent lattice areas. Their distribution may be statistic or discrete. The distortions caused by the impurity atoms fade out quickly. The intrinsic stresses connected with these distortions are the third-order intrinsic stresses.

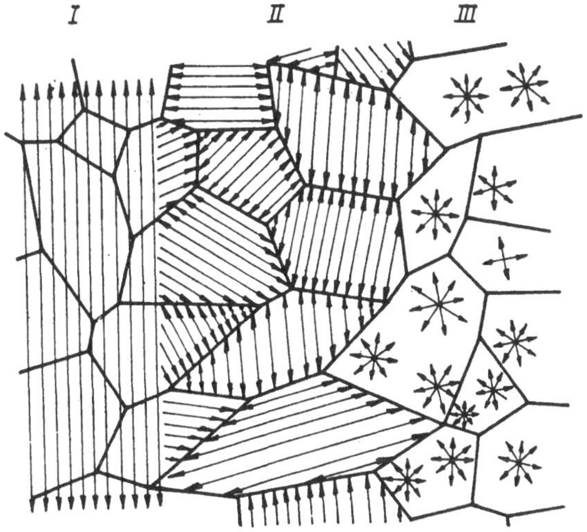

Fig. 3.23. Model of the self-contained stresses I, II, and III in a crystal consisting of coherently spreading ranges [72].

3.2.7. *On the concept of the "frozen-in lattice vibrations"*

The atoms in the lattice vibrate around their normal position with a thermal vibration amplitude corresponding to the temperature. This thermal vibration leads to a reduction of the X-ray interference intensity, expressed in the DEBYE-WALLER factor. The deplacements of individual lattice components according to the intervals of these vibration amplitudes cause a further reduction of the X-ray interference intensities. "Lattice distortions" come into being which according to [78] are comparable with "frozen-in thermal vibrations". According to the method of FRICKE and GWINNER [79] the second power of the magnitude of this deplacement (disturb amplitude) may be derived from the integral intensity. The term "frozen-in thermal vibrations" dates back to the 1930's—1940's. It had been observed that the DEBYE-SCHERRER line become broader and weaker through plastic deformation. It is known from evaluations according to [80] that the reduction of the intensity by the mentioned deplacement might only amount to 0.14 per cent. This is within the measuring error limits in the hitherto existing measuring techniques. The experimentally determined reductions of the intensity can be attributed to other defects. It follows from more recent detailed investigations [81, 68] that the total reduction of the intensity which can occur with the duration of the treatment may be attributed to an amorphization of the solid so that lattice distortions similar to thermal vibration cannot be supposed.

3.2.8. *The influence of the structural imperfections on the solid reactivity*

The main reason for the increase in the chemical reactivity is the increased enthalpy of the solid caused by the variation of the real structure. Correlations between the stored energy and reactivity have been stated in numerous publications [56, 82]. Of special interest is the proportion of the individual defects in the energy content of the mechanically treated solids [83] (see [58]). The specific course of the reaction depends on the nature and the distribution of the structural defects. There are some universally valid correlations between individual structural imperfections and their reactivity which will be discussed in detail in 5.9. Thus it is generally true that point defects, dislocations, and grain boundaries accelerate the transport of matter. The nucleation probability is increased by the presence of dislocations. Size and structure of the surface determine above all the surface processes.

3.2.9. *Methods for the measurement of imperfections*

The methods for the determination of the imperfection are many-sided. This section can only give a survey of some important methods.

Generally several defects occur simultaneously with mechanical treatment, and a division into single defects is usually very difficult. Therefore a defect or an impurity state is frequently characterized. In this way it is possible to give the impurity state by the decrease of the crystallinity or by the increase of the X-ray amorphous part. By this is meant those parts in a powder sample which are no longer diffractionable by X-rays. A further appropriate parameter for characterizing the influence of the mechanical activation is the degree of disorder. It is a summary parameter for all the factors leading to a broadening of the X-ray crystallographic line profile.

A summary of the methods for the characterization of the mechanically activated solid can be found in [84], (see Tab. 3.2.).

Table 3.2. Methods for measuring the imperfection.

1. Investigation of the zero-dimensional imperfection:
1.1. Atomic lattice defect
X-ray diffraction [85], neutron diffraction, electron diffraction, absorption [86], resistence measurement, density, diffusion measurement.
1.2. Electronic imperfection
Exoelectron emission [89], measurement of luminescence [90], electron spectroscopy [91, 92], EPR [93, 94, 95, 96].

2. Investigation of one-dimensional imperfection:
Light microscopy, scanning electron microscopy, transmission electron microscope [97, 98, 99], X-ray topography [55], X-ray diffraction [100,101].

3. Investigation of two-dimensional imperfection:
3.1. Surface
Sorption technique, method for investigation of canal rays, photometric surface measurement [64].
3.2. Grain boundaries, twin boundaries
light and electron microscopy [97, 98].
3.3. Stacking faults:
X-ray diffraction [102].
3.4. Particle sizes
Sieve screen analysis $> 40\ \mu m$, sedimentation analysis $60 - 1\ \mu m$, air-jet sieve $> 10\ \mu m$, counting method (lightmicroscopic) $> 1\ \mu m$, electron microscopic $< 1\ \mu m$ [103–106].

4. Investigation of three-dimensional defects:
4.1. Amorphous structure
X-ray diffraction [107, 108], electron diffraction [109, 110], neutron diffraction [111, 112].
4.2. Pores
Porosimetry [113].
4.3. Inclusions
X-ray topography [55], electron microscopy [97, 98].

5. Investigation of the state of defects:
5.1. Dimension of the primary crystallite size
X-ray diffraction [75, 76, 77].
5.2. Lattice distortions
X-ray diffraction [114–122].
5.3. Crystallinity
X-ray diffraction [123], Infr-red spectrocopy [124], luminescence [125], differential thermal analysis [126], determination of solubility [127].
5.4. Degree of disorder
X-ray diffraction [128].

6. Determination of the energy content:
Solution calorimetry [129], thermal techniques [130].

3.3. Physical elementary processes in the mechanical activation of solids

3.3.1. Introduction and formulation of the problem

Physical elementary processes following the action of mechanical energy on solids have in the same way as the interaction with high-energy radiation or irradiation with high-energy particles, increasingly been the object of experimental and theoretical investigations during recent years.

The field of science described internationally as Tribophysics, is concerned with the relationship between mechanical interactions and physical phenomena among solid interfaces and/or between the latter and their environment including the microscopic and submicroscopic processes occurring with mechanical interference in the structure of solid interfaces as investigated by tribomechanics [131]. Centres of this field of science are in the Soviet Union inter alia in Moscow, Leningrad, Sverdlovsk, Novosibirsk (Chapt. 1., [9]), in the UK in Cambridge [132], in the Federal Republic of Germany in Karlsruhe [133], in Australia in Melbourne [134], in the GDR in Berlin (Chapt. 1., [1]). In addition there are numerous research teams dealing with special problems of tribophysics.

The supply of mechanical energy in treatment and fracture processes causes changes in morphology, electronic structure, and chemical composition of the solid as already described in 3.1. and 3.2. Contrary to other non-thermal excitation processes in which a defined energy supply takes place, e.g. by irradiation or bombardment with high-energy particles, and in which the properties of the activated solid are investigated, the mechanical interference, the nature of the energy absorption as well as the dissipation behaviour of the energy supplied to the solid are in the case of mechanical activation processes object of investigations and have to be taken into consideration in comparative examinations.

Today there is knowledge of a great number of atomic, microscopic, and macroscopic physical processes which occur with mechanical activation and by interaction of the solids involved (tool/workpiece), (Tab. 3.3.).

Table 3.3. Physical processes proceeding during tribo-mechanical treatment.

- Emission of photons
- Emission of electrons
- Static electrification
- Electrostatic discharges
- Change of the electrical conductivity
- Emission of lattice components
- Excitation of lattice vibrations
- Formation and migration of lattice and electron defects
- Local heating-up of the solid
- Formation of fresh surfaces
- Enlargement of the surface
- Fracture processes
- Material abrasion
- Material transitions between solids
- Amorphization
- Inserting of impurities
- Plastic deformation

In many cases a phenomenological approach still dominates the investigation of these processes, which is connected with the specifics of mechanical activation processes:

- Generally the activation energy applied to the solid cannot be determined, because the mechanisms of the energy transmission and dissipation are not yet known and the energy feed during the treatment proceeds irregularly in time and space;
- from the technical point of view the energy losses and the physico-chemical interactions with the tool are only hardly, or not at all, calculable;
- the mechanical activation always includes disturbances of the surface so that an overlapping of surface and bulk processes occurs;
- the processes occurring during the treatment mutually influence each other in many ways. They often occur simultaneously and cause different secondary processes;
- the processes depend in manifold ways on the activation energy and on the treatment parameters, such as contact surface, gas pressure, temperature;
- in order to detect the events occurring during the mechanical treatment, processes are used which themselves are still the objects of research in the field of tribophysics (emission of exoelectrons, triboluminescence, fracture processes).

Below having discussed some principle deformation processes, basic physical excitation processes such as the development of high temperature, the emission of photons and electrons, the electrostatic charge, and the release of lattice components during the mechanical activation shall be dealt with, which are closely connected with the tribochemical processes treated in this book. Special attention is given to the effect of different treatment conditions and the correlations of the activation processes. Depictions of experiments shall be dealt with in terms of their characteristics only, and original literature will be referred to for details.

3.3.2. Mechanical activation

In addition to the integral treatment techniques used in tribochemical investigation (see 3.1.) the research of physical basic processes requires temporally and spatially defined action on the solids under clean environmental conditions. For this purpose falling ball or indentation apparatures, the production of defined fractures, and strain devices with variable strain velocity are used. Both the latter techniques are distinguished by the possibility of on extensive degree of contact-free excitation of the solid. The interaction with the tool remains limited to the impact point of the wedge and the fix point of the sample respectively, which are outside the investigated sample volume. On account of intensity, e.g. in spectroscopic examinations, continuously working milling devices and arrangements according to the lathe principle are also used.

The experimental effort and the variable possibilities of the treatment parameters (velocity, power, tension) depend to a large extent on the required high-vacuum and ultrahigh vaccum conditions, the thermal regime, and the detectors used (see Fig. 3.24.). One of the main problems in the investigation of tribomechanical activation processes as well as in the evaluation and optimization of technical processes is the quantitative approach to the process of energy transmission and dissipation. Some model concepts and possibilities for the correlation of treatment parameters will be discussed below; the arrangement follows methodical aspects. Fluent transitions between the kinds of stresses exist within the scope of the theory of plastic deformation and strength.

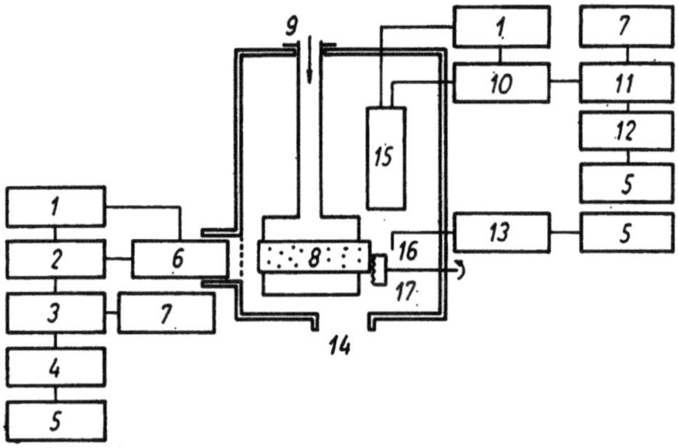

Fig. 3.24. Experimental setup for the measurement of luminescence, electron emission, and crystal charging [136].

1 – high-voltage powder supply unit, 2 – cathode follower, 3 – pulse amplifier, 4 – integrator, 5 – recorder, 6 – photomultiplier, 7 – oscilloscope, 8 – crystal, 9 – cooling device, 10 – pre-amplifier, 11 – linear amplifier, 12 – counting-rate meter, 13 – electrometer, 14 – to the vacuum pump, 15 – multiplier, 16 – charge measuring probe, 17 – milling cutter.

3.3.2.1. *Punctiform stress*

Mechanical treatment processes by which a punctiform stress occurrs through a static compressive load or dynamically through an impact can be dealt with by the equations of the HERTZIAN theory of elasticity [137]. These indicate the relationship between the material and energetic parameters of the pressure and impact partners involved, and the contact surface and contact time.

In recent years, numerous experiments in the field of the morphology of the deformation structure have been carried out with etching techniques and hardness measurements [138–141] by help of which it was possible to prove the validity of the theoretical statements.

As static load a ball is pressed under a certain load onto the surface of a plane sample. With a very low load the deformation remains purely elastic. The sample goes back to the initial state after removing the load. The contact surface is calculated as a circle with radius r by the HERTZIAN theory

$$r = \left\{\frac{3}{4} F \cdot g \cdot R_K \left(\frac{1 - \nu_K^2}{E_K} + \frac{1 - \nu_p^2}{E_p}\right)\right\}^{1/3} \tag{1}$$

(F – load, g – gravitational acceleration, R_K – radius of the ball, ν_K – POISSON's transverse contraction of the ball, ν_p – POISSON's transverse contraction of the sample, E_K – elastic modulus of the ball, E_p – elastic modulus of the sample.)

The contact surface in the elastic sphere is proportional to $F^{2/3}$, the average pressure p_m acting onto the sample corresponds proportionally to $F^{1/3}$.

Examining the basic processes account must be taken of the fact that the pressure p does not have uniform effect on the contact surface. According to [142]

$$p = p_0(1 - x^2/r^2) \tag{2}$$

(p_0 – pressure in the centre, x – position co-ordinate, r – radius of the indentation area.)

If the load pressing the ball onto the target is increased, the elastic deformation turns into plastic deformation. The area of the largest stress is at a distance of $r/2$ below the surface. If $p_{max} > 1.1\sigma_0$ (p_{max} — maximum pressure, σ_0 — limit of eleasticity of the solid) then plastic deformation starts. It further continues at an increase of p_m until complete plastic flow starts around the indentation body at $p_m \approx 3\sigma_0$.

In similar manner to static deformation several stages of deformation can be distinguished in dynamic stress by impact: With a small length of drop and small ball weight the deformation remains elastic. The maximum radius r of the contact surface changes with the density ϱ_K and the velocity v_K of the ball [132]:

$$r = R_K \left[\left\{ \frac{5}{4} \pi \varrho_K \left(\frac{1 - \nu_K^2}{E_K} + \frac{1 - \nu_p^2}{E_p} \right) \right\}^{1/5} \right] v_K^{2/5} \tag{3}$$

(R_K — radius of the ball, ν_K — POISSIN's transverse contraction of the ball, ν_p — POISSIN's transverse contraction of the sample, E_K — elastic modulus of the ball, E_p — elastic modulus of the sample.)

If the deformation energy rises, plastic deformation starts and proceeds to a completely plastic flow as soon as the pressure has exceeded the value p_F, the dynamic flow pressure. The mean flow pressure is greater in dynamic than in static stress and can be evaluated from energetic considerations [132]:

$$p_F = m_K g \frac{h_1 - 3/8 \cdot h_2}{V_S} \tag{4}$$

$$V_S = \frac{r}{4R_K}$$

(h_1 — height of the ball, h_2 — height of recoil, V_S — virtual indentation volume, r — radius of the indentation, R_K — radius of the ball, m_K — mass of the ball, g — gravitational acceleration.)

It is presumed that the energy dissipated during the propagation of elastic waves is low and the rise of temperature around the indentation continues to be small and has no important effect on the strength properties of the material.

Upon the penetration of the impact body into the solid, one part of the kinetic energy is transformed in heat and latent plastic deformation, while another part is preserved for the body as recoil work over the elastic energy.

$$W_K = W_{es} + W_s \tag{5}$$

$$W_s = \frac{m_K}{2} (v_K^2 - v_K'^2) \tag{6}$$

(W_K — kinetic energy of the ball, W_{es} — elastically stored energy, W_S — energy of the latent plastic deformation and heat, m_K — mass of the ball, v_K — impact velocity of the ball, v_K' — recoil velocity of the ball.)

In order to consider the energy balance and to evaluate the possibilities for reaction during deformation, the duration of the solid excitation caused by the impact is important.

The time interval up to the deepest indentation of the ball into the plate may be t_S for the impact (see Fig. 3.25.) and the time up to the recoil t_0. During the time $t_S > 0$ kinetic energy is continuously transformed into heat at the impact point. The value t_S is derived from the HERTZIAN equations in connection with the impact velocity v_K as follows:

$$t_S = \left[2.94 \left\{\frac{5}{4} \pi \varrho_K \left(\frac{1-\nu_K^2}{E_K} + \frac{1-\nu_p^2}{E_p}\right)\right\}^{2/5}\right] v_K^{-1/5} R_K \tag{7}$$

(Meaning of the symbole see equ. (3).)

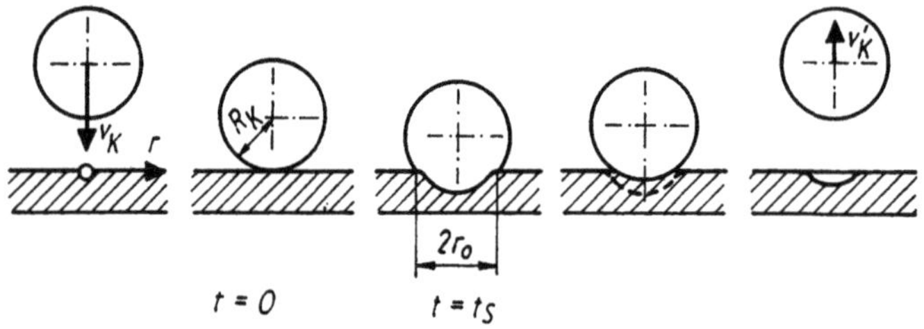

Fig. 3.25. Schematic illustration of the impact process onto a plane surface after [143].
R_K — ball radius, r_0 — radius of the deepest indentation, v_K — impact velocity, v'_K— recoil velocity, t — time, t_S — impact time, r — space co-ordinates.

With the radius r_0 of the deepest indentation of the ball on the plate and the impact time t_S at the moment of impact of steel balls having a diameter of 1 mm the values represented in tables 3.5. result with the material parameters for NaCl, Ni, and Fe given in Tab. 3.4. [144].[1]

From aknowledge of these parameters it is possible to evaluate the heat accumulation and the rise of temperature in impact processes, and these will be discussed in 3.3.3.

Table 3.4. Material parameters for impact and heat conduction [144].

	NaCl	Ni	Fe
ϱ density, g/cm³	2.167	8.910	7.876
E modulus of elasticity N/cm²	$4.107 \cdot 10^6$	$1.933 \cdot 10^7$	$2.114 \cdot 10^7$
K thermal conductivity coefficient J/cm · s · K	0.0578 (300 K)	0.9211	0.7327
$\varkappa$ coefficient of thermometric conductivity cm²/s	0.0298	0.235	0.202
c specific heat J/g · K	0.8960 (273 ... 373 K)	0.4396	0.4606
ν lattice deformation	0.207	0.31	0.30

[1] It has to be taken into consideration in this connection that by the plastic deformation neglected here the actual contact time is longer than the time given in Tab. 3.5.

Table 3.5. Values for the radius of the deepest indentation r_0 and the impact time t_S during the impact of steel balls having a diameter of 1 mm and with different velocities v_K [144].

	v_K, m/s	r_0, cm	t_S, s
NaCl	2	$4.031 \cdot 10^{-3}$	$2.4 \cdot 10^{-6}$
NaCl	40	$1.336 \cdot 10^{-2}$	$1.3 \cdot 10^{-6}$
Ni	40	$1.068 \cdot 10^{-2}$	$8.4 \cdot 10^{-7}$

3.3.2.2. *Plastic deformation*

Clear conditions for the characterization of treatment processes exist in strain stress for the sphere of plastic deformation.

The plastic deformation is the irreversible change of the shape of a body under the influence of stresses. It mainly occurs through the development and motion of dislocations in the solid.

Having passed through n dislocations with a free path length $l \ll L$ the plastic deformation $\varepsilon_p^{(n)}$ of the lattice [145] according to Fig. 3.26. is

$$\varepsilon_p^{(n)} \approx n \frac{b}{L} \frac{l}{L} = \varrho_b \cdot b \cdot l \tag{8}$$

(b — BURGERS vector, L — dislocation length, l — free path length of the dislocation, $\varrho_b = \frac{n}{L^2}$ — density of the mobile dislocations.)

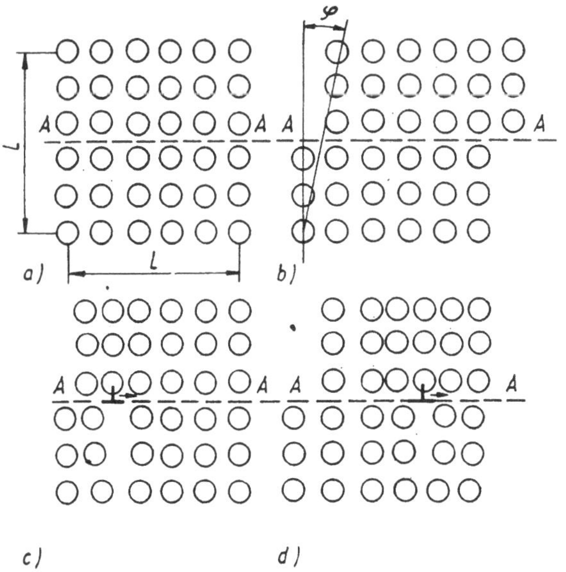

Fig. 3.26. Plastic deformation as motion of dislocations [145].

a) cross-section of the initial sample with the length L, b) shift in the plane AA by the lattice constant a, c) Formation of an edge dislocation with the BURGERS vector $b = a$, d) Motion of the dislocation in the plane AA, ⊥ — edge dislocation.

If v as mean velocity of the dislocations is introduced, there results from equ. (8) for the relationship between deformation velocity $\dot{\varepsilon}_p$ and the mobile dislocation density[1]

$$\dot{\varepsilon}_p = \varrho_b(t)\, bv \tag{9}$$

($\varrho_b(t)$ — density of the mobile dislocations at the time t.)

The deformation velocity $\dot{\varepsilon}_p = 3 \cdot 10^8\ \text{s}^{-1}$ can be evaluated for $v = 10^5$ cm/s ($\leqq$ sound velocity), $b = 3 \cdot 10^{-8}$ cm, $\varrho = 10^{11}\ \text{cm}^{-2}$ for stresses where nearly all the dislocations become mobile. That is to say, the dislocation mobility is theoretically able to elongate the body by a third within 10^{-9} s. Since plastic deformations caused by tension apparatuses take place in metals within the range of about $\dot{\varepsilon}_p = 10^{-5}$ to $10^{-2}\ \text{s}^{-1}$ and high-speed deformations reach about $\dot{\varepsilon}_p = 10^4\ \text{s}^{-1}$ it can be concluded that usual dislocation densities are entirely sufficient for guaranteeing the fastest plastic deformations observed.

Only with very high temperatures $T \geqq 0.9 T_S$ (T_S — melting temperature) and with very small stresses $\sigma \leqq 0.2\ \text{N/cm}^2$ does a vacancy mechanism begin to have effect.

Dislocations develop internal tensile and compressive stresses through the destruction of the short-range order of the solid and represent elementary sources of self-contained stress. The energy content W_V of a screw dislocation can be evaluated by means of the continuum theory from the energy, surplus of a crystal having dislocations with respect to the same crystal volume of an ideal crystal [145].

$$W_V \approx \frac{Gb^2L}{2} \tag{10}$$

(G — shear modulus, b — Burgers vector, L — dislocation length.)

The energy of a dislocation is proportional to its length and to the square of its Burgers vector. In this connection the contribution of the extensive stresses of a dislocation to the energy content is essentially larger than the share of the dislocation core so that the utilization of the linear elasticity theory seems to be justified.

A quantitative evaluation of the energy of a dislocation for a shear modulus $G = 3 \cdot 10^{10}\ \text{N/m}^2$ results according to equ. (10) in a value $\gg 10$ eV for $L \gg b$. Thus dislocations are not defects in the thermodynamic equilibrium but in principle the consequence of a previous power action on the material. The high energy barrier leads to the fact that the dislocation density does not drop to zero in recovery processes either. Under the influence of external mechanical stresses numerous reactions (climbing, multiplication, splitting-up, association) of dislocations among one another and with foreign atoms are possible, whereby the mechanisms of the energy exchange processes and their quantitative part are still unclear in many cases, but they can become effective in many ways in the release of elementary processes of the mechanical activation.

Recently the role of zero-dimensional defects in the plastic deformation has also been subject to increasing investigation. As the point defects described in 3.1. vacancies

---

[1] The total dislocation density $\varrho(t)$ results from the part $\varrho_b(t)$ of the mobile dislocations and from the stationary dislocations $\varrho_u(t)$ stopped immediately after the formation.

in particular occur under the influence of mechanical energy. Their formation energy $E_V{}^f$ is small and is well correlated with the melting temperature T_S [145] for the majority of metals.

$$E_V{}^f \approx 9kT_S \tag{11}$$

For aluminium with $T_S = 933$ K there results for the formation energy of vacancy $E_V{}^f = 0.68$ eV.

The equilibrium concentration n_{V0} of the vacancies is calculated according to the BOLTZMANN-statement

$$n_{V0} = N_0 \exp\left(-E_V{}^f/kT\right) \tag{12}$$

(N_0 — number of the lattice site.)

so that about 10^{10} vacancies per cm^3 exist at room temperature in aluminium. The vacancy concentration increases by 7 orders of magnitude with a temperature rise of $0.7T_S$, which is the boundary of the field of application in engineering. It follows from equ. (11) and (12) that the concentration of vacancies is nearly the same for all crystals at melting temperature.

$$n_{V0}(T_S) = N_0 \exp(-9) \approx 10^{18}\ \text{cm}^{-3} \tag{13}$$

From this derives the theory of the melting of a solid upon reaching a defined vacancy concentration.

Vacancies are the only defects which can produce real physical effects in equilibrium concentration. For this the mobility of these defects is of a decisive importance. The activation energy for the migration of a vacancy $E_V{}^m$ resulting from the relaxation processes of the neighbouring atoms during the jump of a vacancy is approximately

$$E_V{}^m \approx 0.9E_V{}^f \approx 8kT_S \tag{14}$$

so that a thermal fluctuation process must be required for the reversible work of a vacancy migration. The velocity of the vacancy migration v_V can be calculated [145] from equ. [14]:

$$v_V = ap_V = an_K p_a = an_K \omega_0 \exp\left(-E_V{}^m/kT\right) \tag{15}$$

(ω_0 — frequency of the atomic vibration $= 10^{12}\ s^{-1}$, n_K — number of the atoms of the first co-ordination shell, a — lattice constant, p_a — probability of the jump of any atom of the first co-ordination shell, $p_V = n_K \cdot p_a$ probability of the jump for a vacancy.)

For temperatures near melting point there are migration velocities of about 10^2 cm/s. A diffusion of basic and foreign atoms occurs through vacancy jumps, which outweighs the influence of other diffusion processes. For low temperatures the motion of vacancies and foreign atoms along the dislocations pipes and grain boundaries (pipe diffusion) should be taken into consideration.

The formation and migration of zero-dimensional defects is closely connected with physico-chemical responses during and upon completion of a mechanical treatment. In this connection an unambiguous attribution is still not possible in many cases. The different conditions for formation and propagation of the defects prompt investigative methods with a systematic variation of the treatment conditions.

Zero-dimensional defects with higher formation energy may come into existence under intensive mechanical treatment and in interaction with dislocation motions, e.g. interstitial atoms, insertion of foreign atoms, crowdions (see 3.2.2.). The latter have also been observed in impact deformation. Their importance for the investigation of mechanical excitation states lies in the extent of the disturbances which may extend far beyond the deformation region through dislocation motion.

In the field of tribophysics, plastic deformation by strain as a defined treatment process on metals generally appears through an active load with constant deformation velocity $\varepsilon =$ const. (more rarely with constant load velocity $\dot{\sigma} =$ const) or with the stress of $\sigma =$ const (see Fig. 3.27.) in the case of creep. Generally a temperature of between $T = 0.2T_S$ and $0.8T_S$ is used as a parameter. In addition to investigations of strength and deformability in the analysis of stress-strain diagrams in the field of materials science, in the investigations of basic processes, the correlation is sought between static and dynamic processes in the variation of the dislocation density, and the formation and motion of defects.

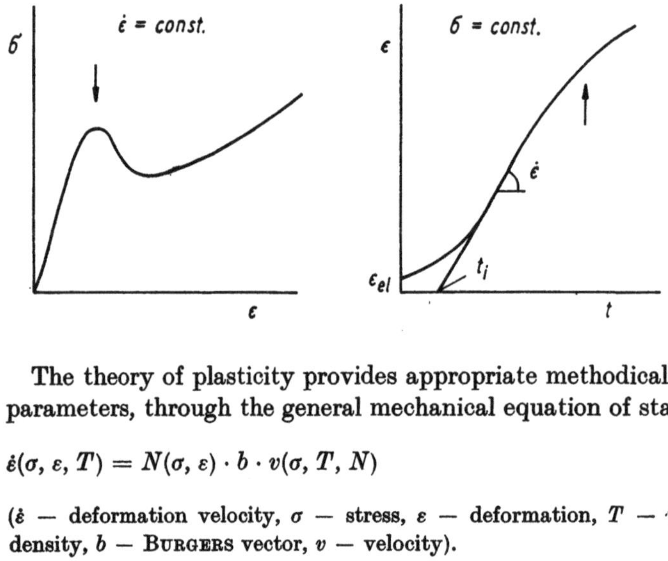

Fig. 3.27. Schematic work hardening and creep curve [146].
σ — tension, ε — strain, t — time,
↓ — tensile yield strength,
↑ — hardening,
t_i — incubation period,
$\dot{\varepsilon}$ — strain velocity.

The theory of plasticity provides appropriate methodically measurable experimental parameters, through the general mechanical equation of state [145]:

$$\dot{\varepsilon}(\sigma, \varepsilon, T) = N(\sigma, \varepsilon) \cdot b \cdot v(\sigma, T, N) \tag{16}$$

($\dot{\varepsilon}$ — deformation velocity, σ — stress, ε — deformation, T — temperature, N — dislocation density, b — Burgers vector, v — velocity).

The limits of this approach consist in the collective treatment of the dislocation structures and stress fields in the zone of deformation. The transition to fracture features shows the necessity of taking into consideration high local stresses which are also the cause of the spatial and temporal anisotropy of numerous elementary processes under mechanical activation.

3.3.2.3. *Development of fractures*

The occurrence of fractures is one of the basic phenomena in mechanical activation processes. The fracture is the physical process of the formation, association, and growth of cracks, which is terminated by the loss in stability of one of the cracks. Micro-cracks and sub-micro-cracks are elements of the defect structure similar to dislocations and vacancies, and can be formed during the plastic deformation. Although here too the quantitative treatment of atomic processes is still at the very beginning, the methods of tribophysics helpfully augment the phenomenologic examination of fractures and allow the suitable choice of experimental conditions. Fracture mechanics uses two criteria for the formation and propagation of fractures; force and energy conditions [147].

Force conditions

At the fracture front the molecular binding forces must be surmounted. This means in the case of the brittle fracture, that local tensile stresses attain the molecular tensile strength. Experiments show that local tensile strengths exceed the average applied tensile stress by 2 to 3 orders of magnitude (see Fig. 3.28.) as a result of the notch effect of existing micro-cracks.

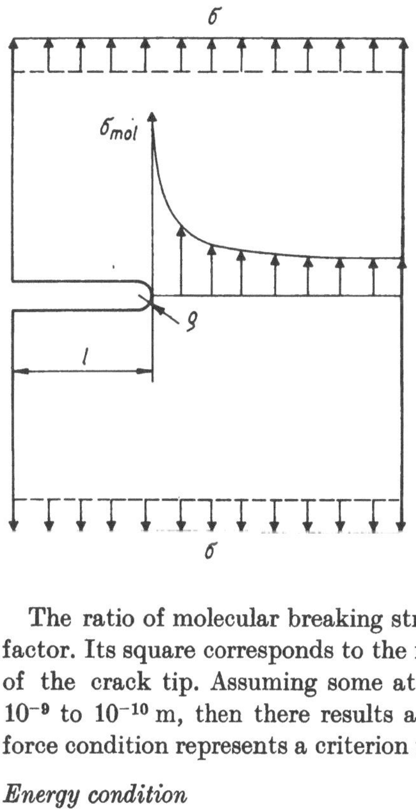

Fig. 3.28. Force condition for the brittle fracture during the tensile test; notch effect at the straight crack [147].
l – crack length, ϱ – radius of curvature of the crack tip, σ – tensile stress, σ_{mol} – molar breaking strength.

The ratio of molecular breaking strength to tensile strength is described as the notch factor. Its square corresponds to the ratio of the crack length to the radius of curvature of the crack tip. Assuming some atomic spacings as radius of the curvature, such as 10^{-9} to 10^{-10} m, then there results a value of a few μm as minimum crack length. The force condition represents a criterion for the mechanical stability of the solid.

Energy condition

The energy condition is a criterium for thermal stability. It requires that the differential crack propagation is energetically advantageous. This will be the case if the released crack propagation energy U is greater or equal to the energy required for the formation of the crack area during the propagation of the crack, the so-called effective fracture area energy E_B

$$\frac{dU}{dl} \geqq \frac{dE_B}{dl} \tag{17}$$

where dl is the differential crack propagation.

The crack length l necessary for equilibrium is designated as the GRIFFITH length [148]. The original GRIFFITH equation only considers the elastic deformation work and the free interfacial energy of the developing fracture planes. It was modified later [149, 150] when the energy supply and the energy uptake during the crack propagation was analysed more exactly.

The kinds of energy contributing to the crack propagation are represented in Fig. 3.29 taking account of the transformation processes. G_i are the specific energies (l — crack front, dl — crack propagation) related to the differential areas ($l \cdot$ dl).

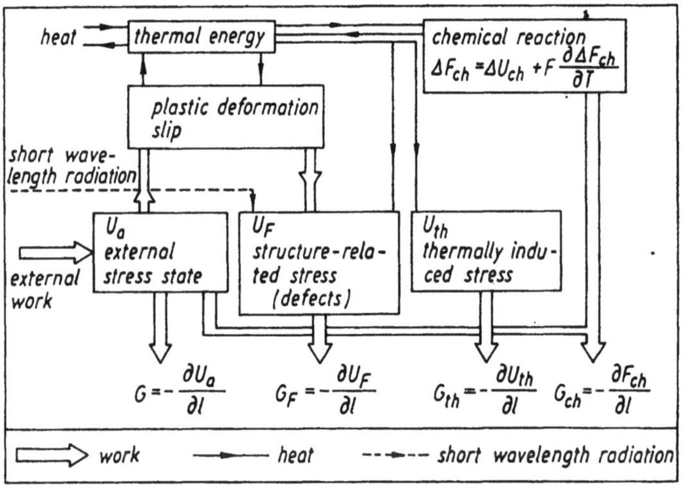

Fig. 3.29. Energy supply during the propagation of the crack [147].
G_i — crack propagation forces, l — crack length;

The energies taken up during the crack propagation are schematically summarized in Fig. 3.30. The free specific interfacial energy is described there as $2\gamma_0$, whereas $2\gamma = \mathrm{d}E_B/\mathrm{d}l$ characterizes the specific fracture plane energy.

Furthermore there is an energy term $\mathrm{d}E_k/\mathrm{d}l$ for the kinetic energy of the fracture plane related to the crack front.

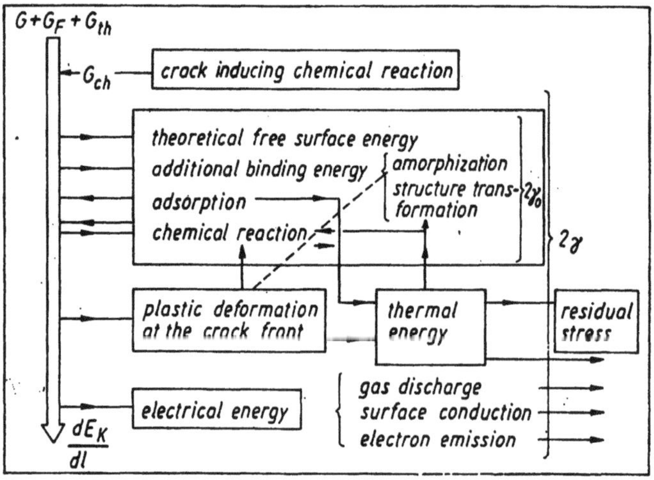

Fig. 3.30. Energy absorption during the propagation of the crack [147].
G_i — crack propagation forces (see Fig. 3.29.), l — crack length, γ — specific fracture-area energy, γ_0 — specific interfacial energy, E_K — kinetic energy of the fragments.

The general energy condition for the crack propagation then, is

$$\sum G_i = 2\gamma + \frac{dE_k}{dl} \tag{18}$$

(G_i — forces of the crack propagation, 2γ — specific fracture plane energy, E_k — kinetic energy, l — crack length.)

In recent papers the equivalence of force and energy conditions here used with the introduction of the GRIFFITH crack length is substituted by the dislocation-theoretical treatment of the instability condition in the crack tip [151] and by the theory of thermal fluctuation [152].

The transformation of energy does not remain constant in the course of the crack formation. Both the elastically stored energy and the absorption of energy by the crack vary as a function of the crack length. In this a static and a dynamic phase of the crack propagation can be distinguished. In the static phase the energy stored in the solid is only sufficient for meeting the energy requirement in the crack propagation. The dynamic phase starts with a critical crack length. The stored energy exceeds the energy demand and is mainly transformed into kinetic energy of the relieved solid regions.

This transformation causes an increase of the crack propagation velocity, but at the same time, the probability for the course of several energy transformation processes also changes. These usually very rapidly running processes are difficult to approach experimentally, so that until now an exact determination of the energetic exchanges in the fracture has not been possible.

One of the basic problems of fracture mechanics is the formation and propagation of micro-cracks up to the critical crack length [153, 154]. In numerous papers the stress field around the crack tip and the conditions of the unstable crack propagation in brittle bodies are calculated by the methods of the linear-elastic continuum theory. As a result it has been possible to derive technically relevant parameters for materials research in good approximation for the field of small tensile strengths and low fracture velocities. The degree of validity of linear-elastic fracture mechanics decreases to the same degree as the plasticity at the crack tip increases.

Experiments have shown that the value of the crack propagation energy at the beginning of the unstable crack propagation exceeded the surface energy by the factor 10. The occurrence of plastic deformations can be attributed to the high stress concentration in the crack tip and thus to the high local energy denstiy, which leads to irreversible deformations, that is to say, plastic deformations, in the nearer environment of the crack tip and to the occurrence of high temperatures (see 3.3.3.). The size of the plastic zone depends inter alia on the fracture velocity, the dimension of the samples, and the temperature.

Attempts have been made to formulate the plastic part as an additional value to the surface energy [155] and as crack resistance [∂56], through extending the linear-elastic fracture mechanics. The continuum-mechanical treatment of the solid based on this approach, encounters fundamental difficulties in the treatment of processes in the atomic dimensions of the fracture tip. Therefore an attempt has been made to describe the fracture propagation as a function of external stress fields [151—154] by means of micro-structural mechanisms of the displacement theory. The micro-cracks are regarded as discontinuities in the atomic structure, that is as lattice distortions simi-

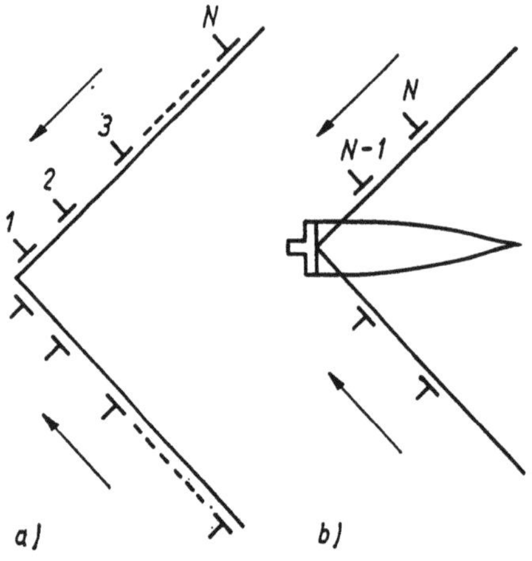

Fig. 3.31. Pattern of the crack formation by piled-up groups of dislocations [154].
a) initial agglomerates of dislocations,
b) formation of a dislocation crack, ⊥ — edge dislocation.

lar to dislocations. The occurrence of cracks may be explained by the pile-up of intersecting dislocations (see Fig. 3.31.).

Brittle crack propagation occurs if the dislocations are for instance blocked by defects. In this model the transition from the brittle to the ductile fracture is explained by the release of dislocations at higher temperatures and the subsequent processes of mobility and multiplication of dislocations. The theoretical formulations permit the interpretation of the temperature-dependence of the specific crack propagation energy and the treatment of the influence of lattice distortions, and generally of initial stress fields, on the crack formation.

We shall only refer here to further theoretical formulations describing crack formation and fracture propagation in solids from the atomic aspects with interaction potentials of the lattice atoms [157] or dynamically by the LAGRANGIAN formalism [158]. The disadvantage still inherent in the atomic models consists in the difficulty of connecting the atomic interaction forces with microscopic quantities in order to obtain quantitatively verifiable evidence.

The crack propagation velocity is an important parameter for the correlation of fracture-physical quantities with elementary processes, such as temperature rise, luminescence, electron emission, electrostatic charging. The influence of the external temperature, surrounding medium, and mechanical stress can be so selected through experimental conditions that isotropic and brittle fracture behaviour may be assumed for appropiate samples with corresponding sample geometry. The deviation of the average and differential fracture propagation velocity from the theoretically possible fracture velocity, the sound velocity, permits conclusions with regard to the transformations of energy which will be correlated with the observed physical and morphological changes. These investigations are still in their infancy.

3.3.3. *Occurrence of high temperatures*

The mechanical treatment processes which were discussed in the preceding section and which proceed in a complex form in technical comminution and activation devices, lead to a temperature rise in the vicinity of the deformation zone. The temperature rise can fundamentally influence the material behaviour, the occurrence of physical elementary processes, and the release of chemical reactions (see 5.3.). In the plastic deformation of metals the part dissipated as thermal energy runs to 95 percent of the deformation energy [159]. Numerous examinations have been carried out to calculate and register experimentally the temperature rise. These take into consideration different deformation mechanisms and are closely connected with the material properties of the mechanically activated solids.

3.3.3.1. *Friction and impact temperatures*

It is known from immediate practical experience that the surface temperature of rubbing and sliding solids increases.

The mechanical work directed against the frictional force is mostly converted into heat so that the following may be stated approximatively for a metal stick sliding with its front surface over a plane [132]:

$$Q = \mu \cdot F \cdot g \cdot v \tag{19}$$

(Q — thermal energy, μ — coefficient of friction, F — load of the cylinder, g — gravitational acceleration, v — velocity.)

Having regard to the heat conduction processes the following rise in temperature at the interface between the rubbing partners results

$$T - T_0 = \frac{\mu \cdot F \cdot g \cdot v}{2\pi r} \sqrt{\frac{1}{2\beta \cdot K \cdot r}} \tag{20}$$

(β — heat transfer coefficient, r — radius of the cylinders, K — thermal conductivity (for the meaning of the symbols see equ. (19).)

For a constantan cylinder with a diameter of 1 mm and a load of 200 g a kinetic friction resistance of $\mu = 0.3$ results with friction on steel and with a sliding velocity of 200 cm/s. As rise in temperature of the constantan surface results $\Delta T \approx 200$ K from $k = 0.2$ J/cm · s · K and $\beta = 0.004$ J/cm^2 · K. This value represents a minimum evaluation, since in particular the true contact surface is much smaller than the macroscopic front surface.

Experiments with several rubbing partners formed as thermoelectric couples provided temperature peaks between 600 K and 1,000 K in the course of periods smaller than 10^{-4} s. Similar temperatures were measured by means of temperature radiation and infra-red measurements in non-conducting materials. Melting phenomena could be observed during the scratching of a glass plate by a steel needle and a temperature of 900 K could be calculated [160].

These investigations in rubbing systems form the basis of the hot-spot theory [132], (see 5.3.).

In distinction to the theories to be dealt with later, the upper limit is the melting temperature of the reactants involved[1]. The intensive local heating on the surface of rubbing bodies plays an important role for a series of engineering processes. Among these latter are friction, lubrication, and wear of metals, the welding of organic materials,

---

[1] At times higher temperatures were observed which could be attributed to the contribution of the chemical heat of reaction to the oxidation of the fresh metal surface.

and the release of chemical reactions, such as oxidation and decomposition during friction and impact stresses.

During the impact treatment under the conditions of the HERTZIAN impact equation (see 3.3.2.1.) the quantity of heat Q_0 per unit of volume and time formed by the kinetic energy of the impact body is considered which spreads concentrically from the impact area $0 \leqq r < r_0$ to the surrounding material $r > r_0$ (r_0 — radius of the deepest indentations in the plate) (see Fig. 3.25.).

This process can be described as a problem of heat conduction with source by subsidiary equations [144]

$$\frac{d^2(r\bar{u})}{dr^2} - q(r\bar{u}) = -\frac{rQ_0}{Kp} \quad \text{für } 0 \leqq r \leqq r_0 \tag{21}$$

$$\frac{d^2(r\bar{u})}{dr^2} - q^2(r\bar{u}) = 0 \quad \text{für } r > r_0 \tag{22}$$

where $\bar{u} = \int\limits_0^\infty e^{-pt}u(t)\,dt, \; u = 0 \quad \text{für } t = 0$

$q^2 = \frac{p}{\varkappa} \qquad \varkappa = \frac{K}{\varrho_P c_P} \qquad p =$ LAPLACE-multiplicator

(r — space co-ordinate, u — temperature, K — thermal conductivity coefficient, $\varkappa$ — temperature conductivity coefficient, t — time, ϱ_p — density of the plate, c_p — specific heat of the impact plate.)

If the latent deformation energy is neglected the heat quantity Q_0 in the impact volume

$$\Delta V \approx \frac{2}{3}\pi r_0^3 \tag{23}$$

can be calculated from the transmitted energy E_{kin}

$$Q_0 \approx E_{\text{kin}} = \frac{m_K}{2}(v_K^2 - v_K'^2) \cdot \frac{1}{\Delta v \cdot t_s} = \frac{3m_K(v_K^2 - v_K'^2)}{4\pi r_0^3 t_s} \tag{24}$$

(t_s — time up to the deepest penetration of the ball.)

The temperature of the impact process can be determined from equ. (21) for the point $r = 0$ at the time $t = t_s$. The temporal heating up and the storage, respectively, of the impact zone can be also given by equ. (21) for the time $t > t_s$.

Rises in temperature $\Delta T = 720$ K and $\Delta T = 475$ K, respectively, are calculated with recoil velocities $v_K' = 0.75 v_K$ and $v_K = 40$ m/s by the impact of a steel ball ($R_K = 0.5$ mm) onto a NaCl and Ni plate, respectively, with those values for r_0 and t_S given in Tab. 3.5. and with material parameters given in Tab. 3.4.

The values conform well with those results obtained in friction processes.

Duration and attenuation of the energetic excitation are important for the time response of mechanically activated processes. After the impact processes the energy density q results for the time $t > t_S$ from the heat conduction equation without source [143]

$$q = \frac{Q}{(4\pi\varkappa t)^{3/2}} \cdot e^{-\frac{r^2}{4\varkappa t}} \tag{25}$$

(Q — heat quantity released at the time $t = 0$ in the point $r = 0$.)

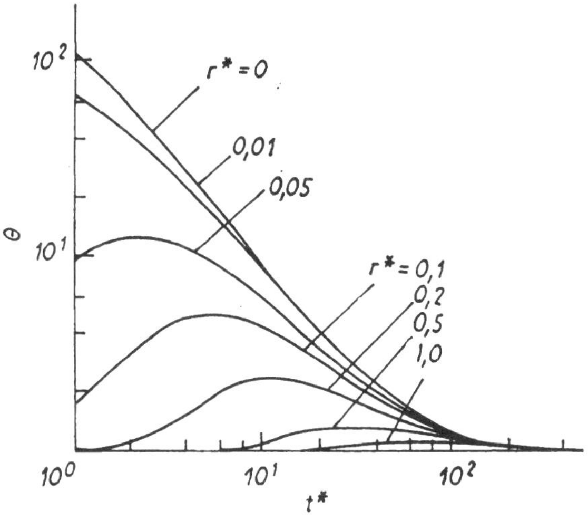

Fig. 3.32. Time behaviour of the excitation on the surface during the ball impact [143].

$\theta = \frac{q}{q_0}$ excitation quantity, $r^* = \frac{\pi r^2}{\pi r_0^2}$ quantity of distance, $t^* = \frac{t}{t_S}$ quantity of time

(q — energy density of the excited solid, q_0 — energy density of the non-excited solid, r — distance to the impact centre, r_0 — radius of the indentation area, t, t_S — see Fig. 3.25.).

Fig. 3.32. shows an evaluation according to equ. (25) by means of the HERTZIAN equations where the solid nearly reaches the normal state in times $t \approx 10^2\, t_S$, that is after about 10^{-4} s in our example.

At some distance from the impact point, that is for growing values of r^*, the excitation is delayed because of the finite propagation velocity of the energy in the solid.

These considerations are directly applied inter alia in adsorption reactions in activated solids. The time available for the incident gas molecules after lifting the impact body to react with the "thermally" initiated solid amounts to fractions of a second only.

3.3.3.2. *Temperatures in dislocation motions*

Mobile dislocations have also been regarded as causes of temperature rises in the dislocation area during the plastic deformation. A motion of dislocations with the velocity v occurs under the effect of a mechanical stress σ causing a rise in temperature ΔT after [161]

$$\Delta T \approx \frac{b \cdot \sigma \cdot v}{2\pi K} \cdot n \cdot \ln \frac{n\Lambda}{\lambda} \qquad \text{für } \lambda \ll \Lambda \tag{26}$$

$$\Delta T \approx \frac{b \cdot \sigma \cdot v}{2\pi K} \cdot n \cdot \left(\frac{2\pi\Lambda}{\lambda}\right)^{1/2} \qquad \text{für } \lambda \gg \Lambda \qquad \Lambda = \frac{2\varkappa}{v} \tag{27}$$

(b — BURGERS vector, σ — mechanical stress, v — dislocation velocity, K — heat conduction coefficient, n — number of the mobile dislocations, $\varkappa$ — temperature conduction coefficient, λ — average spacing of the dislocations.)

It results from this that rises in temperature in the range of some 100 K can only be expected in unusually high dislocation densities, occurring for example in high-speed rolling of ductile bodies. With aluminium, temperatures near melting point can be found.

The rise in temperature caused by the mechanically activated sliding of the molecules during the plastic flow may be significantly larger for non-crystalline materials having a high viscosity, such as polymers. Temperature increases of some 1,000 K were observed under special conditions [162]. The mechanism of the transformation of dislocation energy into heat is not yet clarified. Four models are discussed:

- The kinetic energy of rapidly mobile dislocations is transformed into heat, when the dislocation comes to a standstill;
- the energy transformation is caused by damping of dislocations and scattering of sound waves;
- formation and annihilation of dislocations;
- annihilation of point defects which are produced by the dislocation motion.

Although experimental results do not exist up until now the annihilation of point defects seems to be the most probable explanation [163].

3.3.3.3. *Fracture temperatures*

When fractures appear, a large part of energy dissipated during the fracture propagation is also transformed into heat. The irreversible plastic deformation work described as the crack resistance R and generally considerably exceeding the free interfacial energy is to a large extent transformed into heat energy Q

$$Q = 0.6 \cdots 0.9R \tag{28}$$

Whereas the free interfacial energy of solids totals approximately 0.1 to 1 J/m², values of about 10 J/m² for glass, 10^3 J/m² for polymers, and 10^5 J/m² for steel were determined for the crack resistance R and for the fracture area energy γ, respectively.

The upper limit for the temperature to be expected can be evaluated for the adiabatic case with high fracture velocities from knowing Q:[1]

$$T - T_0 = \frac{Q}{c \cdot \varrho \cdot 2\delta} \tag{29}$$

(c — specific heat, ϱ — density, 2δ — width of the fracture zone.)

Fracture velocity, the thermal properties of materials, the size of the irreversibly deformed zone, and the distance to the fracture have to be taken into consideration for the local distribution of temperature in the surrounding of the fracture. By a miniature thermocouple [161] a rise in temperature of 130 K could be proved in steel at a distance of 30 μm from the fracture front. Several theoretical expressions have been

[1] Accordingly the appearance of tensile stresses at the front edge of the plastic zone leads to the cooling in the adiabatic case which can run to 10 to 100 K.

suggested for the calculation of the temperature distribution in which the fracture tip is regarded as mobile heat source.

Assuming an idealized point-formed heat source at the crack tip the heat quantity dQ/dt produced by the heat source during the time unit ist for $\gamma \approx \gamma_w \gg \gamma_0$ [165]

$$\frac{dQ}{dt} = \frac{dQ}{dl} \cdot \frac{dl}{dt} = 2\gamma_w \cdot v_B \approx 2\gamma \cdot v_B \tag{30}$$

(γ_w — thermal energy of the fracture propagation, γ — fracture area energy, γ_0 — free interfacial energy, v_B — fracture propagation velocity, l — crack front.)

It can be evaluated for quartz glass with $2\gamma = 10$ J/m^2 that the melting temperature is exceeded at a distance of 10^{-9} m from the fracture front (see Fig. 5.10.).

The limits of this model are determined by the fact that

— the times for reaching a thermodynamic balance will become too short and the volume ranges do not allow a statistic definition of the temperature. Therefore the term "equivalent temperature" has been introduced for many of these considerations as a measure for the locally concentrated kinetic energy. This limitation is also indicated in the attribution of high temperatures in the conclusions of the magma-plasma model (see 5.1.);

— the heat conduction does not take place with infinitely high speed but is connected to the physical process of the transmission of vibrations and does not exceed the sound velocity;

— the heat source is not exactly point-like but covers the plastically deformed area in front of the fracture front.

Therefore in [166] a rectangular heat source was considered with the width 2δ and the length d moving with the constant fracture velocity v_B (see Fig. 3.33.). For the sake of simplification, the physical material properties, such as thermal conductivity and thermal diffusivity are assumed to be constant and linear heat conduction laws (Fourier law) are used. Then for the heat source

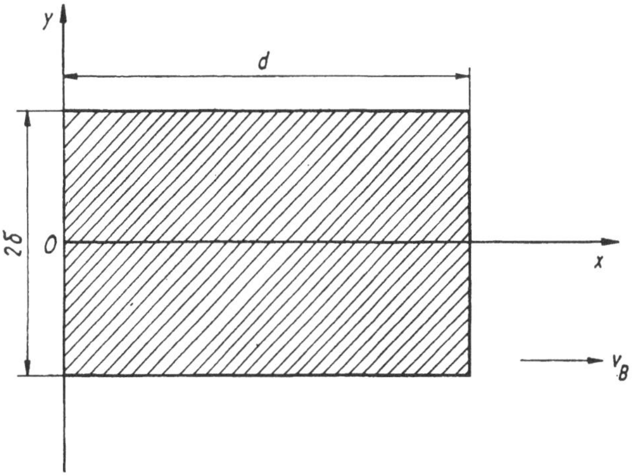

Fig. 3.33. Rectangular heat source at the crack front [166].
2δ — width of the zone, d — length, v_B — fracture velocity.

shown in Fig. 3.33. results the temperature dependent on position

$$T(x, y) = T_0 + (Q/2\delta \cdot c \cdot \varrho)\, \vartheta_3 \left(\frac{x}{\delta}, \frac{y}{\delta}, \frac{d}{\delta}, \psi\right) \tag{31}$$

with $\psi = v_B\delta/2\varkappa$ as parameter of velocity, where 2δ — width of the fracture zone, d — length of the fracture zone, v_B — fracture velocity, c — specific heat, ϱ — density, $\varkappa$ — temperature conduction coefficient, ϑ_3 — normalized temperature function.

Temperature profiles for ϑ_3 in which the maximum value rises to increasing extent, and is no longer situated in the centre but at the back of the mobile heat source, result for increasing fracture propagation velocities ($0.01 \leqq \psi \leqq 100$).

Three different stages of thermal diffusion can be distinguished in Fig. 3.34.:

1. $\psi < 0.1$ isothermal,
2. $0.1 \leqq \psi \leqq 10$ transition stage,
3. $\psi > 10$ adiabatic stage.

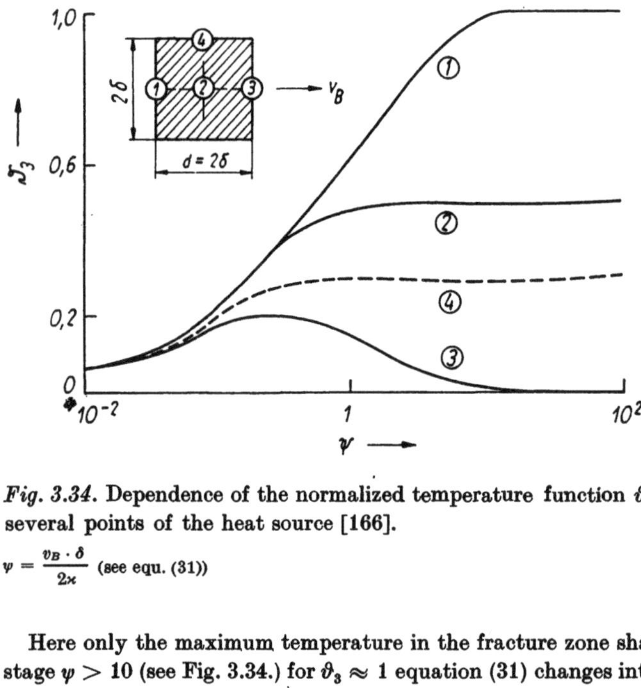

Fig. 3.34. Dependence of the normalized temperature function ϑ_3 on the velocity parameter ψ at several points of the heat source [166].

$\psi = \frac{v_B \cdot \delta}{2\varkappa}$ (see equ. (31))

Here only the maximum temperature in the fracture zone shall be dealt with. In the adiabatic stage $\psi > 10$ (see Fig. 3.34.) for $\vartheta_3 \approx 1$ equation (31) changes into equation (29):

$$T_{\max} - T_0 = Q/2\delta c\varrho.$$

Fig. 3.35. shows temperatures calculated for glass, polymethyl methacrylate, and steel for different values of 2δ in the adiabatic case which were also experimentally examined by photometry [167]. The relative intensities of 4 different wave lengths of the luminescence appearing during the fracture of glass and quartz, and obtained by a spectrometer were compared with the radiation of a black body. Shorttime temperature peaks of 3,200 K for glass and 4,700 K for quartz could be proved. Investigations interpreting similar spectra by other model concepts of triboluminescence will be discussed in 3.3.4. From the investigations of the temperatures resulted $2\delta \approx 2 \cdot 10^{-9}$ m

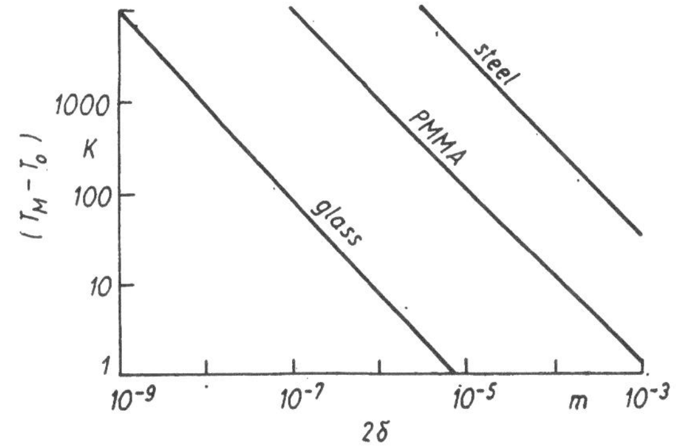

Fig. 3.35. **Maximum temperature for quickly moved fractures ($\psi \gg 1$) in dependence on the width of the fracture zone 2δ [166].**

as the quantity of the irreversibly deformed zone at the fracture front. By using similar approaches of the elasticity theory, rises in temperature of 1,000 K were calculated for NaCl in a volume of 10^{-15} cm^3 at the crack tip for the period of 10^{-9} s [168].

3.3.4. *Tribo-induced luminescence*

The emission of light as an accompanying phenomenon and result of mechanical activation processes was known as early as the 16th century. For instance a luminescence can be observed during the cracking of sugar crystals without any electric or optical means. Triboluminescence in several forms of manifestation have subsequently been proved by sensitive photomultipliers in more than 500 substances [169]. The term triboluminescence introduced in 1895 [170] embraces luminous phenomena occurring during the mechanical treatment (friction, fracture, compression) of solids and which are not in thermal equilibrium with the radiation system.

Systematic researches in recent years have proved that the mechanical energy frequently does not lead directly to the emission of photons in these processes (like the emission of electrons dealt with in 3.3.5., too) but excitational states develop to which the laws of tribo-induced electroluminescence, thermoluminescence, chemoluminescence and others can be applied [171, 172]. True triboluminescence and triboemission are present if an immediate excitation of lattice atoms by mechanical treatment, e.g. by the release of strong lattice vibrations, leads to an emission of photons and electrons. Investigations on tribo-induced luminescence have been carried out during impact stress, dilatation, and compression of monocrystals, in comminution devices, in scratch experiments, and during milling. A summary of the results on tribo-induced luminescence was published in 1977 [169].

In order to record the emission of photons today, photomultipliers combined with monochromators and filtering optics for the recording of the spectrum are exclusively used. Glow curves are obtained by devices for a heating-up linear in time [173].

3.3.4.1. *Electroluminescence*

If crystalline dielectric solids, such as sugar, ionic crystals, mica, are broken in air or comminuted in mills, an intensive luminescence appears. The luminescence can be attributed to electric fields which are formed under the influence of the mechanical treatment (see 3.3.6.). The spectrum of LiF during the compression in air and the gas-discharge luminescence by way of comparison are shown in Fig. 3.36. The bands belong concurrently to the second positive group of the N_2-spectrum. The bands at 310 nm also occurring in the argon atmosphere indicate a residue of OH in the surrounding gas. The pressure-dependence of the intensity of luminescence has a clear maximum in the gas discharge region at 10^3 Pa (see Fig. 3.37.). Differences arising for the various treatment conditions can be explained by the pressure-interval dependence of gas discharge processes [171]. The increase in intensity in the high vacuum during milling treatment is discussed below. The correlation between charge and light impulses during the scratching of LiF in the pressure range of 10^4 Pa and 10^{-5} Pa is represented in Fig. 3.38.

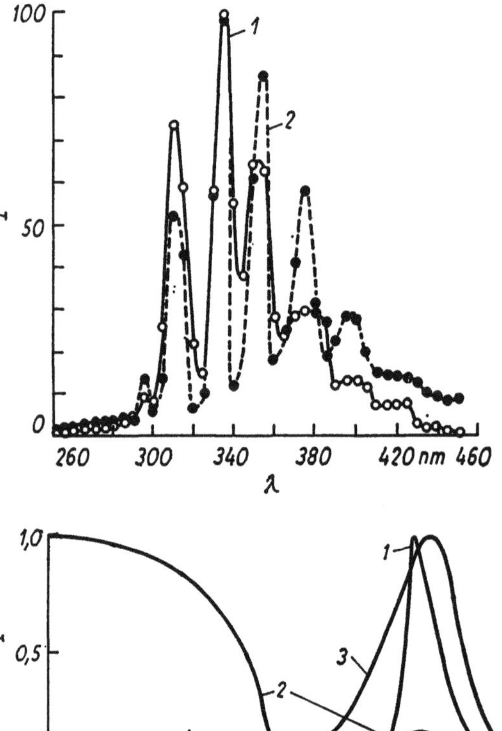

Fig. 3.36. **Luminscence spectrum during the fracture of LiF-crystals [174].**

1 – in air, 2 – gas discharge spectrum in air, *I* – intensity, *λ* – wave length.

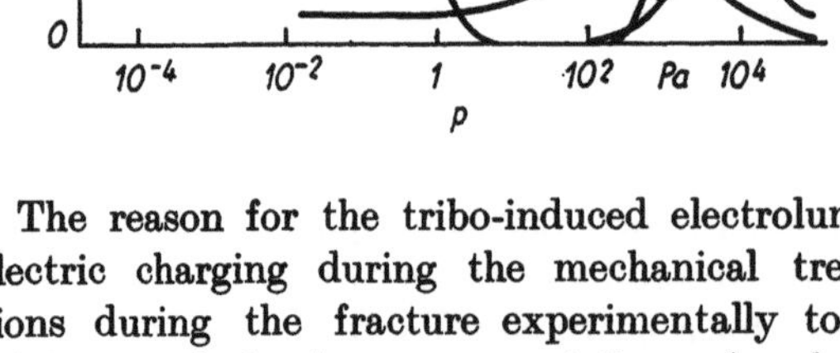

Fig. 3.37. **Pressure-dependence of the LiF-luminscence under different treatment conditions [171].**

1 – impact, 2 – cutter, 3 – scratching device, *I* – intensity, *p* – pressure.

The reason for the tribo-induced electroluminescence of the alkali halides is the electric charging during the mechanical treatment. Contact processes and reactions during the fracture experimentally to a large extent still unexplained (see 3.3.6.), result in charge areas at the surface having charge densities of $5 \cdot 10^{10}$ elementary charges per cm^2 where the field strength is about 10^5 V/cm [175]. The voltage is calculated from the product of field strength and spacing of the fracture surfaces. According to the PASCHEN law [176] an ignition of self-maintained gas discharge appears at the interface with the surrounding gaseous atmosphere for a determined ignition voltage with a fixed value of the product of $p \cdot d$ (p-gas pressure, d – spacing of the electrodes). Taking into consideration the geometry of the treatment ranges the validity of the Paschen law could be demonstrated for tribo-induced luminescence in the range 1 to 10^5 Pa [174, 177]. The decay after the end of the treatment can be attributed to the discharge of the surfaces. Non-self-maintained kinds of gas discharge luminescence are possible through the immediate excitation of gas molecules by electrons emitted and after-accelerated [172].

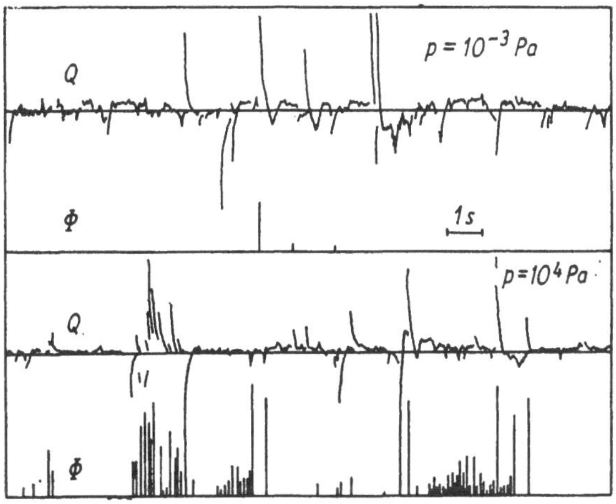

Fig. 3.38. Synchronous appearance of charge (Q) and light pulses (ϕ) during the scratching of LiF [171].

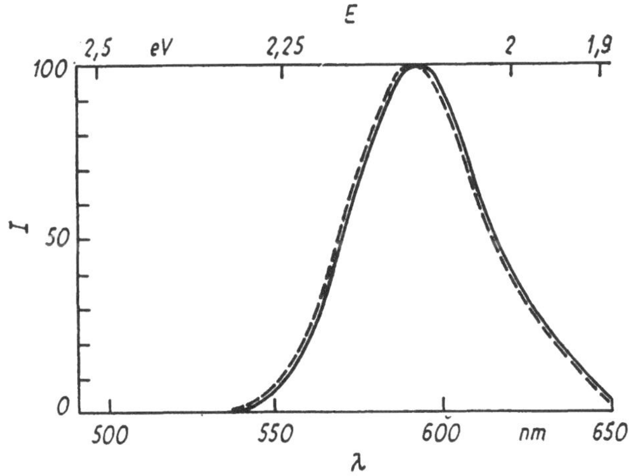

Fig. 3.39. Comparison of spectra of the tribo-induced luminescence and electroluminescence of ZnS : 5 · 10^{-3} Mn [171].
I – intensity of luminescence,
λ – wave lenght,
—— tribo-induced luminescence,
E – energy of the photons,
– – – electroluminescence.

The deformation process and the intensity of the mechanical treatment determine to a considerable extent the density of the electric charging and the geometry of the discharge area in a given system so that time of discharge, the pulse amplitude, the duration of the emission of photons vary to a major degree, in line with the treatment conditions. Tribo-induced electroluminescence in the form of gas discharge luminescence was also observed in, in addition to crystalline substances, glass, mercury (see 5.4.3.), and organic systems [169].

A solid luminescence which can be attributed to laws of electroluminescence has been proved [171] in doped zinc sulfides (ZnS : Cu, ZnS : Mn, ZnS : Cd). The correspondence with the solid luminescence can be shown by the spectra (see Fig. 3.39.).

The band at 590 nm refers to the transition of the ^4G-state to the basic state ^6S of Mn^{2+}. The shift in the spectrum (see Fig. 3.39.) depends on the mechanical pressure on the sample and can be used as a measure for the non-axial compression occurring in mechanically initiated processes [171]. The shift value of 1.5 nm for ZnS: Mn found for the treatment in the jet mill conforms with a pressure of $2.5 \cdot 10^8$ Pa and amounts to about one third of the fracture stress.

Further investigations with differently doped samples of ZnS have shown that luminescence occurs both in the increase and in the decrease of the mechanical stress. In this way a mechanical light generator with periodic triboluminescence can be constructed with piezoelectric ceramics stimulated by alternating voltage. Fig. 3.40. based on a oscillogram shows these relationships for the system ZnS : 10^{-3} Cu, J.

We refer to the details in the original literature [179]. Similar results have also been obtained for SiC [180].

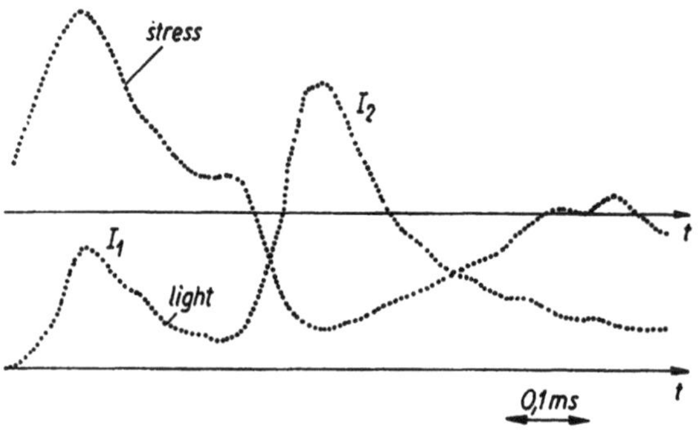

Fig. 3.40. Pressure and luminescence pulse of ZnS : 10^{-3} Cu, J [171].

The piezoelectricity plays the dominating role as the cause of the triboinduced electroluminescence of ZnS. Because of this, in this system field strength already appear upon elastic dilatation and compression of the lattice, which lead to a solid luminescence. Moreover the possible involvement of charged dislocations was also pointed to [181].

3.3.4.2. *Luminescence by re-combination of mechanically excited states*

In Fig. 3.37. a) clear increase of the luminescence in the high vacuum was shown. Today it is certain that fracture processes in the high vacuum, exclusive of gas-discharge phenomena, are accompanied by triboluminescence processes. Here, too, the exposure of spectra, the dependence on material, the temperature behaviour can give important indications as to the mechanism and the connection with tribomechanical activation processes.

Fig. 3.41. shows spectra for monocrystals of LiF, NaF, and NaCl.

The spectra are crystallo-specific and differ significantly from the gas-discharge spectra as illustrated in Fig. 3.36. The spectra consist of a continuum and show in some cases the resonance lines of the alkali-metal atoms in the long-wave part of the spectrum. A clear increase of luminescence during the cutting is registered in the sample cooling period (see Fig. 3.42.).

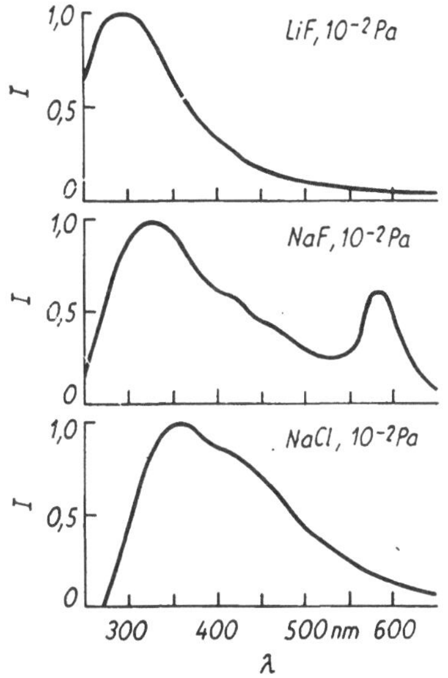

Fig. 3.41. Spectra of the tribo-induced luminescence in the high vacuum corrected in consideration, of the spectral apparatus sensitivity and the self-absorption [171].
I — intensity, λ — wave length.

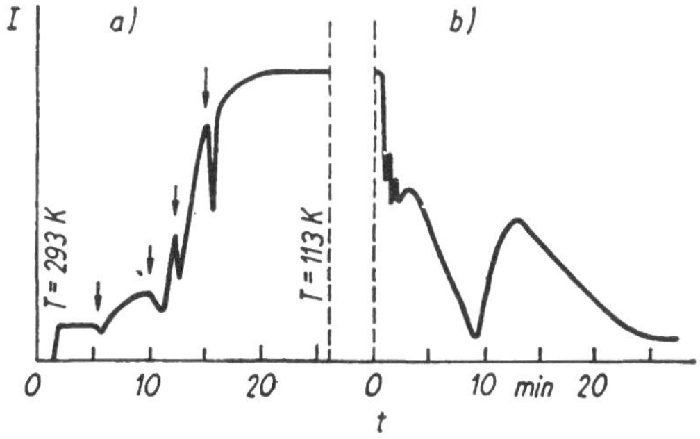

Fig. 3.42. Luminescence in the treatment system LiF/steel cutter with cooling (a) and reheating (b) of the crystal [136].
I — intensity, t — time
↓ — application of the cooling agent.

If the sample is heated up, the intensity decreases again. The intermediate minimum occurring after 10 minutes has to do with the desorption of gases, particularly water, from the cooling trap. It could be proved by measurement of the frictional force in high vacuum, that the intensity of the mechanical treatment severely decreases for a brief period through friction oxidation of the used steel cutter. If LiF is treated with an aluminium cutter, a decrease of the luminescence appears during the cooling. Morphological examinations of the abrasion showed that the crystal, which becomes harder during the temperature decrease, continues to be treated by the aluminium cutter to a small degree only and the wear of the cutter strongly increases.

In order to interpret the excitation mechanism, the spectrum of the emitted photons (see Fig. 3.41.) and the thermoluminescence curve for the centre-identification to LiF were recorded [182]. The correspondence with investigations of X-rayed LiF crystals [183] pennils the conclusion for mechanical treatment, that valence electrons are excited by fracturing in LiF and a V_K-centre comes into being in the form of $(F_2)^-$ also described as a self-trapped hole (see Fig. 3.43.). By the attachement of an electron to this complex, for instance by F- or F'-centre ionization or by capture of the valence electron, a self-trapped exciton state appears disintegrating radiantly and non-radiantly, predomiantly at low temperatures and higher temperatures, respectively. The observed luminescence can be attributed to the recombination of this excited state. The proposed model explains a number of other physical effects, such as electron emission and release of lattice components occurring simultaneously with the luminescence, which will be discussed in the following sections. It is still impossible to develop a model

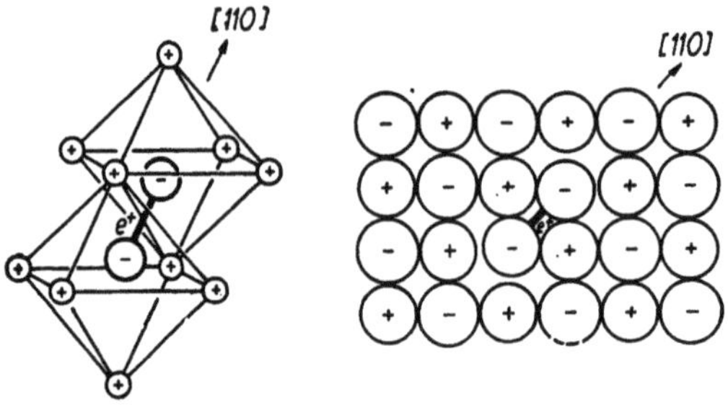

Fig. 3.43. Atomic structure of the V_K-centre [184].
e+ — defect electron.

of the exact mechanism of the formation of the V_K-centre by dissipation of the mechanical energy.

Apart from the V_K-centre some more excited states have been proved after mechanical treatment in alkali halides, oxides, sulphates [185]. These centres can be characterized and their energetic position and frequency factor can be given by an analysis of the glow curves (see 3.5.5.) and by the decay behaviour of the thermo-stimulated luminescence in which the stimulation temperature may even be below the room temperature.

True triboluminescence could appear as a result of an interaction of phonons and photons. However the theoretical consideration meets with great difficulties, since the observed photon energies require an annihilation of a great number of phonons and the lattice imperfection really occurring can only be described as phonons at a greater distance to the disturbation centre. However, the luminescence is determined by processes in the centre of the defect. Therefore it would seem to be more advantageous to start out from a statistical description of possible disturbation terms upon lattice deformation. Already small deflections from the equilibrium state of the lattice components in alkali halides result in an excitation of F-centre electrons on energy levels in the conduction band from where the recombination is possible. This excitation mechanism was used for the explanation of the triboluminescence with slow compression in alkali halides containing many F-centres. It is assumed [186] that such excitations can also occur from the valence band during stronger disturbances. Such defects should for instance exist in the rupture front of a fracture [181].

In [187] the excitation is attributed to the interaction of mobile dislocations with trapped electrons and to the release of disloction energy at luminescence centres.

The annihilation of mobile dislocations have been regarded as excitation mechanisms in the plastic deformation of metals [188]. It leads to the excitation of electrons and to the formation of holes in quasi-local d-states of the precious metals (Cu, Ag, Au) in close proximity to the dislocations. The radiant recombination between the electron and hole states occurs with energies of 1.7, 2.1, and 3.9 eV and conforms well with the experimental results.

3.3.4.3. *Temperature radiation*

As shown in 3.3.3., mechanical treatment processes, particularly fracture processes, are connected with the appearance of high temperatures, so that triboluminescence phenomena have also been attributed to the temperature radiation and the exposured spectra have been compared with thermal radiation sources. In the case of the local thermodynamic equilibrium the spectral distribution of the radiant energy density $P(\lambda, T)$ for the black body is given by the PLANCK radiation formula

$$P(\lambda, T) = \frac{2hc_0}{\lambda^5 \exp(hc_0/\lambda kT)} \tag{32}$$

(h — PLANCK elementary quantum of action, c_0 — light velocity, λ — wave length, T — temperature, k — BOLTZMANN constant.)

By doing so the maximum of $P(\lambda, T)$ shifts according to the WIENS displacement law

$$\lambda_{max} \cdot T = \text{const.} \tag{32a}$$

(λ_{max} — wave length of the emission maximum, T — temperature.)

If the radiant body is not a black body then $P(\lambda)$ still has to be multiplied by the emission capacity dependent on the wave lengths.

For transparent bodies the transmission capacity τ and the reflectance r are taken into consideration [189] as correction factors so that the radiation $E(\lambda)$ of the non-black body results

$$E(\lambda) = P(\lambda) \cdot (1 - \tau) \cdot \frac{1 - r}{1 - r \cdot \tau} \tag{33}$$

Temperatures between 1,000 K and 10,000 K are given for the fracture front [166, 171] and for friction processes [132] in literature but it must be added that the problems of the definition of temperature over very short periods and in very small volumes were referred to in 3.3.3.

Luminescence phenomena have also been inferred from the assumption of high equivalent temperatures in the magma-plasma model (Chapt. 1., [1]).

It has also been shown that the comparison of non-structured continuous spectra which existed in the cases presented here is highly imprecise and does not permit an unambiguous formulation with respect to the mechanism [182].

3.3.4.4. *Chemoluminescence*

The recombination radiation of free dissociated lattice components of the alkali-halide crystals can be deduced from the occurrence of the resonance line of the alkali atoms in the spectrum during the fracture of alkali halides (see Fig. 3.41.). Lattice components are clearly released during the mechanical treatment (see 3.3.6.). They exist as alkali-halide molecules and partially as ions in the vapour phase. The undissociated molecules can be dissociated by electron impact in the electric field. During recombination, taking place within 10^{-8} s after excitation they deliver a defined quantity of the chemical reaction energy as radiation to the environment.

Some of the free metal atoms are excited by the electron impact so that the resonance line of the atom appears in the spectrum. In Fig. 3.44. the energy-level scheme for an NaCl-complex is shown in several states as they can occur in chemoluminescence.

Assuming only neutral halogen and alkali metal atoms are present in the initial state, for great intermolecular distances the interaction of a complex of atoms is deter-

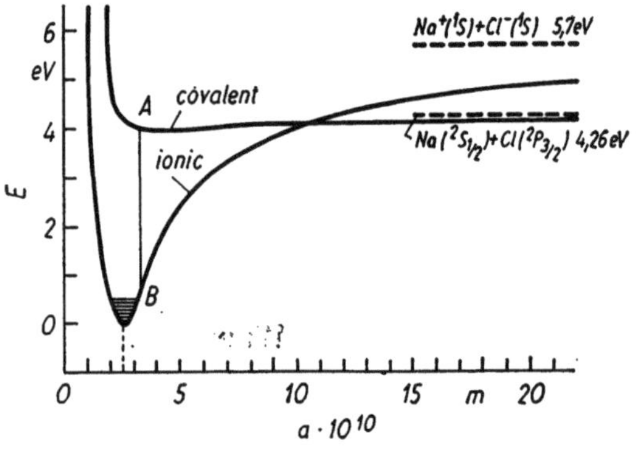

Fig. 3.44. Energy pattern of a NaCl-complex in various states [190]. *E* — energy, *a* — nucleus distance.

mined by the potential curve which conforms with the covalently bound molecule. Greater vicinity can appear due to the thermal motion of the particles (with NaCl $< 10.5 \cdot 10^{-10}$ m) in which the lowest energy state for the complex of atoms is given by the potential curve for the ionically bound molecule. The transition takes place by the transfer of one electron from metal to halogen atoms. The energy difference of these transfers is not limited towards the low energies. The upward boundary is given by the dissociation energy of the molecules in neutral atoms (Tab. 3.6.). The short-wave boundary of the spectra (Fig. 3.41.) has been used as confirmation of this relationship [190].

Table 3.6. Dissociation energy of the alkali halides [190].

	Dissociation energy		Cut-off wavelength
	eV	kJ/mol	nm
LiF	6.6	636	188
NaF	5.3	511	234
NaCl	4.26	410	292

The intensity maximum of the band is determined by the transition between the minimum of the upper potential curve and the potential curve of the ionically bound molecule [191]. A quantitative interpretation of the course of the spectra is difficult since the course of the potential curve of the covalent molecule, the interation with quenching gas components and the transition probabilities occur very complexly and are generally not exactly known.

The tribo-induced chemoluminescence could also be observed in fresh surfaces coming into existence during the mechanical treatment. Recently luminescence phenomena have been increasingly examined in systems like Al, Mg, O_2 [192], Al/O_2 + H_2O [193], NiO, MgO [194], and W/O_2, CO, NO [195] in connection with molecular beam experiments. Fig. 3.45. shows the tribo-induced luminescence during the continuous treatment of Al with a cutter, at different temperatures. Often it is assumed that the respective adsorption heat and reaction energies, occurring in the examined system lead to the occupation of excited states in the surface bond complex and recombine with radiation.

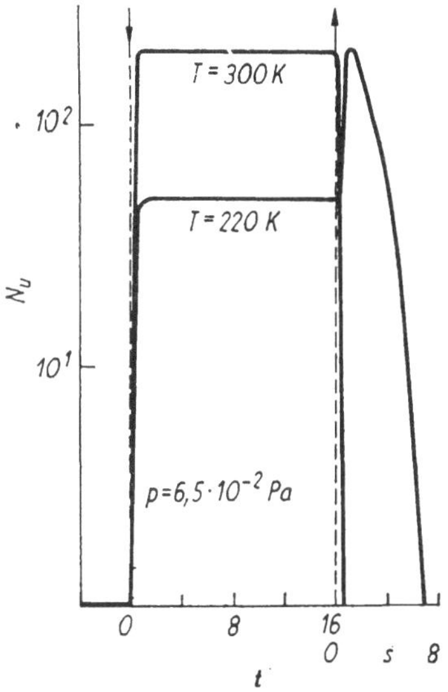

Fig. 3.45. Luminescence during the cutting treatment of aluminium [193].
N_u — pulses/revolution, t — time, — ↓ — beginning of the treatment, ↑ — end of the treatment.

3.3.5. *Tribo-induced electron emission*

Similar to the luminescence, the emission of electrons (and charged particles) during and after a mechanical treatment is investigated as a physical elementary process and as a method for the characterization of excited states. The term exoelectron emission (EEE) for this field originates from investigations of the emission in freshly treated metals which were interpreted as a consequence of exothermal transformation processes of the surface (re-crystallization, phase transformation) [196]. Today the phenomenological term EEE is used as a comprehensive term for emission phenomena which do proceed insteadily, with changed electron work function, and after excitation by non-thermal energy. Recently several comprehensive surveys have been published [197—203].

As in 3.3.4. an attempt shall be made here to given a classification according to the basic mechanisms and to express the relationship to the excitation by mechanical energy by the term "tribo-induced". In electron emission processes a stimulation step is nearly always necessary for overcoming the energy barrier at the surface, mainly by addition of optical or thermal energy (optically stimulated electron emission — OSEE, thermally stimulated electron emission — TSEE). Therefore the electron emission is influenced to a much greater degree by surface processes than the luminescence.

3.3.5.1. *Emission from mechanically excited states*

TSEE has mainly been used for defect analysis in mechanically activated ionic crystals, oxides, and on metals covered by oxides. Assuming the emission appears through the conductivity band the energy level and the frequency factor of the traps could be determined by interpretation of the glow curves [204], (see Fig. 3.46.).

Similar calculations can also be carried out during the isothermal decay of the emission. The postemission generally overlapping from several traps can be divided into

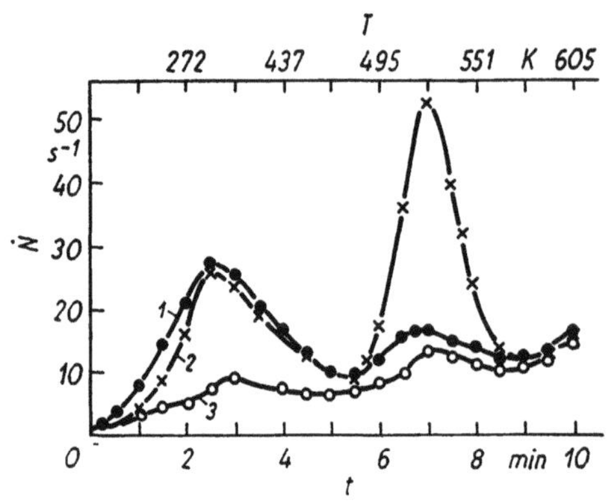

Fig. 3.46. Thermally stimulated electron emission of an aluminium sample [216].

1 – after the mechanical treatment,
2 – after electron excitation (Sr^{90}-β-radiation),
3 – after X-ray radiation (29 keV),
$\dot{N}$ – intensity, T – temperature, t – time.

single exponential decay curves as shown in Fig. 3.47. In many cases the examinations of the TSEE have been correlated with examinations of the thermoluminescence so that it has been possible to obtain information with respect to the characteristic relationship of both fundamental processes [185]. The interpretation of the tribo-induced TSEE proceeds from the band model (see Fig. 3.48.). Defects are produced through mechanical activation in the crystal lattice, and the additional energy states thus obtained are partially occupied by electrons.

This process corresponds with the production of new localized terms in the forbidden band and with the partial occupation by electrons. Localized states immediately beneath the conduction band are described as traps. The activator terms, e.g. the F-centres, lie deeper in the forbidden band. The electrons of the activator terms can be energetically excitated by the activation process and occupy the traps. Generally then, a very low thermal or optical stimulation is sufficient to raise the electrons into the

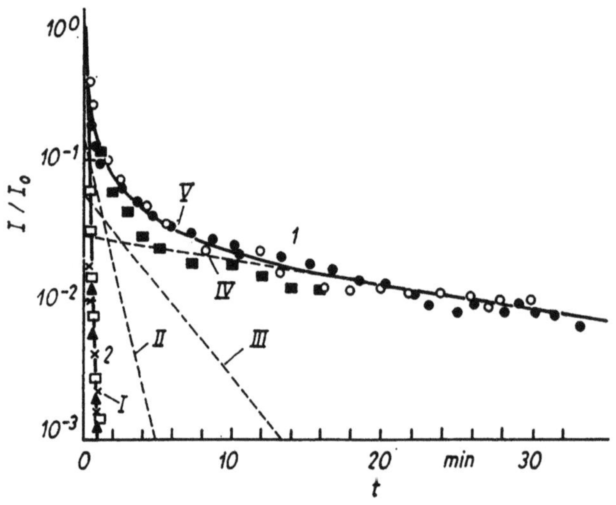

Fig. 3.47. Electron emission decay of air-grown and vacuum-grown monocrystals after the mechanical treatment [206].

I_0 – intensity during the treatment, I – intensity, t – decay time.
Curve V represents the superposition of the curves I to IV (I–IV : $I/I_0 = A_n e^{-B_n t}$)

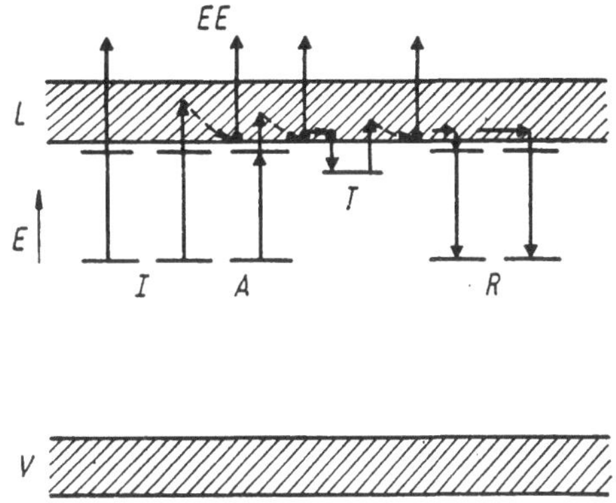

Fig. 3.48. Schematic band model for an alkali-halide crystal [207].
E – energy, V – valence band, L – conduction band, I – Ionization of a colour centre, A – excitation of a colour centre, T – trap, R – recombination with fluorescence, EE – electron emission.

conduction band from where they can emit into the vacuum after overcoming the energy threshold, generally very low in insulators. One part of the electrons in the conduction band recombines with previously emptied activator terms accompanied by the emission of photons (thermostimulated luminescence, see 3.3.4.2.).

The mechanism of the respective mechanical excitation and ionization of colour centres, has not yet been completely clarifield. It must be supposed that thermal energy releases lead to excitations through interaction of the F-centres with migrating dislocations [208] or through immediate ionization by displacement of the ion configuration in alkali halides during the fracture and the migration of dislocations [186].

The OSEE on dielectric solids is attributed to the same causes [209]. The stimulation with photons in the OSEE enables the addition of very precise energy quantities to the mechanically excited solid. In the case of dielectric solids with broad forbidden band the results of the thermal centre analysis have been confirmed by OSEE to a large extent. Differences [200] have to be attributed mainly for the involvement of surface process and physical differences in the stimulation with photons and thermal energy, for instance in the influence on adsorption layers. In recent years the OSEE has been widely applied in the field of materials science with the plastic deformation of real metal surfaces covered by oxides.

The first information on the influence of vacancies, micro-cracks, and dislocations on the electron emission on mechanically deformed surfaces was already published in 1961 [210]. Excluding fresh surface reactions (see 3.3.5.2.) in the ultra-high vacuum it could be calculated from results with different strain velocities that the emission is closely connected with the diffusion of vacancies and dislocations and the annihilation processes at the surface (see Fig. 3.49.).

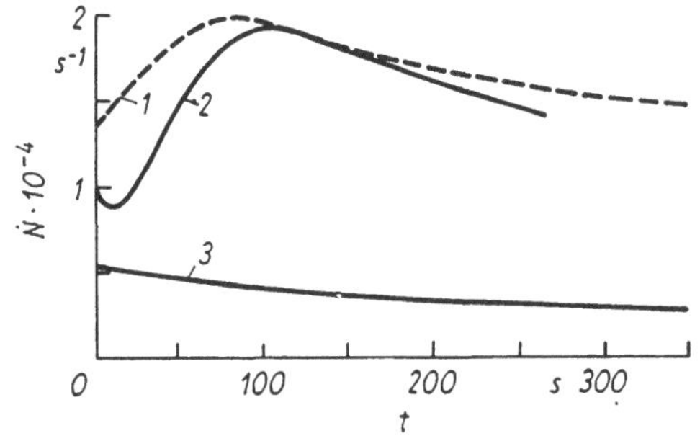

Fig. 3.49. Electron emission decay of aluminium (99,999 per cent) after the plastic deformation with different strain velocities $\dot{\varepsilon}$ [211]. 1; 2 – $1.2 \cdot 10^{-2}\,s^{-1}$, 3 – $3 \cdot 10^{-4}\,s^{-1}$; with strain ε 1; 3 – 7%, 2 – 15%, $\dot{N}$ – intensity, t – time.

The emission from mechanically excited centres also without any external stimulation has been observed in some papers. Thus not only the variation of the energetic position of the emission centres but also the overcoming of the work function as a consequence of the mechanical excitation has to be discussed in this spontaneous or dark emission processes. Experimental results are available for the phase changes [212], for the plastic deformation, and for the fracture of the ionic crystals [136, 213, 214]. Auger processes can be released through the formation of exciton states, for instance during the mechanical excitation of V_K-centres (see 3.3.4.2.) according to Fig. 3.50. in the non-radiative recombination of the defect, resulting in the emission of electrons from F-centres at the surface area.

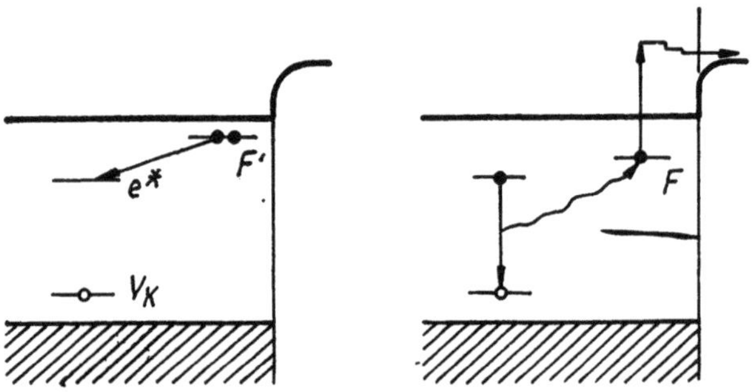

Fig. 3.50. Electron emission in the non-radiant exciton decay [215].
e^* — disturbed exciton state, V_K — V_K-centre, F — F-centre, F' — F'-centre.

3.3.5.2. *Photoemission with changed work function*

In the examination of metals and semiconductors the OSEE can in many cases be attributed to the changed work function of mechanically treated surfaces. It can be regarded as a special case of the external photoeffect [215—217].

The mechanical activation removes top layers at the surface and in dependence on the environmental conditions leads to a metallic clean surface for a short period. This surface is characterized by its work function $e\psi$. This energy threshold which an electron has to surmount when it leaves the solid is varied by the adsorption of polar molecules at the fresh surface. Electronegative gases lead to increases in work function, electropositive gases decrease the work function.[1]

According to the HELMHOLTZ equations the variation in the work function $e\Delta\psi$ is calculated by

$$e\Delta\psi = 4\pi n_0 \mu \theta \text{ [eV]} \quad (34)$$

(n_0 — number of the adsorption sites per cm^2, θ — degree of coverage, μ — dipole moment.)

The theoretically determined value according to equation (34) is not reached due to polarization effects and interaction between neighbouring molecules on the surface. At the most 3 eV [220] were observed for the system Al/H_2O.

[1] Numerous exceptions as the very low variation in the work function in the O_2-adsorption to aluminium are attributed to the special structure of the adsorbate complexes for instance by umklapp processes [219].

Under these circumstances some metals and semiconductors are capable of emitting electrons during the exposure to visible and ultra-red light. A maximum of the emission is observed (see Fig. 3.51.) in dependence on the coverage of the freshly treated surface in the system Al/H_2O which appears simultaneously with the development of the work function minimum in monolayer coverage [218]. The further growth of the adsorption layer then leads to an increase of the work function and to a decay of the emission. In addition the decreasing electron transmission of the growing reaction layer reduces the photoemission process, and traps and other defects of the grown up reaction layer cause the tribo-induced photostimulated emission. Only few investigations of this transition process have been published until now.

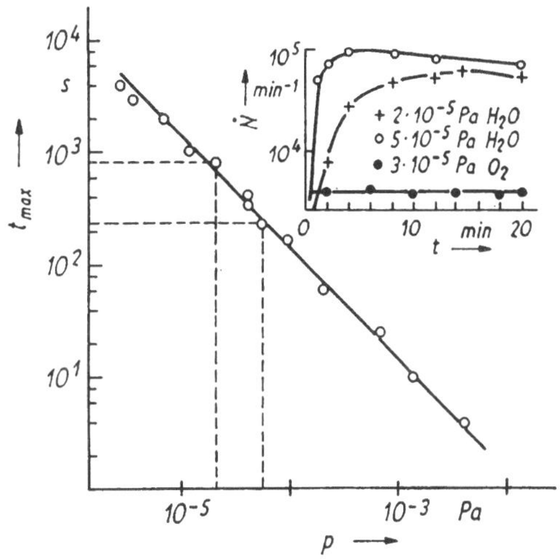

Fig. 3.51. Time up to reaching the maximum of the photoemission t_{max} after the mechanical treatment of aluminium in dependence on the partial pressure p of H_2O [218].
t — time after the treatment, $\dot{N}$ — pulse rate.

Together with other surface methods the abrasive mechanical treatment has been successfully used in metal systems, such as Al, Mg, which are very difficult to prepare by evaporating or thermal desorption processes under UHV-conditions [172].

The photostimulated EEE has been used as a contrast-forming elementary process for the identification of micro-cracks and fatigue fractures, in special emission microscopes [221—223]. The emission profile of an aluminium sample deformed by tensile stress and registered by a scanning method UV light probe (diameter = 70 μm) is shown in Fig. 3.52., [224]. Differences in the emission intensity during the strain and the compression of the samples are explained with the different arrangement of the glide steps compared with the light source (see Fig. 3.53.).

Under UHV-conditions glide steps represent surface areas free of oxides, the work function of which is said to be lower compared with the areas covered by oxides. However, measurements of the work function in abrasive treatment show contrary situations [220].

The different decay of the emission after abrasive treatment and plastic deformation is explained by the oxidation velocity smaller by the factor 10 of the (111)-glide step surface formed during the strain compared with the polycrystalline surfaces [225].

Attributing the OSEE on metals to the external photoelectric effect with changed work function also includes examinations at clean metal surfaces without any in-

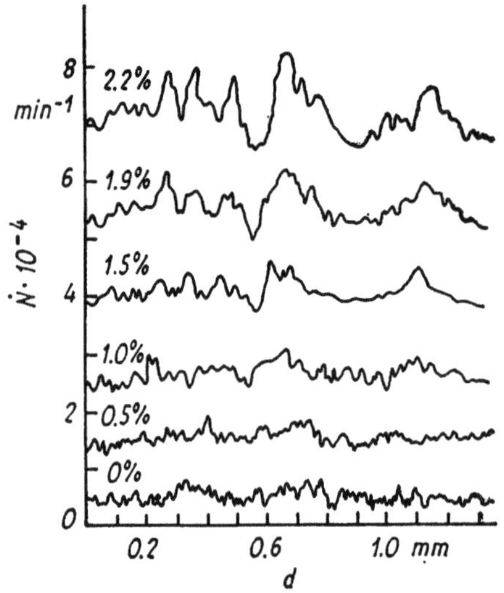

Fig. 3.52. Electron emission during the strain (0 → 2.2 per cent) of (1100)-aluminium [224].
Diameter of the light probe – 70 μm, Ṅ – pulse rate, d – position of the light probe on the sample. The successive measurements are vertically displaced by an ordinate unite in each case.

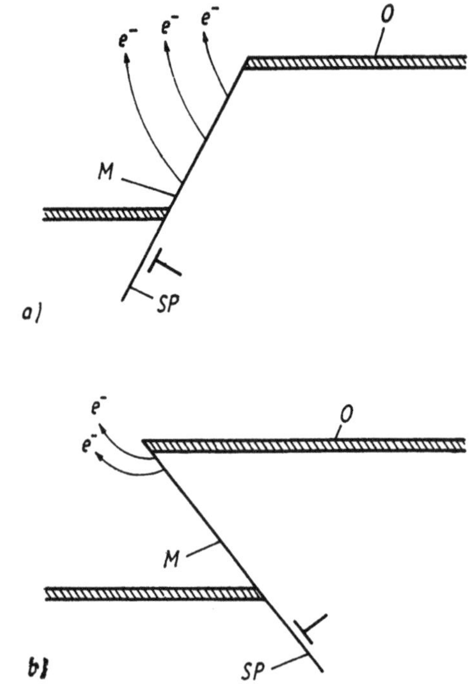

Fig. 3.53. Schematic representation of the effect of the glide step geometry on the electron emission during the action of tensile (a) and compressive stresses (b), [225].
O – oxide, M – fresh metal surface, e^- – emitted electrons, SP – glide plane, ⊥ – edge dislocation.

fluence of adsorbates. The evidence of the influence of a mechanical treatment on the work function and/or the band structure of the clean metal surfaces has not yet been unambiguous.

3.3.5.3. *Chemoemission*

Recent papers also deal with the possibility of an emission of free electrons at freshly treated surfaces by adsorption processes. Fig. 3.54. shows as an example the spontaneous emission of an aluminium surface during continuous milling treatment with various rotational speeds and with pressures between 10^{-2} and 10^{-4} Pa. The emission is initiated by oxygen or water adsorption at the produced fresh surface and reaches its maximum intensity if the formation of a monolayer is possible between two successive treatments [226]. This is the case for the system Al/H_2O with an influence of 80 Langmuir ($\approx 10^{-2}$ Pa · s) and accords with the minimum of the work function. The pressure-dependence points to the influence of the chemical reaction intensity. An electron yields of 10^{-9} to 10^{-10} per chemisorption act can be evaluated. The emission yields of some chemoemission processes published in literature are summarized in Tab. 3.7. The photon processes occurring simultaneously have already been referred to in 3.3.4.4. The

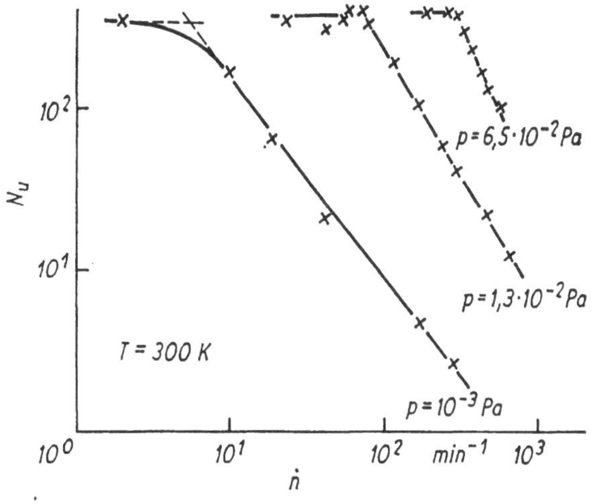

Fig. 3.54. Spontaneous emission during the milling of aluminium with various rotational speeds [226].

$\dot{n}$ – speed of revolution, N_U – pulses/revolution;

more rapid decay of the photon emission before reaching the monolayer coverage excludes that the electron emission is optically "self-stimulated" [193, 230].

The mechanism of the transfer of chemical reaction heat on individual electrons has not yet been clarified. Models are discussed of the emission of AUGER electrons during the

Table 3.7. Chemoemission.

System	Pressure Pa	Emission probability electrons per incident gaseous molecule	Ref.
$Ni-O_2$		$10^{-9} \ldots 10^{-10}$	[227]
$Mg-H_2O-O_2$	$10^{-8} \ldots 10^{-4}$	10^{-6}	[228]
$Al-H_2O-O_2$	$10^{-8} \ldots 10^{-4}$	no emission	
$Cs-O_2$	10^{-6}	10^{-2}	[229]
$K-O_2$	$> 10^{-3}$	10^{-5}	[272]
$Al-O_2$	$10^{-6} \ldots 10^{3}$	no emission	
$Mg-O_2$		no emission	
$Na-Cl_2$	10^{-4}	$5 \cdot 10^{-8}$	[230]
$Na-J_2$			
$Na-Br_2$			
$Al-O_2-H_2O$	$10^{-4} \ldots 10^{-2}$	$10^{-9} \ldots 10^{-10}$	[226]
$Mg-O_2-H_2O$			
$Mg-O_2$	$8 \cdot 10^{-5}$	10^{-8}	[231]
$Mg-Cl_2$		10^{-8}	
$Al-O_2$		no emission	
$Al-Cl_2$		no emission	

junction of solid and gaseous atoms [228], in the formation of surface states during the adsorption by ionic shifting [231], and in the initiation of lattice vibrations (phonons, plasmons) during chemisorption processes with a large free reaction enthalpy [226]. Further progress can be expected in clarifying the role of surface and volume states at real surfaces and in interaction with gases, through the examination of the electronic surface structure by means of work function measurements and sophisticated electron spectroscopic techniques.

In [232] the thermostimulated chemoemission during adsorption and desorption processes is attributed to the recombination of active states (radicals, ions) within the adsorption layer. Tab. 3.8. shows a pattern for the surface reaction with O_2 which mainly regards the emission process widely independent from the solid substance and in connection with surface processes and the interaction with dislocations and point defects in the reaction layer. The high proton mobility of the adsorbed water favours the transmission of the excitation energy. In addition to electrons the emission of ions and neutral particles has also been observed in these examinations.

Table 3.8. Electron emission at the oxygen desorption [232].

$$O_2^- \rightarrow O^- + O_{ads}$$
$$2\,O_{ads} \rightarrow O_{2(ads)} \rightarrow O_{2(gas)}$$
$$2\,O^- \rightarrow O_{2(ads)} + 2\,e\uparrow$$

3.3.5.4. *Thermal emission*

Proceeding from the equivalent temperatures estimated in 3.3.3. the emission of thermal electrons is possible. Electrons are predicted as components of the microplasma immediately during the impact in the magma-plasma model (Chapt. 1., [1]. Attempts have been made to determine the equivalence temperature from the energy of the electrons during the treatment. But the influence of the electric fields and the non-equilibrium state of the treatment zone has to be taken into consideration in all the energy determinations.

The simultaneous emission of electrons, ions, and photons during the mechanical treatment of aluminium and nickel samples covered by oxides have also been attributed to thermal causes [233].

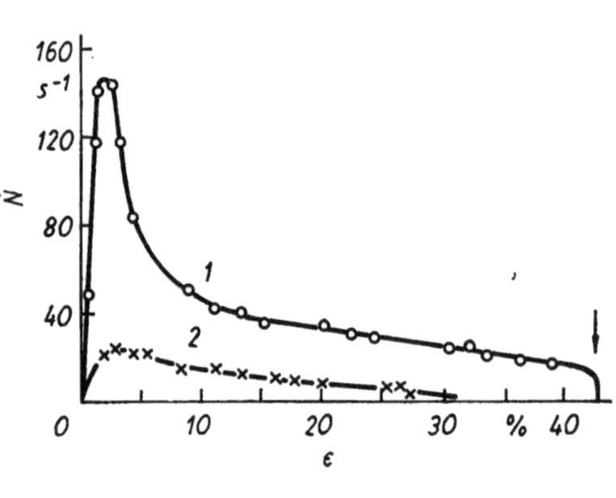

Fig. 3.55. Tribo-induced emission of electrons (1) and negative ions [2] during the strain of nickel with thick oxide layer (200 nm), [233].

Deformation velocity $\dot{\varepsilon} = 3.3 \cdot 10^{-4}\,s^{-1}$, $\dot{N}$ – pulse rate, ε – strain, ↓ – crack

Fig. 3.55. shows the emission rate for electrons and negative ions in dependence on the strain. It could be proved by parallel examinations with the scanning electron microscope that the emission depends on the propagation and morphology of micro-cracks in the oxide layer (thickness $\geqq 5 \cdot 10^{-8}$ m). A thermal emission mechanism at temperatures of 3,000 K was estimated by help of the RICHARDSON equation [153] from the intensity and taking into account the geometric conditions in the crack front. The electrons and ions are emitted from the oxide layer. The recombination of the excited vaporized lattice components is assumed to be the reason for the emission of photons.

3.3.5.5. *Field emission*

In the discussion of the reasons for the spontaneous emission of charged particles from mechanically activated surfaces the influence of electric fields has been referred to at an early stage [234, 235]. As described in detail in 3.3.6. electrostatic charges come into existence through different contact and excitation processes during the treatment of solids. If the field strength reaches

$$E = \frac{\sigma}{2\,\varepsilon_0} \tag{35}$$

(E — field strength, σ — charge density, ε_0 — dielectric constant.)

values of 10^6 V/cm and more, then electrons can be emitted by field emission.[1]

The field emission represents a quantum mechanical tunnel effect. The surface-potential barrier is "deflected" by the strong electric field so that the electrons can tunnel through the wall with finite probability. In aluminium having oxide layers $\geqq 10^{-8}$ m the heterogeneous charging of the crater walls during the formation of fis-

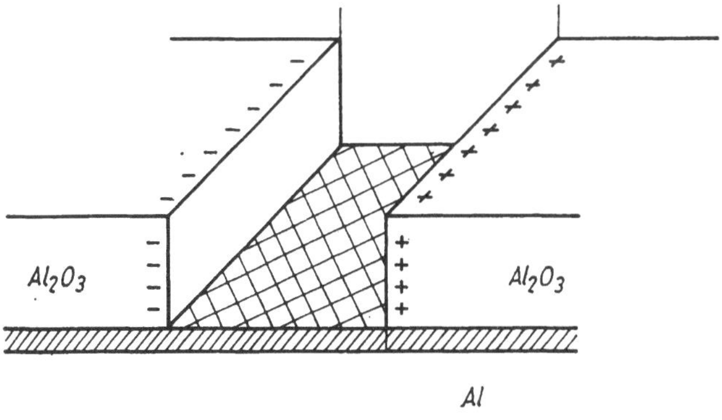

Fig. 3.56. Model for the charging of micro-cracks during the strain of metals covered by oxides [236].

[1] Whereas relatively small average charge densities in metals may already lead to appropiate field strengths by the effect of tips and edges with atomic dimensions the tip effect cannot play any role in dielectrics because of the small charge carrier mobility.

sures in the oxide after tensile stress is asummed to be a reason for the field emission of electrons [236], (see Fig. 3.56.). The position of the maximum depends on the thickness of the oxides, the intensity additionally on the deformation velocity and the sample preparation. In air atmosphere the electric field leads to the emission of charged particles by the ignition of a gas discharge.

The field emission has been experimentally proved and theoretically evaluated [237] by the separation of adhesive systems polymer/solid and by the cleavage of mica and quartz samples. The relationship between adhesion and electron emission is shown for the systems guttapercha — glass and nitrocellulose — glass in Tab. 3.9. Fig. 3.57. shows the course of the emission during the cleavage of mica. The discontinuous cleavage process and the dependence of the electric field strength on the charge density of the cleavage face leads to impulse-like emission peaks with electron energies in the keV range. The electrons appearing are described as mechano-electrons (DERJAGIN-KROTOVA electrons) and release secondary processes, such as X-ray radiation, high-frequency fields, luminescence, and chemical processes (see 5.4.).

Table 3.9. Dependence of the adhesion A and of the emission $\dot{N}$ on the break-away velocity v [237].

System	v cm/s	A J/cm²	$\dot{N}$ Imp/s
Guttaperchaglass	10^{-3}	$2.8 \cdot 10^{-3}$	350
	10^{-2}	$4.2 \cdot 10^{-3}$	$8 \cdot 10^{3}$
	10^{-1}	$2.06 \cdot 10^{-2}$	$4 \cdot 10^{4}$
	1	$2.94 \cdot 10^{-2}$	$5.5 \cdot 10^{4}$
Nitrocelluloseglass	10^{-3}	$1.9 \cdot 10^{-4}$	251
	10^{-2}	$2.3 \cdot 10^{-3}$	836
	10^{-1}	$5.3 \cdot 10^{-3}$	$5.5 \cdot 10^{3}$
	1	$7.1 \cdot 10^{-3}$	$1.01 \cdot 10^{4}$

In a more recent work a chain reaction occurring in the lattice defect annihilation with a yield of 10^{-7} is discussed as cause of the mechanoelectrons [238].

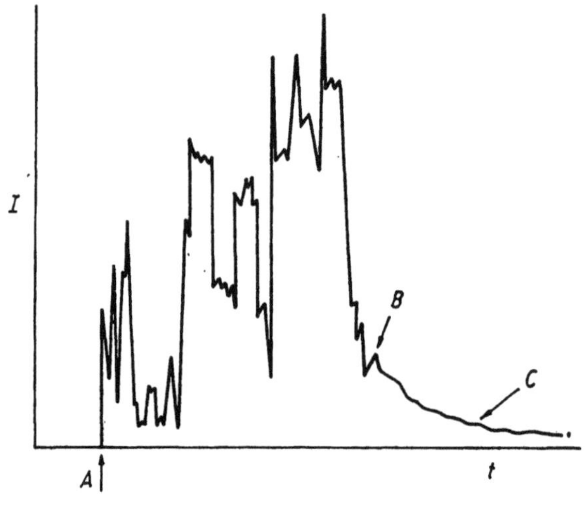

Fig. 3.57. Electron emission during the cleavage of mica [237].
I — pulse rate, t — time, A — beginning, B — end of the cleavage, C — postemission.

3.3.5.6. Energy of the electrons

During recent years through a series of improved methods energy analyses of the emitted electrons have formed part of the consideration of the tribo-induced emission [223, 240, 242]. The energy distribution and its dependence on excitation and stimulation conditions permit to draw conclusions for TSEE and OSEE which relate to the mechanism of the electron emission. The energies are in the range of 10^{-1} to 10 eV [239]. The evidence of the MAXWELLIAN distribution is regarded as pointing to emission via the conduction band. The influence of electric fields is mentioned in few cases only, but it could be the explanation for the discrepancies [239] occurring.

A very widely composed model of an emission influenced by the electric field is shown in [243]. The electric field is built up by the bombardment of SiO_2 and Al_2O_3 surfaces, with electrons. Having been stimulated thermally and optically the electrons can emit from the depths of 10^{-7} m and possess energies of up to 100 eV.

Electrons with energies in the keV-range have been proved in the cleavage of ionic crystals [214] and in the separation of adhesive compounds [244]. The maximum energy of the electrons was determined by deflection experiments and by retarding in thin foils. The energy spectrum could be recorded by a scintillation counter [242]. The energy distribution during the fracture of NaF-crystals is represented in Fig. 3.58. (the maximum is simulated through the reduced senstitivity of the scintillation counter for energies smaller than 20 keV).

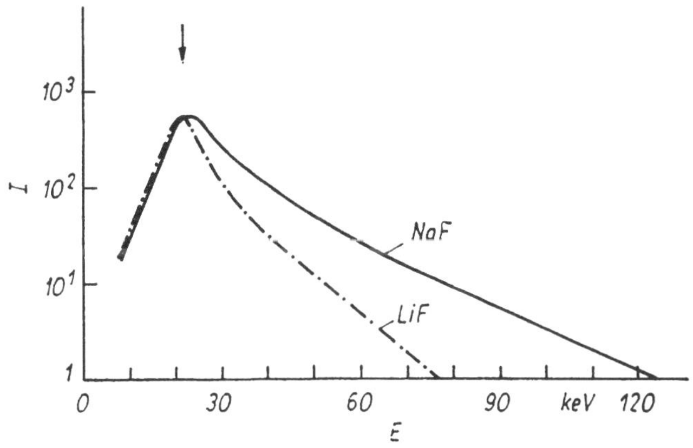

Fig. 3.58. Energy spectrum of the high-energy electron emission during the cleavage of LiF and NaF [242].

I — intensity (electrons per energy interval 1 keV), E — electron energy, ↓ — break at $E < 20$ keV by reduced sensitivity of the scintillation counter.

The number of high-energetic electrons during the compression of a NaF-crystal amounted to about 100, the total number of the emitted electrons is 4 to 5 orders of magnitude above it. It has been estimated from geometric examinations of the crystal fragments and of the distance to the recording instrument that the measured charge density leads to potentials of 100 kV so that some electrons can be accelerated to the energies shown in Fig. 3.58. By far the greatest part of the emitted electrons is deflected to contrary charged crystal domains or metallic parts of the holding components.

3.3.6. *Electric charging processes*

The close correlation of numerous tribo-induced basic processes with electrostatic charges has been referred to in the preceding sections. The appearence of electric charges and the variation of conductivity during the mechanical treatment was mentioned for the first time in 1833 [245] and since 1928 has also been investigated experimentally in connection with the formation of vacancies and the motion of dislocations mainly in disturbed and doped alkali halides [246].

Examples for the occurrence of electric charges and surface potentials partially in correlation to other physical processes are represented with different mechanical treatment conditions in Fig. 3.59. to Fig. 3.64. and in Fig. 3.38.

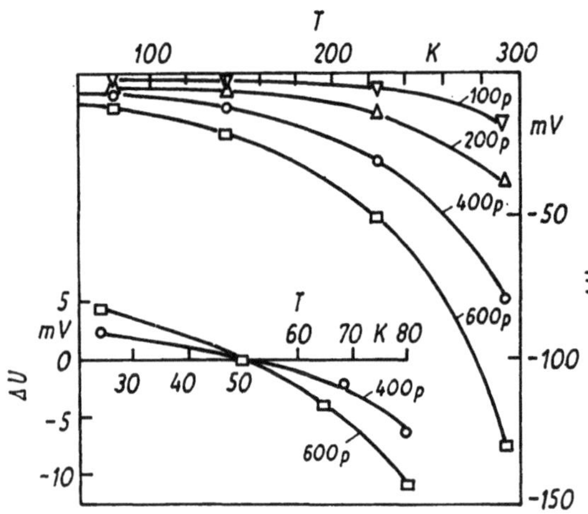

Fig. 3.59. Potential difference ΔU in dependence on the temperature T on NaCl after the impact with different load [248].

The measurements of the charges have generally been carried out with probe methods [175, 248, 250]. Fig. 3.63. and 3.64. show recordings of the decoration of charge regions by agglomerates of organic molecules and dielectric powders. It may be pointed out that the various methods record different charge structures and charge regions. Whereas decoration techniques are capable of determining qualitatively dipole arrangements with high local resolution directly on the surface, influence probes record quantitative measurements of the charge density as mean values of the space charge in a surface region. The horizontal extent of this surface region depends on the probe diameter and on the probe spacing ($\approx$ 50 μm [249]) and the vertical dimension on the dielectric constant of the sample matter. The differences in the results obtained at the mirror surfaces of LiF and shown in Fig. 3.60. and 3.63. are attributed to these methods.

Probes being in contact with the sample (see Fig. 3.62.) can lead to uncontrollable interactions between probe and sample and mostly only permit a determination of surface potentials.

The factors leading to an accumulation of charges on the surface during the mechanical treatment are of varying kind and shall be mainly discussed here for crystalline inorganic solids with insulating properties (alkali halides, oxides). Examinations of polymer systems are represented elsewhere [252].

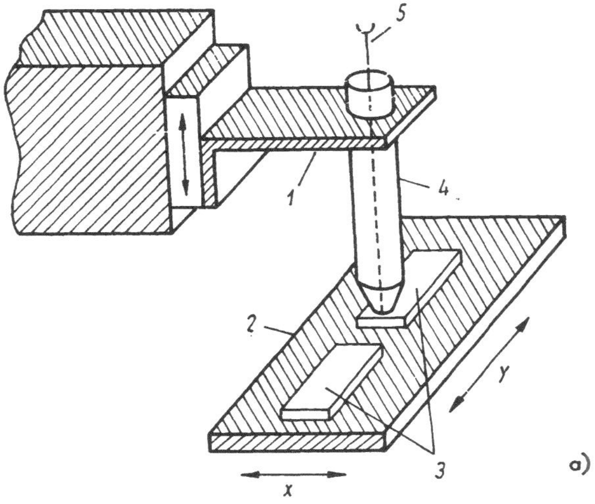

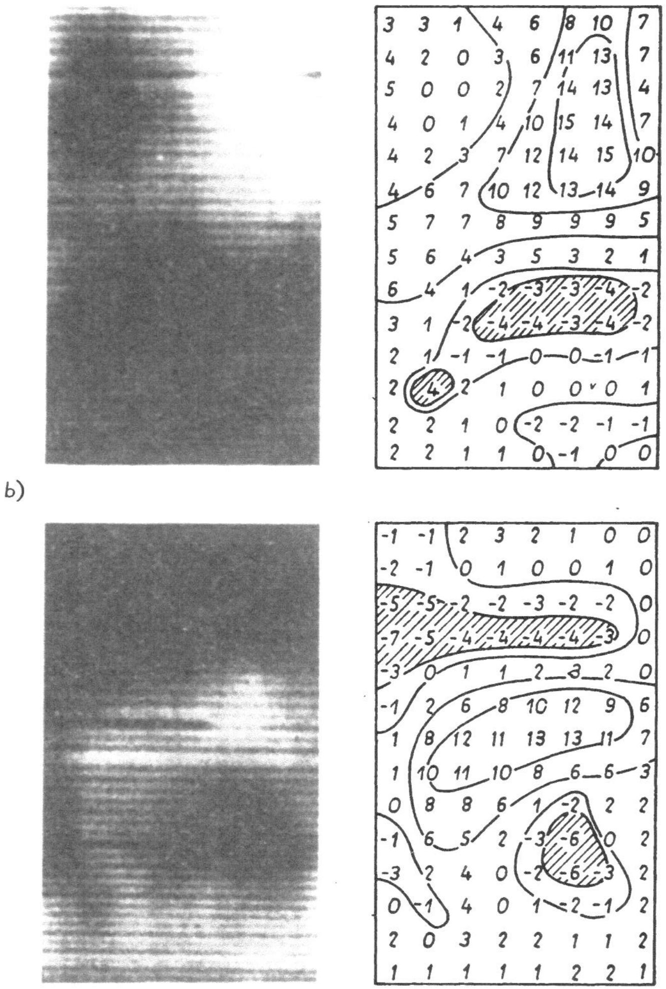

Fig. 3.60. Electrostatic charging of alkali-halide cleavage surfaces [178].

a) arrangement of the measurement: 1 – probe mount, 2 – desk (movable in the directions x and y), 3 – crystal halves, 4 – shield, 5 – probe;
b) qualitative (left) and quantitative (right) charge relief of corresponding cleavage surfaces of LiF (5.0×8.4 mm^2). Cleavage edge in the middle of the Fig., numbers in 10^9 elementary charges/cm^2.

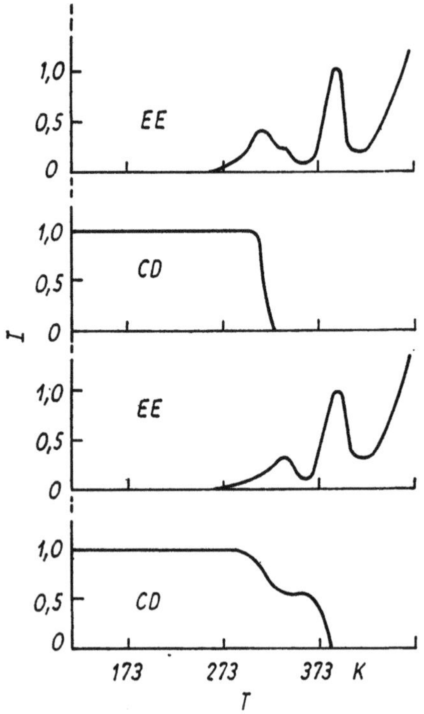

Fig. 3.61. Thermally stimulated electron emission (EE) and charge density (CD) on cleavage surfaces of vacuum-grown (above) and air-grown (below) LiF [249].
I — intensity, *T* — temperature.

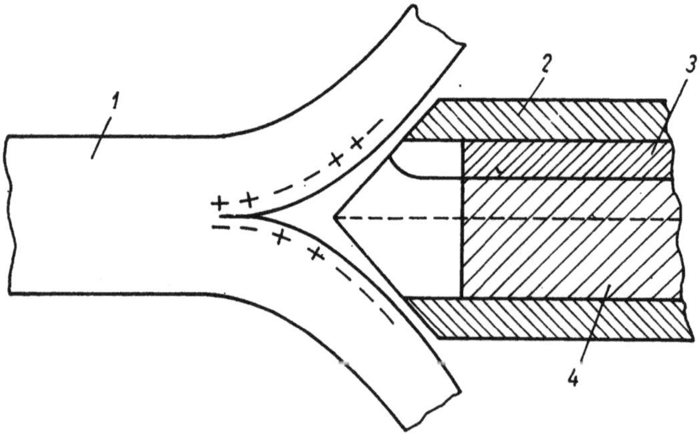

Fig. 3.62. Diagram of a mobile charge measuring probe for cleavage tests with mica [250].
1 — mica crystal, 2 — mount of the probe, 3 — probe, 4 — insulator.

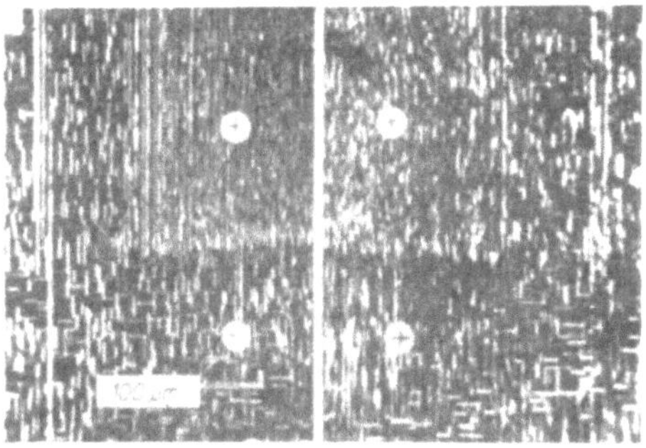

Fig. 3.63. Positively and negatively charged regions at corresponding cleavage surfaces of LiF (0.001% Mg) decorated with anthraquinone [251].

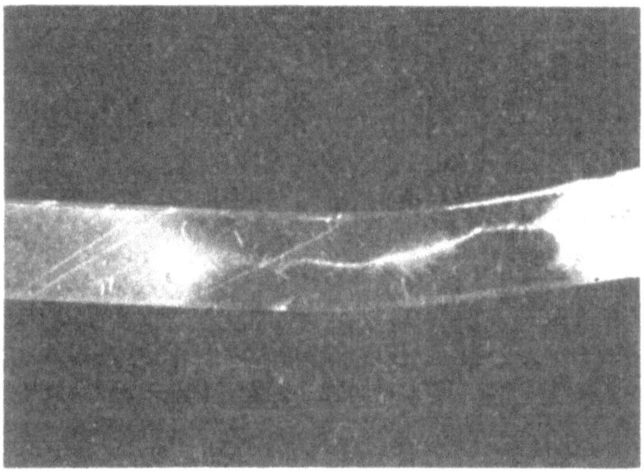

Fig. 3.64. Charged regions during the bending of a LiF crystal decorated with dielectric powder (red lead) [207].

3.3.6.1. Charging by piezoeffect

A polarization appears in the ionic lattice during the introduction of a mechanical stress in piezoelectric crystals

$$P_i = d_{ijk} \cdot \sigma_{jk} \tag{36}$$

(d_{ijk} — piezoelectric modulus, P_i — polarization, σ_{jk} — mechanical stress.)

Conversely, the introduction of an electric field E_i results in a strain ε_{jk}:

$$\varepsilon_{jk} = d_{ijk} \cdot E_i . \tag{37}$$

Precondition for the piezoelectric behaviour is a dissymmetry in the lattice structure as possessed for example by SiO_2 and ZnS. The triboinduced electroluminescence of ZnS appearing during plastic and elastic deformations could be attributed to the piezoelectric charging (see 3.3.4.1.).

In more recent works piezoelectric potential differences have also been measured in dependence on the introduced stress in symmetrically structured crystals, such as alkali halides.

Mobile charged dislocations are discussed as the cause [254]. During the one-sided stress onto NaCl, LiF, and KCl, charges were found in the order of magnitude of 10^{-9} C in deformations of 15 per cent. From this a potential of 10^4 V on the surface and an electric field of 10^5 V/cm were calculated under experimental conditions. The piezoelectric effect is attributed to quadrupole and dipole displacements in the crystal [255].

3.3.6.2. *Contact charging*

The mechanism of the contact charging is not connected with a defined material class or crystal structure. Cause of the charging is the contact potential of the materials touching each another. These processes have been dealt with most clearly for metals in which the contact potential is given by the difference of the work function of the electrons [256]. The charge transfer takes place until the compensation of the FERMI energy of the contact partners. If one of them is an insulator, a charge flow in the sense mentioned above is not possible because of the restricted mobility of the charge carriers.

However, charge transfers may also appear at insulators upon contact by the existence of electric surface states with quasiband structure and electron or molecule terms having low ionization energy [257–259]. Here electrolytic processes play an important role in water films at real surfaces.

The small distances between the contact partners ($< 25 \cdot 10^{-10}$ m) necessary for the charge transfer are obtained by the mechanical treatment, so that the tool shape or the direction of the effective force is often substantially more important than the intensity of the treatment. The return flow of the charges may be prevented by the sudden separation of the contact partners. The electric field E (for homogeneous conditions with small spacings) results as:

$$E = \frac{U}{d} \quad \text{where} \quad U = \frac{Q}{C} \tag{38}$$

(Q — transferred charge quantity, C — idealized capacity between the contact partners, d — spacing of the charged faces.)

The conductivity both in the gaseous medium and within the solid, and the electron emission processes which may contribute to charging and discharging also influence the charge density and charge distribution in contact processes. These processes are additionally influenced by the appearance of high temperatures in the contact zone.

3.3.6.3. *Charged fracture surfaces*

The possibilities of the appearance of charged regions at fracture planes during the mechanical treatment shall be discussed by the example of the alkali halides.

Fig. 3.65. shows the (111)-plane of the NaCl-lattice to have charge carriers of one sign. Ions with different signs appear alternately in all the other crystal faces, so that the total charge equals zero.

The theoretical surface charge density σ' for the (111)-plane can be calculated

$$\sigma' = 4 \cdot 10^{16}/\sqrt{3} \cdot a_0^2 \tag{39}$$

(elementary charges/cm²)

(a_0 — lattice constant.)

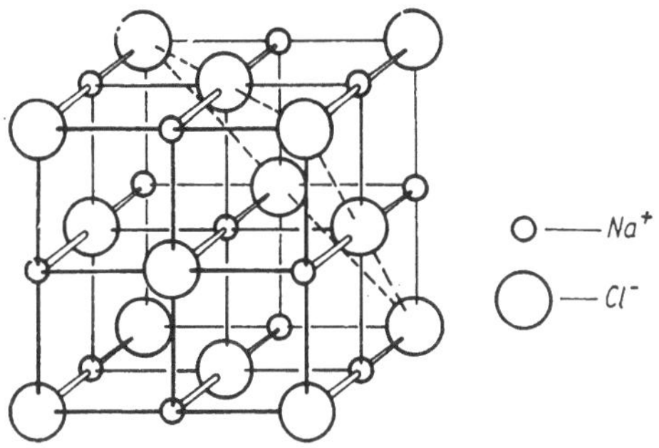

Fig. 3.65. Elementary lattice cell of NaCl with an indicated (111)-plane [190].

The charge is partially neutralized by the contrary charged (111)-plane lying immediately below it so that

$$\sigma = \frac{\sigma'}{2} \tag{40}$$

can be assumed to be the effective charge density. The field strength E (in the homogeneous case for the infinitely extended area) results

$$E = \frac{\sigma}{\varepsilon_0} \tag{41}$$

(σ — charge density in As/cm^2, ε_0 — dielectric constant.)

The effective charge densities are calculated for some alkali halides in Tab. 3.10.

Table 3.10. Charged fracture surfaces [190].

	Lattice constant $a_0 \cdot 10^{10}$ m	Charge density on [111] $\sigma \cdot 10^{-14}$ elementary charges/cm^2
LiF	4.02	7.0
NaF	4.26	5.5
NaCl	5.65	3.6
KCl	6.28	3.0

It can be deduced from this that small (111)-regions on fracture surfaces already lead to high field strengths at the surface and cause gas discharges, field emission, and after-acceleration of electrons. However, for energetic reasons, with alkali halides the formation of (111)-cleavage surfaces is very improbable in comparison to the (100)- and the (110)-area, since the surface energy is higher by the factor 6 and 2.5, respectively. So far a portion of (111)-fracture surfaces could not be proved experimentally either. In the charge image, such regions would be seen through very high charge densities and a mirror-symmetrical charge accumulation with opposite signs on the corresponding fracture surfaces.

3.3.6.4. *Charged dislocations*

Every irregularity in an ionic crystal, such as edges, surface steps, intersection points of dislocations, are potentially charged [260], (Fig. 3.63.). Charged dislocations [248, 260] have been discussed as essential causes of the electric charging in plastic deformation without any tool contact (see Fig. 3.59.).

The edge dislocations in alkali halides of the NaCl-lattice type formed during the plastic deformation have BURGERS vector $a_0/2$ $[\bar{1}\bar{1}0]$. The sliding occurs in the (110)-plane.

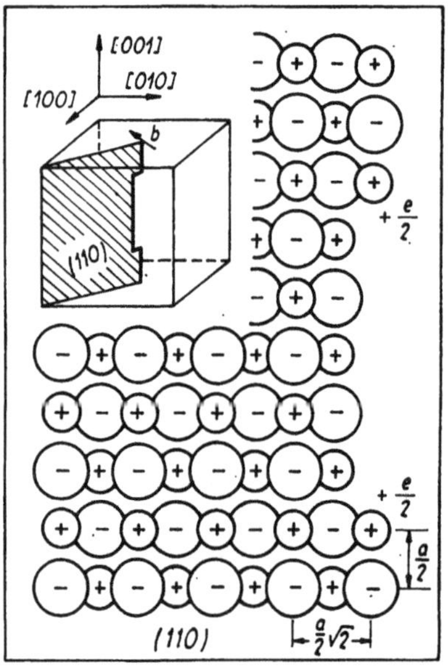

Fig. 3.66. Geometry of a charged edge dislocation [171].
a — lattice constant, e — elementary charge.

Fig. 3.66. shows the dislocation line running in the crystal which limits the pushed plane. Ions of different signs appear in turn on an ideally running dislocation line, an excess charge does not exist. However, if the dislocation line has jogs, as shown in Fig. 3.66., excess charges come into existence which may be positive or negative. If the dislocation as a whole is to be charged jogs of one sign have to appear with priority. This is the case if the formation energy for cationic and anionic vacancies is varyingly great. The dislocations collect vacancies on their way through the crystal and take them to the surface. Hence nature and concentration of the impurities in the crystal as well as their local distribution are of essential importance to the formation of the charged regions. The formation of positive and negative charged regions on a fracture surface could be connected with the irregular distribution of impurities in the crystal caused by the growth conditions.

3.3.6.5. *The formation of electric double layers*

Finally reference shall be made to the possibility of a local charge accumulation by mechanical separation and the formation of electric double layers, respectively.

Anionic and cationic vacancies immigrating from the surface to the crystal result in a dipole layer near to the surface and in the vicinity of screw dislocations because of different formation and place exchange energies [261].

In other papers adsorption layers leading to the occupation of surfaces by dipoles [262] or compensating inside charges [263] are held responsible for the charging. Also differences in the contact potentials coming into existence inside the oxide layers and resulting in high field strengths because of the small distances [264] can be attributed to charge displacements.

Proof of these charge structures has mainly been achieved means of decoration techniques (Figs. 3.63., 3.64.), whereas a recording by means of influence techniques is only possible with distances of the charged regions > 100 μm.

A mechanical charge separation is obtained by tearing off adhesion layers from metallic and semi-conductive substrates [237].

In the electrostatic adhesion theory the formation of electric double layers at the interface of amorphous bodies through donor-acceptor bonds determines the strength of the adhesive bond. The displacement of electric charges is proved quantitatively by the variation of the adhesion partners involved and by electric measurements, e.g. of the varied conductivity of semiconductor substrates. The electric processes observed as secondary effects (see Tab. 3.9.) were used for the characterization of the adhesion systems.

A consistent interpretation cannot be given for the charged regions shown in Fig. 3.60., [175] on fracture surfaces of alkali halides by means of the mentioned mechanisms. The mean charge density of surfaces of alkali halides up to a maximal 10^{11} elementary charges/cm² could be determined by a scanning method with high local resolution of the measuring probe (diameter of the recorded area ≧ 65 μm). In general the fracture surfaces have a positive excess charge. In the range of 110 K to 425 K the maximum charge density does not depend on the crystal temperature. The dimensions of the

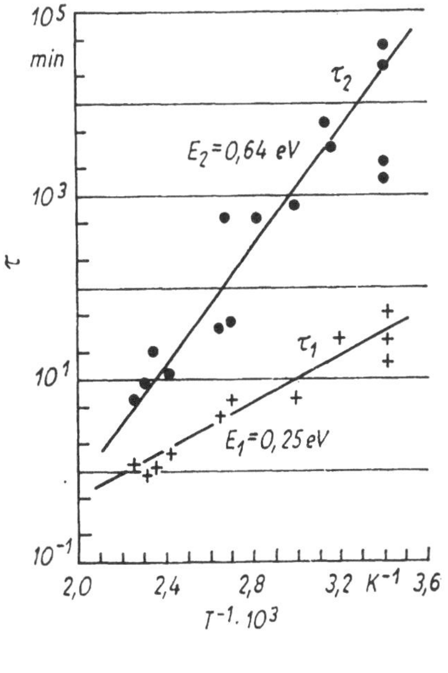

Fig. 3.67. ARRHENIUS plot of the time constant τ_1 and τ_2 for the thermally stimulated discharge of cleavage surfaces of LiF at different temperatures T [249].
E_1, E_2 — activation energy.

homogeneous charged regions are in the mm-range. The charge accumulations on the mirror surfaces have no correlation.

In the Arrhenius representation the activation energies for discharging of 0.64 and 0.25 eV respectively [249] result from the temperature dependence of the time constant for the discharging process (Fig. 3.67.). This indicates the contribution of electron processes, as also expressed in Fig. 3.61. Obviously dynamic processes during the fracture propagation contribute to the charge distribution.

3.3.7. *The release of lattice components*

The "vaporization" and the release of gases during mechanical treatment processes have also been referred to in the description of causes and influencing quantities of physical elementary processes and chemical reactions.

From the thermodynamic point of view the increase of the vapour pressure in strongly deformed micro-ranges can be deduced as a possibility for the separation of lattice components. The Gibbs-Thomson equation [265] includes the relationship between vapour pressure and particle radius r

$$p_r = p_\infty \exp\,(2\gamma_0 M/RTr\varrho) \tag{42}$$

(P_r — vapour pressure for crystal with the radius of curvature r, p_∞ — saturated vapour pressure of the compact crystal, $2\gamma_0$ — surface energy, M — molecular weight, T — temperature, ϱ — density.)

A vapour pressure increase of several orders of magnitude can be deduced for NaCl with particle sizes having a radius $< 0.1\ \mu$m. The vapour pressure increase is additionally influenced by locally varying surface energies and the occurrence of high temperatures during the treatment. The emission of ions and neutral particles in crack tips dealt with in 3.3.5.4. and the triboluminescence during the fracture (see 3.3.4.4.) may be attributed to these causes.

The "evaporization" of lattice components is also included as a consequence in the magma-plasma model. With a treatment period of 40 to 80 hours, intensive friction processes including stick-slip motions of the scratching tool on a zinc sample (deformation and work hardening alternate intermittently) lead to a micro-analytically detectable precipitation of zinc atoms on a quartz plate mounted above the sample (tribosublimation) [266].

The release and the dissociation of molecules of the lattice could be proved by careful mass-spectrometric examinations during the cleavage of alkali halides (LiF, NaF, KCl). Fig. 3.68. [267] shows the mass spectrum corrected to take into consideration the various sensitivities. Noticeable here is the non-stochiometric relation of the released lithium and fluorine ions and the different kinetics.

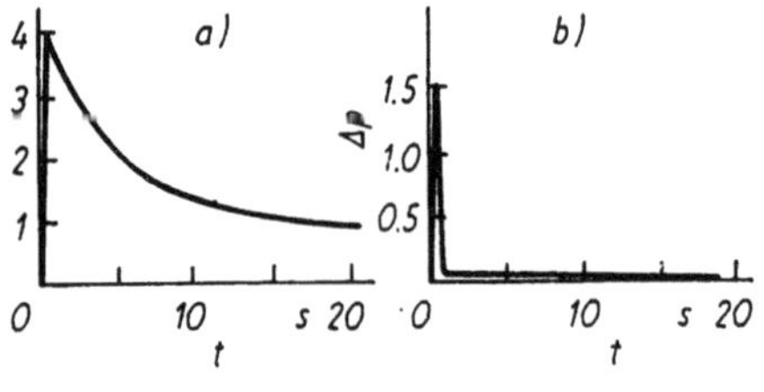

Fig. 3.68. Partial pressure increases Δp of fluorine (a) and lithium (b) after the cleavage of LiF crystals [267].
t — time, Δp — arbitrary units.

A possible explanation is based on the V_K-centre model, as already discussed in 3.3.4.2. Fig. 3.43. shows how strong repelling forces of neighbouring halogen ions appear with the recombination of the V_K-centre in alkali halides. The formation of impact sequences in the direction of densest halogen spacings (110-direction) can lead to a preferential emission of halogen ions in close proximity to the surface.

The directional non-stochiometric release of lattice components by V_K-centres recombination could also be proved experimentally with the bombardment of alkali halides with electrons and X-rays [268]. In connection with the appearance of gaseous discharge spectra in the high vacuum, enclosed gases which can be released from the crystal lattice by the cleavage have been assumed to be the cause of the triboluminescence [174].

The amount of gas released during the fracture is shown for several alkali halides in Fig. 3.69. The average amount of 10^{11} gaseous molecules/cm^2 of the fracture surface is released. Similar experiments are described in [269] in which mainly H_2O and CO_2 have been proved. The enclosed amount of gas depends on the growth conditions and the crystal structure. During the cleavage of mica, 10^{13} molecules per cm^2 surface appear [270].

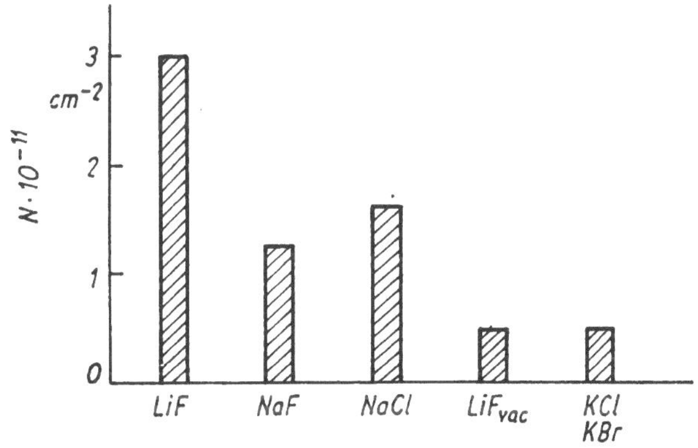

Fig. 3.69. Gas release during the fracture of ionic crystals [172].
N — number of gaseous molecules per cm^2 fracture plane.

The release of gas molecules during the abrasion of adsorption layers has to be taken into consideration for examinations at mechanically cleaned surfaces in the vacuum, since it enables an adsorption of the treated surface or of the freshly produced fracture surface. In this predominantly hydrocarbons, CO, and H_2O have been proved in the residual-gas spectrum [271].

3.3.8. *Summary*

The application of mechanical energy releases a great number of physical elementary and micro-processes in solids which can be the cause and accompanying phenomenon of tribochemical reactions. The influence of different mechanical treatment processes on the formation and interaction of excitation states was represented by examples on the appearance of high temperatures, on luminescence, and electron emission, on the formation of surface charges, and on the release of lattice components. It has been shown that these activation processes cannot generally be attributed to a uniform mechanism but that close correlations exist between them. Some of the theoretical approaches presently being discussed were considered. In comparison with other non-thermal excitation techniques, further advances in fundamental research on the mechanical activation of solids depend in particular on the more comprehensive qualitative and quantitative understanding of the transmission and dissipation of mechanical energy.

References

[1] Thiessen, P. A., Heinicke, G., Meyer, K.: „Forschen und Wirken", Festzeitschrift zur 150-Jahrfeier der Humboldt-Universität, II, p. 119, Berlin 1960

[2] Thiessen, P. A.: chapt. 1., [1], p. 15

[3] Thiessen, P. A.: chapt. 1., [1], p. 16

[4] Puchegger, F.: Naturwiss. **39** (1952), 428

[5] Smekal, A.: "Proc. Int. Symp. Res. of Solids" p. 152, Gothenburg 1952

[6] Smekal, A.: "2 Plansee-Seminar, De Re Metallica", p. 28, Reutte, Tirol 1955

[7] Thiessen, P. A.: „II. Intern. Pulvermetallurgie-Tagung, Eisenach 1961", p. 23, Berlin 1962

[8] Meyer, K.: chapt. 1., [1], p. 33

[9] Stejn, R. P.: J. appl. Phys. **32** (1961), 1951

[10] Duwell, E. J.: J. appl. Phys. **33** (1962), 2691

[11] Wentorf, R. H.: J. appl. Phys. **30** (1959), 1765

[12] Lin, T. S., Li, C. H.: J. appl. Phys. **35** (1964), 3325

[13] Duwell, E. J., Butzke, H. C.: ASLE Transactions **7** (1964), 101

[14] Amelinckx, S., Strumane, G.: „Siliconcarbide", Oxford 1960

[15] Thiessen, P. A., Heinicke, G., Meyer, K.: chapt. 1., [4], p. 501

[16] Meyer, K., Linke, E.: Wissenschaft u. Fortschritt **23** (1973), 326

[17] Breckel, H.: Techn. wiss. Ber. MPA Stuttgart **10** (1968), 92

[18] Brendel, H.: „Wissensspeicher Tribotechnik", p. 308, Leipzig 1978

[19] Meyer, K.: „Phys. chem. Kristallographie", p. 247, Leipzig 1968

[20] Meyer, K.: Dissertation Berlin (1960), 9

[21] Rumpf, H.: Aufber.-Techn. **7** (1966), 427

[22] Meyer, K.: chapt. 1., [1], p. 52

[23] Rumpf, H.: Chem. Ing. Techn. **31** (1959), 333

[24] Rehbinder, P. A.: Z. Physik **72** (1931), 91

[25] Gutman, E. M.: „Mechanochimija i zaščita ot korrozii", Moskva 1974

[26] Ref. [23], p. 323

[27] Baumgardt, S., Buss, B., May, P., Schubert, H.: Aufber.-Techn. **16** (1975), 399

[28] Thiessen, K. P.: chapt. 1., [7], p. 15

[29] Doerschel, J.: Kristall u. Technik **6** (1971), 663

[30] Horiguchi, Y., Nomura, Y.: Chem. and Ind. (1965), 1791

[31] Graham, R. A., Jones, O. E., Holland, J. R.: J. Phys. Chem. Solids **27** (1966), 1519

[32] Horiguchi, Y., Nomura, Y., Fussayama, T.: Kogyo Kagaku Zasski **68** (1965), 910

[33] Horiguchi, Y., Nomura, Y.: Less-Common Metals **11** (1966), 378

[34] Horiguchi, Y., Nomura, Y.: Bull. chem. Soc. Japan **36** (1963), 486

[35] Horiguchi, Y.: J. Am. Ceram. Soc. **49** (1966), 519

[36] Adadurov, G. A., Barkalov, I. M., Goldanskij, V. I., Dremin, A. N., Ignatovič, T. N., Michailov, A. M., Galroze, V. L., Jampolskij, P. A.: Vysokomol. soed. **7** (1965), 180.

[37] Heinicke, G.: chapt. 1., [1], p. 116
[38] Lauer, O.: Aufber.-Technik **15** (1974), 378
[39] Batel, W.: Chem. Ing. Techn. **30** (1958), 567
[40] Žirnov, E. N., Jusupov, T. S.: chapt. 1., [9], p. 9
[41] Chint, I. A.: chapt. 1., [9], p. 12
[42] Chint, I. A.: Osnovy proizvodstva silikal'citnych izdelij" Leningrad, Moskva 1962
[43] Hint, J.: Silikattechnik **21** (1970), 116
[44] Hint, J.: Wiss. Z. der Hochschule f. Architektur u. Bauwesen **17** (1970), 340
[45] Hint, J.: Aufber.-Techn. **12** (1971), 96
[46] Ref. [34], p. 571
[47] Muschelknautz, E., Giersiepen, G., Rink, N.: Chem. Ing. Techn. **42** (1970), 6
[48] Korda, P.: Aufber.-Techn. **2** (1961), 239
[49] Bradley, A. A.: „III. Europäisches Symposium Zerkleinern", Dechema-Monogr. **69** (1971), 781
[50] Bernhardt, C., Heegn, H.: „Viertes Europäisches Symposium Zerkleinern", Prepints, I. p. 213, Nürnberg 1975
[51] Seeger, A.: „Handbuch d. Physik" Bd. VII, 1, p. 383, Berlin 1955
[52] Paufler, P., Leuschner, D.: „Kristallographische Grundbegriffe der Festkörperphysik", Berlin 1975
[53] Seeger, A.: „Moderne Probleme der Metallphysik", Bd. 1, Berlin 1965
[54] Smirnow, A. A.: „Metallphysik", Berlin 1974
[55] Anleytner, J.: "X-Ray Methods in the Study of Defects in Single Crystals", Warschau 1967
[56] Chapt. 1., [4], p. 19–106
[57] Ref. [19], p. 11–320
[58] Schmalzried, H.: „Festkörperreaktionen, Chemie des festen Zustandes", Berlin 1973
[59] Schmalzried, H., Navrotzki, A.: „Festkörperthermodynamik: Chemie des festen Zustandes", Weinheim 1975
[60] Williams, J. O.: "Structural imperfections in solid state Chemistry, Sci. Prog. Oxf. **64** (1977), 247
[61] Skakow, Ju. A.: Kristall u. Technik **12** (1977), 283
[62] Kittel, Ch.: „Einführung in die Festkörperphysik", Leipzig 1971
[63] Messerschmidt, U., Appel, F.: Spektrum **10** (1978), 27
[64] Koglin, B., Leschonski, K., Alex, W.: Chem. Ing. Techn. **46** (1974), 984
[65] Schrader, R.: Dechema-Monogr. **41** (1961), 303
[66] Schneider, U.: Aufber.-Techn. **9** (1968), 567
[67] Hübner, W., Thiessen, K. P.: Wear **55** (1979), 305
[68] Steinike, U., Müller, B., Ebert, I., Hennig, H.-P.: Kristall u. Technik **14** (1979), 1469
[69] Dubinin, M. M.: Pure and Appl. Chem. **48** (1976), 407
[70] Macherauch, E.: Acta Phys. Austriaca **18** (1964), 364
[71] Hartmann, R. J., Macherauch, E.: Z. Metallkunde **54** (1963), 161
[72] Schmidt, C., Löwer, H., Peger, F., Ho van Oi, Hannemann, M., Liebscher, H.: XIII. Intern. Wiss. Koll. TH Ilmenau 1968
[73] Macherauch, E.: Z. Werkstofftechn. **10** (1979), 97
[74] Kochendörfer, A.: Chem. Ing. Techn. **48** (1976), 601
[75] Klug, H., Alexander, L. E.: "X-ray diffraction procedures", New York 1954
[76] Neff, H.: „Grundlagen u. Anwendung d. Röntgen-Feinstruktur-Analyse", Oldenburg/München 1962
[77] Glocker, R.: „Materialprüfung mit Röntgenstrahlen" Berlin/Göttingen/Heidelberg 1959
[78] Fricke, R., Lohrmann, O., Wolf, W.: Z. phys. Chem. **37** (1937), 60
[79] Fricke, R., Gwinner, E.: Z. phys. Chem. **183** (1938), 165
[80] Heimendahl, M. V., Weyerer, H.: Z. Metallkunde **51** (1960), 573

[81] Brückner, W., Hase, W., Fritzsch, E., Pietzsch, C.: Kristall u. Technik 7 (1972), 41
[82] Wagendristel, A., Berger, R.: Z. Naturforsch. **26**a (1971), 1101
[83] Heegn, H.: Banicke listy-Revue des Institutes d. SAV, Košice (CSSR) Sonderheft (1976), 58
[84] Autorenkollektiv, zusammengestellt Juhász, Z.: „Untersuchungsmethoden zur Charakterisierung mechanisch aktivierter Festkörper", Budapest 1978
[85] Jost, K. H.: „Röntgenbeugung an Kristallen", p. 139, Berlin 1975
[86] Greenwood, N. N.: „Ionenkristalle, Gitterdefekte und nichtstöchiometrische Verbindungen", Weinheim 1968
[87] Kröger, F.: "The Chemistry of Imperfects Crystals", Amsterdam 1964
[88] Van Bueren, H. G.: "Imperfections in Crystals", Amsterdam 1960
[89] Proceedings of the 5th International Symposium on "Exoelectron Emission and Dosimetry", Zvikov 1976
[90] Walton, A. J.: Adv. Physics **26** (1977), 887
[91] Ibach, H.: "Electron Spectroscopy for Surface Analysis", Berlin 1977
[92] Carlson, T. A.: "Photoelectron and Auger Spectroscopy", New York 1975
[93] Abragam, A., Bleaney, B.: "Electron Paramagnetic Resonance of Transition Ions", Oxford 1970
[94] Wertz, J. E., Bolton, J. R.: "ESR-Elementary Theory and Practical Applications", New York 1972
[95] Ayscough, P. B.: "Electron Spin Resonance in Chemistry", London 1967
[96] Atkins, P. W., Symons, M. C. R.: "The Structure of Inorganic Radicals. An Application of Electron Spin Resonance to the Study of Molecular Structure", Amsterdam/London/New York 1966
[97] Schneider, H., Woltersdorf, J.: „Strukturen kristalliner Phasengrenzen, Elektronenmikroskopischer Bildkontrast", Leipzig 1976
[98] Petzow, G.: Microscopica Acta **78** (1976), 1
[99] Urban, K.: „Die Anwendung der Hochspannungselektronenmikroskopie in der Materialforschung", 1978
[100] Wilkens, M.: Kristall u. Technik **11** (1976), 1159
[101] Read, W. T.: "Dislocations in Crystals", New York 1953
[102] Faninger, G., Schmidt, R., Sturm, F.: Acta Phys. Austriaca **29** (1969), 112
[103] Leschonski, K., Alex, W., Koglin, B.: ref. [64], p. 23, 101, 821
[104] Alex, W., Koglin, B., Leschonski, K.: ref. [64], p. 387, 477, 729
[105] Ref. [64], p. 563
[106] Gregg, S. J., Sing, K. S. W.: "Adsorption, Surface Area and Porosity", London/New York 1967
[107] Ratnasamy, P., Leonard, A. J.: Cat. Rev. **6** (1972), 293
[108] Warren, B. E., Krutter, H., Morningstar, G.: J. Am. ceram. Soc. **19** (1936), 202
[109] Steeb, S.: "Springer Tracts in Modern Physica", Vol. 47, Berlin/Heidelberg/New York 1968
[110] Jagodzinski, H., Moritz, W., Wolf, D.: Surface Science **77** (1978), 233
[111] Bacon, G. E.: "Neutron Diffraction", Oxford 1962
[112] Krivoglaz, M. A.: "Theory of X-ray and Thermal Neutron Scattering by Real Crystals, New York 1969
[113] Mikhail, R. S., Brunauer, S., Boda, E. E.: J. Colloid Interface Sci. **32** (1970), 367
[114] Fricke, R., Heinle, K.: Z. Elektrochem. angew. physik. Chem. **55** (1951), 261
[115] Kochendörfer, A., Wolfstieg, U.: Z. Elektrochem. angew. physik. Chem. **61** (1957), 83
[116] Warren, B. E., Averbach, B. C.: J. appl. Phys. **21** (1950), 595
[117] Warren, B. E. u. Averbach, B. C.: "Imperfections in Nearly Perfect Crystals", p. 152, New York 1953
[118] Trömel, H., Urmann, E.: Berichte d. Bunsenges. phys. Chem. **72** (1968), 580

[119] STURM, F.: Z. angew. Phys. **27** (1969), 100
[120] KOCHENDÖRFER, A.: Chem. Ing. Techn. **48** (1976), 601
[121] OETTEL, H.: Kristall u. Technik **12** (1977), 275
[122] RÖLL, H.: Kristall u. Technik **12** (1977), 383
[123] OHLBERG, S. H., STRICKLER, D. W.: J. Am. ceram. Soc. **45** (1962), 170
[124] STEINIKE, U., GEISSLER, H.: Kristall u. Technik, in Vorber.
[125] JEDAMZIK, J., BODEN, G., HENNIG, H.-P., EBERT, D., GEISSLER, H., STEINIKE, U., RICHTER, E., SCHOLZ, G.: Z. phys. Chem., **261** (1980), 258
[126] BARTA, R., BRUTHANS, Z.: Silikaty **6** (1962), 9
[127] DUSDORF, W.: Kristall u. Technik **1** (1966), 59
[128] PATZAK, I.: Ber. dtsch. kerm. Ges. **43** (1966), 77
[129] MASING, G.: J. techn. Physik **6** (1925), 569
[130] WAGENDRISTEL, A., POLAK, R.: Aluminium **47** (1971), 195
[131] Ref. [16], p. 326
[132] BOWDEN, F. P., TABOR, D.: „Reibung und Schmierung fester Körper", Berlin 1959
[133] RUMPF, H.: Aufber.-Techn. **14** (1973), 59
[134] CSIRO, Division of Tribophysics, Divisional Report 1973—1975, Melbourne
[135] THIESSEN, P. A.: chapt. 1., [1], p. 11
[136] HOFFMANN, K., LINKE, E.: Kristall u. Technik **11** (1976), 835
[137] HERTZ, H.: Ges. Werke, Bd. 1, p. 155, Leipzig 1895
[138] MEYER, K., GRAGERT, E.: phys. stat. sol. **6** (1964), 803
[139] DAVIDGE, R. W., WITHWORTH, R. W.: Phil. Mag. **6** (1961), 217
[140] BOJARSKAJA, Ja. S., VAL'KOVSKAJA, M. I.: Fiz. tv. tela **5** (1963), 2324
[141] URUSOVSKAJA, A. A., TJAGARAL'ŽAN, R., KLASSEN-NEKLJUDOVA, M. V.: Kristallografija, **8** (1963), 625
[142] TABOR, D.: "The Hardness of Metals", p. 45, Oxford 1951
[143] SCHÖNE, R.: Chem. Ing. Techn. **41** (1969), 286
[144] SPANGENBERG, H. J.: Mber. Dt. Akad. Wiss. **13** (1971), 130
[145] VLADIMIROV, V. I.: „Einführung in die Physikalische Theorie der Plastizität und Festigkeit", Leipzig 1976
[146] HAASEN, P.: „Festkörperprobleme", III, Braunschweig 1964
[147] RUMPF, H.: „1. Europäisches Symposium Zerkleinern", p. 1, Weinheim 1962
[148] GRIFFITH, A. A.: Phil. Trans. Roy. Soc. **A 221** (1920), 163
[149] OROWAN, E.: Rep. Progr. Phys. **12** (1949), 185
[150] RUMPF, H.: Materialprüfung **3** (1961), 253
[151] KOCHENDÖRFER, A.: Z. Metallkunde **62** (1971), 1, 71, 173, 255
[152] ŽURKOV, S. N.: Intern. Journal of Fracture Mechanics **1** (1965), 311
[153] STROH, A. N.: Adv. Physics **6** (1957), 418
[154] COTTRELL, A. H.: "Fracture", p. 20, New York/London 1959
[155] KERKHOF, F., SOMMER, E.: Materialprüfung **14** (1972), 289
[156] IRWIN, G. R.: „Handbuch Physik VI", p. 551, Berlin 1958
[157] SINCLAIR, J. E., LAW, B. R.: Proc. Roy. Soc., London, **A 329** (1972), 83
[158] BURNS, S. J., WEBB, W. W.: J. appl. Phys. **41** (1970), 2078, 2086
[159] TAYLOR, G. I., QUINNEY, H.: Proc. Roy. Soc., London, **A 107** (1934), 307
[160] MADELUNG, E.: Naturwiss. **14/15** (1942), 223
[161] ESHELBY, J. D., PRATT, P. L.: Acta metallurgica **4** (1956), 560
[162] EIRICH, F. R., TABOR, D.: Proc. Cambr. Phil. Soc. **44** (1948), 566
[163] NICHOLAS, J. F.: Acta metallurgica **7** (1959), 544
[164] SCHÖNERT, K., WEICHERT, R.: Chem. Ing. Techn. **41** (1969), 295
[165] ROSENTHAL, D.: Trans. ASME **68** (1946), 849
[166] WEICHERT, R., SCHÖNERT, K.: J. mech. phys. Solids **26** (1978), 151
[167] WEICHERT, R.: Dissertation, Karlsruhe 1976

[168] Urakaev, F. Ch.: Izv. Sib. otd. AN SSSR, ser. chim. nauk (1978), 5
[169] Walton, A. J.: Adv. Physics **26** (1977), 887
[170] Wiedemann, E., Schmidt, G.: Ann. Phys. Chemie, Neue Folge **54** (1895), 604
[171] Meyer, K., Obrikat, D., Rossberg, M.: Kristall u. Technik **5** (1970), 5, 181
[172] Linke, E.: Dissertation, Berlin 1970
[173] Chen, R.: J. mat. Sci. **11** (1976), 1521
[174] Beljaev, L. M., Martyšev, Ju. N.: phys. stat. sol. **34** (1969), 57
[175] Wollbrandt, J., Linke, E., Brückner, U.: Exp. Techn. Physik **23** (1975), 65
[176] Dosse, J., Mierdel, G.: „Der elektrische Strom im Hochvakuum und in Gasen", Leipzig 1945
[177] Polly, F., Obrikat, D., Meyer, K.: Kristall u. Technik **2** (1967), 425
[178] Chudáček, J.: Czech. J. Physics **B 17** (1967), 34
[179] Obrikat, D.: Dissertation, Berlin 1970
[180] Sodomka, L.: phys. stat. sol. (a) **7** (1971), K 65
[181] Beljaev, L. M., Golovistikov, A. N., Nabatov, V. V.: fiz. tv. tela **10** (1968), 3733
[182] Hoffmann, K., Linke, E.: Kristall u. Technik **12** (1977), 495
[183] Podgorsak, E. B.: J. appl. Phys. **42** (1971), 2761
[184] Pick, H.: "Optical Properties of Solids", p. 653, Amsterdam/London 1972
[185] Bohun, A.: phys. stat. sol. **3** (1963), 779
[186] Butler, C. T.: Phys. Rev. **141** (1966), 750
[187] Alzetta, G., Chudáček, I., Scarmozzino, R.: phys. stat. sol. (a) **1** (1970), 775
[188] Molockij, M. I.: fiz. tv. tela **20** (1978), 1651
[189] Mc Mahon, H. O.: J. opt. Soc. Amer. **40** (1950), 376
[190] Rossberg, M.: Dissertation, Berlin 1970
[191] Levi, H.: Dissertation, Berlin 1934
[192] Kasemo, B.: Phys. Rev. Letters **32** (1974), 1114
[193] Grabo, G.: Dissertation, Berlin 1976
[194] Roginskij, S. E., Rufow, Ju. N.: kinetika i kataliz **11** (1979), 383
[195] Mc Carrol, B.: J. Chem. Phys. **50** (1969), 4758
[196] Kramer, J.: „Der metallische Zustand", Göttingen 1950
[197] Brotzen, F. R.: Phys. stat. sol. **22** (1967), 9
[198] Scharmann, A.: „Festkörperprobleme" VI, p. 106, Braunschweig 1967
[199] Ramsay, J. A.: Progress in Surface and Membrane Science **11** (1976), 70
[200] Huster, E.: Naturwiss. **64** (1977), 448
[201] Minc, R. I., Kortov, V. S./izd.: „technika i metodika izmerenija ekzoelektronnoij i akustičeskoj emissii", Sverdlovsk 1973
[202] "Proceedings 5 th Int. Symp. Exoelectron Emission and Dosimetry, Zvikov 1976", Gießen 1976
[203] Kelly, P., Himmel, L.: Comm. sol. stat. Phys. **7** (1976), 81
[204] Holzapfel, G.: Vacuum **22** (1972), 467
[205] Schlenk, W.: phys. stat. sol. (a) **33** (1976), 217
[206] Wollbrandt, J., Likne, E.: ref. [202], p. 274
[207] Wollbrandt, J.: Dissertation, Berlin 1973
[208] Seitz, F.: Rev. Mod. Phys. **26** (1954), 7
[209] Holzapfel, G.: ref. [202], p. 19
[210] Ku, T. C., Pimbley, W. T.: J. appl. Phys. **32** (1961), 124
[211] Soršorov, M. Ch., Meščerjakov, V. N., Aleksandrov, L. S.: ref. [202], p. 247
[212] Kramer, J.: Z. Physik **133** (1952), 629
[213] Zakrevskij, V. A., Pachotin, V. A., Vaitkevič, S. K.: fiz. tv. tela **21** (1979), 723
[214] Krotova, N. A., Linke, E., Chrustalev, Ju. A.: Wollbrandt, I., Sipovskij, V. I.: Dokl. AN SSSR **208** (1973), 138
[215] Bičevin, V. V., Kjaembre, Ch. F.: phys. stat. sol. (a) **4** (1971), K 235

[216] SCHARMANN, A., SEIBERT, G.: Z. Physik **183** (1965), 249
[217] RAMSAY, J. A.: J. appl. Phys. **37** (1966), 452
[218] LINKE, E., MEYER, K.: Surface Science **20** (1970), 304
[219] ELEY, D. D., WILKINSON, P. R.: "Structure and Properties of Thin Films", p. 508, New York 1959
[220] BORN, D., LINKE, E.: ref. [202], p. 265
[221] BAXTER, W. J., ROUZE, S. R.: Metall. Trans. A **7 A** (1976), 647
[222] VEERMANN, C. Chr.: Mat. Sci. Engin. **4** (1969), 329
[223] MINC, R. I., KORTOV, V. S.: „Issledovanie poverchnosti konstrukcionnych materialov metodom eksoelektronnoj emissii", Sverdlovsk 1969
[224] BAXTER, W. J.: Vacuum **22** (1972), 571
[225] BAXTER, W. J.: J. appl. Phys. **45** (1974), 4692
[226] LINKE, E., GRABO, G.: ref. [202], p. 255
[227] DELCHAR, T. A.: J. appl. Phys. **38** (1967), 2403
[228] GESELL, T. F., ARAKAWA, E. T., CALLCOTT, T. A.: Surface Science **20** (1970), 174
[229] MOUCHARATIEH, N., OLMSTEDT, J.: J. Phys. Chem. **75** (1971), 1928
[230] KASEMO, B., WALLDEN, L.: Solid state comm. **15** (1974), 571
[231] FERRANTE, J.: ASLE Trans. **20** (1976), 328
[232] KRYLOVA, I. V.: Z. Chem. **13** (1973), 415
[233] ROSENBLUM, B., BRÄUNLICH, P., HIMMEL, D.: J. appl. Phys. **48** (1977), 5262
[234] KARASEV, V. V., KROTOVA, N. A., DERJAGIN, B. V.: Dokl. AN SSSR **89** (1953), 113
[235] GIEROSZYNSKI, A., MADER, J., SUJAK, B.: Acta Phys. Pol. **26** (1964), 1033
[236] SUJAK, B., GIEROSZYNSKI, A., GIEROSZYNSKA, K.: Acta Phys. Pol. A **46** (1974), 3
[237] DERJAGIN, B. V., KROTOVA, N. A., SMILGA, B. P,: "Adgezija tverdych tel", Moskva 1973
[238] MOLOCKIJ, M. I.: Dokl. AN SSSR **234** (1978), 1438
[239] KORTOV, V. S., MINC, R. I., MJASNIKOV, I. E., ŠEVČENKO, Ju. A.: phys. stat. sol. (a) **2** (1970), 55
[240] SCHLENK, W.: phys. stat. sol. (a) **29** (1975), K 151
[241] BRUNSMANN, U., SCHARMANN, A.: phys. stat. sol. (a) **15** (1973), 525
[242] WOLLBRANDT, J., LINKE, E., MEYER, K.: phys. stat. sol. (a) **27** (1975), K 53
[243] FITTING, H.-J., GLAEFEKE, H., WILD, W., NEUMANN, G.: J. Phys. D: Appl. Phys. **9** (1976)
[244] CHRUSTALEV, Ju. A., USTINOVA, V. I., KROTOVA, N. A.: Dokl. AN SSSR **186** (1969), 381
[245] FARADAY, M.: "Experimental Researches in Electricity", Vol. 1, p. 82, London 1839
[246] GYULAI, Z., HARTLY, D.: Z. Physik **51** (1928), 378
[247] STEPANOV, A. V.: Z. Physik **81** (1933), 560
[248] FRÖHLICH, F., SEIFERT, P.: phys. stat. sol. **25** (1968), 303
[249] WOLLBRANDT, J., LINKE, E.: "V. Polish Seminar on EEE and Related Phenomena", Karpacz 1978
[250] MEČIK, M. S.: „Fizika rasščeplenija sljud", Irkutsk 1967
[251] DISTLER, G. I., i dr.: „Dekorirovanie poverchnosti tverdych tel", Moskva 1976
[252] a) ELSDON, R., MITCHELL, F. R. G.: J. Phys. D: Appl. Phys. **9** (1976), 1445
b) SCHEIN, L. B., CRANCH, J.: J. appl. Phys. **46** (1975), 5140
[253] KITTEL, Ch.: „Einführung in die Festkörperphysik", p. 491, München/Wien 1969
[254] ROBINSON, W. H.: Phil. Mag. **25** (1972), 355
[255] GLADKIJ, V. V., KIRIKOV, V. A., MAGATAEV, V. K.: fiz. tv. tela **21** (1979), 932
[256] HARPER, W. R.: "Contact and Frictional Electrification", Oxford 1967
[257] TAMM, J.: Phys. Z. Sov. **1** (1932), 733
[258] SHOCKLEY, W.: Phys. Rev. **56** (1939), 317
[259] LOEB, L. B.: "Static Electrification", Berlin 1958
[260] WHITWORTH, R. W.: Adv. Physics **24** (1975), 203
[261] KLIEWER, K. L., KOEHLER, J. S.: Phys. Rev. **140** (1965), A 1226, A 1241
[262] SUJAK, B., LESZ, J.: ref. [202], p. 66

[263] Kornfeld, M. I.: J. Phys. D: Appl. Phys. **11** (1978), 1295
[264] Schaafs, W.: Z. angew. Physik **11** (1959), 220
[265] Ref. [19]. p. 211
[266] Thiessen, P. A., Franke, E., Florek, H.: Mber. Dt. Akad. Wiss. **7** (1965), 371
[267] Gallon, T. E., Higginbotham, I. G., Prutton, M., Tokutaka, H.: Surface Science **21** (1970), 224
[268] Elliot, D. J., Townsend, P. D.: Phil. Mag. **23** (1971), 249
[269] Vanderslice, T. A., Whetten, N. R.: J. Chem. Phys. **37** (1962), 535
[270] Mercer, P. D.: Vacuum **17** (1967), 267
[271] Edelmann, Ch., Schneider, H. G.: „Vakuumphysik und -technik", Leipzig 1978
[272] Saadeh, N. K., Olmsted, J.: J. Phys. Chem. **79** (1975), 1325

4. Kinetics and thermodynamics of tribochemical reactions

4.1. Tribochemical excitation model

The first papers dealing with tribochemical processes predominantly took the view that mechanical energy is able to cause chemical reactions only via thermal energy. This view proceeded from the experience that mechanical energy is nearly completely transformed into thermal energy during the action of mechanical energy onto solids (see 4.4.2.). This view was supported by investigations of sliding processes where it could be observed that temperature peaks occur temporarily during the mechanical interactions of solids in the micro-ranges of the contact zones, reaching 1300 K and more with suitable metal pairing ("hot-spot" model, lightning-temperature hypothesis) [1]. Even at the beginning of the sixties the view was still held that "the sequence of tribochemical reactions always has to be taken into consideration if mechanical forces act onto a solid matter system in which reaction possibilities are available in thermodynamic terms" [2]. At the same time tribochemical reactions were reported the development of which could not be interpreted on the basis of a purely thermal excitation. Numerous papers were subsequently published which demonstrated that tribochemical reactions show a divergent kinetic and thermodynamic behaviour in comparison with thermally related reactions. Some characteristic differences are shown in the following:

1. A number of solids respond differently during mechanical treatment to other reaction products than with the application of thermal energy. Thus a decomposition of the compounds takes place during the pestling of sublimate and silver halides.

On the other hand $HgCl_2$ sublimes during the heating, whereas, the silver halides melt in an undecomposed state [3]. The decomposition of bromates may serve as an additional example [4]:

$$2\,NaBrO_3 \xrightarrow{\text{tribochem.}} Na_2O + \frac{5}{2}\,O_2 + Br_2 \quad (1)$$

$$NaBrO_3 \xrightarrow{\text{therm.}} NaBr + \frac{3}{2}\,O_2 \quad (2)$$

2. In comparison with a thermally stressed solid, mechanically treated solids show a reactivity often increased by orders of magnitude, particularly in the low temperature range. In this way tarnish layers which would need 10^{17} years for their growth without mechanical stress come into existence on the stressed metal surfaces during the rolling action in a period of some minutes [5]. The tribochemical development of nickel carbonyl from CO and metal also takes place under technological conditions with a reaction

velocity increased by nearly 2 orders of magnitude compared to thermal plants [6], (see 7.2.2.1.).

3. In comparison with thermally released reactions tribochemical reactions show a varied dependence on temperature and temperature coefficients towards smaller values with the limit zero [7], (see 4.5.).

4. Compared to thermal reactions the dependence on pressure is decreased in tribochemical reactions. Very often it has the value zero. For instance the thermally induced formation of nickel carbonyl follows the second order reaction but with tribochemical guidance of reaction in dependence on the treatment intensity and on the temperature, an order in the range of 0.5 ··· 0 [8], (see 4.6.).

Upon mechanical excitation, chemical reactions are observed which do not run in the thermal way, for instance the reaction [9]:

$$Au + \frac{3}{4} CO_2 \rightarrow \frac{1}{2} Au_2O_3 + \frac{3}{4} C \tag{3}$$

For tribochemical reactions only limited use can be made of the relationship of classical thermodynamics. Thus reactions with a positive free enthalpy (ΔG) also appear during mechanical treatment, such as [10, 11]:

$$SiC + 2\,H_2 \rightarrow Si + CH_4 \quad \Delta G_{238} = +59\ \text{kJ/mol} \tag{4}$$

$$2\,Cu + CO_2 \rightarrow 2\,CuO + C \quad \Delta G_{298} = +140\ \text{kJ/mol} \tag{5}$$

7. The "tribochemical equilibria" measured during mechanical treatment are essentially different from the thermal equilibria, which is expressed inter alia in the equilibrium constant. The CO_2 decomposition pressures of several carbonates in the thermodynamic equilibrium and upon reaching the "tribochemical equilibrium" are compared by way of example in Tab. 4.1., [12].

Table 4.1. CO_2 decomposition pressure p_{CO_2} in thermal and tribochemical equilibrium for $T = 298$ K [12].

Carbonate	Li_2CO_3	$MgCO_3$	$CdCO_3$	$ZnCO_3$
p_{CO_2} in thermal equilibrium, Pa	10^{-26}	10^{-7}	10^{-4}	10^{2}
p_{CO_2} in tribochemical equilibrium, Pa	10^{5}	10^{5}	10^{5}	10^{5}

A deformation model for the interpretation of the results which in terms of equilibrium thermodynamics seem to be contradictory was made in [13—15] which permits an explanation of the chemical and high-energy processes according to the principle of the smallest contradictions and also to predict heuristically additional phenomena while taking special account of the processes during the impact stress of solid surfaces.

The mentioned deformation model proceeds on the basis that a quasiadiabatic energy accumulation occurs at solid surfaces with impacts of sufficient intensity for a very short period and in very small spaces, caused by the finite speed of propagation of disturbances in the real discontinuum. This energy excess can be evaluated from the energy application to be calculated according to HERTZ, from the impact braking [16] as well as from the experimentally observed deformation volume [17, 18], (see 3.3.). According to the model the highest energy states called triboplasmas appear immediately during the impact. They are characterized by an extremely disturbed solid structure as well as by non-steady high-excited fragments of the solid and of the surrounding chemical reaction sphere in the form of lattice components, ions, electrons, etc. The kinetic temperatures are very high in this phase ($> 10^4$ K). The separation of lattice components from the impact region can be directly proved by a tribomechanically released molecular sublimation [19].

The short life of the triboplasma causes no MAXWELL-BOLTZMANN distribution so that an equilibrium temperature cannot be given and the chemical process taking place in this excitation phase cannot be described by the laws of thermodynamics. The conversions in triboplasmas are of a stochastic nature. The highest statge of energy excitation changes dynamically into the next stage characterized by the relaxation of the plasma states and termed "edge-plasma" and "post-plasma". A step diagram of the energy dissipation was composed for the total process in the form of a "hierarchy" of the energy states, which is reproduced in Fig. 4.1., [20].

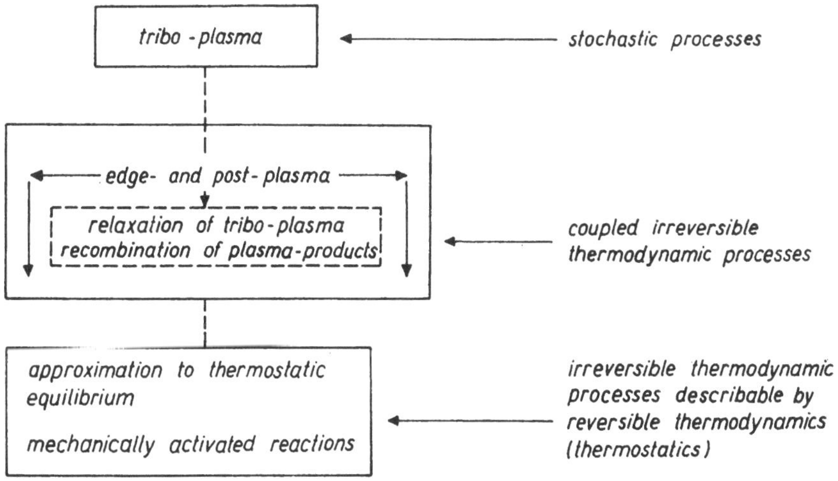

Fig. 4.1. Step diagram of the energy dissipation on solids stressed by impact [20].

A number of physical processes (see 3.3.) take place in this step, such as the recombination of plasma products, the propagation of dislocations, fracture processes, the propagation of photons, the emission of electrons and photons which have important functions for the initiation of chemical reactions. Since the individual processes have an essentially longer life than the tribo-plasma, conditions can be generally expected for this state which lead to an equidistribution of the energy to the molecules and permit the use of the thermodynamic definition of entropy [21]. Hence it is possible to use the thermodynamics of irreversible processes for the reactions occurring in the second step and to connect these with the dissipation processes (see 4.8.2.). Since the reaction cross-section of the peripheral and post-plastmatic processes is higher by about 2 orders of magnitude than the triboplasmatic sequences this state plays the essential role for the sequence of tribochemical reactions and has become a main field of tribochemical research, as shown in Chapt. 5.

In the network of the dissipation processes the above-mentioned "hot-spots" also occur and represent one of the possible excitation states. Thus their contribution in the excitation of tribochemical reactions has to be taken into consideration in the analysis of the causes releasing the reactions.

However, the "hot-spots" are not the only excitation states to be taken into account, as it was formerly often simplistically supposed. Neither do they belong to the highest excitation states in the hierarchy of dissipating energy. In this respect there is no contradiction between the original "hot-spot" model and the excitation model shown here. The latter, however, takes into account with respect to time and energetic nature the complexity of the dissipation process occurring during the mechanical treatment and thereby enables an explanation of the initially discribed apparent contradictions [20].

The approximation to the thermostatic equilibrium of the system has to be regarded as the last stage of the energy dissipation hierarchy. A part of the dissipating energy remains in the solid in the form of the so-called excess free energy. This is bound up with morphological and electronic defects and has a long life especially with low temperatures. The flowing of excess free energy is an irreversible process which occurs so slowly, however, in relation to the chemical reaction in the solid, that in many cases the activated state adapting itself can be regarded as quasistatic and described sufficiently approximatley by the use of classic thermodynamic relationships (see 4.8.1.), [22].

In addition to the energy step scheme shown in Fig. 4.1. the processes occurring during the tribomechanical impact treatment can be illustrated by the ball model represented schematically in Fig. 4.2. The processes taking place during the penetration of a tool into an undisturbed solid are schematically shown in column I, and the decay processes taking place during the dissipation of the energy accumulated in the impact

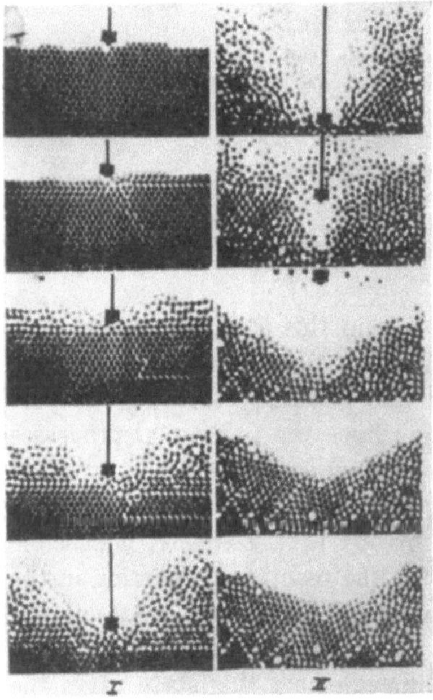

Fig. 4.2. Different stages of the impact stress schematically shown by the ball model [7].

Column I: Penetration of the tool into the lattice of the solid (↓); column II: Decay phase up to the condition of frozen lattice distortions (↑).

space in column II. From the flow diagram the strong defect state immediately during the treatment leading to a loosening of the lattice can be clearly discemed and also a disturbance remaining upon conclusion of the treatment.

A number of important elementary processes resulting from the dissipation action are discussed in section 3.3. As shown in Tab. 4.2. differences in the magnitude exist during the period of appearance and for the duration of elementary excitation states. If the mechanical treatment process is separated from the chemical process in time, for instance by addition of the reactant after completion of treatment, then a number of highly excited states are already decayed and are no more available for the release of the chemical reaction (see Fig. 4.11.).

Table 4.2. Relaxation times of excitation processes in mechanically stressed solids.

Excitation process	Relaxation time	Ref.
Impact process	$> 10^{-6}$ s	
	(impact time after HERTZ)	[23]
Triboplasma	$< 10^{-7}$ s	[20]
Gaseous discharge	$\approx 10^{-7}$ s	[24]
"Hot spots"	$10^{-3} \cdots 10^{-4}$ s	[1]
Electrostatic charging	$10^{2} \cdots 10^{5}$ s	[25]
Emission of exoelectrons	$10^{-6} \cdots 10^{5}$ s	[26]
Triboluminescence	$10^{-7} \cdots 10^{3}$ s	[26, 27]
	(fluorescence,	
	phosphorescence)	
Lattice defects	$10^{-7} - > 10^{6}$ s	[28]
(e.g. V_k centres in LiF with		
different temperatures)		
Dislocation motion	$< 10^{5}$ cm/s	[29]
Lattice vibrations	$10^{-9} \cdots 10^{-10}$ s	[30]
	(at 0 K)	
Fracture formation	$10 \cdots 10^{3}$ cm/s	[31]
	(fracture velocity)	
Fresh surface	at $1{,}3 \cdot 10^{-4}$ Pa:	[32]
	$1 \cdots 10^{2}$ s	
	at 10^{5} Pa: $< 10^{-6}$ s	
Life time of excited	10^{-3} s	[33]
metastable states	10^{-2} s	[34]

Furthermore it can be seen from Tab. 4.2. that the duration of certain excited states depends not only on the physical properties of the solid but also to a large extent on external parameters such as pressure, temperature, and gas atmosphere. Thus the external conditions often decisively determine the moment at which certain excitation states become effective in the network of the dissipation processes, and their duration.

Changes in the solid influenced by mechanical energy are generally described by the term "mechanical activation". If a constant field takes effect, for instance through the application of a constant stress in the elastic range, it is called a "static activation" and a "dynamic activation" with a non-constant stress [35].

The tribochemical excitation model discussed above applies to the theoretically as well as practically important case of the impulse treatment of sufficient intensity, and in practice is applicable to friction and impact processes as they occur in bearings and gears as well as in crushing devices. If the stress of the solid with lower intensity follows through the decrease in velocity of a stress pulse, then certain high excitation states such as triboplasma do not longer occur and the hierarchy of the energy propagation starts with lower excitation steps in each case. Such stresses are represented by processes like dilatation and compression in the form of press-working, expanding, and bending, leading in a border-line case to a simple elastic deformation and influencing the reactivity of the solid without thereby passing through the entire energy step pattern (see 5.2.).

In the following sections some kinetic and thermodynamic aspects of tribochemical reactions shall be discussed by using the tribochemical excitation model, whereby the macroscopic kinetics in dependence on external conditions such as intensity of treatment, temperature and pressure are comprehensively explained in Chapt. 4., whilst the kinetics of the elementary processes are discussed together with a discussion of the reaction mechanisms in Chapt. 5.

4.2. Experimental arrangement for kinetic and thermodynamic measurements

According to the statements in Chapt. 3.1. the mechanical energy may be transmitted to the solid in various forms. In particular the so-called "vibration vessel" has proved suitable for kinetic and thermodynamic investigations. The filled-in material is subjected to a pulse-like impact treatment during the vibration of such a vessel. Thus according to the model represented in 4.1. high-energy quantities are applied to the solid for a short time in narrowly confined spaces. If the treated material takes the form of granules or powder then overall an integral treatment spectrum with a statistical distribution of the individual impact energies is obtained by the vibration of the reaction vessel [36]. Under constant external mechanical conditions, for instance with constant amplitude and frequency of the vibration, the respective treatment spectrum can be held constant for a long period and started again after an interruption of the treatment. This creates conditions by which kinetic examinations can be strictly reproduced and quantitative relationships comprehended. In practice, 2 experimental arrangements have particularly proved their worth:

1. The solid to be activated consists exclusively of coarsegrained material, e.g. of SiC, Al_2O_3, SiO_2 grains or metal flakes. In addition to the impacts to the wall of the vessel a mutual stress of the grains mainly takes place during the vibration treatment. Frequently an additional hard substance, for instance corundum, is added to the grains of the solid to be activated. If the proportion of this hard substance is sufficiently high then its impacts onto the solid to be activated are decisive for the progress of the tribochemical reaction (see Tab. 4.7.). A constancy of the tribochemical reaction velocity could be maintained for more than 1000 hours with the arrangement described above and with constant external condictions [10].

2. The solid to be treated forms a finely distributed powder. The mechanical treatment takes place through impact stress of the powder by means of glass, porcelain,

corundum, or metal balls in the manner of a vibration milling or through hard substance particles having grain diameters of about 10^2 times larger than the powder particles. Using balls as treatment tools a rapid comminution of the grains mostly takes place at the connencenent of the stress until the HÜTTIG milling equilibrium is feached. The steady state appearing on this occasion can also be maintained constantly ror a period of more than 1,000 hours [38].

The experimental setup shown schematically in Fig. 4.3. is particularly suitable for the investigation of tribochemical reactions in which the gaseous phase participates [37].

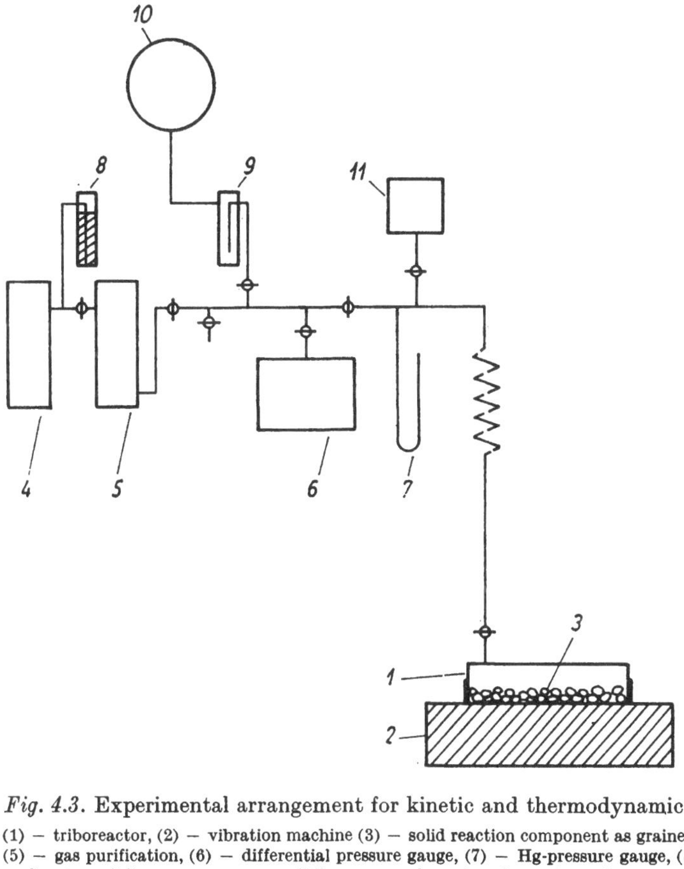

Fig. 4.3. Experimental arrangement for kinetic and thermodynamic investigations [37].

(1) – triboreactor, (2) – vibration machine (3) – solid reaction component as grained substance, (4) – gas supply vessel, (5) – gas purification, (6) – differential pressure gauge, (7) – Hg-pressure gauge, (8) – Hg-pressure relief valve, (9) – cooling trap, (10) – vacuum pump, (11) – gas analyzer (gaschromatograph, mass spectrometer).

Centrepiece of the apparatus is the vibrating vessel in which the intensity of the mechanical stress can be varied by changes of frequeny and amplitude. A temperature range of 77 to 1,200 K [10] and a pressure range of 10^{-3} to 10^5 Pa were obtained by such an arrangement [39, 40].

Kinetic examinations on tribochemical reactions may also be carried out to a limited extent by the help of the single impact treatment method, if the repeatability of the impact action is guaranteed under the same energetic conditions (see Fig. 4.15.) by a suitable arrangement [41].

4.3. General course of tribochemical reactions

The general course of a tribochemical reaction is described during the steady-state effect of mechanical energy on a chemical system by the scheme shown in Fig. 4.4. In detail, 4 different reaction phases can be distinguished:

1. reaction on the untreated solid,
2. rise phase at the beginning of the treatment,
3. steady-state reaction phase with constant application of energy,
4. decay phase after the interruption of the treatment.

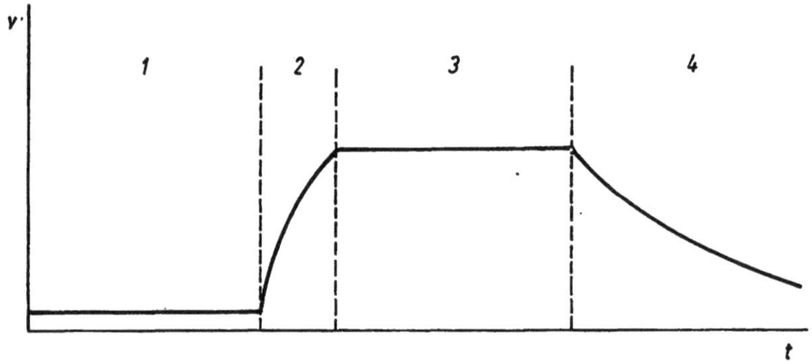

Fig. 4.4. Diagram of a tribochemical reaction course [42].
v – reaction velocity, t – time; (1) – reaction on the untreated solid, (2) – rising reaction, (3) – steady-state reaction stage, (4) – decaying reaction.

The reaction before the beginning of the treatment (phase 1) is determined by the thermal excitation and is therefore a function of the reaction temperature. However, at room temperature most solid reactions proceed so immeasurably slowly that this reaction phase can serve for a comparison with tribochemical reactions in a few exceptional cases only, such as in the formation of nickel carbonyl from nickel and carbon oxide [43].

The application of mechanical energy generally results in a significant increase of the solid reaction velocity, often amounting to some orders of magnitude (see Tab. 4.3.). After having passed through a rise phase a constant velocity appears under external constant conditions. This velocity can be maintained for a longer period ($> 10^3$ h) and can be reproduced again after the interruption of the treatment. This reaction phase has been proved in numerous types of reaction.

Table 4.3. Velocity v of the nickel carbonyl formation on reduced and mechanically treated nickel surfaces at 298 K [44].

	v in the steady state mole/h
Effect of CO on a nickel surface reduced by H_2 (phase 1 in Fig. 4.4.)	$5 \cdot 10^{-7}$
Nickel carbonyl formation during tribomechanical treatment (phase 3 in Fig. 4.4.)	$3 \cdot 10^{-5}$

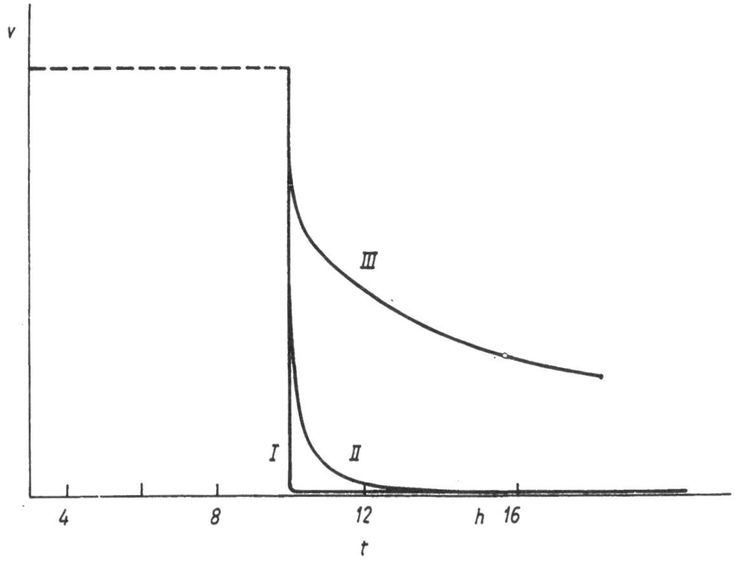

Fig. 4.5. Decaying reaction for 3 different types of reaction [45].
v – reaction velocity, t – time; I – formation of methane, II – formation of nickel carbonyl, III – hydrogenation of benzol; – – – during the treatment, —— after interrupting the treatment.

Distinct differences appear in the individual types of reaction after the interruption of the treatment in the decay phase. Fig. 4.5. schematically shows the course of three reactions in this phase as examples of different types of reaction.

The formation of methane according to equation (4) belongs to that group of chemical reactions, the velocities of which immediately return to unmeasurable values if the treatment is interrupted. These reactions are bound to the highly active states

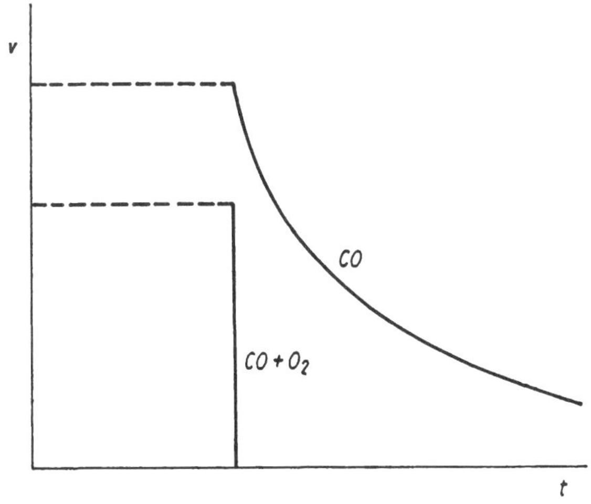

Fig. 4.6. Influence of O_2 on the decay behaviour of the mechanically activated formation of nickel carbonyl [47].
v – reaction velocity, t – time; – – – during the treatment, —— after interrupting the treatment.

which occur in the plasma and post-plasma phases and decay again in fractions of a second. All reactions which do not proceed voluntarily in thermodynamic terms (free reaction enthalpy $\Delta Gr^0 >$ excess free enthalpy of the solid) (see 4.8.1.) belong to this group such as the formation of methane from SiC and hydrogen. In addition all the reactions with a high activation energy also belong, such as the formation of methane from carbon and hydrogen [45] and finally also such reactions which are spontaneously inhibited by reaction poisons in the surrounding atmosphere during the interruption of the treatment, such as the formation of nickel carbonyl in the presence of traces of oxygen [47], (Fig. 4.6.).

In a number of reactions the velocity does not decrease to unmeasurably small values immediatly upon interruption of the treatment, it rather decreases according to a decay funktion. Such reactions which proceed with measurable velocity in the first reaction phase, that is to say without treatment, belong to this group, such as the formation of nickel carbonyl. However since longlife active regions are built-up in the solid by the treatment process, the reaction yields are usually higher than before the beginning of the treatment. In the case of the formation of nickel carbonyl the decrease of the reaction velocity occurs both through the decay of the active state (see 4.8.1.) [48] as well as through consumption of the active solid regions by the chemical reaction.

The tribo-catalytic reactions, such as benzene hydrogenation, belong to the 3rd group. Because there is no consumption of the activated lattice regions by the chemical reaction, but slowly a degeneration of the active state, the decay process only takes place very slowly.

As shown in Fig. 4.7. the reaction decay after the interruption of the treatment is little changed by the intensity of the foregoing treatment, in contrast to the reaction during the treatment. On the other hand it is influenced to a considerably greater

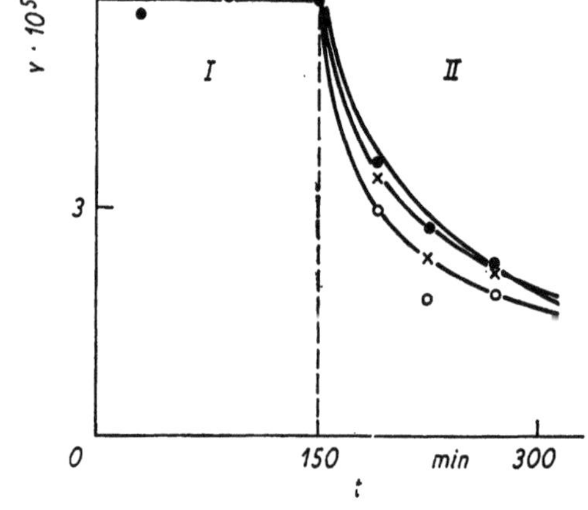

Fig. 4.7. Decay curves after the treatment with different frequency [43].

v – reaction velocity, t – time, ⊙ – 10.4 Hz, x – 9.6 Hz, • – 8.4 Hz; I – during the treatment, II – after interrupting the treatment.

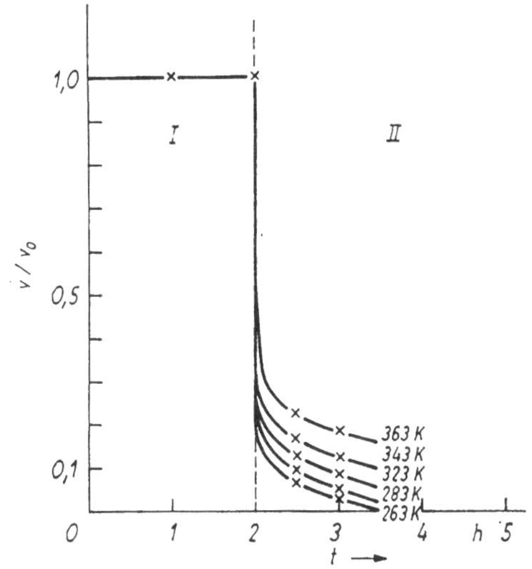

Fig. 4.8. Temperature dependence of the decay curves for the formation of nickel carbonyl [7].
v – reaction velocity during the treatment (I) and after interrupting the treatment (II), v_0 – v during the mechanical treatment, t – time

degree by the existing reaction temperature than the steady-state reaction occurring during the treatment.

As shown in the example of the formation of nickel carbonyl in Fig. 4.8., the decay curves always drop more steeply with decreasing temperature and are no longer experimentally measurable with a sufficiently low temperature.

The rise phase represented in Fig. 4.4. is not measurable in a number of reactions, but it extends to many hours in other reactions. If a tribochemical reaction is exclusively released by highly-excited states then a steady-state level is reached immediately after the beginning of the treatment. So-called rectangular curves result from the interruption of the treatment under these conditions [40]. The slow rise of the reaction curve up to the steady-state level leading to an increase of the catalytic activity by 5 orders of magnitude in the course of 60 minutes in the case of the ethylen hydrogenation on iron contacts [50] can be attributed to several causes dependent on the system, being considered such as the formation of long-lived active lattice-disturbed regions (see 5.9.), to the increase of the solid surface until the milling equilibrium is obtained (see 5.5.), and to the removal of the top layers inhibiting the reaction (see 5.6.). Consequently the velocity measured in this reaction phase is essentially determined by non-steady state physical and physico-chemical processes.

A quantitative description of the reaction course on surfaces treated by impacts was made in [51]. As a basis for the mathematical setup the active state of the crystal surface was assumed to be characterized by the existence of active centres. These are characterized by the fact that the tribochemical reaction mainly takes place in them. For the formation and the decay of such active centres the general relation is valid:

$$A \underset{w'}{\overset{w}{\rightleftharpoons}} A^* \qquad (6)$$

(A – non-active centre, A^* – active centre, w and w' – probability factors for the formation and the decay of the active centres.)

Then the velocity equation for the variation of the active centres is:

$$\frac{dN_A}{dt} = w \cdot N_A - w' N_{A*} = w N_0 - (w + w') \cdot N_{A*} \tag{7}$$

(N_{A*} and N_A — number of the active and of the non-active centres, respectively, N_0 — number of the altogether changeable centres, t — time.)

If the activity of the surface Φ is defined as the ratio N_{A*}/N_0 then the following differential equation results:

$$\frac{d\Phi}{dt} + (w + w')\,\Phi = w. \tag{8}$$

Taking into account the time dependence of w and w' one obtains the equation:

$$\frac{d\Phi}{dt^*} + Z_1(t^*)\,\Phi = Z_2(t^*) \tag{9}$$

(Z_1 and Z_2 — coefficients including the potential barriers for the formation and decomposition of the active centres, $t^* = \frac{t'}{t_0}$, t' — decay time and t_0 — impact duration.)

The values calculated from equation (9) are contrasted with the experimentally measured values in the example of the formation of nickel carbonyl in Fig. 4.9. Except for one constant factor, the mean degree of activity and the chemical reaction yield can be compared with each other with the help of the described mathematical model. The decay behaviour is correctly described in its temporal progress.

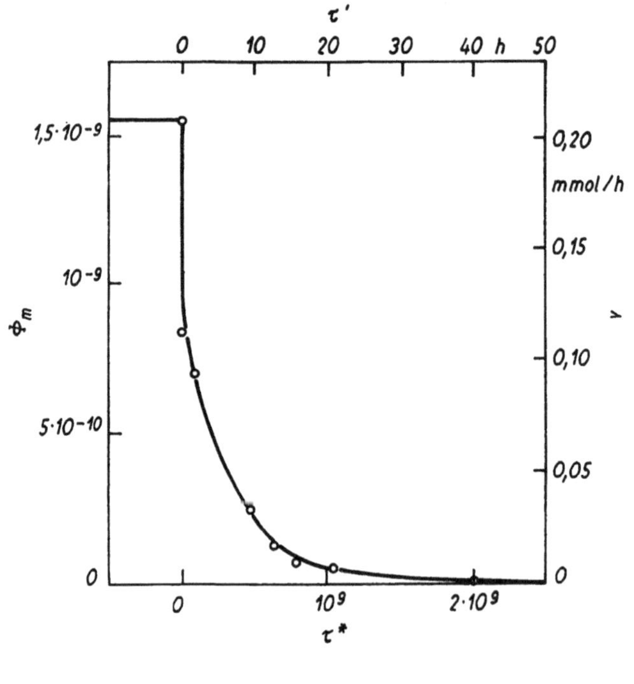

Fig. 4.9. Comparison between the theoretically calculated decay behaviour of the mean activation degree Φ_m (unbroken line, left side ordinate) [51] and the experimentally measured decay reaction of the formation of nickel carbonyl (points after [52], right side ordinate).

v — reaction velocity, τ' — time, $\tau^* = \tau'/t_0$, t_0 — duration of impact.

The consumption of the active centres by the chemical interaction with the molecules of the surrounding gas is included in the mathematical model in a similar statement [53]. Furthermore it is assumed that the velocity of the formation of active centres is proportional to the velocity of the formation of the new surface and that the active centres are of the same value in the first approximation.

Then the velocity of the formation of active centres is:

$$\frac{\mathrm{d}N_{A^*}}{\mathrm{d}\tau} = n\frac{\mathrm{d}S}{\mathrm{d}t} = W_0 \tag{10}$$

(N_A, N_{A^*}, t — see symbols equ. (7), S — specific surface area, n — concentration of the active centres on the surface.),
the velocity of the decay of the active centres is:

$$-\frac{\mathrm{d}N_{A^*}}{\mathrm{d}t} = \frac{1}{t_0}\cdot N_{A^*} \tag{11}$$

(t_0 — lifetime of the active centres.)
and the velocity of the interaction with the molecules of the gas:

$$-\frac{\mathrm{d}N_{A^*}}{\mathrm{d}t} = \alpha \cdot K \cdot N_{A^*} \cdot p \tag{12}$$

(p — pressure of the gas, α — probability of the chemical reaction during the impact, $K = w/\sqrt{2\pi \cdot mkT}$, $w = 10^{-15}$ cm^2, m — mass of the gaseous molecules.)

The total change of the active centres results, taking account of the processes [10]—[12]:

$$\frac{\mathrm{d}N_{A^*}}{\mathrm{d}t} = W^0 - \frac{1}{t_0}N_{A^*} - \alpha K N_{A^*}p = W_0 - \frac{1}{t}\cdot N_{A^*}, \tag{13}$$

where

$$t_p = (t_0^{\ 1} + \alpha K \cdot p)^{-1} \tag{14}$$

By integration of equ. (13) results:

$$N_{A^*} = W_0 t_p \left(1 - \exp\left(-\frac{t}{t_p}\right)\right) \tag{15}$$

In the steady-state range is $t \gg \tau_p$. Then

$$N_{A^*} = N_{st} = W_0 t_p = \frac{W_0 t_0}{1 + \alpha \cdot t_0 \cdot K \cdot p} \tag{16}$$

If chemosorption is regarded as a simple chemical process then its velocity v is in the steady-state range

$$v = \frac{\mathrm{d}N_{A^*}}{\mathrm{d}t} = \alpha \cdot K \cdot N_{st} \cdot p = \frac{W_0 \cdot \alpha \cdot K \cdot pt_0}{1 + \alpha \cdot t_0 \cdot K \cdot p} \tag{17}$$

or

$$\frac{W_0}{v} = 1 + \frac{1}{\alpha \cdot t_0 \cdot K}\cdot\left(\frac{1}{p}\right)$$

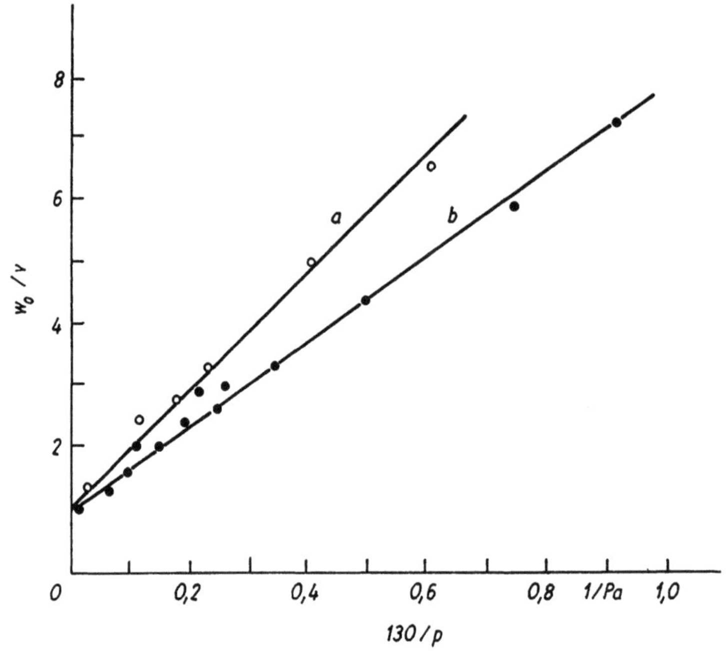

Fig. 4.10. Dependence of the sorption velocity of O_2 (a) and of CO (b) on the gas pressure during the milling of MoS_2 (a) and Fe_3C (b) at $T = 300$ K [54].

W_0 — velocity of the formation of active centres, v — reaction velocity, p — gas pressure.

Equation (17) could be successfully applied to the mechanically activated chemosorption of oxygen to molybdenum disulphide and of carbon dioxide to iron carbide. As Fig. 4.10. shows the chemical processes follow equ. (17) in both cases.

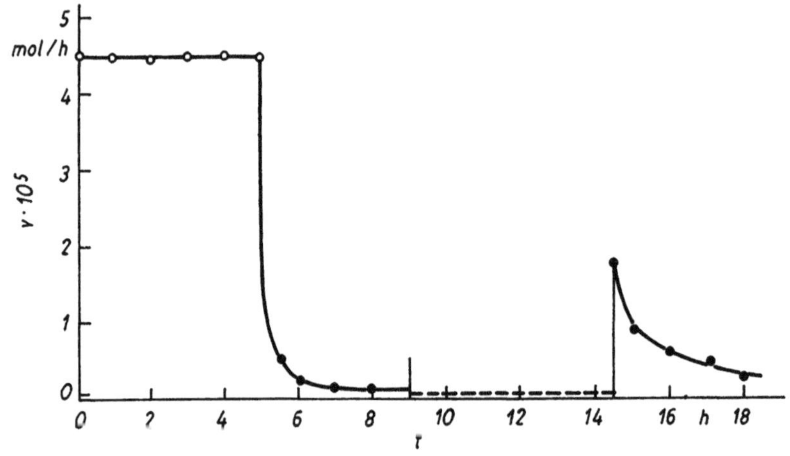

Fig. 4.11. The course of the decay curves for the formation of nickel carbonyl after the treatment in atmospheres of CO and N_2 [55].

v — reaction velocity, t — time; ○ — reaction during treatment in a CO-atmosphere, ● — decay reaction in a CO-atmosphere, — — — treatment in a N_2-atmosphere.

However, in many cases the above-mentioned assumption, that one kind of active centres exists and influences the reaction does not apply. So the formation of nickel carbonyl in the steady-state phase is caused by more highly active, rapidly decaying states and by less active, long-lived states. If for instance the mechanical treatment of nickel is carried out in an inert-gas atmosphere then as shown in Fig. 4.11. the reactivity in CO measured after the interruption of the treatment is only slightly higher than after the treatment in CO. The more highly active states produced during the treatment in N_2 decayed before the subsequent application of CO and now only the less active states are still available. In these cases an overlapping of two or more kinetic processes comes into being.

The constancy of the steady-state levels in Fig. 4.4. is only guaranteed for a longer period if gaseous reaction products come into being, such as nickel carbonyl.

During the formation of solid reaction products the resulting solid phase influences with a growing concentration the reaction kinetics so that they show a steady state course only in the beginning. Thus the following expression is valid for the decomposition of pyrite to iron sulphide and sulphur by using a planetary mill [56]

$$\alpha' = K' \left[\tau - \frac{1}{k_2} (1 - e^{-k2t}) \right] \tag{18}$$

(α' — degree of reaction, k_2 — constant of the mill. K' — constant, τ — duration of the treatment.)

With brief treatment times, a linear relation exists between the degree of reaction and duration of the treatment by a simplification of equ. (18). The following equation is valid for the reaction of two solids in a planetary mill, such as in the reaction of pyrite with metallic iron to iron sulphide [56]:

$$\alpha' = 1 - \exp \left\{ -K' \left[\tau - \frac{1}{k_2} (1 - e^{-k2\tau}) \right] \right\} \tag{19}$$

In all the investigations show that the course of the tribochemical reactions is not only determined by the type of reaction but also by the nature and intensity of the applied mechanical energy, since these factors also determine the formation of the active centres responsible for the reaction.

4.4. The influence of the treatment intensity on tribochemical reactions

4.4.1. Relationships between reaction velocity and intensity of the treatment

The influence of mechanical energy on a solid generally changes its properties very comprehensively. In most cases several parallel processes are involved in the dissipation of the mechanical energy which depend to varying degree on the kind of mechanical energy application, on the specific properties of the solid, and on the external boundary conditions. Hence quantitative data on how the mechanical energy determines the velocity of the chemical reaction via the elementary processes is extremly difficult to furnish, up till now only in individual cases (see 5.3.). However, simple relationships are obtained for a large number of tribochemical reactions, if the mechanical energy applied to the solid from outside is related to the reaction velocity.

If the impact stress of the grained reaction material is carried out in a vibration vessel (see 4.2.), then in such a system the acting mechanical energy E_s can be represented by the amplitude a and the frequency f of the vibration of the reaction vessel [57]:

$$E_s \sim a^2 \cdot f^2 \tag{20}$$

As can be seen in Fig. 4.12. and Tab. 4.4. the velocity v of numerous tribochemical reactions is related with the amplitude a and the frequency f by the following equation

$$v \sim a^2 \cdot f^2 \tag{21}$$

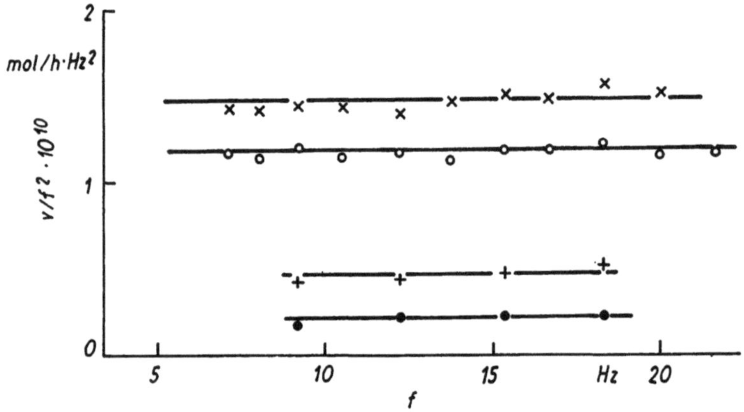

Fig. 4.12. The tribochemical formation of metallic oxides in dependence on the intensity of treatment [57].
v — reaction velocity, f — frequency; x — iron, o — nickel, + — copper, • — Aluminium.

Table 4.4. The dependence of tribochemical reaction velocity v on the intensity of mechanical stress.
f — frequency, a — amplitude.

Chemical and physico-chemical process	Measured relation	Measuring range	Ref.
$Ni + 4CO \rightarrow Ni(CO)_4$	$v \sim f^2$	$f = 3.3 \cdots 11.6$ Hz	[11]
$SiC + 2H_2 \rightarrow Si + CH_4$	$v \sim f^2$	$f = 9.7 \cdots 11.6$ Hz	[11]
Fe oxidation	$v \sim a^2$	$a = 2.9 \cdots 7.4$ cm	[57]
Tribodesorption of noble gases from nickel	$v \sim f^2$	$f = 4.2 \cdots 7.2$ Hz	[58]

From the equations (20) and (21) results

$$v \sim E_s \tag{22}$$

Thus the vibration energy as an expression of the intensity of the mechanical stress is combined in a determined range by a linear relation with the reaction velocity. Equation (22) expresses the fact that causal correlations exist between the flow of the dissipating mechanical energy and the chemical reaction released by it. Equation (22) is called the "tribochemical energy relation" [62].

The stress with a high impact frequency takes place according to a statistical distribution in a grained reaction material mechanically treated by vibration motions. The linear relation between tribochemical reaction and treatment intensity can also be proved under conditions of defined individual impacts. For this the treatment intensity can be determined from the kinetic energy E_z of the pushing bodies in first approximation

$$E_z = m \cdot g \cdot h \tag{23}$$

(m — mass, g — acceleration due to gravity, h — height of fall.)

The connection between the tribochemical process and the quantities given in equ. (23) can be seen from Fig. 4.13.—4.15.

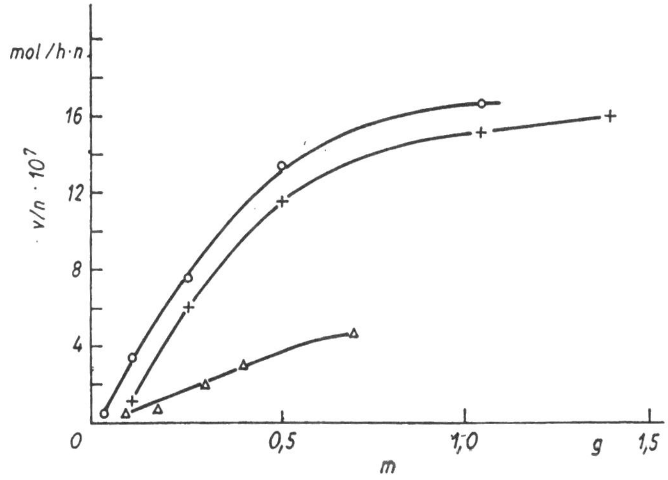

Fig. 4.13. Dependence of the tribochemical formation of nickel carbonyl on the mass of the impacting balls [59].

v — reaction velocity, m — ball mass, n — number of balls, ○ — steel balls in the fall down pipe, + — steel balls in the vibration vessel, Δ — glass balls in the vibration vessel.

The behaviour of the curves is similar in all three cases: a nearly linear slope at the beginning approximating to a boundary value with higher energy values. Fig. 4.13. shows differences resulting from the use of glass and steel balls as impact tools. In this it has to be taken into account that the density of glass is smaller by about the factor 3 than steel, and this alters the impact cross-section. Apparently for reaching high local excitation states the energy density also plays an important role. The use of treatment tools having different densities, such as steel or ceramic balls, also leads to a different reactivity of the stressed solids in ball and planetary mills [62—64], (see Tab. 4.5.).

In equ. (23) g is the gravitational acceleration and is valid for the free fall. The acceleration b in planetary mills can be varied within wide limits by the modification of the mill parameters (up to over 60 times the amount of the g value; Fig. 4.14.).

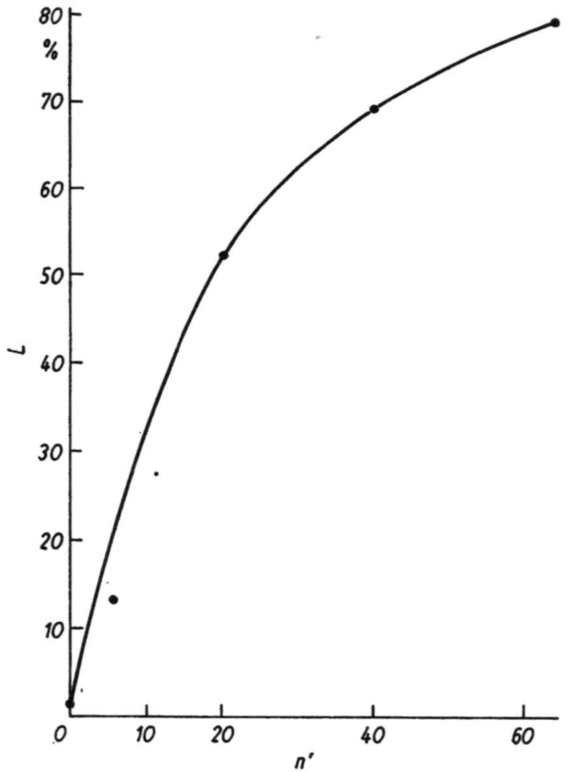

Fig. 4.14. Activation of the Marocco-phosphate in a planetary mill in dependence on the acceleration ratio n' [60].

L — solubility of P_2O_5 in alkine ammonium citrate solution (%), $n' = b/g$, b — acceleration, g — gravitational acceleration.

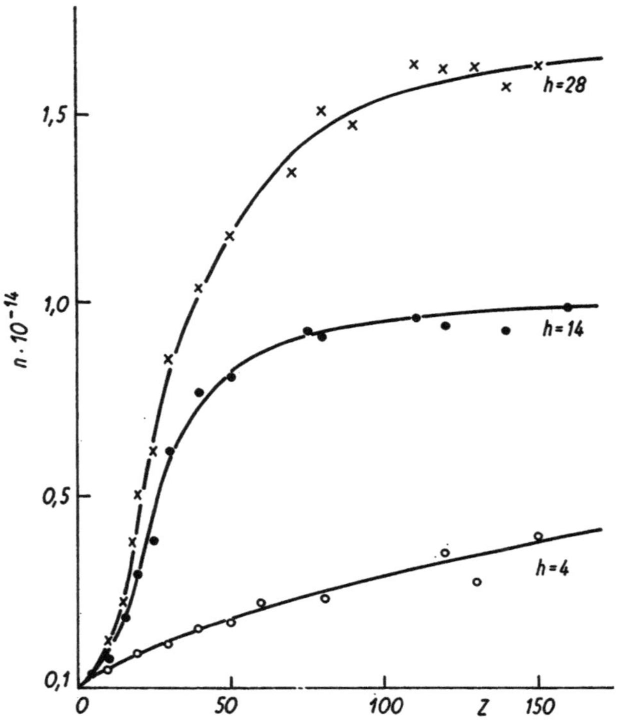

Fig. 4.15. Tribosorption of $^{14}CO_2$ molecules n on silver in dependence on the number of impacts z and the height of fall h [61].

Table 4.5. Dependence of the activation degree on the material of the planetary mill for Marocco-phosphate [64] time of treatment 30 min.

Ball and wall material	Density ϱ, g/cm^3	P_2O_5-solubility, %	
		in citric acid solution	in alkaline ammonium citrate solution
Nephrite	2.9	47	35
Steel	7.8	0	80

Fig. 4.15. shows the tribosorbed quantity n' of CO_2-molecules on silver in dependence on the hight of fall h and on the number of single impacts z. The following is applicable in the range of the linear slope of the sorption curve [61]:

$$n' \sim E_z \cdot z \sim h \cdot z \tag{24}$$

(E_z — kinetic energy of the single impact.)

Until now only few results exist for the causes of the deviations appearing with a higher treatment intensity. For instance the deviation in the system shown in Fig. 4.15. can be attributed to the CO_2 saturation of the treated boundary layer, that is to the appearance of a tribosorption equilibrium (see 4.8.3.), [61].

In vibration mills the dissipation flow of mechanical energy and the tribochemical reaction released by it are directly measured by electrical means. The following equation could be derived [65] for the power input N:

$$N = a \cdot M_c \cdot \omega^3 \cdot r \tag{25}$$

(a — experimentally measurable quantity characterizing the position of the centre of gravity of the grinding load, M_c — total mass of the milling material, ω — angular velocity of the circular vibration, r — amplitude and vibration radius, respectively.)

Since according to Fig. 4.16. the velocity of tribochemical reaction changes proportionally to the vibration radius, according to equ. (25) a direct proportionality exists between the power input and the reaction velocity:

$$v \sim N \tag{26}$$

Equ. (26) could also be confirmed for simple reaction systems, such as the formation of carbonyls from metals and CO, for reactions in planetary mills (see Tab. 4.6.).

Table 4.6. Dependence of the velocity $d\alpha/dt$ of the tungsten hexacarbonyl formation on the power input N in planetary mills at 313 K [66].

N, kJ/min	0.6	1.5	2.8
$\frac{d\alpha}{dt} \cdot 10^2$, h^{-1}	3	9	14

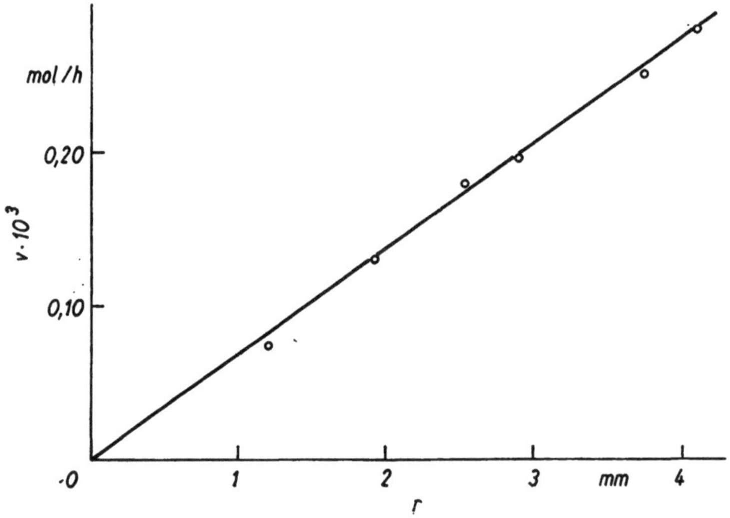

Fig. 4.16. Dependence of the velocity v of the reaction $2\,Cu + CO_2 \rightarrow 2\,CuO + C$ on the vibration radius of the circular vibration r [65].

To sum up it can be said that the simple dependences of the reaction velocity on the intensity of the stress represented in the equations (21)—(26) was mainly measured during the reaction of gases with solids, especially with metals. Where the systems are more complicated, such as the reaction of pyrite and iron resulting in iron sulphide or the decomposition of pyrite accompanied by the formation of iron sulphide and sulphur, then the relationships between power input N and reaction yield can no longer be described by simple linear relations. The following equation could be derived for the latter reactions [56]:

$$K = C \cdot \exp\left(\frac{E_{def.}}{\alpha'' \cdot t_{rel} \cdot N}\right) \tag{27}$$

(K — constant of the reaction velocity, C — constant, $E_{def.}$ — kinetic deformation energy leading to the release of a chemical reaction, α'' — coefficient, t_{rel} — time for the stress relaxation.)

In order to explain the phenomenological equation a connection is required between the chemical elementary processes and those processes in the solid occurring with the dissipation of the mechanical energy (see Chapt. 5.).

The intensity of the treatment may also be influenced by the application of grained or powdery substances of different hardness. As Tab. 4.7. shows the reaction velocity increases by orders of magnitude, if silicon carbide is used instead of KCl as the treating matter for the activation of nickel. The reason for this can above all be seen in the fact that the transferred energy during the impact stress is mainly utilized for the deformation of the pushing particles, if substances are used which have a smaller hardness than nickel. But if hard substances such as Al_2O_3 and SiC are used, then the roughness and therefore the ability to a preferred local stress in micro-ranges is maintained even after a longer treatment period. The acceleration of the reaction by the application of SiC to iron powder which is milled in the presence of water in a vibration vessel filled with steel balls can also for example be attributed to this (see Fig. 4.17.).

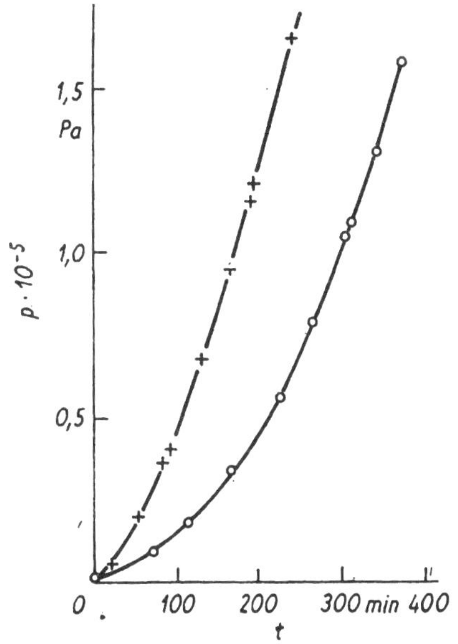

Fig. 4.17. The formation of hydrogen during the treatment of iron powder in presence of water vapour [68].
p_{H2} — hydrogen pressure, t — time of treatment,
o — without addition of SiC, + — with addition of SiC.

Table 4.7. Velocity v of the nickel carbonyl formation in dependence on the hardness of the used material [67].

Hard material	$v \cdot 10^5$ mole/h	MOHS hardness	Micro-hardness $H_M \cdot 10^{-7}$, Pa at 50 g test load
Silicon carbide	7.6	9	5,200
Quartz	2.4	6	1,475
Fluorite	0.7	4	178
Marbel	0.2	3	133
Potassium chloride	< 0.01	2	22

4.4.2. *On the efficiency of tribochemical reactions*

In view of its technological importance, the effect of the mechanical energy in comminution processes in respect to the work to be done for the production of new surfaces is the object of numerous researches [69, 70, 72]. The following defining equation is given for the efficiency of the comminution η_z [70]:

$$\eta_z = \frac{A_0}{A_G} \cdot 100 \tag{28}$$

(A_0 — effective work for the surface enlargement of the milling substance, A_G — total milling work used for the drive of the mill.)

The value A_G consists of a part available for the actual milling and from that part of the energy used up in the springs and bearings by friction losses.

If the specific surface energy of the solid σ is known then A_0 can be calculated taking into account the surface alteration ΔO:

$$A_0 = \Delta O \cdot \sigma \cdot M_c \tag{29}$$

(M_c — total mass of the milling material.)

The total energy A_{tr} taken up by the solid has to be taken into consideration in the formulation of an efficiency η_{tr} with respect to the course of tribochemical reactions [70]:

$$\eta_{tr} = \frac{A_{tr}}{A_G} \cdot 100 \tag{30}$$

A_{tr} consists of the excess energy stored in the volume phase, and the surface energy.

An example of the efficiency of a vibration mill is shown by the use of calcite in Fig. 4.18. If it is assumed that the energy alterations occurring through the treatment in the solid essentially causes the respective release and acceleration of tribochemical reactions (see 4.8.), then according to Fig. 4.18. the chemically effective part of energy A_{tr} is very small in comparison to the total energy expended. The largest part of the mechanical energy applied dissipates into thermal energy.

The efficiency η_{tr} can vary by orders of magnitude if different machines are used, so that comprehensive millingtechnical examinations are necessary in the case of a technological utilization of tribochemical reactions for an energetic optimization. As an

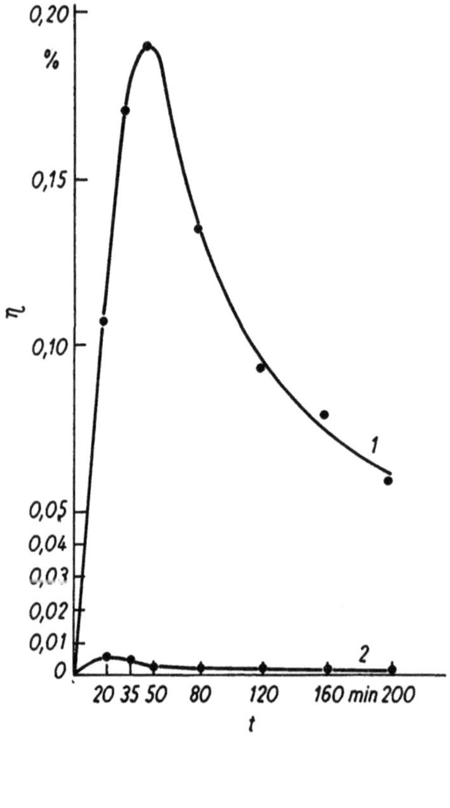

Fig. 4.18. Efficiency η of a laboratory vibration mill for calcite in dependence on the milling time t [70].

1 — efficiency η_{tr} related to the total energy taken up by the solid, 2 — efficiency η_s related to the surface energy.

example in Tab. 4.8. the energy input (A_G) is represented which is required to obtain a defined reactivity released by the treatment process. This value varies by orders of magnitude if several tools are used.

Table 4.8. Efficiency of various activation sets.
Comparison of electric energy for the production of a defined amount of an activated apatite per unit of time having an ammonium citrate solubility of 6% [71].

Activation set	Installed driving power, kWh	Throughput amount per unit of time for getting an ammonium citrate solubility of 6%, kg/h	Input of specific energy, kWh/kg
Jet-O-Mizer	2.0	0.03	67
Micronizer	2.0	0.045	44
Disintegrator	5.0	3.0	1.7
Vibration mill I	0.8	0.08	10
Vibration mill II	1.0	0.5	2
Double-pipe vibration mill I	2.2	8.0	0.28
Double-pipe vibration mill II	100	1,000	0.10
Vibration mill III	100	300	0.33

The efficiency η_{tr} for tribochemical reactions has been defined for a certain range of negative affinity in the following way [72]:

$$\eta_{tr} = \frac{-v \cdot A}{N_{\text{stat}}} \tag{31}$$

(v — reaction velocity, A — affinity of the chemical reaction, N_{stat} — power input in the steady-state, see 4.3.)

Here the reaction work $v \cdot A$ done in the time unit is directly related to the power input. As Tab. 4.9. shows, for a variety of tribochemical reactions the efficiency of $10^{-3} \cdots 10^{-2}$ per cent calculated from equ. (31) is smaller by $1 \cdots 2$ orders of magnitude than the value calculated from equ. (30). This difference is based on the fact that tribochemical reactions having a negative affinity run with a smaller cross section than those with a positive affinity. However, the examples discussed here show unanimously that tribochemical reactions usually run with a small efficiency.

As Fig. 4.19. shows, the efficiency remains constant during the alteration of the power input in the investigated measuring range. If $-A$ is maintained in a constant state then a proportionality results from equ. (31) which corresponds to equ. (26):

$$v \sim N_{\text{stat}} \tag{32}$$

From this the conclusion can be drawn that the above-mentioned proportionalities between energy input and reaction velocity are only valid if the tribochemical efficiency and the chemical affinity are constant.

Table 4.9. Efficiency η_{tr} of tribochemical reactions with negative affinity [12].

Reaction	η_{tr}
$Ni + 4\,CO \rightarrow Ni(CO)_4$	$3.3 \cdot 10^{-5}$
$CdCO_3 \rightarrow CdO + CO_2$	$1.3 \cdot 10^{-5}$
$2\,NiO + C \rightarrow 2\,Ni + CO_2$	$1.7 \cdot 10^{-4}$
$MgO + C \rightarrow Mg + CO$	$1.2 \cdot 10^{-4}$
Calcite → aragonite	$2.4 \cdot 10^{-5}$

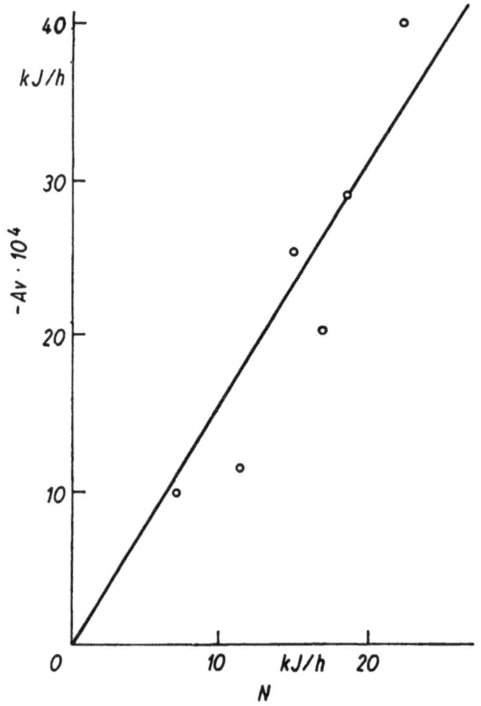

Fig. 4.19. Dependence of the product $-A \cdot v$ for the reaction $Ni + 4\,CO \rightleftharpoons Ni(CO)_4$ on the respective power input N in the range of negative affinity [12].

4.5. The dependence of tribochemical reactions on temperature

The activation energy E_A in the thermally released reactions can be determined by the following equation:

$$k = C \cdot e - \frac{E_A}{R \cdot T} \tag{33}$$

(k — constant of velocity, C — constant.)

Examinations of solid reactions have shown that the heterogeneous kinetics can also be described by the equation which initially was derived for homogeneous reactions [73]. However, the use of equ. (33) presupposes that the activation energy required for the reaction is exclusively taken up from the heat supply of the environment. On the other hand, if the energy required for the release of a chemical reaction is applied exclusively in a non-thermal manner, e.g. by light, or by fast electrons, then chemical reaction can no longer be described by equ. (33). Non-thermally activated reactions frequently show no, or a lowered, dependence on temperature in comparison to completely thermally activated reactions. If the ARRHENIUS equation is used formally, activation energies result the values of which deviate negatively from those of the thermally conducted reaction and are zero in the borderline case.

In comparison with the thermally activated reaction, an altered temperature dependence has also been found in tribochemical reactions. Numerous reactions are independent of temperature in a wide measuring range [46, 74]. For instance, the velocity of the formation of stannanes from tin and benzyl chloride is constant in the range of 277 ··· 353 K [75]; (see Fig. 4.20.).

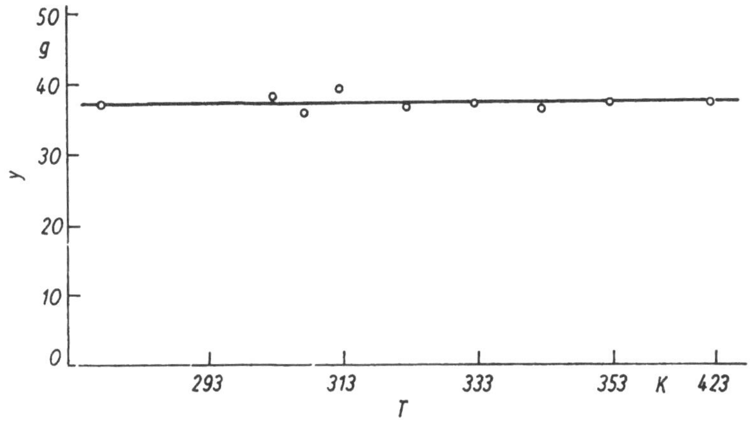

Fig. 4.20. Temperature dependence of the tribochemical reaction: $Sn + 2\ C_6H_5CH_2Cl \rightarrow (C_6H_5CH_2) \cdot SnCl_2$ [75].
y – reaction yield, T – temperature.

Further examples are given in Tab. 4.10. where the activation energies are compared during mechanical treatment and during thermal activation. It appears that the activation energies always have smaller values with tribo-mechanical treatment than with exclusively thermal stress.

The occurrence of a reaction velocity independent of temperature means that the energy required for releasing the reaction is not drawn from the heat supply of the environment but is applied to the solid through the mechanically activated regions only. This mainly applies to reactions with higher activation energies released by correspondingly high activation states and showing no measurable decay reaction after the interruption of the treatment (see Fig. 4.8.). As shown in Fig. 4.21. the chemosorption at short-lived surface centres also occurs as a thermally non-activated reaction [53].

A pre-condition for the temperature independence of a tribochemical reaction is that the appearance of states responsible for the reaction is not, or only to a limited extend influenced by the surrounding temperature or that the efficiency of the tribo-

Table 4.10. Apparent activation energy E_A of tribochemical and thermal reactions

Reaction	Tribochemical reaction				Thermal reaction			
	E_A, kJ/mole	Temperature T, K	Contact	Ref.	E_A, kJ/mole	Temperature T, K	Contact	Ref.
$Ni + 4\,CO \rightarrow Ni(CO)_4$	8.8	273 ··· 283	—	[89]	43	273 ··· 283	—	[76]
	3.5	233 ··· 273	—	[89]				
	0	233	—	[43]				
	neg.	90 ··· 233	—	[43]				
$Fe + O_2 \rightarrow$ Iron oxide	0.7	90 ··· 298	—	[54]	54	273	—	[77]
$CH_4 \rightarrow C + 2\,H_2$		313	Corundum, SiC	[46]	497	1673 ··· 1973	Corundum, C,	[78]
$5\,Ni + CO_2 \rightarrow Ni_3C + 2\,NiO$	1.3	293 ··· 368	—	[90]				
$CH_3OH \rightarrow CO + 2\,H_2$	13 ··· 21	291 ··· 318	Fe	[91]	146	573 ··· 673	ZnO	[79]
$C_6H_6 + 3\,H_2 \rightarrow C_6H_{12}$	25	263 ··· 473	Ni	[50]	44,4	298 ··· 373	Ni	[80]
$C + H_2O \rightarrow CO + H_2$		298	C	[92]	146	1273 ··· 1433	C	[81]
Formic acid decomposition		303	Ni	[89]	48	393 ··· 473	Ni	[82]
$3\,H_2 + N_2 \rightarrow 2\,NH_3$		298	Fe act.	[93]	58		Fe	[83]
H_2 chemosorption		298	SiC, C	[92]	92 ··· 145		C	[84]
CH_4 chemosorption		313	Corundum, SiC, C	[92]	222	973	C	[85]
Nitrate decomposition	0	77 ··· 293	—	[94]				
$Sn + 2\,C_6H_5CH_2Cl \rightarrow (C_6H_5CH_2) \cdot 2\,SnCl_2$	0	277 ··· 353	—	[75]				
O_2, H_2 sorption to SiO_2	0	78 ··· 370	SiO_2	[34]				
$C(SiC) + 2\,H_2 \rightarrow CH_4 + Si$	0.03	77 ··· 195	—	[95]				
	1.3	195 ··· 300	—	[95]				
	2.1 ± 1.3	300 ··· 350	—	[95]				
	17.2 ± 2.1	400 ··· 700	—	[95]				
	72	730 ··· 930	—	[95]				
	108	700 ··· 950*)	—	[95]	126	923 ··· 1143	C	[86]

Table 4.10. (Continuation)

Reaction	Tribochemical reaction				Thermal reaction			
	E_A, kJ/mole	Temperature T, K	Contact	Ref.	E_A, kJ/mole	Temperature T, K	Contact	Ref.
$2\,C(SiC) + 3\,H_2 \rightarrow C_2H_6 + (2\,Si)$	1.7 ± 0.4	77 ··· 300	—	[95]				
	9.7 ± 0.8	300 ··· 350	—	[95]				
	22.2 ± 1.7	400 ··· 650	—	[95]				
$3\,C(SiC) + 4\,H_2 \rightarrow C_3H_8 + (3\,Si)$	23.8 ± 0.8	400 ··· 640	—	[95]				
Fat hydrogenation	27*)	—	Ni	[87]	84	—	Ni	[87]
Dissolution of apatite	3.4*)	288 ··· 313	—	[88]	29.8	288 ··· 323		[88]

*) After interrupting the treatment.

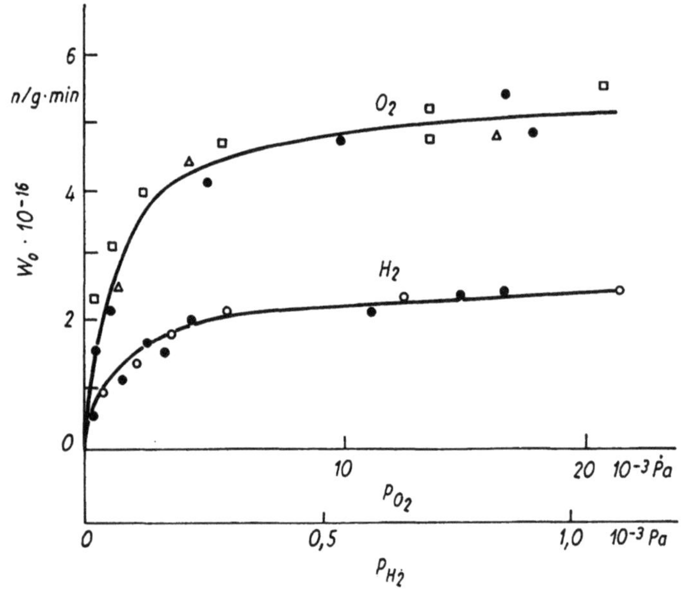

Fig. 4.21. Dependence of the sorption velocity v of hydrogen and oxygen molecules on milled quartz on the gas pressure with different temperatures [53].
⊙ – 78 K, Δ – 195 K, • – 300 K, □ – 370 K.

chemical reaction (see equ. (31)) is independent of temperature. This condition for instance is met in first approximation in the mechanically activated formation of nickel carbonyl (Tab. 4.11.).

Table 4.11. Efficiency η_{tr} of the reaction $Ni + 4\,CO \rightarrow Ni(CO)_4$ in dependence on temperature T [12].

T, K	$\eta_{tr} \cdot 10^5$
298	3.5
308	3.3
323	3.0
343	3.0

However, if an increasingly faster relaxation of the active states takes place with the rising temperature so that they are available to increasingly smaller degree for the respective release and acceleration of chemical reactions, then a negative temperature coefficient is measured. This behaviour has been observed particularly in mechanically activated catalytic reactions above the TAMMANN temperature [96–98] as shown in Fig. 4.22.

As Fig. 4.23. shows a negative temperature gradient also occurs during the formation of nickel carbonyl with temperatures < 233 K [43]. The same dependence has also been observed during the formation of wolfram carbonyl [66]. In order to explain these results it is assumed that the active centres decay more rapidly than CO diffuses

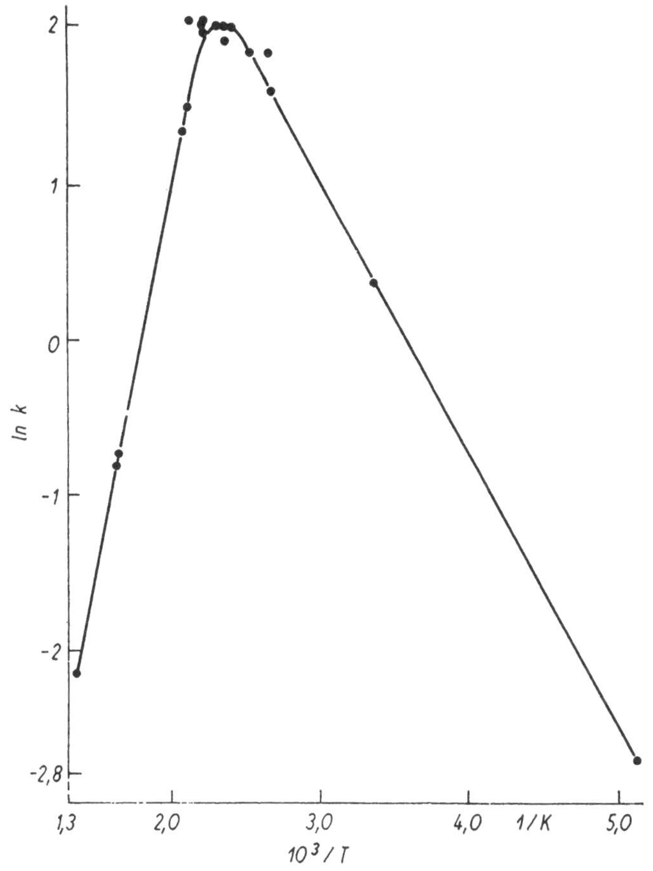

Fig. 4.22. Temperature dependence of the ethylen hydrogenation during the mechanical treatment at the armco-iron catalyzer [96].

k — constant of velocity,
T — temperature.

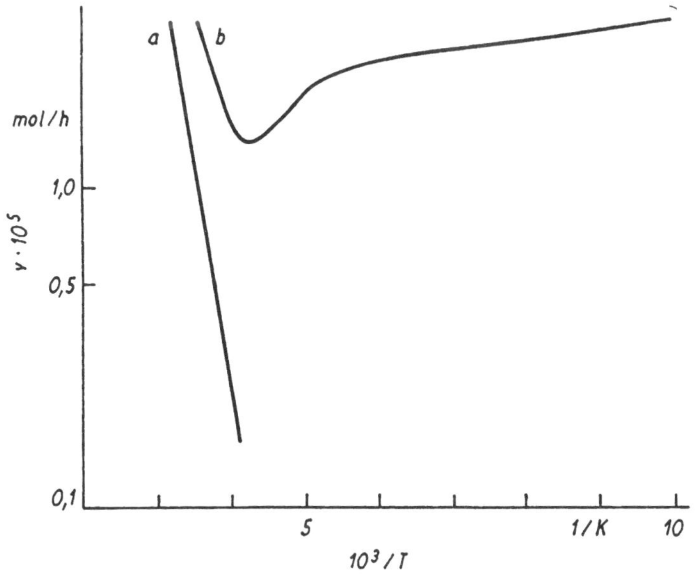

Fig. 4.23. Dependence of the reaction velocity v of the nickel carbonyl formation on the temperature T [43].

a — for mechanically pretreated nickel, b — during the mechanical treatment.

from the gaseous atmosphere to the reaction place. Therefore only the pre-adsorbed molecules are available for the tribochemical reaction (see 4.6.). Then the negative temperature gradient results from the increasing CO-coverage with decreasing temperature [99, 100]. As can also be seen in Fig. 4.23. curve *b*, the formation of nickel carbonyl has a positive temperature gradient above 233 K. In this range the formation of nickel carbonyl also runs with measurable velocity after the interruption of the treatment (curve *a* of Fig. 4.23.) and follows the ARRHENIUS equation.

Table 4.12. Activation energy E_A of the reaction.
$Ni + 4\,CO \rightarrow Ni(CO)_4$ before, during, and after mechanical treatment [101].

Treatment	E_A, kJ/mole
Before the mechanical treatment	54
During the mechanical treatment	14.6
After the mechanical treatment	25

As Tab. 4.12. shows an activation energy may be calculated from this which is only half as large as that measured on the untreated nickel. Apparently the velocity of the formation of nickel carbonyl is influenced by at least 2 kinds of active surface centres: by more highly active centres exciting the formation of nickel carbonyl independently on the surrounding temperature and by centres with lower activity accelerating the formation of nickel carbonyl, accompanied however by a simultaneous application of thermal energy from the surrounding heat supply. The latter centres appear in a great number of tribochemical reactions and often lead to an increase of the reaction velocity by orders of magnitude due to the decreased activation energy (see Tab. 4.10.). With special regard to the degradation reactions of organic polymers a modified ARRHENIUS equation has been derived for this [102]:

$$k = k_0 \cdot \exp\left(-\frac{U_0 - \gamma\sigma}{RT}\right) \qquad (34)$$

(k — constant of velocity for the tearing, for instance of polymeric foils, fibres, or crystals being stressed by strain accompanied by a constant elastic stress σ until the fracture; k_0, U_0, γ — constants).

The derivation of this equation was based on the assumption that by the effect of an external elastic stress σ the binding energy U_0 decreases by the quantity of the deformation energy $\gamma\sigma$. Since the bursting of the bond represents the decisive activation step for the tearing process the difference $U_0 - \gamma\sigma$ is included as activation energy in the ARRHENIUS equation. If the sample is destroyed its life span t_L is inversely proportional to the velocity constant k. Consequently equ. (34) can also be written in the following form [103]:

$$t_L = t_0 \cdot e^{\frac{U_0 - \gamma\sigma}{RT}} \qquad (35)$$

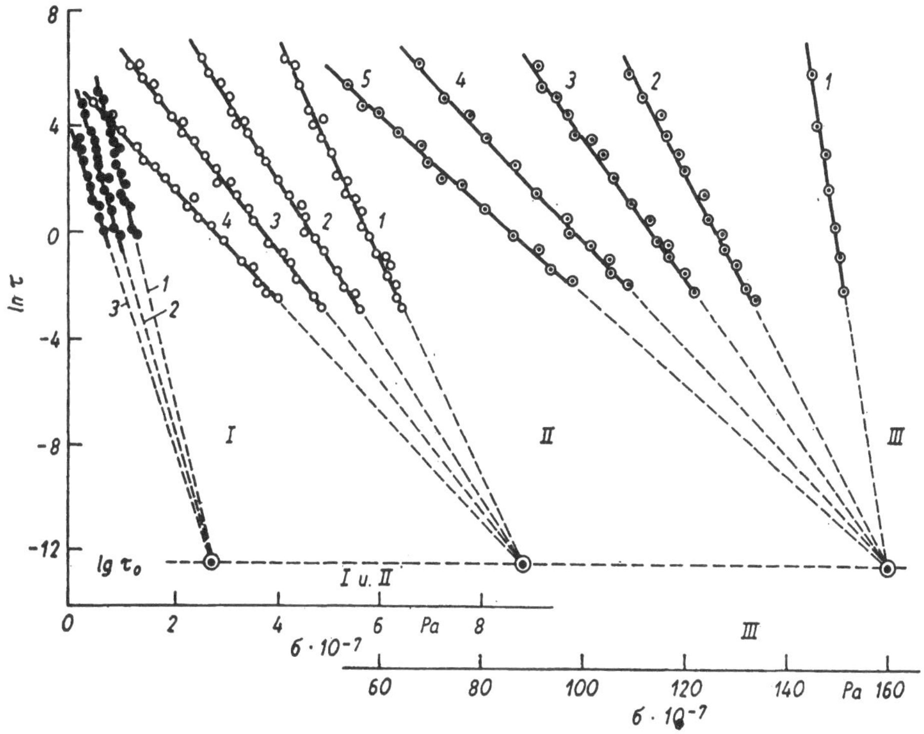

Fig. 4.24. Dependence of the time τ until the destruction of several substances on the external stress σ during uniaxial tensile strain and with different temperatures [103].

I – rock salt: 1 – 673 K, 2 – 773 K, 3 – 873 K; II – aluminium, polycrystalline: 1 – 291 K, 2 – 373 K, 3 – 473 K, 4 – 573 K; III – polyamide fibres: 1 – 93 K, 2 – 153 K, 3 – 198 K, 4 – 293 K, 5 – 353 K.

In Fig. 4.24. the time up to the appearance of the destruction of NaCl, polycrystalline aluminium, and caproic fibres is shown in dependence on the applied unidirectional tensile stress for several temperatures. The experimental results can be described for a wide range of σ and T by equ. (35). The intersection point of the straight lines is in $t_0 = 10^{-12}$ s, which is the duration of an elementary oscillation. Thus t_0 has the character of a molecular constant. According to Tab. 4.13. the same also applies for U_0. The values partly calculated from Fig. 4.24. are posed by way in Tab. 4.13. of comparison with the heats of sublimation of the metals and the activation energies of the thermal degradation of the polymers, respectively. In the first approximation U_0 does not deviate very strongly from the value of the energy of the interatomic linkage.

The equations (34) and (35) are derived in the case where external stresses are applied to a solid. The production of active regions in a mechanical way is connected with the formation of lattice defects which influence energetically the linkages in the solid through the appearance of inner stresses. For instance an elastic lattice deformation exists in the environment of a dislocation. Therefore equ. (34) is also applicable in a modified form to such tribochemical processes which run on mechanically disturbed solids. Consequently the generally observed reduction of the activation energy by the mechanical treatment of the solid results from the attenuation of the lattice bonds due to inside stresses.

Table 4.13. Comparison of the constant U_0 with the heat of sublimation of metals E_S and the activation energy of the thermal degradation of polymers E_A, respectively [103].

Substance	U_0, kJ/mole	E_s and E_A, respectively kJ/mole
Platinum	500	530
Iron	418	405
Copper	340	335
Aluminium	222	230
Zinc	126	130
Rock salt	285	—
Polytetrafluorethylene	314	318
Polypropylene	234	242
Caprone	188	180
Polyvenyl chloride	146	134

As can be seen further in Tab. 4.12., the activation energy of the nickel carbonyl formation continues to be reduced at room temperature during the treatment in comparison with the value measured on the pre-activated nickel. Taking into accout the appearance of at least two kinds of centres only the higher-active centres of which lead to a reaction independent of temperature, the value given as 14.6 kJ/mol is assumed to be a quantity consisting of the activation energies of the reactions on the individual kinds of centres. Since the activation energies of thermally conducted solid reactions are always higher than the corresponding mechanically activated reactions, the influence of the thermal excitation on the reaction grows more and more with in-

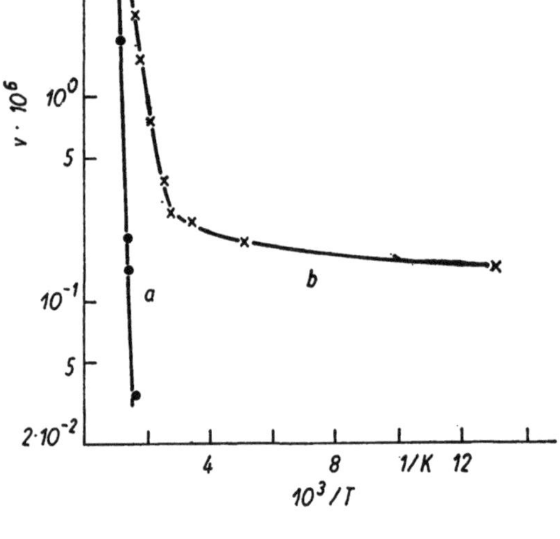

Fig. 4.25. Tribodesorption velocity v of CH_4 in the SiC/H_2 system in dependence on the temperature T [10].
a — for mechanically pretreated SiC,
b — during the mechanical treatment.

creasing temperature so that the differences in the reaction velocity decrease (see Fig. 4.25.). The temperature at which the reaction velocity on the non-treated solid is equal to that on the treated solid at room temperature is termed "corresponding temperature". For instance the corresponding temperature runs to 900 ··· 1,000 K (Fig. 4.25.) in the formation of methane from SiC and C and hydrogen, respectively [86, 104].

However, the conditions of excitation are very different in both cases. Whereas the solid has a unifromly high activity in the entire volume in an exclusively thermal activation within the scope of the Boltzmann distribution, only one small part in each case is highly activated and reactive for a short period at room temperature during the treatment. As will be shown in chapter 5. more highly excited states exist during treatment in comparison to the thermal activation (see Fig. 5.16.), [105].

4.6. The pressure-dependence of tribochemical reactions

The measurements of the pressure-dependence for the determination of the reaction order plays an important role in kinetic examinations of gas-solid reactions. Usually heterogenous reactions are of lower order compared to gaseous phase reactions. The reaction is a first-order reaction in many cases if the gas, due to a weak sorption, only covers a part of the surface and the coverage density is proportional to the gas pressure. If, however, the interaction forces are very high between gas and solid then a nearly complete coverage density already takes place with small gas pressures so that the solid surface concentration of the reactants does not further change with a variation of the pressure, and the reaction velocity remains thereby constant, too. This is a zero-order reaction in this case. Examples for this are the decomposition of formic acid and of NH_3 on wolfram. Often the conditions are extremely complicated in heterogenous reactions and broken orders are measured.

If heterogeneous reactions on solids occur under the influence of mechanical energy e.g. during milling stress, then considerable deviations in the pressure dependence often appear in comparison to non-mechanically influenced reactions. In general the reaction order is reduced in tribochemical reactions compared with thermal reactions and has even been the value zero in several cases [106]. Thus for example the thermally excited formation of nickel carbonyl was measured to be a second-order reaction [76]. The value drops to 0.2 ··· 0.5 during mechanical treatment and also depends to a small degree on the type of the hard materials used for the activation of nickel, as Fig. 4.26. shows. Thus apart from a slower reaction velocity a higher order is found if glass balls are used than if sharp-edged and harder particles are used [43]. The reason for the reduction of the order of reaction during the mechanical energy influence on the solid, consists above all in the fact that the reaction velocity is determined by the velocity at the more highly activated impact centres which have a strongly increased absorptive and reactive capacity in comparison to the non-stressed solid surfaces (see 4.7.). Because of this the concentration of the reactants at the solid surface and generally the order of reaction are changed as well.

Proceeding from the postulate that at least two kinds of centres determine the total velocity in the formation of nickel carbonyl, the dependence of the order of reaction on the type of treatment shown in Fig. 4.26. is easy to understand in so dar as with less intensive treatment, for instance with the use of glass balls, fewer active regions come into existence resulting in a lower reactivity and in a higher order.

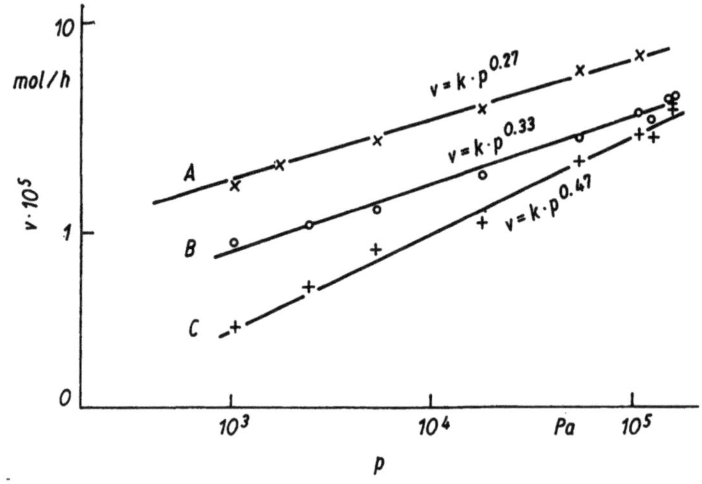

Fig. 4.26. Dependence of the velocity v of the tribochemical formation of nickel carbonyl on the CO pressure p by using different hard substances [45].
A — SiC, *B* — refuse glass, *C* — glass balls.

The tribochemical formation of iron and nickel oxides from the elements does not depend on the pressure through 5 orders of magnitude ($1 - 10^5$ Pa) [57], whereas the relation $v \sim p^{0.28}$ has been found for the thermally activated formation of iron oxide [77].

Due to the sorption of the gas to the mechanically activated solid surface as prestage of the reaction the velocity v of the gas consumption is no longer independent on the pressure p with pressures below 1 Pa, but it complies with the proportionality in first approximation [107] as Fig. 4.27. shows:

$$v \sim p \tag{36}$$

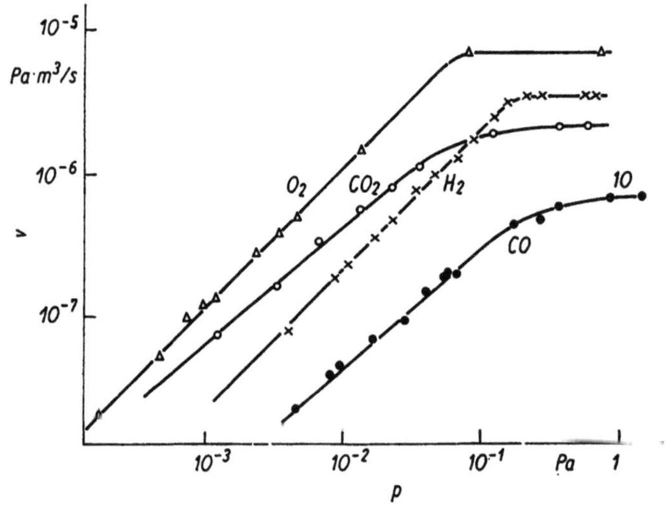

Fig. 4.27. Velocity v of the tribosorption of several gases on mechanically stressed nickel surfaces in dependence on gas pressure p [107].

Consequently the reaction with small pressures occurs according to the first order. By this the pressure-dependence of the gas intake corresponds for the whole pressure range of $10^{-4} \dots 10^{-5}$ Pa with the LANGMUIR adsorption isothermes:

$$\theta = \frac{k_1' \cdot p}{k_2' + p} \tag{37}$$

(θ— degree of coverage, k_1', k_2' — constants.)

and analogy can be described by the equation:

$$v = \frac{k_1 \cdot p}{k_2 + p} \tag{38}$$

(k_1, k_2 — constants.)

The tribomechanically induced sorption of CO_2, O_2, and CO to silver complies with equation (38) as well [108].

In order to explain the pressure-dependence it can be assumed that an equally large number of sites capable of sorption is produced by steady-state treatment in a unit of time. If, however, these sites exist for a short period then a sorption takes place only if the gas molecule interacts with the solid by collision during the lifetime of the activated state. The number of collisions is sufficiently large with high pressure so that the impinging gas molecules saturate all the mechanically activated centres by sorption. Since there is not enough time for a complete saturation below a defined gas pressure the active centres partially die out again before they can begin to react. Under these conditions equation (38) is identical with equation (17) derived in 4.3. Information on the maximum number of the mechanically activated centres can be gained from the height

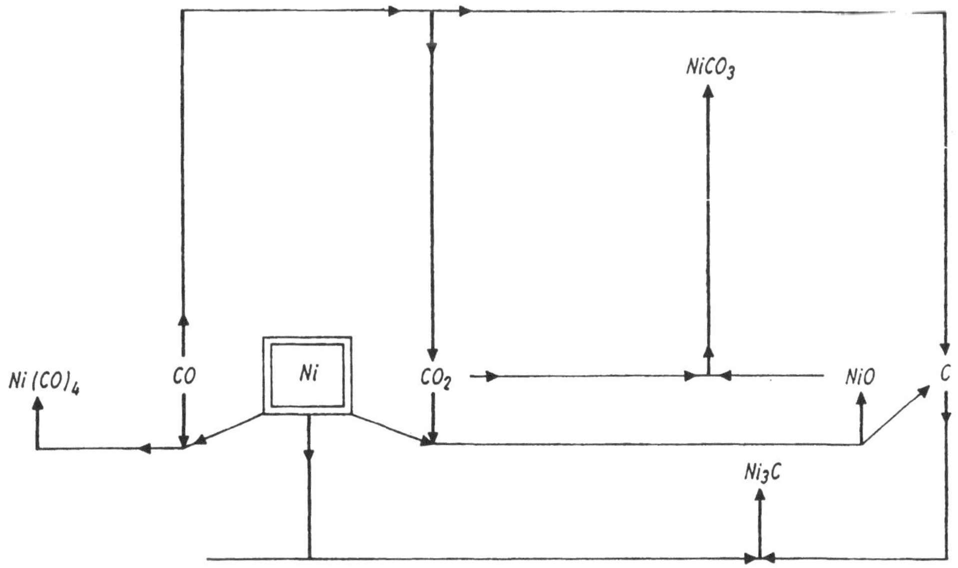

Fig. 4.28. Scheme of the tribochemical reactions of nickel in carbon oxide atmosphere [110].

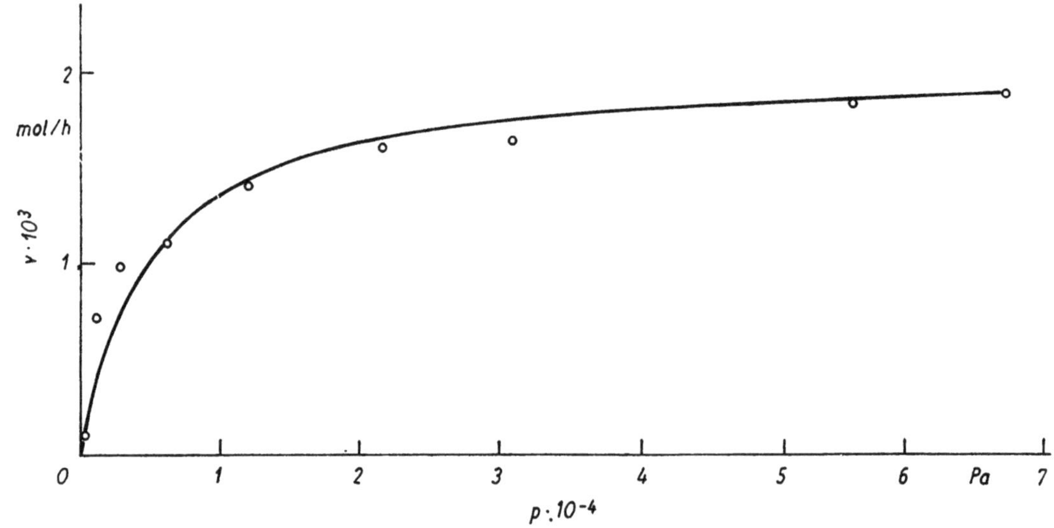

Fig. 4.29. Dependence of the reaction velocity v on the CO-pressure p for the reaction [112].
$2\,Fe + 3\,CO \rightarrow Fe_2O_3 + 3\,C$.

of the pressure-independent course of the curves shown in Fig. 4.27. and information on the lifetime of the centres from the position of the transition of the curve course from the zero order to the first order. Assuming a sticking probability of ≈ 1, values of $10^{-2} - 10^{-3}$ s were calculated for the lifetime of the sorption centres during the sorption of CO to treated nickel [109]. A value of $\approx 10^{-2}$ s was also found for the sorption of O_2 and H_2 to quartz [53].

However, the sorption of gases is only the first step in a series of continuing tribochemical reactions of the treated solid. As shown schematically in Fig. 4.28. for instance, carbonylization, oxidation, and carburization reactions are released during the treatment of nickel in a CO atmosphere running with different velocities and frequently overlapping in a complicated manner [90, 110, 111]. The reactions also comply with equation (38) as far as they appear as gas-solid reactions [112]. A typical curve behaviour is shown in Fig. 4.29. The constants k_1 and k_2 of equation (38) are compiled for a series of tribochemical reactions in Tab. 4.14.

A comparison of the curves represented in Figs. 4.27. and 4.29. shows that important differences exist in the quantitative values with a qualitatively similar behaviour.

Table 4.14. Summary of the constants k_1 and k_2 of the equation $v = \dfrac{k_1 \cdot p}{k_2 + p}$ for the number of tribochemical reactions [112].

Reaction	$k_1 \cdot 10^3$, mole/h	$k_2 \cdot 10^{-2}$, Pa
$2\,Al + 3\,CO \rightarrow Al_2O_3 + 3\,C$	4.45 ± 0.06	19.0 ± 4.5
$4\,Al + 3\,CO_2 \rightarrow 2\,Al_2O_3 + 3\,C$	3.74 ± 0.10	22.4 ± 9.1
$2\,Fe + 3\,CO \rightarrow Fe_2O_3 + 3\,C$	2.02 ± 0.04	50.2 ± 7.2
$4\,Fe + 3\,CO_2 \rightarrow 2\,Fe_2O_3 + 3\,C$	0.75 ± 0.03	170 ± 17.2
$2\,Ni + CO_2 \rightarrow 2\,NiO + C$	0.27 ± 0.01	92.0 ± 18.8
$CdO + CO_2 \rightarrow CdCO_3$	4.58 ± 0.38	30.3 ± 9.8

For instance the constant k_2 of the curve in Fig. 4.29. is larger by about 4 decimal powers than the constant k_2 of the curve in Fig. 4.27. Accordingly the pressure-dependent part of the curve in Fig. 4.29. extends in excess of 10^3 Pa. From this the conclusion is drawn that gases which are less strongly adsorbed decisively determine the course of the tribochemical reaction. This is also shown by the temperature-dependence of the reaction represented in Tab. 4.15. and in Fig. 4.30.:

$$2\,Ni + CO_2 \rightarrow 2\,NiO + C \tag{39}$$

Table 4.15. Dependence of the constants k_1 and k_2 in equation $v = \frac{k_1 \cdot p}{k_2 + p}$ on temperature T for the reaction $2\,Ni + CO_2 \rightarrow 2\,NiO + C$ [112].

T, K	k_1, Pa/s	$k_2 \cdot 10^{-2}$, Pa
298	1.06 ± 0.102	5.6 ± 0.95
308	0.98 ± 0.107	6.5 ± 1.14
323	0.90 ± 0.154	8.9 ± 2.14

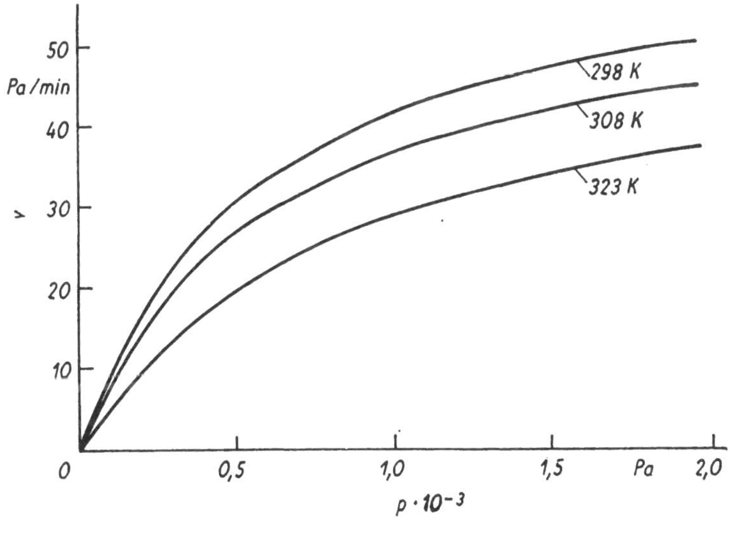

Fig. 4.30. Dependence of the velocity v of the reaction $2\,Ni + CO_2 \rightarrow 2\,NiO + C$ on the CO_2-pressure p for 3 different temperatures [112].

Whereas k_1 continues to be constant with increasing temperature, k_2 grows with the temperature. Therefore, the decrease of the concentration of weakly adsorbed molecules with growing temperature is the reason for the negative temperature coefficient of the reaction in question.

As Fig. 4.31. shows, analogous conditions exist in the tribochemical formation of nickel carbonyl. A complete coverage of the Ni-surfaces ($\theta = 1$) by CO already appears with very small CO-pressures during the reduction of the temperature to a proximity of the condensation level of CO, so that the reaction is pressure-independent with $T = 100$ K in the measured pressure range compared with examinations at room temperatu-

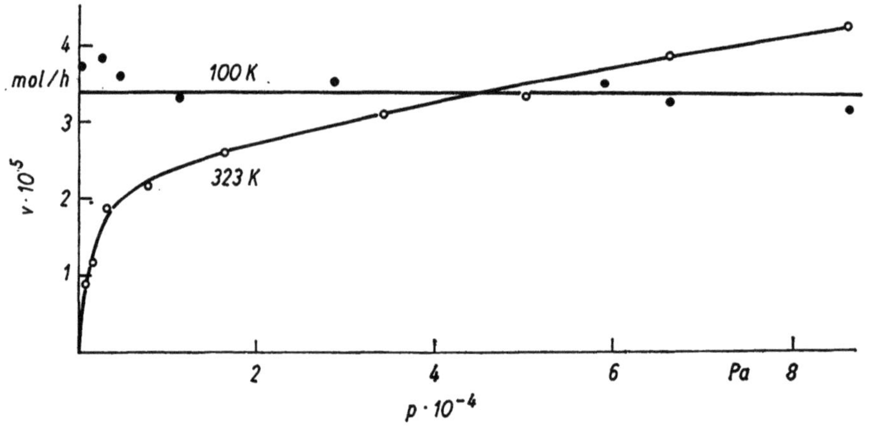

Fig. 4.31. Dependence of the velocity v of the nickel carbonyl formation on the CO pressure p for 2 different temperatures [112].

res. As represented in Fig. 4.23., the reaction also has a negative temperature gradient with low temperatures. Such a temperature gradient was also observed in the tribosorption of O_2 to MoS_2 [53].

The following fundamental result can be derived from the pressure and temperature-dependence of tribochemical reactions for their kinetics:

The tribochemical reaction of the mechanically stressed solid with gases depends on the concentration of the gases to be adsorbed to the solid surface whereas their binding state is of secondary importance [112].

Under these condition is

$$v \sim \theta \tag{40}$$

By this equation (38) directly results from equation (37) in which $k_2' = k_2$. If the intensity of the treatment is varied with temperature being constant, e.g. by an increase of the reaction vessel's amplitude of vibration (see 4.4.), then the constant k_1 in equ. (38) increases [112], (Tab. 4.16.).

Since k_2 does not, or only to a small degree, vary within the scope of the measuring accuracy, the specific adsorption on the treated solid surface is not, or to a small degree, changed by intensity variations. However, the portion of the adsorbed gas continuing to react increases proportionally to the intensity of the dissipation flow. As Fig. 4.32.

Table 4.16. Dependence of the constants k_1 and k_2 in equation $v = \frac{k_1 \cdot p}{k_2 + p}$ for the reaction $2\,Ni + CO_2 \rightarrow 2\,NiO + C$ on the radius r of the circular vibration [112].

r, mm	k_1, Pa/s	$k_2 \cdot 10^{-2}$, Pa
1.09	0.48 ± 0.085	9.7 ± 2.5
1.69	0.76 ± 0.110	8.8 ± 1.9
2.18	1.20 ± 0.225	14.4 ± 3.3
2.72	1.38 ± 0.200	14.3 ± 2.4

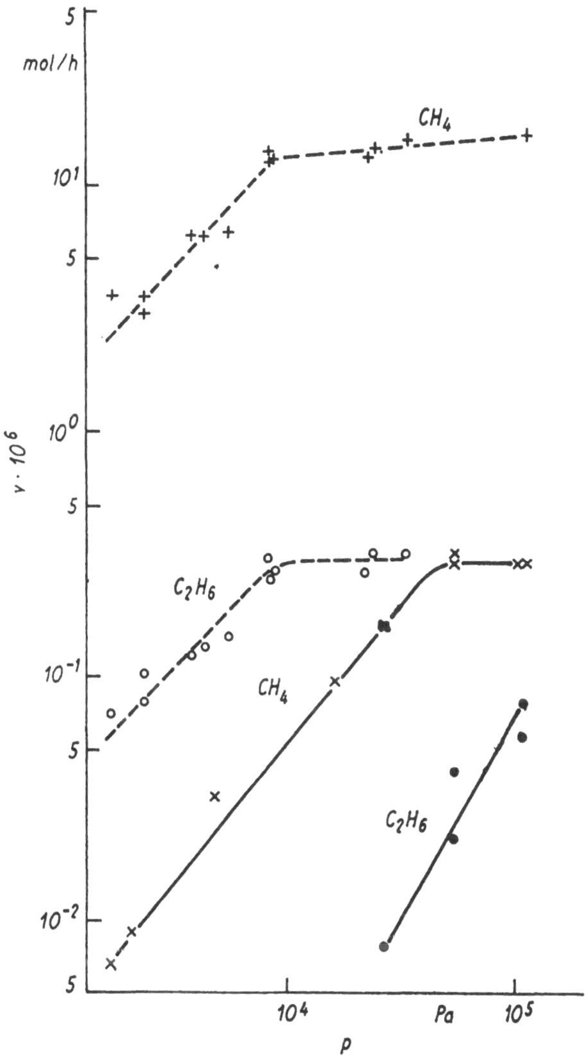

Fig. 4.32. Dependence of the Alkan-tribosorption velocity v on the H_2 gas pressure p with and without addition of iron [10].
—— SiC/H_2, - - - SiC + Fe/H_2.

shows, the formation of methane and ethene from SiC and H_2 also takes place according to equ. (38). A shift of the curves takes place upon application of Fe, which is apparently caused by the changed sorption behaviour of hydrogen towards iron with catalytic effect [113].

The relationships between the velocity of the tribochemical reaction and the gas pressure explained here correspond with the activation model discussed in 4.1. Proceeding from the supposition that the highly active regions in the post-plasma caused by impacts are stable for fractions of a second only, then a chemical reaction can only take place if the gases have the possibility for an reactive interaction during this time. This condition is met especially if the gases are already pre-adsorbed before the impact or if the reaction runs with higher pressures ($> 10^4$ Pa) and this guarantees a sufficiently large number of gas molecule impingements onto the solid. The pre-adsorbed molecules however represent the essential part in the tribochemical reaction for low and medium pressures as can be seen in Fig. 4.33. The curve I of the nickel carbonyl formation turns into curve III with the transition from 300 K to 100 K caused by the higher adsorption rate with smaller pressures [115] but it does not show the behaviour II as would be the case with predominance of the gas-kinetic part [114].

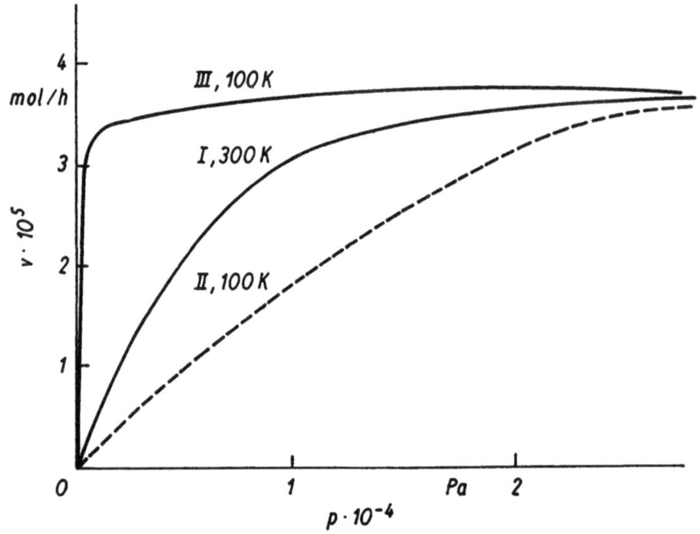

Fig. 4.33. Velocity v of the nickel carbonyl formation in dependence on the gas pressure p with different temperatures [114].
—— experimental curve, – – – calculated curve with predominance of the gas-kinetic part.

4.7. Kinetics of the tribosorption

As described in 4.6., an increased sorption of gases onto solids occurs under the influence of mechanical energy. This process, called tribosorption, both leads to an increased adsorption to the surface of the solid (tribo-adsorption) and to an inserting of the gases into the mechanically disturbed lattice [116]. This process, called triboabsorption, does not usually occur in mechanically non-stressed solids. It is a specific characteristic for the influence of non-thermal energy on solids and brings about a material penetration of the solid with molecules of the surrounding medium. From this possibilities arise both for an altered phase structure in the mechanically treated boundary layer (see 6.4.) and for altered kinetics of tribochemical reactions in respect of that of thermally activated reactions (see 4.7.6.). In the following the relationships will be discussed between triboadsorption and triboabsorption as well as between triboabsorption and triboreaction, under consideration of the kinetic of the triboabsorption.

4.7.1. The detection of the triboabsorption

The kinetic progress of triboadsorption and triboabsorption has been studied by repeated single impact treatment on metal surfaces.

In Fig. 4.34. the quantity of the CO_2-molecules tribosorbed per unit of area is represented for a series of metals in dependence on the number of ball impacts acting

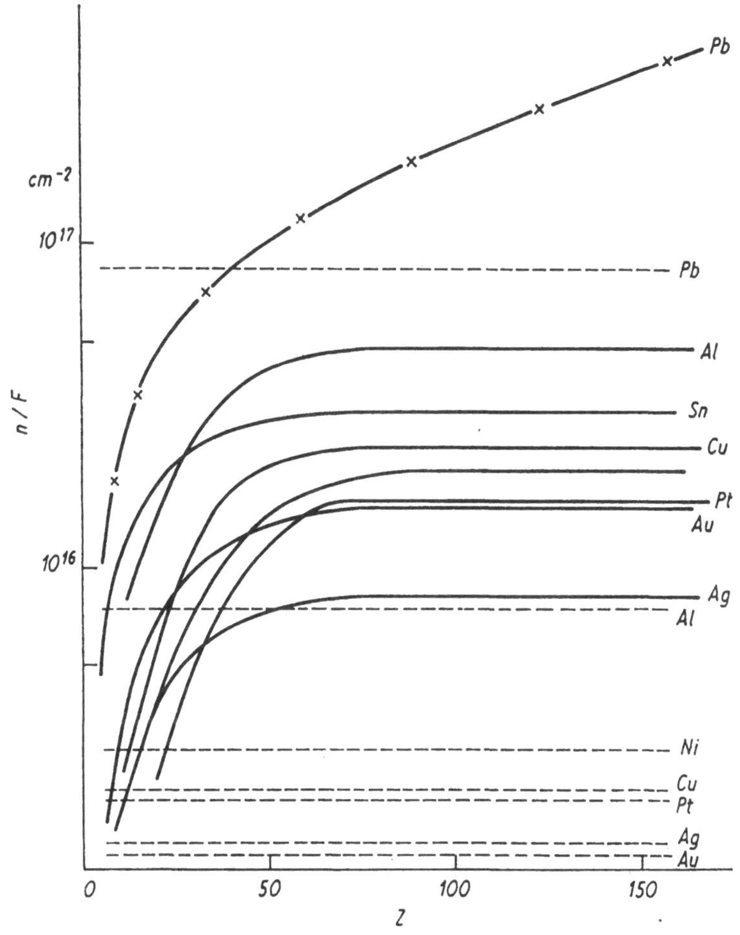

Fig. 4.34. Tribosorption of CO_2 molecules n per surface F at different metals in dependence on the number of impacts z [117].
—— tribosorption, - - - triboadsorption, —x— triboabsorption.

on the metal surface. If the gas is added after the finished treatment then the triboadsorption is independent of the number of ball impacts but is strongly dependent on the chemical nature of the metal. If labelled gases are used, such as $^{14}CO_2$, then the tribosorption on metal surfaces can also be observed by radioautographs. As Fig. 4.35.a) shows, a detectable sorption exclusively takes place on the mechanically treated parts of the metal surface. If the treatment of metal grains is carried out statistically in a vibration reactor then the tribosorbate spreads evenly over the entire surface of the grained substance, see Fig. 4.35.b).

At the beginning of the mechanical stress the quantity of the tribosorbate increases proportionally to the time and the number of impacts, respectively. The saturation state, the so-called tribosorption equilibrium, appears after a treatment time dependent on the system (see 4.8.3.). According to Fig. 4.34. the amounts of the tribosorbed gas quantity are approximately the factor 10 larger than the amounts of the triboadsorbed gas in the tribosorption equilibrium. As Fig. 4.36. shows the values for the

Fig. 4.35. Autoradiographs of copper mechanically treated in a $^{14}CO_2$-atmosphere [61].
a — treatment of a copper surface by different numbers of ball impacts, number of the impacts z starting at x and running in the direction of arrow 1; 5; 10; 15; 25; 50; 100; 125; 150; 200; b — polished sections of treated copper grains.

tribosorption of methane, hydrogen, and acethylene are also higher by up to one order of magnitude than the triboadsorption calculable from the surface. This result can be explained only by the fact that a part of the tribosorbed gas is inside the solid.

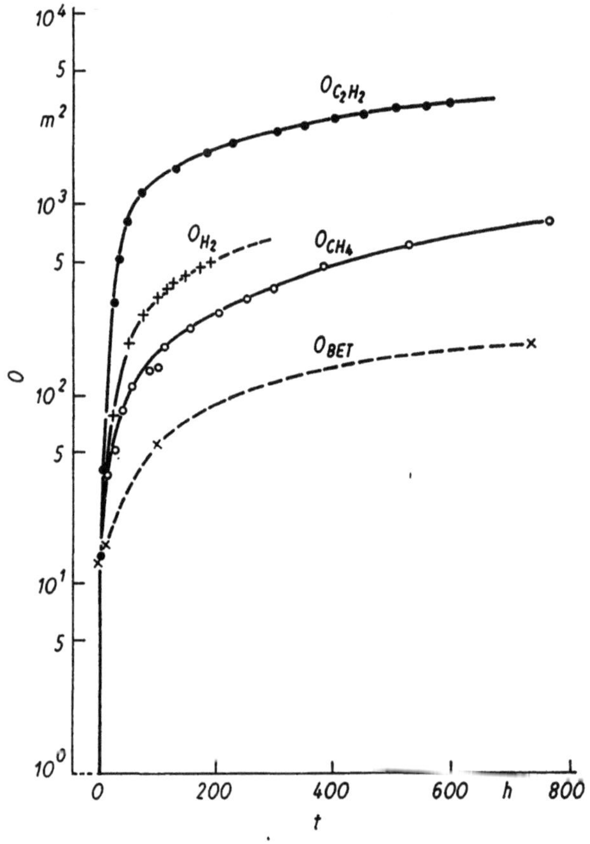

Fig. 4.36. Measured SiC surface area O_{BET} and calculated areas $O_{C_2H_2}$, O_{CH_4}, and O_{H_2} assuming exclusive adsorption of gases onto SiC in dependence on the treatment time *t* [46].

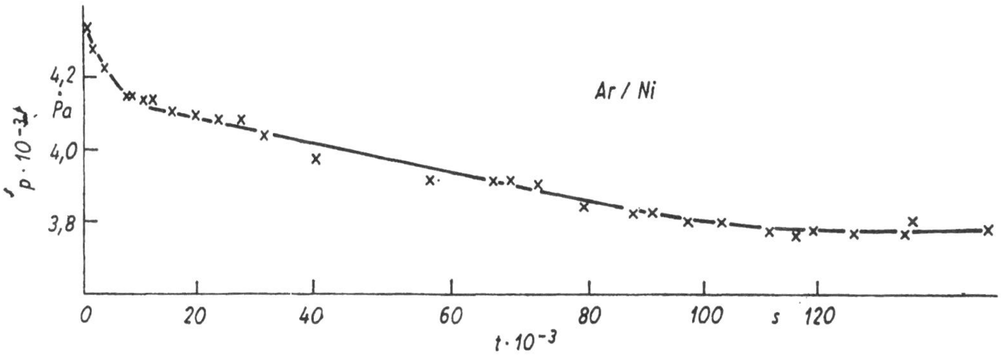

Fig. 4.37. Variation of the pressure p in dependence on the treatment time t in the argon/nickel system [121].

4.7.2. *Kinetics of the triboabsorption*

Inert gases are especially suitable for the investigation of the kinetics of the tribosorption, because the sorption processes are not overlapped by simultaneously running chemical actions in the inert-gas-solid system. Furthermore the triboadsorption of inert gases on metals is extremely small at room temperature and can be neglected in comparison to the triboabsorption [118—120].

As Fig. 4.37. shows for the nickel-argon system, the triboabsorption runs in analogy to the curves represented in Fig. 4.34.: Constancy of the tribosorption velocity (linear part of the curve) for a longer period of treatment at the beginning of the treatment after the passage of a non-linear part of the curve and transition to a saturation state.

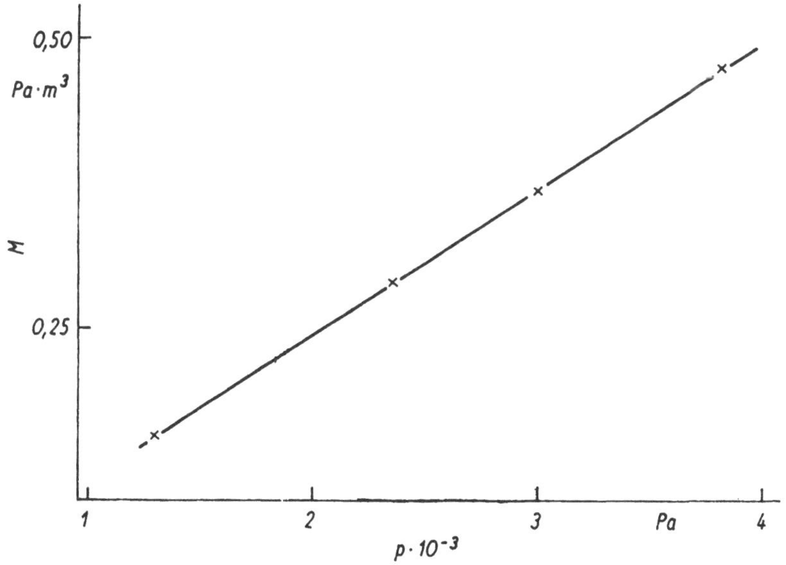

Fig. 4.38. Dependence of the amount of triboabsorbed gas M in the tribosorption equilibrium *TSG* on the gas pressure p in the argon/nickel system at 83 K [112].
argon/nickel system [121].

The amount of the tribosorbate in the tribosorption equilibrium M_{TSG} depends linearly on the gas pressure (see Fig. 4.38.):

$$M_{TSG} \sim p. \tag{41}$$

Since the measurements were carried out in the linear range of the LANGMUIR isotherms a direct proportionality apparently exists between the degree of coverage θ and the tribosorbate in the saturation state:

$$M_{TSG} \sim \theta. \tag{42}$$

An analogous relation results from the gas pressure for the dependence of tribosorption velocity v_{TS}.

According to Fig. 4.39. this relation may be described by equ. (38) and a relation corresponding to equ. (40) may be derived from this [112]:

$$v_{TS} \sim \theta. \tag{43}$$

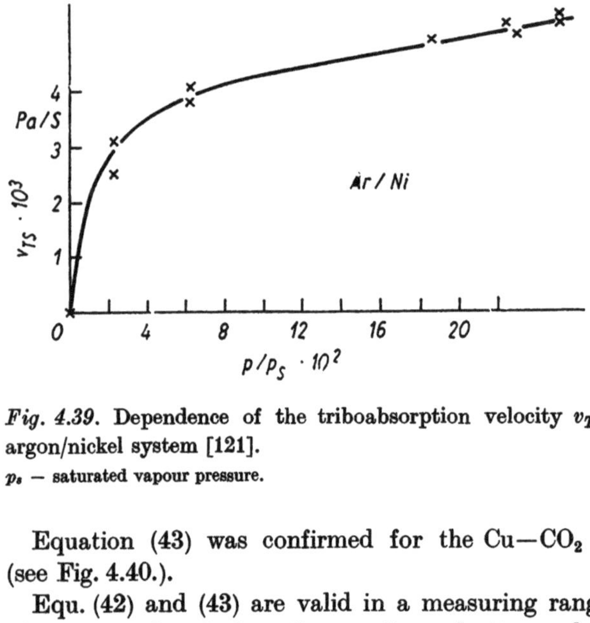

Fig. 4.39. Dependence of the triboabsorption velocity v_{TS} on the relative pressure p/p_s in the argon/nickel system [121].
p_s — saturated vapour pressure.

Equation (43) was confirmed for the $Cu—CO_2$ system directly in the experiment (see Fig. 4.40.).

Equ. (42) and (43) are valid in a measuring range up to $\theta \approx 1$. Above the degree of coverage $\theta \approx 1$ the tribosorption velocity no longer grows proportionally with θ (see Tab. 4.17.).

Table 4.17. Dependence of the CO_2 tribosorption velocity v_{TS} in copper on the degree of coverage θ [112].

θ	$v_{TS} \cdot 10$, Pa/min
0.5	1.8
4	4
1,500	9

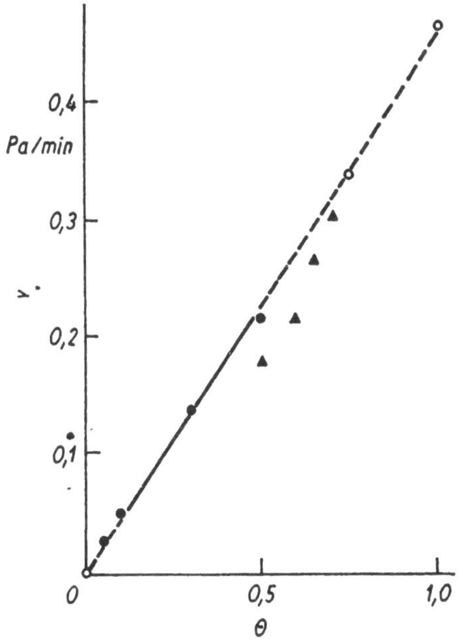

Fig. 4.40. Velocity of the triboabsorption v_{TS} in dependence on the degree of coverage θ in the CO_2/copper system [112].

• – measuring points in the pressure-dependent range, ○ – the same obtained by extrapolation, △ – measuring points in the nearly pressure-independent range.

By this the triboabsorption is directly proportional to the concentration of the gas adsorbed on the mechanically treated solid surface both with respect to the saturation value and with respect to the velocity (for $\theta < 1$), and independent from the chemical nature of the gas molecules.

Apart from the variation of the gas pressure the degree of coverage can also be varied by temperature changes through orders of magnitude. As Tab. 4.18. shows, the triboabsorption of noble gases is very small at room temperature due to an extremely small physisorption. Only in the range of the saturation pressure of the noble gases we observe relatively high triboabsorption. Thus, for instance, a well measurable triboabsorption occurs at 80 K with the use of Kr and Ar. By contrast, helium is in practice not triboabsorbed at this temperature due to its much lower boiling point [34].

Table 4.18. Dependence of the tribosorbate n of ^{85}Kr at copper on the temperature T after 30 h meehanical treatment [112].

T, K	n, ^{85}Kr atoms/cm^2
296	$\approx 10^{11}$
83	$\approx 5 \cdot 10^{15}$

4.7.3. *Factors influencing the tribosorption*

If one proceeds from the fact that the velocity of the triboabsorption is determined by the concentration of gas molecules on the solid surface then it can be supposed that in a homologous series the velocity of the tribosorption increases either by growing physisorption trends or by increasing chemisorption trends of the individual substances of this series.

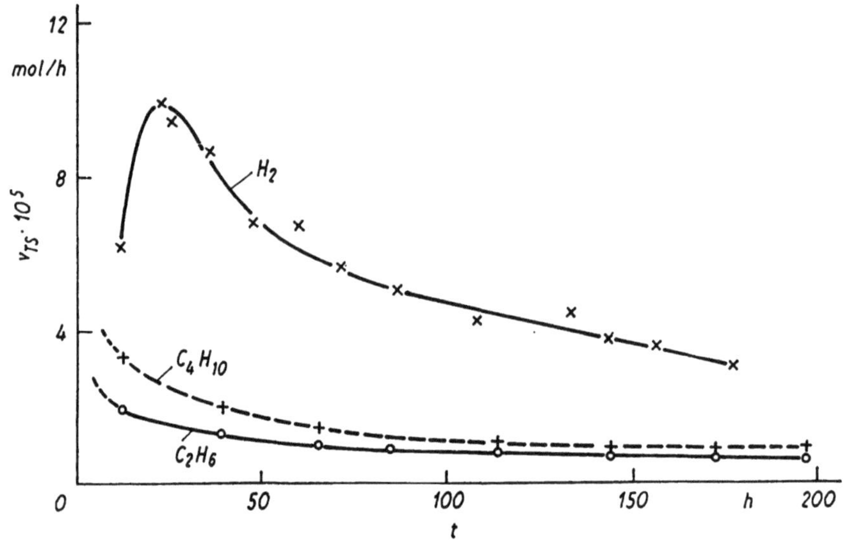

Fig. 4.41. Velocity of the tribosorption v_{TS} of several gases at the mechanically stressed SiC in dependence on the treatment time t [10].

This tribosorption can also be observed in gaseous paraffins. There, substances with different molecular weights show very similar kinetics of the tribosorption. As follows from Fig. 4.41., the tribosorption velocity of *n*-butane (boiling point 273 K) on average is higher by a factor of 1.6 than that of ethane (boiling point 184 K), evidently caused by the higher physisorption trend of the butane [10].

According to Fig. 4.41. the triobsorption of hydrogen (boiling point 20 K) takes place with a considerably higher velocity than the alkane tribosorption and is a consequence of the increased chemical interaction with the treated solid compared to the paraffins.

According to Fig. 4.42. the unsaturated hydrocarbons ethylene (boiling point 189 K) and acetylene (boiling point 169 K) also show a tribosorption velocity increased by up to one order of magnitude in comparison with the paraffins. With these gases it is not the physical adsorption properties of gases but the chemisorption processes which determine the tribosorption velocity as is the case with hydrogen [10].

From a comparison of the tribosorption velocities of several gases on two different solids, it can be seen that apart from the physical properties the chemical properties of the gases also determine the tribosorption onto the solid to a high degree (see Tab. 4.19.).

As follows from Tab. 4.19. the values measured for the individual gases on SiC and corundum are almost identical. H_2 and O_2 form an exception, and on SiC lead to a chemical reaction accompanied by the development of paraffins and carbonoxides, respectively. Consequently they are much stronger tribo-absorbed than on corundum [123].

The different tribosorption trends of CO_2 on several metals discussed above can be attributed to the same reason. Thus according to Fig. 4.34 the tribosorption velocity of CO_2 is higher by about one order of magnitude on chemically very reactive aluminium than on precious metals, such as silver and gold.

To sum up, from the examinations discussed above, it appears that the tribosorption velocity of the individual gases depends on the surface coverage which is determined both by the physical and the chemical properties of the reacting substances and serves like a diffusion source.

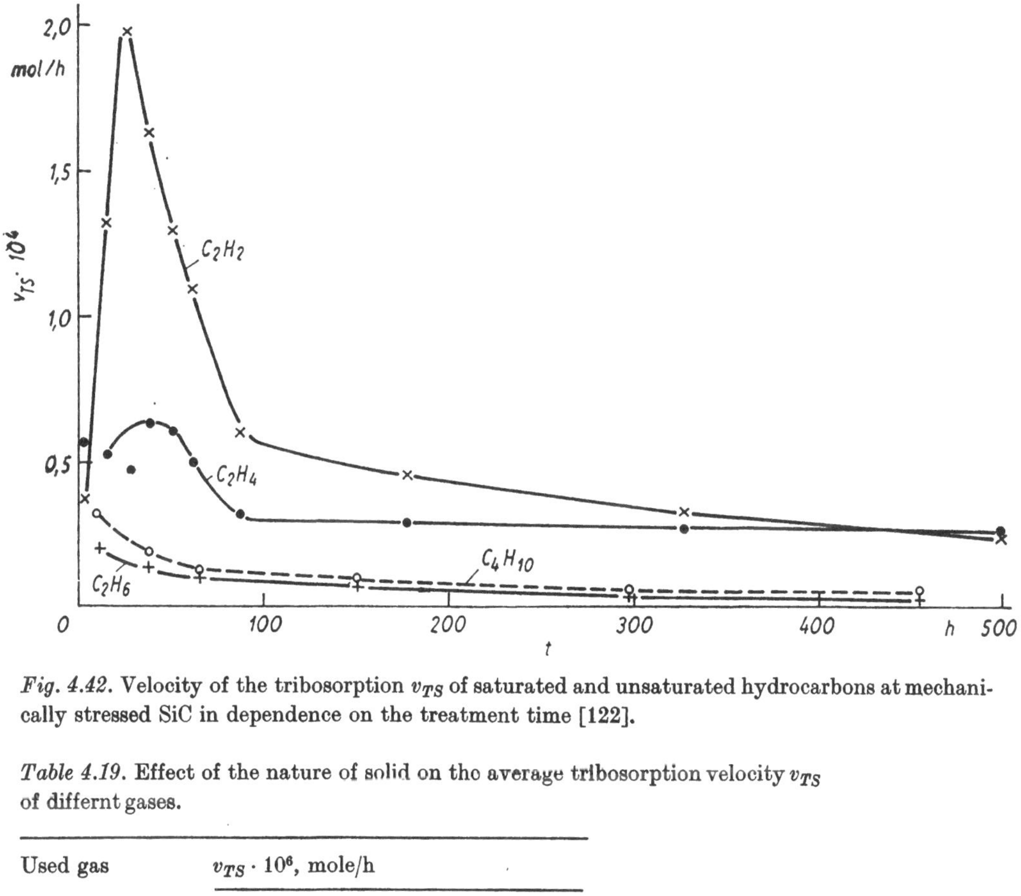

Fig. 4.42. Velocity of the tribosorption v_{TS} of saturated and unsaturated hydrocarbons at mechanically stressed SiC in dependence on the treatment time [122].

Table 4.19. Effect of the nature of solid on the average tribosorption velocity v_{TS} of differnt gases.

Used gas	$v_{TS} \cdot 10^6$, mole/h	
	at SiC	at corundum
N_2	$2 \cdot 10^0$	$3 \cdot 10^0$
O_2	$4 \cdot 10^1$	$3 \cdot 10^0$
H_2	$6 \cdot 10^1$	$1 \cdot 10^1$
CH_4	$1 \cdot 10^1$	$1 \cdot 10^1$
C_2H_6	$1.5 \cdot 10^1$	$1 \cdot 10^1$
C_2H_2	$1 \cdot 10^2$	$6 \cdot 10^1$

4.7.4. *Penetration profile of triboabsorbed gases*

The penetration depth of triboabsorbed gases depends inter alia on the temperature and the duration of the treatment. As an example, the penetration depth of radioactive krypton in copper is shown in dependence on the treatment time in Fig. 4.43. An other type of concentration profile also arises during the tribosorption of CO_2 in copper (see Fig.

4.44.). As Tab. 4.20. shows, a much steeper concentration profile arises for the tribosorption at 83 and 93 K, respectively, than at room temperature. Evidently in addition to mechanical excitation, temperature also plays an important role for the tribodiffusion in the solid.

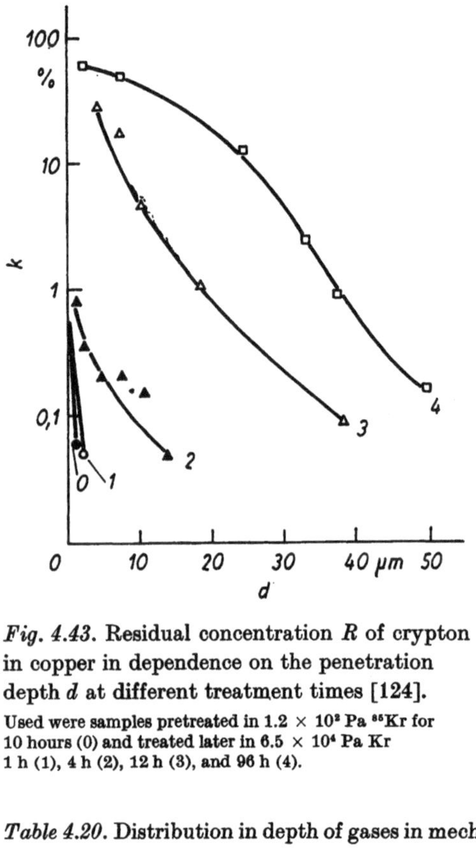

Fig. 4.43. Residual concentration R of crypton in copper in dependence on the penetration depth d at different treatment times [124].
Used were samples pretreated in 1.2×10^2 Pa ^{85}Kr for 10 hours (0) and treated later in 6.5×10^4 Pa Kr 1 h (1), 4 h (2), 12 h (3), and 96 h (4).

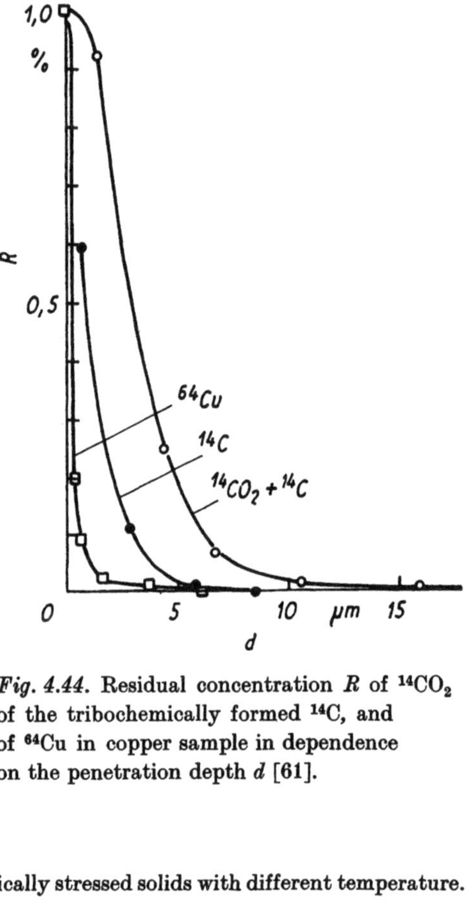

Fig. 4.44. Residual concentration R of $^{14}CO_2$ of the tribochemically formed ^{14}C, and of ^{64}Cu in copper sample in dependence on the penetration depth d [61].

Table 4.20. Distribution in depth of gases in mechanically stressed solids with different temperature.

System	Treatment temperature, K	Part of the triboabsorbed gas (%) in the boundary layer Δd: Δd, nm	%	Experimentally measurable maximum penetration depth mµ	Ref.
$Cu-^{85}Kr$	93	$5 \cdot 10^2$	99	0.5	[124]
$Cu-^{85}Kr$	296	$5 \cdot 10^2$	70	50	[124]
$Cu-^{14}CO_2$	296	10^3	5	≈ 20	[124]
Ni—Ar	83	6	93		[123]

In all, the investigations show that the main part of the triboabsorbed gas exists in the layer very near the interface but simultaneously the gas penetrates into greater depths of some µm. The penetration profile depends to a high degree on the stress parameters (see 6.6).

4.7.5. *Desorption of the tribosorbate*

Information on the tribosorption state can also be obtained by recording thermal desorption spectra after the mechanical treatment in a defined gaseous atmosphere.

As Fig. 4.45. shows, a triboadsorption solely takes place during the application of gas to the pre-treated metal (curve 3). The activation energy of the CO_2-desorption from copper can be calculated by means of the desorption curve amounting to $E_{De} = 146$ kJ/mol and corresponding to the energy of the chemisorption binding on the surface of copper [124]. If the treatment takes place in the presence of the gas then both triboadsorption and triboabsorption will appear. The second and third maximum of the desorption curve can be attributed to the thermal release of the triboabsorbed gas from the metallic and oxidic phase, respectively ($E_{De} = 175$ kJ/mol). They are situated at the temperature of the commencing exchange of sites [124].

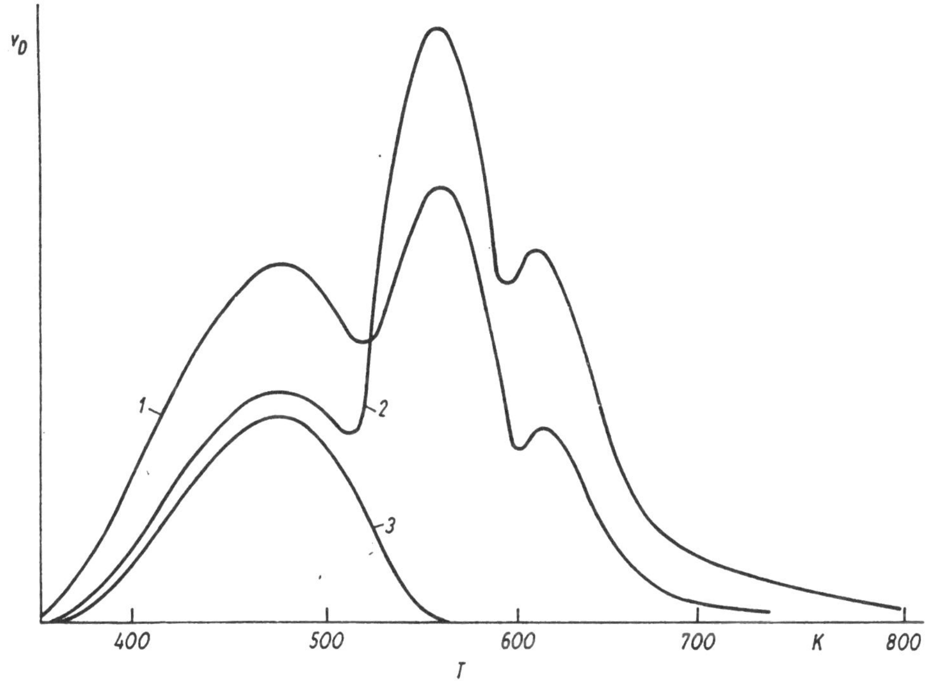

Fig. 4.45. Desorption velocity v_D of $^{14}CO_2$ from copper in dependence on the temperature T [61]. 1 — 10 h pretreated in 13 Pa $^{14}CO_2$, 2 — 20 h pretreated in 13 Pa $^{14}CO_2$, 3 — 3 h pretreated in the vacuum with succeeding addition of 13 Pa $^{14}CO_2$.

If noble gases are used, the desorption from the absorbed state is exclusively observed. As Fig. 4.46. shows for the Ni—Ar system, only one desorption maximum appears under this condition. Since the activation energy ($E_{De} = 136$ kJ/mol) calculable from the spectrum corresponds with the energy ($E_0 = 135$ kJ/mol) required for the migration of mono-vacancies, the release of Ar can be explained as noble gas desorption from the bulk phase by the migration of vacancies occupied by noble gases out for the inner solid to the interface [123], (mechanism see 5.7.).

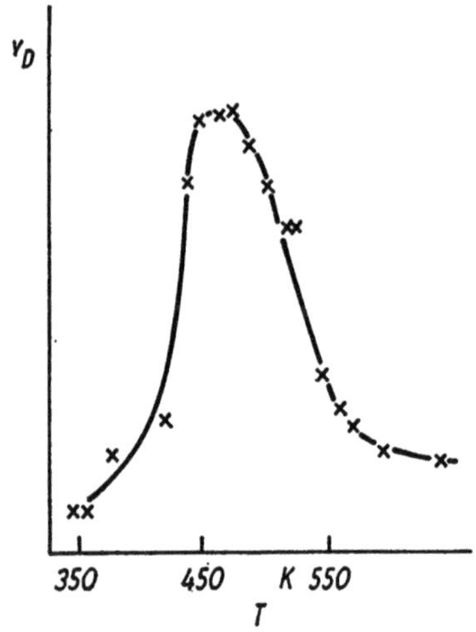

Fig. 4.46. Desorption velocity v_D of argon triboabsorbed in nickel in dependence on temperature T [123].

4.7.6. *Relations between triboabsorption and triboreaction*

As can be seen in Fig. 4.43. not only a distribution of the CO_2 concentration can be determined in the treated layer near the interface of copper but also a similarly occurring carbon distribution. Apart from the triboabsorption and diffusion processes chemical reactions take place simultaneously leading to the formation of free carbon according to the following reaction

$$\mathrm{Cu} + \frac{1}{2}\,\mathrm{CO_2} \rightarrow \mathrm{CuO} + \frac{1}{2}\,\mathrm{C} \tag{44}$$

According to Tab. 4.21. a reaction with a negative affinity is involved, which is exclusively forced by the treatment. As was inter alia also shown for the $H_2 - SiC$ system [10], tribosorption processes usually run considerably faster than tribochemical processes. In these cases the chemical reactions take place mainly inside the solid. Assuming the velocity of the development of carbon $\left(\frac{\delta \tau_c}{\delta t}\right)$ to be directly proportional to the CO_2 concentration of CO_2 existing in the solid, the following relation can be established:

$$\left(\frac{\partial \tau_c}{\partial t}\right) \sim c_{CO_2} \tag{45}$$

Table 4.21. Base state affinities ${}^0A_T$ for the oxydation of copper by CO_2 in dependence on temperature T [116].

Reaction	${}^0A_T$, kJ/mole		
	$T = 298$ K	$T = 1{,}000$ K	$T = 2{,}000$ K
$4\,Cu + CO_2 \rightarrow 2\,Cu_2O + C$	− 102	− 200	− 342
$2\,Cu + CO_2 \rightarrow 2\,CuO + C$	− 269	− 271	− 477

Taking into consideration the CO_2 distribution in the Cu matrix the following relation results [125]

$$\frac{N_c}{N_g} = 1 - \frac{N_g - N_{c_{An}}}{N_g} \cdot \exp\left[-k_4(t - t_{An})\right],$$

$$\text{respec.} \quad -\ln \frac{N_g - N_c}{N_g - N_{c_{An}}} = k_4(t - t_{An}) \tag{46}$$

(N_c — quantity of the C atoms in the Cu matrix, $N_{c_{An}}$ — $N_{c_{An}}$ at the time $t = t_{An}$, N_g — quantity of the C atoms and CO_2 molecules in the Cu matrix, k_4 — factor of proportionality, t_{An} — time at which the steady state appeared.)

In Fig. 4.47. the comparision of equ. (46) with the experimental results is made. By this one can regard as justified the above assumption in a first approximation, that tribochemical reactions can also run as first order bulk reactions. The ratio $CO_2 : C$ in the copper matrix does not remain constant with continual treatment if the constant supply of CO_2 is interrupted through the triboabsorption process. According to Fig. 4.48. a growing degree of reaction of CO_2 to carbon takes place under the given conditions. The quantity of CO_2 which can be desorbed thermally, decreases more and more, to a corresponding extent. The chemical reaction inside the solid is to a large extent determined by the tribosorption and diffusion processes.

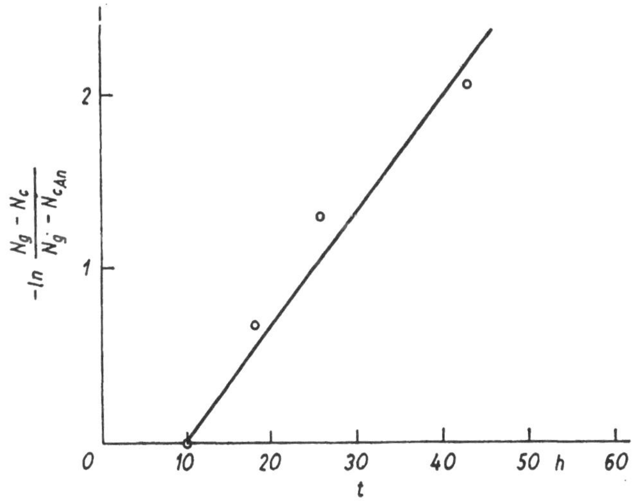

Fig. 4.47. Temporal course of the tribochemical reaction $2\,Cu + CO_2 \rightarrow 2\,CuO + C$ in copper samples saturated by CO_2 at room temperature [125].
+ — time of treatment, N_g, $N_{c_{An}}$, N_c — see text, —— calculated curve, o — experimental values.

If the velocities of the tribosorption and triboreaction have the same order of magnitude, then the place of reaction shifts to the surface to a major degree. Thus, for instance, the mechanically activated reactions of O_2, CO, and CO_2 with Al may occur largely as surface reactions.

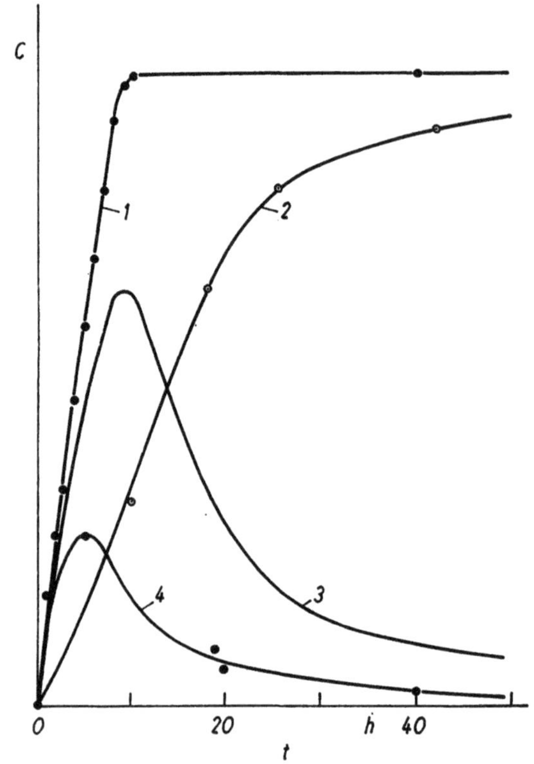

Fig. 4.48. Material total balance of carbon in the $Cu/^{14}CO_2$ system in dependence on the treatment time *t* [61].

I' – impulse rate, 1 – tribosorption of the gas during the decrease of gas pressure from 16 to 0.1 Pa, 2 – triboreaction to ^{14}C, 3 – $^{14}CO_2$ not reacted (difference of the curves 1 and 2), 4 – gas ready for desorption after interrupting the treatment at 296 K.

4.7.7. *Tribokinetic model*

A tribokinetic model has been established for the various processes taking place during the treatment of a solid in the presence of a gaseous phase. It is represented in Fig. 4.49. [124].

According to the pattern given in this figure the gas reaches the metal surfaces by diffusion and adsorbes there. A triboadsorption takes place at treated points. Depending on the boundary conditions a triboabsorption in the triboplasma out of the gaseous phase occurs directly during the treatment. This is combined with a tribochemical reaction or an adsorption and reaction of the gas on the metal surface. Moreover the triboadsorbed gas is absorbed by the metal and only then chemically transformed. In addition the triboabsorbed gas is conveyed further (tribodiffusion) within the solid and remains stored in the deeper boundary layers or reacts with the solid matrix influenced by the treatment. As will be discussed in section 5.7. the triboinduced processes often differ fundamentally from the corresponding thermally conducted actions so that the absorption and diffusion constants derived from thermally activated processes are not suitable for the description of the corresponding tribokinetic processes (see Tab. 4.20. and section 5.7.).

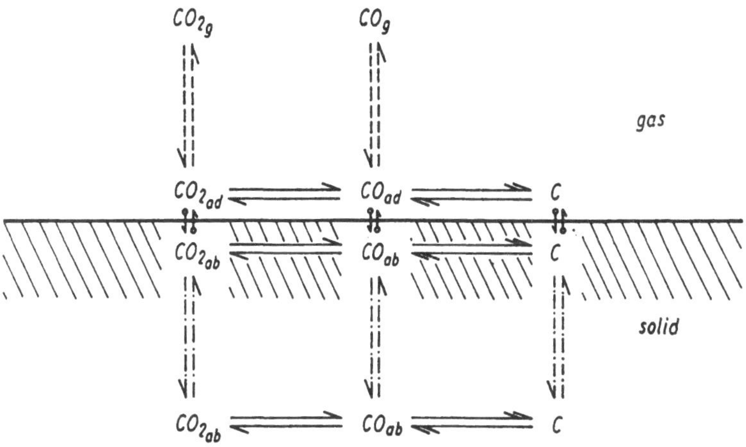

Fig. 4.49. Tribokinetic model for the tribosorption and triboreaction processes in the CO_2/metal system [61].

CO_{2g} – CO_2 in gaseous state, CO_{2ad} – CO_2 in adsorbed state, CO_{2ab} – CO_2 in adsorbed state, for CO the same symbols resp.,
⇌ – chemical reaction: $CO_2 + Me \rightleftharpoons CO + MeO$
⇌ – chemical reaction: $CO + Me \rightleftharpoons C + MeO$

– triboabsorption, – tribodesorption, – tribodiffusion, /// – phase boundary layer dist urbed by mechanical treatment.

4.8. The thermodynamics of tribochemical reactions

4.8.1. Thermodynamic characterization of mechanically disturbed solids

The excitation model represented in section 4.1. is based on the premise that in the dissipation of energy transferred to the solid by the mechanical treatment the highly activated states decay very rapidly and a frozen metastable disturbed state remains. Under normal conditions the irreversible conversion of this disturbed state towards the thermodynamic equilibrium occurs only extremely slowly in most cases and often can not be detected experimentally. This means there are no conditions of equilibrium as are equired by the thermodynamics of reversible processes. Apart from mechanical treatment, solids having irreversible lattice defects may also be produced chemically, by precipitation, reduction, and other chemical conversions as well as by bombardment with elementary particles. They have been named "active solids" [126].

In order to be in a position to describe active substances by the relationship of reversible thermodynamic in spite of the presence of non-equilibrium states, the excess free enthalpy ΔG is used as additional quantity for the active solid description. It indicates the excess content of free enthalpy which the active substance possesses in comparison with its stable state of equilibrium [127]. ΔG is also designated as solid activity and is defined by the following equation:

$$\Delta G = G_T^* - G_T \tag{47}$$

(G_T^* – free enthalpy of the active solid, G_T – free enthalpy of the same substance in the initial state at the temperature T.)

At the same time ΔG indicates the affinity A of the active substance to change into its stable state ($\Delta G = -A$). Since the excess free enthalpy effects both the equilibrium position and the velocity of a solid reaction it represents and important quantity for characterizing the solid reactivity. In the following the ΔG values which can be obtained by the treatment process will be regarded quantitatively and compared with ΔG values of active products prepared in a chemical way. A direct determination of the quantity ΔG can with certain reservations (see 6.6.) be obtained for metals by the measurement of the electromotive force E', which becomes effective with the association of the active and stable states to a galvanic chain. The following relationship exists:

$$\Delta G = n \cdot 96513 \cdot E' \tag{48}$$

(n — valency of the metal ion, ΔG — is expressed in J, E' in V.)

A further direct possibility for determining ΔG which is not limited to metals, is the examination of the equilibrium positions and of their shifting during the transition from equilibrium to the active state. The examination of the reaction

$$Ni + 4CO \rightleftharpoons Ni(CO)_4 \tag{49}$$

has proved particularly suitable, because only one solid phase exists in the system and the deviations from the equilibrium which occur can be unambiguously attributed to the changed activity of the solid.

The following equation can generally be used for the calculation of ΔG [128]:

$$\Delta G = RT(\ln K_p' - \ln K_p) \tag{50}$$

(K_p — equilibrium constant of the non-mechanically activated system, K_p' — equilibrium constant with use of a pretreated solid.)

For equ. (49) for instance $K_p = p_{Ni(CO)_4}/p_{CO}^4$.

In the equilibrium state the basic affinity $°A_T$ of the reaction can be derived from the equilibrium constant:

$$°A_T = RT \ln K_p. \tag{51}$$

Analogously the basic affinity $°A_T{}^*$ of the reaction by use of an active solid is:

$$°A_T{}^* = RT \ln K_p'. \tag{52}$$

From this results:

$$\Delta G_T = °A_T{}^* - °A_T. \tag{53}$$

As shown in Fig. 4.50. the chemical equilibrium at the mechanically preactivated metal can be adjusted reversible whereby the use of relations of reversible thermodynamics, taking account of additional excess quantities, is justified.

For equ. (49) an excess free enthalpy of 10.9 kJ/mol can be calculated for $T = 298$ K according to equ. (53). In Tab. 4.22. this value is compared with results obtained with variously pretreated nickel samples. Both the mechanically pretreated active products and the ones prepared in a chemical way lead to the same activity values in terms of the order of magnitude. This result corresponds with the general experience that active metal

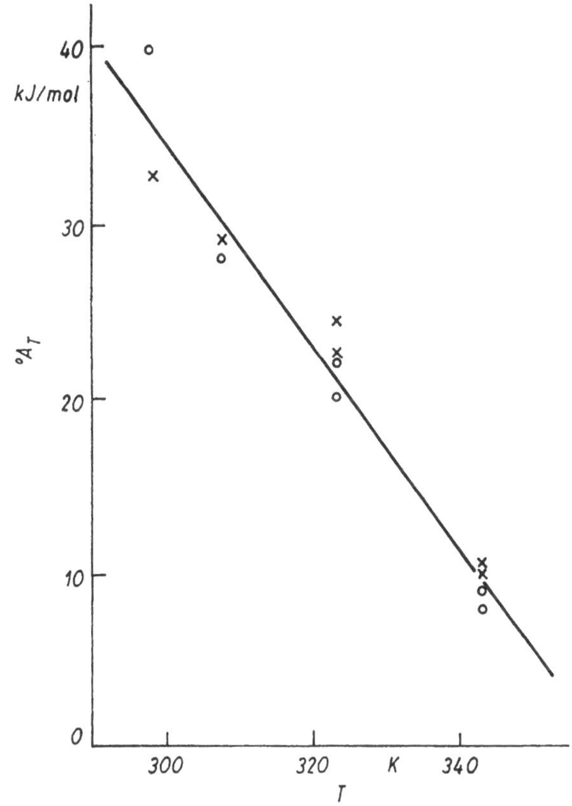

Fig. 4.50. Dependence of the base state affinity ^0A_T of the reaction $Ni + 4CO = Ni(CO)_4$ on the temperature T for pre-milled nickel powder [12].
x — obtained by decomposition, o — obtained by formation.

powders having pyrophoric properties can be prepared by both careful reduction of nickel salts [129] and by intensive mechanical treatment in vibration mills [130]. ΔG values measured in other metals, such as in copper, have the same order of magnitude as in nickel. Still higher values are measured in non-metal systems, e.g. for $CdCO_3$ 36 kJ/mol (see Tab. 4.22.).

Further important methods for the determination of excess free enthalpy are calorimetry and differential thermal analysis. However, they only provide the excess enthalpy ΔH which is related to ΔG in the following manner [12]:

$$\Delta G_T = (H_T{}^* - H_T) - T(S_T{}^* - S_T) \tag{54}$$

$$\Delta G_T = \Delta H_T - T\Delta S_T \tag{55}$$

($H_T{}^*$, $S_T{}^*$ — enthalpy and entropy, respectively, of the active solid, H_T, S_T — enthalpy and entropy, respectively, of the same substance in the initial state at the temperature T.)

In a number of cases the second term of equ. (55) is so small that $\Delta G_T \simeq \Delta H_T$ can be expressed in first approximation. However, it becomes evident that the entropy term should not to be neglected particularly with high defect concentration. ΔS can

Table 4.22. Excess free enthalpy ΔG and excess enthalpy ΔH of different pretreated active solids.

Substance	Pretreatment of the solid	Measuring method	ΔG, kJ/mole	ΔH, kJ/mole	Ref.
Ni	Reduction of a nickel salt	Equilibrium measurement	11.7	—	[76]
Ni	Vibration milling	Equilibrium measurement	10.9	—	[12]
Ni	Rolling of nickel sheets	Potential measurements	13.8	—	[131]
Ni	Reduction of nickel-hydroxide	Galorimetric measurements	—	6.5	[129]
Ni	Reduction of nickel oxide	Galorimetric measurements	—	5.2	[129]
Ni	Reduction of nickel carbonate	Galorimetric measurements	—	1.5	[129]
Ni	Torsion	Galorimetric measurements	—	0.19	[132, 133]
Ni	Milling	Galorimetric measurements	—	0.88	[134]
Al	Torsion	Galorimetric measurements	—	0.12	[133]
Al	Milling	Galorimetric measurements and DTA	—	0.84	[135, 136]
Cu	Vibration milling	Potential measurement	7.5	—	[137]
Cu	Reduction of a copper salt	Potential measurement	5.7	—	[138]
Cu	Torsion	Calorimetric measurement	—	0.06	[139]
Cu	Torsion	Calorimetric measurement	—	0.13	[133]
Fe_2O_3	Dehydration of α-FeOOH	Calorimetric measurement	—	20.0	[140]
Fe_2O_3	Vibration milling	Calorimetric measurement	—	9.2	[141]
$MgCO_3$	Vibration milling	Calorimetric measurement	—	12.6	[142]
$MgCO_3$	Disintegrator milling	Calorimetric measurement	—	2.2	[142]
$CdCO_3$	Vibration milling	Equilibrium measurement	36.0 298 K	—	[12]
$CaCO_3$	Vibration milling	Measurement of the decomposition pressure	9.8 873 K	80	[143, 144]
$CaCO_3$	Vibration milling	Measurement of the decomposition pressure	6.6 973 K	75	[145]
$CaCO_3$	Utilization of a freshly precipitated product	Measurement of the decomposition pressure	3.5 1073 K	—	[146]
$CaCO_3$	Utilization of a freshly precipitated product	Measurement of the decomposition pressure	3.2 1097 K	—	[147]
$CaCO_3$	Vibration milling	Calorimetric measurement	—	5.9	[145]
Tricaclciumsilicate	Vibration milling	Calorimetric measurement	—	47.7	[148]
SiO_2	Vibration milling	Calorimetric measurement	—	6	[149]
Sugar	Vibration milling	Calorimetric measurement	—	21	[150]

be calculated according to the following equation (22):

$$\Delta S_T = S_T^* - S_T = S_0^* + \int_0^T \frac{C_p^* - C_p}{T}\, dT \tag{56}$$

(S_0^* — zero-point entropy, C_p^*, C_p — specific heat of the active product and of the same product in the initial state, respectively.)

Even when C_p^* and C_p hardly differ from each other the difference in entropy may possibly be considerable in consequence of a high zero-point entropy. At room temperature and for little disturbed crystalline substances a comparison of ΔG and ΔH is permissible in evaluations in terms of order of magnitude. However, at high temperatures and for highly disturbed substances, the ΔG and ΔH values differ in the order of magnitude (e.g. $CaCO_3$, Tab. 4.22.). Since $\Delta S_T \geqq 0$ is $\Delta H_T \geqq \Delta G_T$.

In comparison to active products prepared by chemical means or by milling the excess enthalpies of metals mechanically stressed by strain or by torsion are relatively small. Important differences can also be obtained by the use of different milling sets. Thus ΔH values obtained during the vibration milling of $MgCO_3$ are higher by the factor 6 than during the activation by disintegrators (see Tab. 4.22.). The activation of $MgCO_3$ also leads to higher ΔH values in vibration mills compared to ball mills [151].

As Fig. 4.51. shows, with constant treatment time considerable differences in the energy input are also measured, if the same activation set is used but individual parameters of the mill such as amplitude are varied. The determination ΔH values has thus proved an appropriate method for the investigation of the efficiency of activation sets.

The excess enthalpy of nickel measured by calorimetry is smaller by one order of magnitude than the excess free enthalpy of correspondingly activated nickel products.

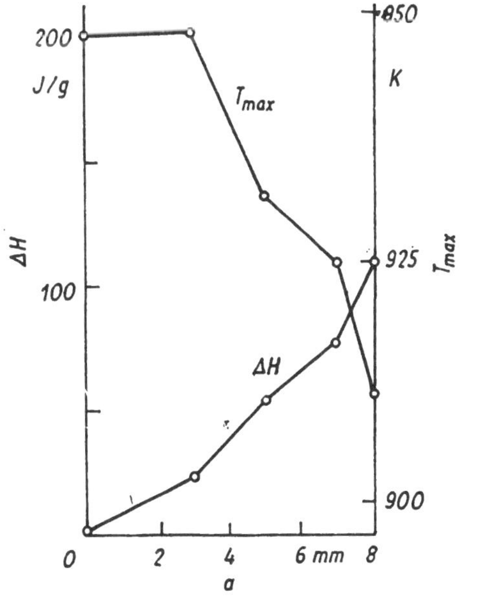

Fig. 4.51. Changes in the excess enthalpy ΔH and in the temperature of the maximum decomposition velocity T_{max} of mechanically activated magnesite in dependence on the amplitude a of the vibration mill [142]. Time of treatment 0.5 h.

Similar differences have also been measured in treated calcite. The reason can mainly be seen in the fact that methods used for the determination of ΔG only cover the particularly active positions of the solid, because they essentially determine the solid reactivity, whereas the values measured by calorimetry cover the entire bulk phase and lead to a medium activation value. Therefore one has to distinguish in each case which of the two quantities characterize the respective solid properties and the process running inside it, in a suitable manner.

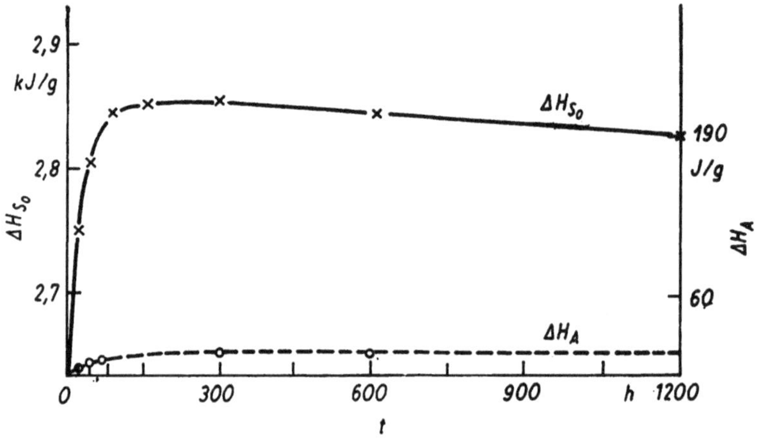

Fig. 4.52. Changes in the solution enthalpy ΔH_{S0} and in the surface enthalpy ΔH_A in dependence on the treatment time t for mechanically activated tricalciumsilicate [149].

The excess enthalpy of a solid is determined by the sum of all the solid defects not being in a thermodynamic equilibrium state. Often the excess enthalpy is divided into portions caused by surface enlargement and into portions caused by bulk defects. As shown in Fig. 4.52. by the example of mechanically activated tricalcium silicate, sometimes only a small part ($\approx$ 10 per cent) of the excess enthalpy ΔH of the activated product accounts for the surface enlarging. The main portion of the excess enthalpy can be attributed to the development of thermodynamically unstable states in the lattice. In sugar too the proportions of enthalpy accounting for the development of fresh surfaces only run to few per cent of the total increase in enthalpy [150]. In other substances, however, such as in metals, the enthalpy value which has to be related to the surface enlargement amounts to an important portion of the total increase in enthalpy. Generally the following is valid for the distribution of excess enthalpy among the various kinds of defects [152]:

$$\Delta H = \sum_{i=1}^{n} \left(\frac{N_i}{N_L}\right) H_i \tag{57}$$

(N_i — number of the lattice components in the ith state, N_L — LOSCHMIDT number, H_i — enthalpy rate per mole for the development of the ith state.)

If in addition to the changes in the surface under the bulk disturbances, changes in primary crystalinity, in dislocation density, and in phase composition, are taken into consideration as decisive quantities for the variation of the excess enthalpy, then the following equation results [152]:

$$\Delta H = \left(\frac{N_A}{N_L}\right) \Delta H_A + \left(\frac{N_K}{N_L}\right) \Delta H_K + \left(\frac{N_v}{N_L}\right) \Delta H_v + \left(\frac{N_p}{N_L}\right) \Delta H_p \tag{58}$$

(ΔH_A — molare surface enthalpy, ΔH_K — molare grain boundary enthalpy, ΔH_v — molar dislocation enthalpy, ΔH_p — molare transformation enthalpy, N — number of the lattice components, N_A — at the surface, N_K — at grain boundaries, N_v — in dislocations, N_p — in the metastable phase.)

Approximations have been developed for the calculation of enthalpy and concentration of the individual defects, which permit a calculation of the individual portions of the enthalpy by means of experimentally determined values [152]. These are listed for selected substances in Tab. 4.23. and compared with the ΔH values determined experimentally. In individual cases a good uniformity was obtained between the values calculated theoretically and these determined experimentally.

Table 4.23. Calculation of the excess enthalpy $\Delta H_{\text{theor.}}$ from the changes in enthalpy of the individual portions of defects in comparison to the experimentally determined ΔH values for selected vibration milled samples [152], kJ/mole (symbols see text).

Substance	$\left(\frac{N_A}{N_L}\right) \Delta H_A$	$\left(\frac{N_K}{N_L}\right) \Delta H_K$	$\left(\frac{N_v}{N_L}\right) \Delta H_v$	$\left(\frac{N_p}{N_L}\right) \Delta H_p$	$\sum \left(\frac{N_i}{N_L}\right) \Delta H_i = \Delta H_{\text{theor.}}$	$\Delta H_{\text{exp.}}$
SiO_2	0.71	0.42	—	7.10	8.23	21.0
MgO	2.54	6.20	0.15	—	8.89	5.1
$CaCO_3$	0.76	5.06	—	—	5.82	5.0
Fe	0.06	0.53	0.025	—	0.62	—

Moreover it can be seen in Tab. 4.23. that the main portion of the excess enthalpy in the individual substances is made up by various defects. For instance in quartz the increase in enthalpy is mainly caused by the amorphization, in MgO, $CaCO_3$, and Fe, however, mainly by the comminution of the primary particles. In some cases the following simple relationship exists between the structural quantities of the solid and the excess enthalpy [153]

$$\Delta H \sim \frac{S}{X_1} \tag{59}$$

(S — specific surface, X_1 — portion of the crystalline phase in the solid.),

and

$$\frac{\Delta H}{S} \sim \frac{1}{X_1}, \quad \text{respectively.} \tag{60}$$

By this the excess enthalpy related to the surface is inversely proportional to the portion of the crystalline phase. Equ. (60) is an expression of the fact that the specific excess enthalpy is essentially caused by the transition of the lattice into the partial-crystalline state. Since the partial-crystalline phase has a considerably increased reactivity in

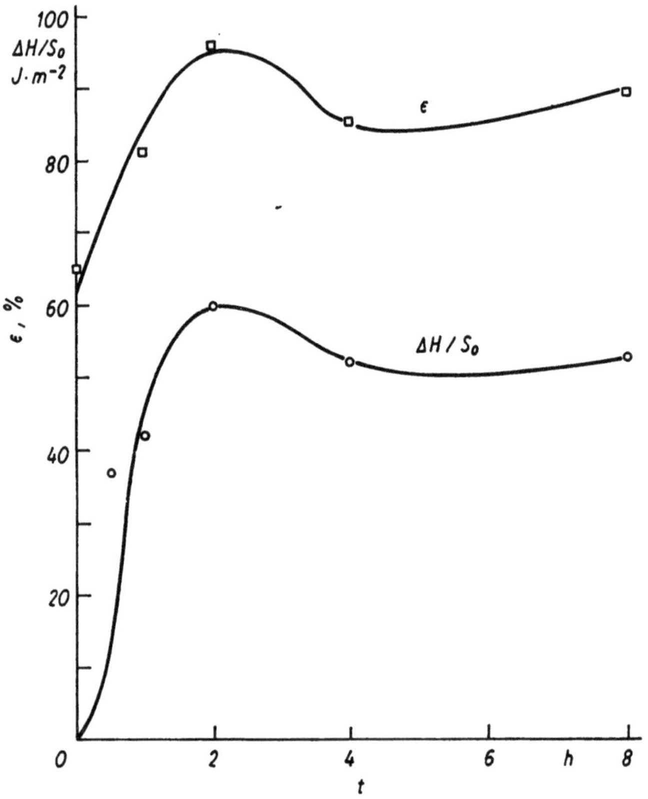

Fig. 4.53. Dependence of the yield ε during the thermal decomposition of magnesite and of the specific enthalpy increase $\Delta A/\Delta S_0$ on the time of treatment t [154].

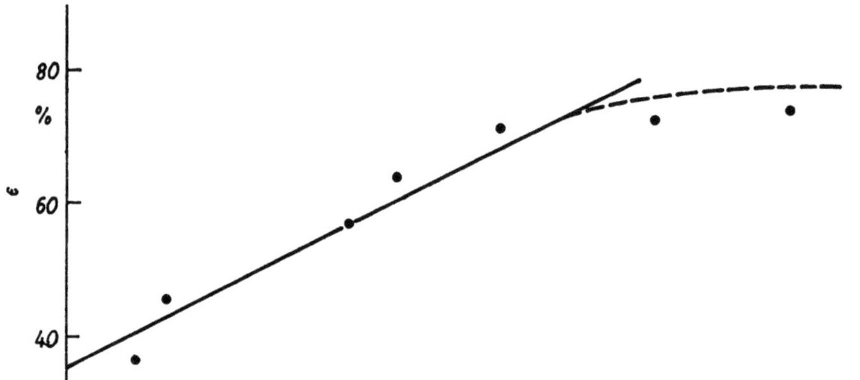

Fig. 4.54. Dependence of the quantity of dissolved magnesite ε on the excess enthalpy ΔH of milling samples treated for differently long times at constant time for dissolution [154].

comparison with the crystalline ones the reactivity is often determined by those portions only. An analogous curve behaviour has been measured for the reaction yield, e.g. in the thermal decomposition of magnesite, and for the specific increase in enthalpy in dependence on the treatment time in such cases, as shown in Fig. 4.53.

According to Fig. 4.54. the solubility of magnesite is, a within broad limits, also directly proportional to the excess enthalpy of the mechanically activated magnesite. Altogether these values would show that the determination of the ΔH-values of mechanically activated substances represents a suitable method both for characterizing the varied structure and the reactivity of the solid.

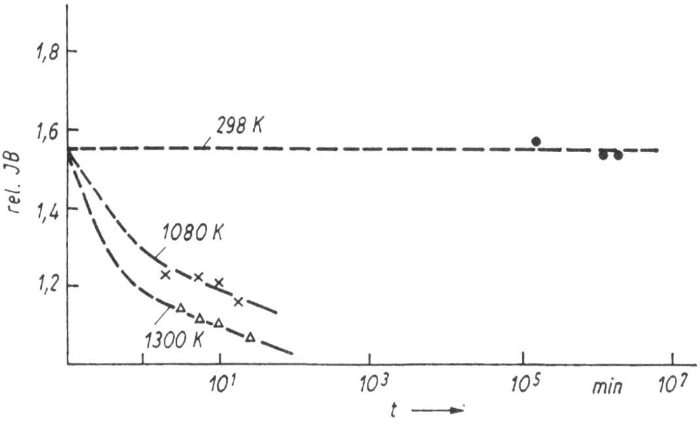

Fig. 4.55. Isothermal recovery curves for mechanically activated MgO [155].
Decay of the relative integral width rel. IB of the 222 reflection in dependence on the time of aunealing *t*.

The metastable states remaining in the solid after the mechanical treatment usually have a long lifetime at room temperature. As Fig. 4.55. shows, the defects developed by mechanical treatment in MgO will be nearly unchanged with $T = 298$ K for 1.5 years.

Upon the commencement of the lattice mobility by an increase in temperature, recovery and recrystallization processes appear which are connected with a reduction of the excess enthalpy. The temperature of the commencing lattice mobility, the so-called TAMANN temperature T_T, amounts to $0.3 \cdots 0.4 T_S$ with metals, to about $0.6 T_S$ with oxides, and to $0.8 \cdots 0.9\ T_S$ with organic carbon compounds (T_S — melting temperature) [235].

In Fig. 4.23. for instance the maximum is directly at the TAMANN temperature of iron, so that the decrease in reactivity above T_M can be attributed to the annealing of disturbed lattice regions. As Fig. 4.55. shows noticable lattice recreactions may also occur in mechanically activated substances below T_M (for MgO $= 1{,}750$ K) already.

In the presence of reactants a measurable decrease in activity already takes place at room temperature. The reason for this can be found in the increased reactivity of the active states, which interact primarily chemically with the surrounding atmosphere as compared with less active regions. In the system Ni/CO this process leads to a permanent alteration of the equilibrium state. As Fig. 4.56. shows the ln K_p value decreases by 0.62 over a longer period and consequently the excess free enthalpy of Ni by 3.4 kJ/mole and by about 30 per cent of the total yield, respectively [156]. The rapid decrease in immersion enthalpy of mechanically activated quartz in the presence of a wet atmosphere is also based on the same cause [157].

The decomposition of mechanically activated lattice regions occurs extremely rapidly in metals at room temperature if the treated metal is in an electrolytic solution.

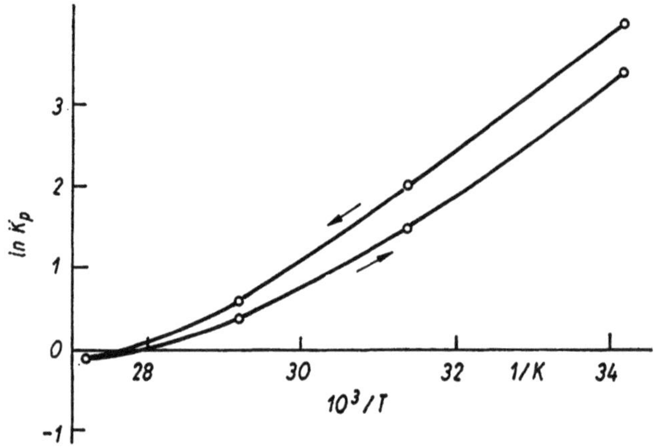

Fig. 4.56. Equilibrium constant K_p of the reaction $Ni + 4CO \rightleftharpoons Ni(CO)_4$ at different temperatures [156].

The temporal succession of the individual determinations took place in the direction of arrow:

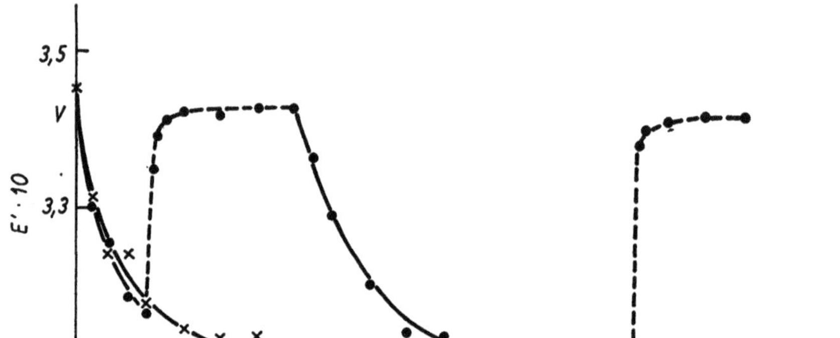

Fig. 4.57. Changes in normal potential E' of a Cu/Cu^{++} halfcell by mechanical treatment of the Cu electrode [137].

t — time, —— during treatment, – – – after interrupting treatment.

As Fig. 4.57. shows, the decrease of the potential characterizing the copper activity drops abruptly to almost the initial value, after interrupting the treatment. This rapid decomposition of activity is caused by the occurrence of local elements and the high local currents combined with it [137] (see also section 6.4 ff.).

The velocity of the decomposition of excess enthalpy and of excess free enthalpy, respectively, remaining in the solid after mechanical treatment is thus also a function of external conditions. In dependence on these parameters the lifetime of states possessing excess energy can vary from 10^{-7} to more than 10^6, s, as shown in Tab. 4.2 by the example of the V_k-centres.

The lifetime of excited states increases in solids with falling temperature. This explains why higher activities are often obtained by mechanical treatment than by chemical pretreatment of the solid as for instance by decomposition or conversion (see Tab. 4.22.). The creation of excited states can occur by means of mechanical energy at lower temperatures as well, whereas temperatures > 400 K are required in most cases for the production of highly active products by chemical means which already lead to the decomposition of the most highly active states in the solid. To sum up it can be said that the relations of reversible thermodynamics can be applied to tribochemical reactions if the lifetime of the mechanically produced lattice distortions is greater than the time of the chemical process in question, e.g. of the equilibrium establishment. In this way excess free enthalpy has proved itself to be quantitative measure for the mechanical activation.

4.8.2. *Application of irreversible thermodynamics to tribochemical processes*

According to the statements in section 4.8.1. the relationship of reversible thermodynamics can be applied to the description of the reactions taking place at mechanically treated solids, whilst taking account of the excess free enthalpy of the treated solids. However, other reaction conditions exist directly during the mechanical stress, which

Table 4.24. Tribochemical reactions with negative affinity A.

No.	Reaction	A_{298K}, kJ/mole	Ref.
1	$C + 2H_2O \rightarrow CO_2 + 2H_2$	− 62.4	[92]
2	$2Cu + CO_2 \rightarrow 2CuO + C$	− 140.2	[11]
3	$Au + 3/4\,CO_2 \rightarrow 1/2\,Au_2O_3 + 3/4\,C$	− 311.5	[9]
4	$SiC + 2H_2 \rightarrow Si + CH_4$	− 58.9	[46]
5	$BaO_2 \rightarrow BaO + 1/2\,O_2$	− 55.2	[57]
6	$Fe_3O_4 + 2C \rightarrow 2CO_2 + 3Fe$	− 330.2	[158]
7	$2MgO + C \rightarrow CO_2 + Mg$	− 744.9	[159]

cannot be described by the reversible thermodynamics. Thus, for instance, reactions having a negative affinity appear under the direct influence of mechanical energy (see Tab. 4.24.). Also the steady gas pressures appearing during the treatment in heterogeneous reactions are shifted from the equilibrium pressure by orders of magnitude (see Tab. 4.25.) and have no direct connection with the values calculated by application of reversible thermodynamics (Tab. 4.1.).

As Tab. 4.25. shows a still higher decomposition pressure appears at $CdCO_3$ during the treatment than after its interruption. From this it can be concluded that consequently still higher *excited* states appear in the solid during the treatment [161].

It could be shown that the chemical reactions running under the direct influence of mechanical energy are coupled with the dissipation processes in the solid, and the application of irreversible thermodynamics to tribochemical reactions is possible [162, 163].

Table 4.25. Effect of the mechanical treatment on the displacement of the CO_2 equilibrium pressure for 2 selected reactions at 298 K.

Reaction	Equilibrium pressure (calculated) p, Pa	Steady-state pressure after the mechanical treatment p, Pa	Steady-state pressure during the mechanical treatment p, Pa	Ref.
$CdCO_3 \rightleftharpoons CO_2 + CdO$	$1.2 \cdot 10^{-4}$	$2.2 \cdot 10^{2}$	$7.8 \cdot 10^{4}$	[12]
$2\,Cu + CO_2 \rightleftharpoons 2\,CuO + C$	10^{29}		10^{5}	[160]

For the dissipation function ψ is valid:

$$\psi = T \cdot \mathrm{d}S_i/\mathrm{d}t = \sum_k J_k \cdot X_k \geqq 0 \tag{61}$$

$\mathrm{d}S_i/\mathrm{d}t$ — generation of entropy per unit of time, J_k — generalized flux, X_k — generalized force.)

Taking into consideration that both scalar and vectorial fluxes and forces can appear, then equ. (61) changes into [164]:

$$\psi = T \cdot \mathrm{d}S_i/\mathrm{d}t = \sum_{k=1}^{n_1} J_k \cdot X_k + \sum_{j=1}^{n_2} \boldsymbol{J}_j \boldsymbol{X}_j \tag{62}$$

(J_k, X_k — scalar generalized fluxes and forces, $\boldsymbol{J}_j \boldsymbol{X}_j$ — vectorial generalized fluxes and forces.)

The left addend is the so-called chemical term and takes into account the coupling of scalar quantities, e.g. of reaction velocities with affinities and of scalar heat fluxes with scalar differences in temperature, occurring particularly in recombination processes of triboplasma.

The right addend is the so-called electro-mechano-caloric term coupling matter fluxes with heat fluxes. Involved here too is the coupling of dislocation fluxes with macroscopic and microscopic stresses, electric currents with the electrochemical affinity, and heat fluxes with the temperature gradient.

The processes represented by the terms are not able to couple with one another. Therefore each term has to be defined positively for itself. If only one single tribochemical reaction is regarded in the system and if the part vA is separated from $\sum J_k X_k$ then results:

$$\psi = vA + \sum_{i=1}^{n_1} J_i X_i + \sum_{j=1}^{n_2} \boldsymbol{J}_j \boldsymbol{X}_j \tag{63}$$

Taking into consideration the condition formulated above then the following is valid:

$$\left(vA + \sum_{i=1}^{n_1} J_i X_i\right) > 0\,; \quad \sum_{j=1}^{n_2} J_j X_j > 0 \tag{64}$$

According to this follows $vA < 0$, if (65)

$$|vA| < \left|\sum_{i=1}^{n_1} J_i X_i\right| \tag{66}$$

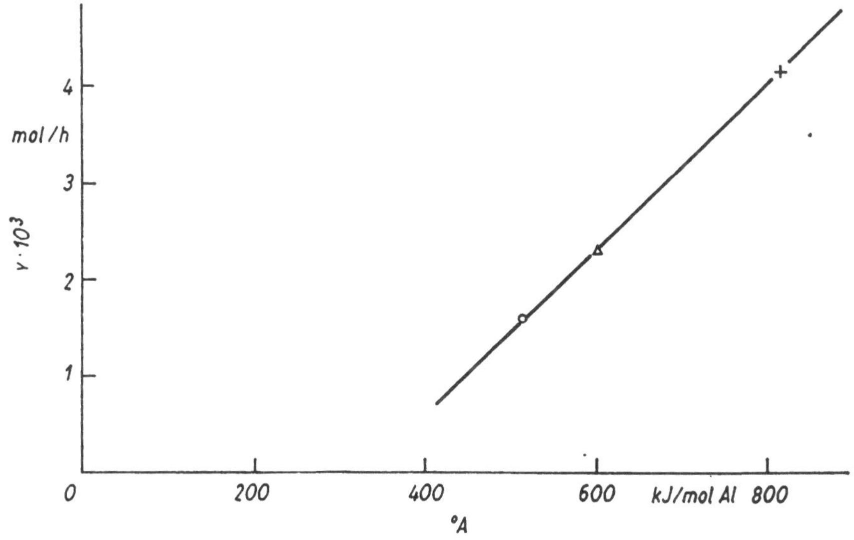

Fig. 4.58. Dependence of the reaction velocity v for three different tribochemical reactions of aluminium on the basic affinity ^0A [165].
+ – reaction with O_2, Δ – reaction with CO, o – reaction with CO_2.

It could be shown that this relation is valid even, in the more intensified form, in numerous tribochemical systems examined experimentally [12]:

$$|vA| \ll \left|\sum_{i=1}^{n_1} J_i X_i\right| \tag{67}$$

Equ. (65) shows that reactions which have a negative affinity can also occur through the coupling, and hence there is nothing preventing a qualitative explanation of a great number of tribochemical reactions, as for instance the reactions mentioned in Tab. 4.24.

In order to connect reaction velocity and affinity with sufficiently small A-values the phenomenological equations of linear irreversible thermodynamics can be used [164]:

$$v = L_{11}A + \sum_{i=1}^{n_2} L_{1i}X_i \tag{68}$$

(L_{11}, L_{1i} – coefficient of linearity.)

In the steady state of treatment the 2nd term is constant, resulting in [165]:

$$v = aA + b \tag{69}$$

(a, b — constants.)

As a rule it is difficult to prove the validity of equ. (69), because kinetic inhibitions appear in most of the chemical reactions, such as the growth of covering layers during starting reactions [166]. However, it is a special characteristic of tribochemical reactions that through the treatment action kinetic inhibitions are, to a wide extent removed

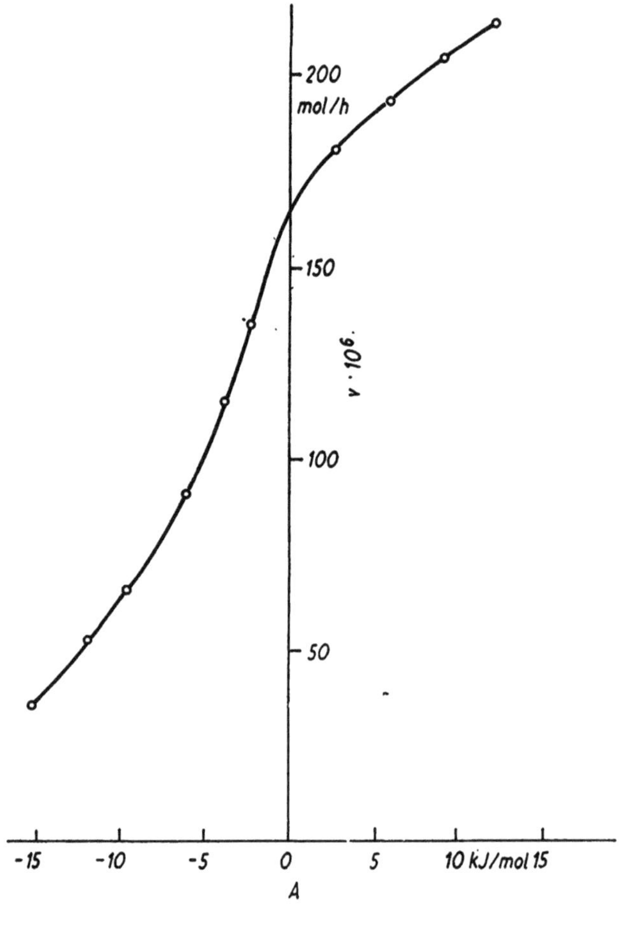

Fig. 4.59. Dependence of the reaction velocity vf for the reaction $Ni + 4CO \rightarrow Ni(CO)_4$ on the affinity at 308 K [12].

and the chemical solid reaction becomes the stage determining the velocity of the total process, e.g. as fresh surface reactions on treated metal surfaces [107]. There is a correlation between reaction velocity and affinity as expected under these conditions.

In Fig. 4.58. the reaction velocity v is represented in dependence on the affinity A as an example for 3 oxidation reactions. It becomes apparent that v depends linearly on A in a very wide range.

Near equilibrium the formation of nickel carbonyl has been examined. As Fig. 4.59. shows, the curve established experimentally can be described qualitatively by equ.

(69) resulting in positive values for both a and b. This conforms to general experience [165]. The importance of the electro-mechano-caloric terms is dealt with separately in section 6.4.

4.8.3. *Tribochemical equilibrium*

In a series of tribochemical reactions, such as the decomposition of cadmium carbonate, a decline of the reaction velocity to zero is observed when a determined concentration of the reactants is reached. If the concentration is varied by the application of one reactant in the system in question, then again a reaction occurs towards the re-establishment of the initial state.

In Fig. 4.60. the dependence of the CO_2 pressure p on the treatment time t is shown as an example for the formation (1) and decomposition (2) of cadmium carbonate. According to this the same steady state appears from both the side of the formation

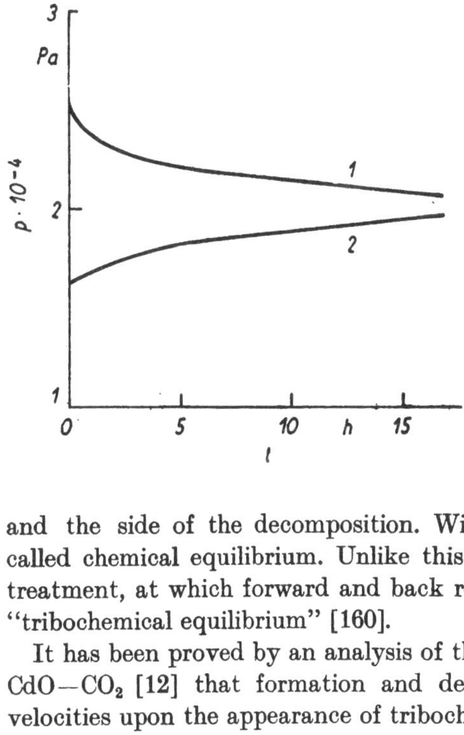

Fig. 4.60. Dependence of the CO_2 pressure p on the treatment time t for the formation and decomposition of $CdCO_3$ [12].

Establishement of a steady state by formation (1) and by decomposition (2) of $CdCO_3$.

and the side of the decomposition. With totally thermal excitation such a case is called chemical equilibrium. Unlike this the steady state appearing with mechanical treatment, at which forward and back reactions run with the same velocity, is called "tribochemical equilibrium" [160].

It has been proved by an analysis of the individual time laws in the system $CdCO_3-CdO-CO_2$ [12] that formation and decomposition reactions happen with the same velocities upon the appearance of tribochemical equilibrium.

In analogy to equ. (38), valid for the kinetics of the reaction is:

$$v_b = \frac{k_1 \cdot p}{k_2 + p} \tag{70}$$

The opposite reaction $CdCO_3 \rightarrow CdO + CO_2$ can be expressed approximatively by the equation:

$$v_d = k_3 + k_4 p \tag{71}$$

(v_b — formation velocity, v_d — decomposition velocity, k_1, k_2, k_3, k_4 — constants, p — gas pressure.)

An expression of the total velocity can be found for the overlapping of decomposition and formation reactions with any mixture ratio by the help of both the equations for pure CdO and pure $CdCO_3$, respectively. Assuming the reactions to become effective proportionally to the relative concentrations of the components there results with

$$x_{CdCO_3} = x \quad \text{and} \quad x_{CdO} = 1 - x \tag{72}$$

for the total velocity v_r:

$$v_r = a \cdot v_d^0 - b(1 - x)\, v_b^0 \tag{73}$$

(a, b — constants of proportionality, x — mole fraction $CdCO_3$, v_b^0, v_d^0 — reaction velocities of the pure phases.)

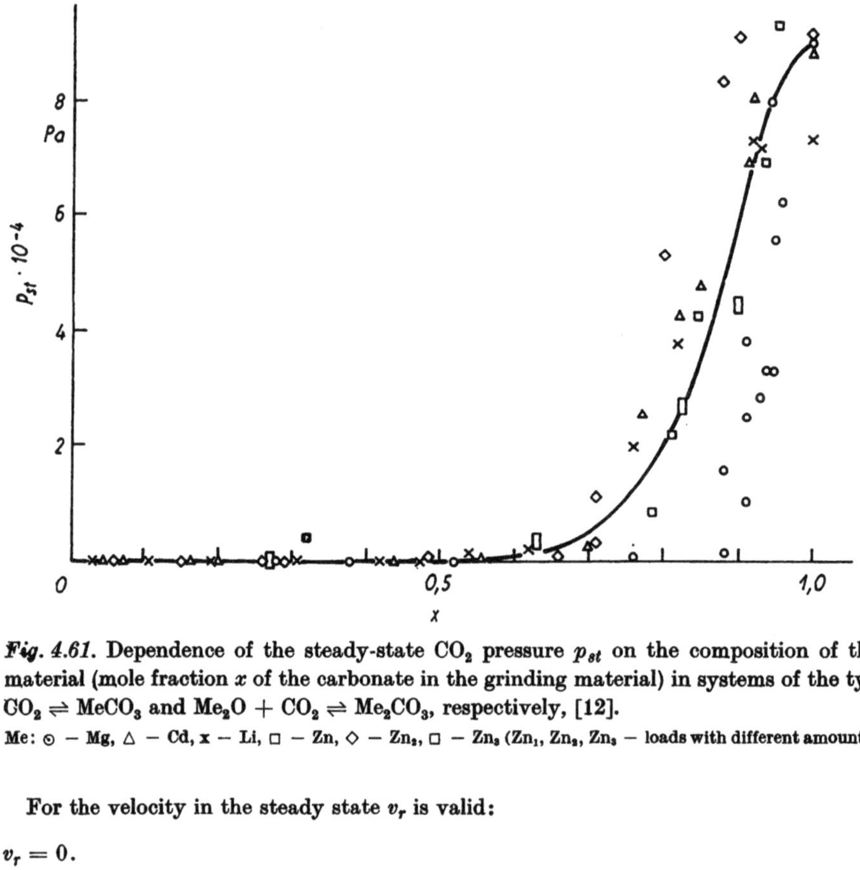

Fig. 4.61. Dependence of the steady-state CO_2 pressure p_{st} on the composition of the grinding material (mole fraction x of the carbonate in the grinding material) in systems of the type $MeO + CO_2 \rightleftharpoons MeCO_3$ and $Me_2O + CO_2 \rightleftharpoons Me_2CO_3$, respectively, [12].
Me: ⊙ — Mg, △ — Cd, x — Li, □ — Zn, ◇ — Zn_2, □ — Zn_3 (Zn_1, Zn_2, Zn_3 — loads with different amounts of products)

For the velocity in the steady state v_r is valid:

$$v_r = 0. \tag{74}$$

From this can be derived the following relationship:

$$x = \frac{k_5 \cdot p_{st}}{k_6 + k_7 \cdot p_{st} + k_8 \cdot p_{st}} \tag{75}$$

(k_5, k_6, k_7, k_8 — constants, p_{st} — steady-state pressure.)

This functional relationship is represented by the unbroken curve in Fig. 4.61. In addition the steady-state pressures obtained experimentally for various mixture relations are plotted for the system

$$CdCO_3 \rightleftharpoons CdO + CO_2 \qquad (76)$$

as well as for a number of other analogous reaction systems. As can be seen, tribochemical equilibria can appear for all the examined reactions of the type

$$MeCO_3 \rightleftharpoons MeO + CO_2 \qquad (77)$$

the position of which can be described by the simple kinetic statement discussed above in dependence on the composition of the grinding material in first approximation.

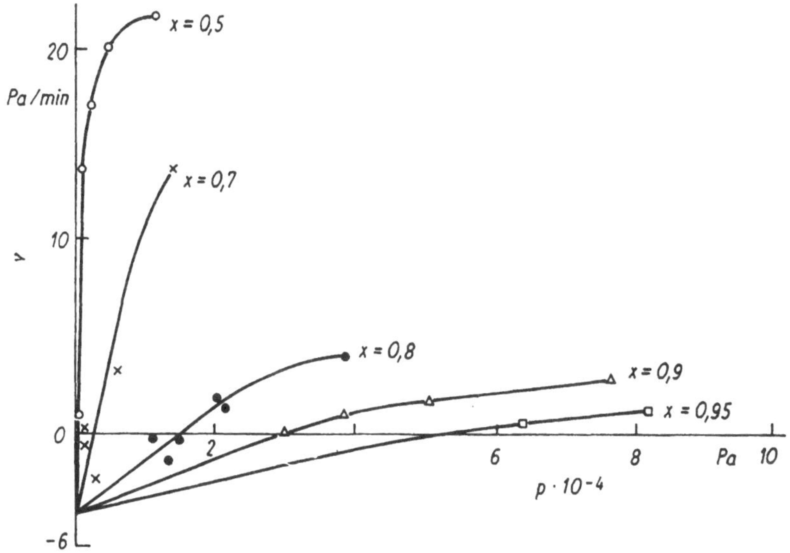

Fig. 4.62. Dependence of the reaction velocity v on the CO_2 pressure p for several compositions of the grinding material phase in the system $Cd + CO_2 \rightleftharpoons CdCO_3$ [160].
x — mole fraction of $CdCO_3$ in the grinding material phase.

Hence tribochemical equilibrium is a dynamic state characterized by the same velocity of the forward and back reactions. There is, however, a principle difference between a reversible chemical equilibrium and the tribochemical equilibrium, as can be seen from equ. (73). If the Gibbs phase law

$$F = B - P + 2 \qquad (78)$$

F — degrees of freedom, B — number of the independent components, P — number of the phases.)

is valid, the CO_2-equilibrium pressure is independent of the carbonat-oxid ratio in the solid phase in the first mentioned case, since the system has one degreee of freedom only.

In contrast to this according to equ. (75), the position of the tribochemical equilibrium is decisively defined by the concentration ratio of the solid phases. The shifting of the steady-state CO_2-pressure (p_{st} with $v_r = 0$) due to the change of the mole fraction x of $CdCO_3$ in the grinding material can also be seen from Fig. 4.62.

In the system $C—Cu_2O—CuO—C/CO_2$ the position of the tribochemical equilibrium is also defined by the portion of the reacted $CO_2x^s_{CO_2}$ related to the total amount of Cu. Consequently the composition of the solid phase also influences the steady-state appearing upon treatment in this example (see Fig. 4.63.).

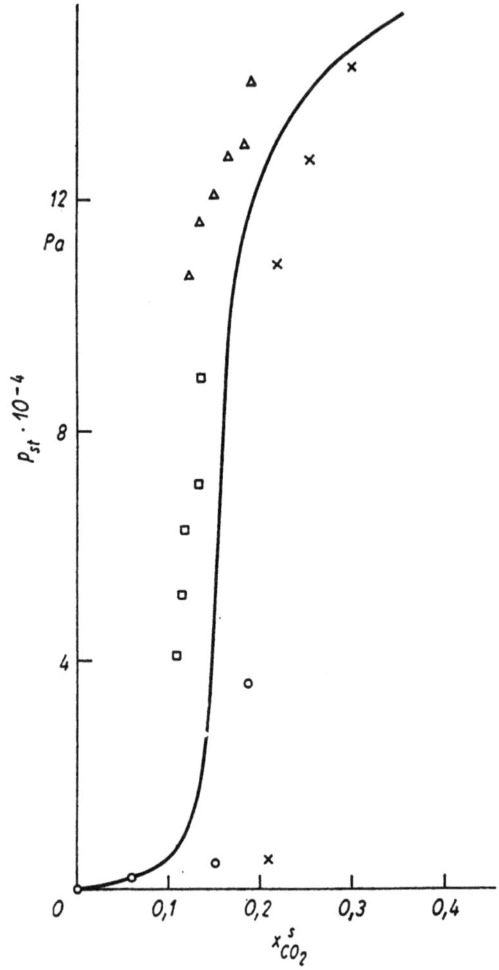

Fig. 4.63. Dependence of the steday-state CO_2 pressure p_{stat} on the composition of the grinding material phase in the system $CuO—C—Cu—Cu_2O$ [160].

$x^s_{CO_2}$ – mole fraction CO_2 in the solid phase initial grinding material: o – Cu, x – 2 CuO + C, Δ – 2 Cu_2O + C, □ – Cu + 2 Cu_2O + C.

These decisive differences can be explained insofar as in chemical equilibrium each partialsystem of the total thermodynamic system can be described by the same state variables at any time. The maintenance of the tribochemical equilibrium, however, is connected with a permanent entropy production in the solid in question and with a permanent energy flux into the reaction space neighbourhood caused by the forces and fluxes forced mechanically. An equilibrium between the partial systems in time and space as defined in thermodynamics does not appear. From this also results the dependence of the steady state on the quantitative ratio of the solid phase.

As defined in irreversible thermodynamics, a steady state appears by virtue of the fact that a number of fluxes drop to zero by maintaining determined forces constantly in time. In the case of mechanical treatment in a closed milling set, such as in a vibration mill, steady states appear after an incubation period (see 4.3.) which lead to the appearence of forces constant in time. If the $(n+1)^{\text{th}}$ flux vanishes with maintenance of n forces then a nth order steady state is present. In consideration of a mechanically released reaction the tribochemical equilibrium represents such a state. This steady state regularly follows the LE CHATELIER-BROWN principle, as can be seen from the preceding representation. By identification of the tribochemical equilibrium as a steady state of nth order, steady-state tribochemical reactions could be proved to run

Table 4.26. Chemical and physico-chemical processes running during mechanical treatment, which can be reversed by the alteration of reaction conditions.

Nature of process	Process			Ref.
Tribochemical reactions	$MeO + CO_2$	$\rightleftharpoons$	$MeCO_3$, Me = Cd, Li, Zn, Mg	[12, 52]
	$Ni + 4CO$	$\rightleftharpoons$	$Ni(CO)_4$	[90]
	$C + 2H_2$	$\rightleftharpoons$	CH_4	[92]
	$3Ni + C$	$\rightleftharpoons$	Ni_3C	[90]
	$Si + C$	$\rightleftharpoons$	SiC	[10]
	$2Me + CO_2$	$\rightleftharpoons$	$MeO + C$, Me = Cu, Fe, Ni, Mg	[111, 167]
Tribosorption	Ar_{gas}	$\overset{Ni}{\rightleftharpoons}$	$Ar_{triboabsorbed}$	[112]
	H_{2gas}	$\overset{Fe/SiC}{\rightleftharpoons}$	$H_{2triboabsorbed}$	[46]
Tribochemical modification transformation	α-PbO_2	$\rightleftharpoons$	β-PbO_2	[168]
	Calcite	$\rightleftharpoons$	aragonite	[169]

isoaffinely. This is evidence that an open thermodynamic system is present in the case of tribochemical reactions during the treatment. The thermal reaction system undergoes a permanent matter and energy feed from the nonthermal higher excitation state system (triboplasma), [164].

Equ. (69) for $v = 0$ changes during the tribochemical equilibrium into

$$-A = \frac{b}{a} = A_0. \tag{79}$$

Consequently the affinity takes on a finite value ($A_0 \neq 0$) in the steady state.

As Tab. 4.26. shows an inversion of the reaction direction was found for a larger number of reaction systems with the alteration of external reaction parameters. Steady-state gas pressures p^{st} in the tribochemical equilibrium were also determined in a series of systems. As Tab. 4.1. and Tab. 4.25. show these deviate by orders of magnitude from the gas pressures determined in the chemical equilibrium and cannot be correlated with them, either.

Steady states will also be obtained if the tribochemical reaction consists in a modification transformation. As an example of "tribochemical modification equilibrium" the steady state in the calcite-aragonite system is represented in Fig. 4.64.

As results from Tab. 4.26., a saturation state the so-called "tribosorption equilibrium" – also appears during the tribosorption of noble gases and hydrogen in solids, in which sorption and desorption of gases occur with the same velocities. As Fig. 4.65.

shows, the direction of the sorption in the range of saturation depends on the adjusted partial pressure. Desorption prevails with smaller pressures and sorption prevails with higher ones. The tribochemical reaction is not influenced by this.

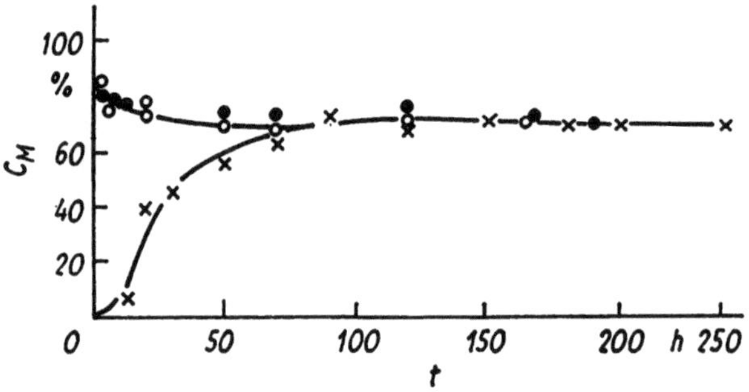

Fig. 4.64. Adjustment of a tribochemical modification equilibrium in the calcite-aragonite system during mechanical stress in a vibration mill [170].

C_M — proportion of aragonite (%), (100 — C_M) — proportion of calcite, t — treatment time, x_* — synthetic calcite, o — natural aragonite, • — synthetic aragonite.

A shifting of the tribochemical equilibrium appears with the changed intensity of the mechanical treatment, as investigations on the modification equilibrium in the system α-PbO_2—β-PbO_2 showed [171]. Thus the reactant concentration in the tribo-

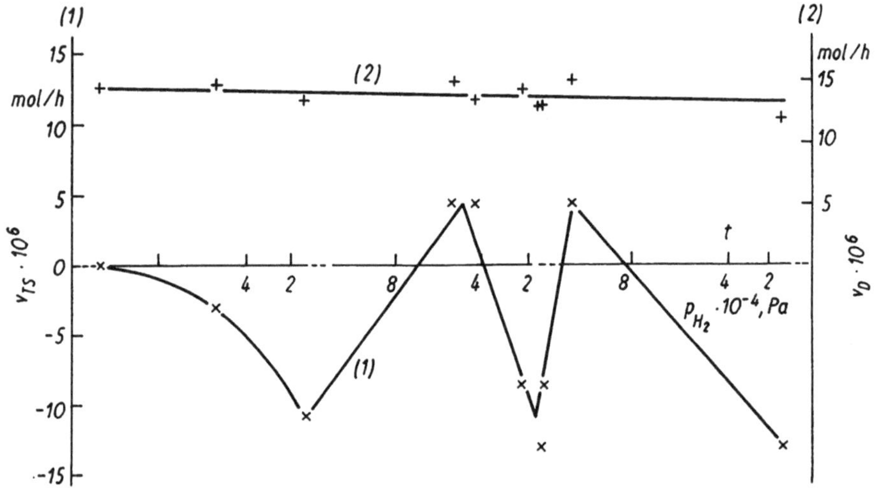

Fig. 4.65. Dependence of the hydrogen sorption in the environment of the tribosorption equilibrium on pressure [46].

v_{TS} — sorption velocity, v_D — desorption velocity, t — time of treatment.
Treatment material: SiC + iron powder saturated with H_2. 1 — tribosorption curve of H_2, 2 — curve for the simultaneously occurring tribodesorption of methane.

chemical equilibrium measured experimentally, apart from the usual parameters, such as pressure and temperature, also depend on the dissipation conditions, which in their turn are determined by the intensity of mechanical treatment (see 4.4.). By this the number of state variables required for the characterization of the tribochemical equilibrium also grows compared to the number of parameters required for the description of a reversible thermodynamic equilibrium.

4.9. Tribochemical modification transformation

Phase transformations can be understood as chemical reactions of the type

$$A_{\text{solid}} \rightarrow B_{\text{solid}} \tag{80}$$

and consequently tribomechanically effected transformations can be regarded as tribochemical reactions of this type. The transformation of the crystalline phase appears with constant material composition.

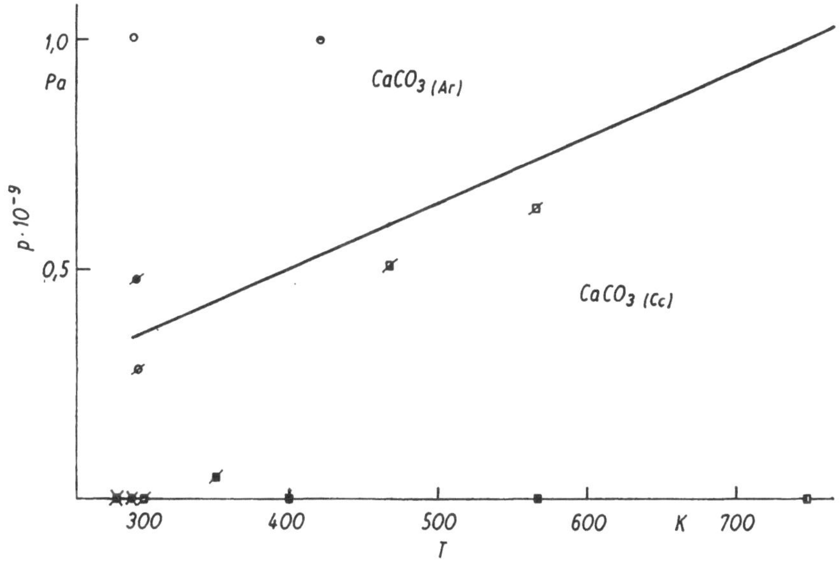

Fig. 4.66. Effect of different treatment on the calcite (Ca)-aragonite (Ar) transformation in dependence on pressure p and temperature T [248].

o – calcite, ■ – aragonite, ⌀ – with additional shear stress, – with intensive stress in a vibration mill, – proportion of the black area shows the transformation into aragonite. – proportion of the white area shows the transformation to calcite.

A considerable number of tribochemical modification transformations examined up until now are listed in Tab. 4.27. In most cases phase transformations without treatment take place at temperatures above 400 K only, so that at room temperature the metastable phase also has a long lifetime. By contrast, under the influence of mechanical energy all the reactions listed in Tab. 4.27. already occur at room temperature.

As can also be seen from Tab. 4.27., modification transformations having both positive and negative free enthalpy can be carried out by treatment. In numerous cases the transformation of the crystalline phases is observed in both directions. As shown in 4.8.3. a tribochemical modification equilibrium arises in such systems.

Table 4.27. Summary of a number of tribochemical modification transformations.

No.	Substance	Phase A	→ Phase B	Affinity	Ref.
1	ZnS	Wurtzite (hexag.)	→ spalerite (cub.)	+	[172–178]
2.1	AgJ	β (hexag.)	→ γ (cub.)	–	[178–180]
2.2		γ (cub.)	→ β (hexag.)	+	[Chapt. 1., [81]), [182–185]
3.1	$CaCO_3$	Calcite (trig.)	→ aragonite (rhomb.)	–	[167], [186–191], [178], [192–195]
3.2		Aragonite	→ calcite	+	
3.3		Vaterite (hexag.)	→ calcite	+	[183, 196, 195]
4.1	PbO	Litharge (tetr.)	→ massicot (orthorh.)	–	[197,–203], [178], [204–207]
4.2		Massicot	→ litharge	+	[194, 198, 208]
5	MnF_2	(rutil)	(orthorh.)	–	
6.1	PbO_2	β (tetr.)	→ (orthorh.)	–	[198, 168, 209, 210]
6.2		α	→ β	+	
7	Sb_2O_3	Senarmonitite (cub.)	→ valentinite (orthorh.)	–	[198]
8	SiO_2	α-quartz (trig.)	→ coesite (monokl.)	–	[198, 211]
9	CaC_2	III (orthorh.)	→ II (dto.)	+	[212, 213]
10	$UO_3 \cdot H_2O$	Bipyramidal	→ rod-shaped	–	[214]
11.1	Co	(cub.)	→ (hexag.)	+	[215–218]
11.2		(Hexag.)	→ (cub.)	–	
12	$Ag_2(HgJ_4)$	Yellow	→ red	–	[141]
13	$Hg_2(CuJ_4)$	Red	→ violet	–	[141]
14.1	ZrO_2	(Monocl.)	→ (tetr.)	–	[219, 220]
14.2		(Tetr.)	→ (monocl.)	+	[220–224]
15	Fe_2O_3	γ (tetr.)	→ (α rhomb.)		
16	Eu_2O_3	C (cub.)	→ β (monocl.)		[225]
17	CdS	Wurtzite-type	→ spalerite-type	–	[226–229], [194], [230]
18	Al_2O_3	Boehmite	→ γ-Al_2O_3		[231]
19	BeF_2	Low pressure	→ high pressure	–	[198]
		Red	→ yellow		[232, 233]
20	$CuFe_2O_4$	(tetr.)	→ (cub.)		
21	TiO_2	Anatase	→ rutile	+	[234, 235, 228, 236, 237, 238]
22	$Na_4P_4O_{12}$				[239]
23	Na_2SO_4	V (orthorh.)	→ III (dto.)		[240]
24	$BaTiO_3$	(tetr.)	→ (cub.)		[241, 178, 242]
25	Ca_2SiO_4	γ-C_2S	→ β-C_2S	–	[243, 244, 149]
26	WO_3	α	→ β	–	[245]
27	C	(hexag.)	→ (rhomb.)	–	[246]
28	CdSe	α	→ β	–	[230]
29	Nb_2O_5	(monocl.)	→ (pseudohexag.)		[247]
30	Ta_2O_5	β (rhomb.)	→ δ (pseudohexag.)		[247]

Tribochemical modification transformations have been investigated in detail in the calcite-aragonite system.

In Fig. 4.66. the respective measured and extrapolated p-T equilibrium curve, is represented for this system. According to this calcite is the stable phase with normal pressure, whereas aragonite represents the thermodynamically stable phase with high pressures above the equilibrium line. Often calcite is described as the high-temperature phase and aragonite as the high-pressure phase. It can be seen from the measures drawn in (■, O) that at room temperature the two phases are metastable and do not undergo any transformation with completely thermal stress in the equilibrium area of the other phase. A measurable transformation of calcite into aragonite occurs at 423 K only (◑). The transformation of aragonite into calcite can be observed at a higher temperature, namely from 750 K (◧). Apparently the transition is more inhibited from aragonite to calcite than the inverse process. Thus a relatively high value of 444 kJ/mole was also found for the activation energy of the aragonite transformation [249].

If the material is additionally treated by shearing stress then the transformation of calcite into aragonite already takes place at room temperature in the equilibrium area of aragonite (●). The inverse reaction takes place at a temperature of 473 K only (◪). Hence also with mechanical stress it is more inhibited than the transformation of calcite. In both the cases the shearing stress effects a reduction of the transformation temperature. A transformation of calcite being in its own equilibrium area (∅) could not be obtained by shearing stress. Evidently the shearing stress only effects an acceleration of the polymorphous transformations towards the establishement of equilibrium due to the activation energy reduction (see 4.5.).

The two transformations can be proved at room temperature with sufficiently intensive mechanial treatment, e.g. in a vibration mill. Intensity and mechanical treatment threshold levels are observed the falling short of which no longer permits the realization of a modification change in both directions.

According to equ. (25) the power input N of a vibration mill is:

$$N = a' \cdot M_c \cdot \omega^3 \cdot r. \qquad (25)$$

It results that the transformation of calcite to aragonite appears with a sufficiently high power input, namely with a milling body mass of $M_c > 1.3$ g (●) and a vibrating radius $r > 4.5$ mm (◙) [178]. In accordance with the thermal exitation here too a higher energy barrier has to be overcome in the impact-induced aragonite transformation than in the inverse reaction.

The transformations carried out under the most varying conditions show in all that to overcome an energy barrier is very important in the formation of the new phase. The appearance of energy barriers was for instance not observed in tribochemical reactions between a solid and a gas. Evidently the local energy concentration must reach a minimum value in the modification transformation in order to enable a new phase to come into excistence. For instance such a threshold can be caused by the nucleation of the new phase [178].

Particularly high energy barriers occur if the various polymorphous forms strongly differ from one another in structure. Thus e.g. γ-Fe_2O_3 has a spinel structure with a great number of vacancies on the B—Fe sites. A diffusion of Fe atoms across oxygen layers is required during the transformation [222]. Thus it happens that in this system the transformation runs only monotropically and γ-Fe_2O_3 is formed from the α-phase by neither thermal nor tribochemical excitation.

In PbO the structure of the atomic layers differs only little in the polymorphous forms so that litharge massicot transformations can be easily realized [205].

Such transformations can also be carried out with a small energy input at which a continuously running transformation occurs through the installation of structure defects by the treatment process. Thus by tribomechanical means the transformation of the wurtzite structure into the spalerite structure takes place relatively easily, since above all a successive transformation of the lattice planes occurs in succession by the mechanically-caused installation of stacking faults.

A tribochemical modification transformation through continuous a alteration of the lattice parameters was also observed during the vibration milling of $BaTiO_3$. As Tab. 4.28. shows, a successive transition to the cubic lattice can be found with low treatment intensity, by matching the dimensions of the elementary cell in the tetragonal lattice, and this with constancy of the elementary cell volume (line 2 and 3). With higher treatment intensity (line 4) an additional lattice expansion takes place whih is said to be connected with the displacement of lattice ions at the interstitial positions [178]. These investigations indicate that a different treatment can lead to qualitatively different disorders.

Table 4.28. Lattice constants of $BeTiO_3$ after a vibration milling by steel balls (determined from the [420] and [422] reflexes).

No.		Amplitude a, mm	Diameter of balls d, mm	Duration of treatment t, h	Lattice constant $a \cdot 10$ nm	$c \cdot 10$ nm	Vol. of the elementary cell $a^2 \cdot c \cdot 10^3$, nm^3	Rel. difference between a and c, %
1	unmilled	—	—	—	3.988	4.038	64.21	1.25
2	milled	4	7	7	4.003	4.006	64.19	0.07
3	milled	4	10	7	4.000	4.013	64.22	0.32
4	milled	9	10	17	4.024	4.039	65.40	0.37

As Fig. 4.66. has shown for the transformation of calcite to aragonite with normal pressure ■, such transformations are also tribomechanically possible which run contrary to the establishment of the thermodynamic equilibrium. According to Tab. 4.27. such modification changes with negative affinity appear very often.

The assumption was manifest that the formation of high-pressure phases occurring with negative affinity come about through the local occurrence of high pressures in the impact treatment of the low-pressure phases. In Tab. 4.29. a selection of inorganic substances is represented with the corresponding pressures at which the transformation to the high-pressure modification takes place. Since a transformation of these substances up to BeF_2 could be realized by the mechanical stress it was concluded that pressures appear under treatment up to about $15 \cdot 10^8$ Pa [198]. However, compressive forces are not assumed to appear in the general sense of a hydrostatic pressure having an all-round effect, under the conditions of vibration milling. One should rather expect compressive stress in determined directions within the grinding material particles, leading to shearing stresses which for their part release deformation processes. These conditions do not conform with these conditions which usually exist during high-pressure phase transformation under thermodynamic equilibrium. Therefore the shearing processes resulting in a bonding fracture, the stored stress energy, and lattice distortions are also included in the discussion of the reasons for the transformation in the high-pressure phase [178, 250].

Table 4.29. Transformation pressures of some inorganic substances at room temperatures [198].

Substance	Transformation pressures $p \cdot 10^{-8}$, Pa
PbO	5.5
MnF_2	9.5
PbO_2	10.0
Sb_2O_3	10.0
SiO_2	13.5
BeF_2	15.5
B_2O_3	18.5
$BaSO_4$	30.0

From the thermodynamic point of view a coupling of the tribochemical reactions with the processes running during the dissipation of mechanical energy is also present here, as has been described in 4.8.2. Consequently one is rather more justified in taking the affinity of the transformation as a basis for the classification instead of the transformation pressure. Four phase transformations are described for the transition from the low-pressure to the high-pressure phase in Tab. 4.30. Therefore the affinities are negative at room temperature and atmospheric pressure. The first transformation can be easily realized by vibration milling, whereas a higher treatment intensity is required for the second one.

Table 4.30. Data on some tribochemical modification transformation [178].

No.	Substance	Transformation	Transformation pressure	Molar transformation volume	Affinity A, J/mole
1	PbO	Litharge → massicot	5.5	− 0.76	423
2	$CaCO_3$	Calcite → aragonite	3.9	− 2.76	1,088
3	Sb_2O_3	Senarmonite → Valentinite	10.0	− 2.20	2,227
4	AgJ	Modif. (II) → halite modification (III)	3.0	− 7.18	2,181

The remaining transformations could not be realized in spite of the increased treatment intensity in the vibration mill. Their affinities become more negative in the same order as the realizability of the tribochemical transformation decreases. It is assumed that the amounts of energy required for the last two transformations are not available under the given stress conditions [178].

In order to determine the affinity of the modification transformations dealt with, the following relationship was established [178]:

$$A \approx p_{Tr} \cdot \Delta V \quad (82)$$

(p_{Tr} — transformation pressure in thermodynamic equilibrium, ΔV — molar variation of volume during the transformation.)

A further argument for the fact that the mechanical treatment not only suspends kinetic inhibition with modification transformations but also creates conditions for the use of the irreversible thermodynamics is provided by the appearance of the negative temperature gradients [200] typical for tribochemical reactions and of tribochemical modification equilibria (see Fig. 4.64.). The latter deviate in their p-T dependence and in their properties from the thermally determined equilibrium values and can be regarded as n^{th} order steady states (see 4.8.3.). Thus for instance the position of the equilibrium and the steady-state composition of the two modifications depend on the intensity of the mechanically induced dissipation flow. A higher intensity leads to larger

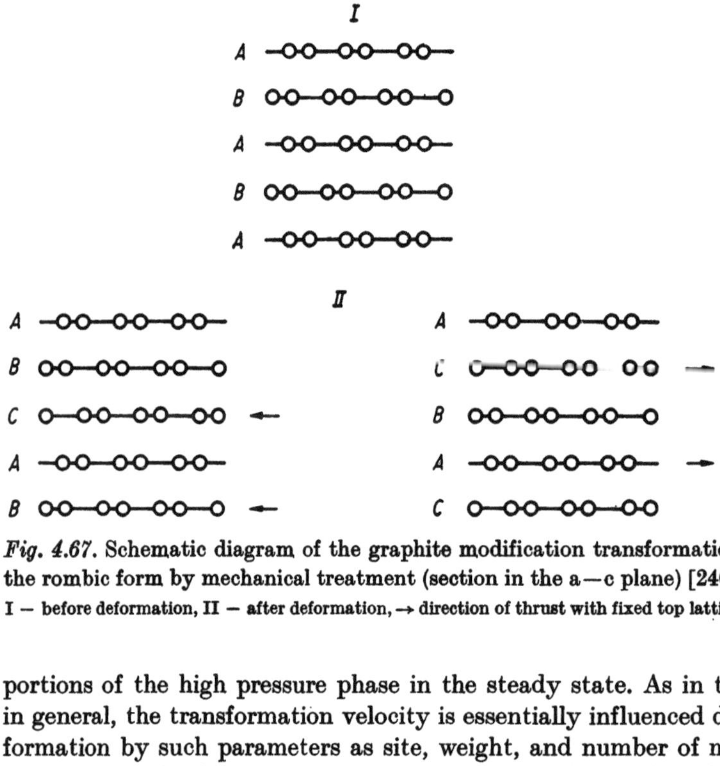

Fig. 4.67. Schematic diagram of the graphite modification transformation from the hexagonal to the rombic form by mechanical treatment (section in the a—c plane) [246].
I — before deformation, II — after deformation, → direction of thrust with fixed top lattice equence.

portions of the high pressure phase in the steady state. As in tribochemical reactions in general, the transformation velocity is essentially influenced during the phase transformation by such parameters as site, weight, and number of milling balls, the filling composition, the dimensions of the mill as well as amplitude and frequency of vibration (see 4.4.). Higher milling capacities bring about higher transformation velocities [224].

It is assumed with respect to the mechanism of modification transformations that the forward and back reaction is released by different excitation states in the solid. In this way the critical shearing stresses of calcite are below the critical rupture stresses so that the transformation of calcite is brought about by the influence of deformation processes. By contrast hot spots or plasmatic-magmatic states are held responsible for the retrograd reaction [251, 252]. The triboinduced formation of the martensitic phase in steel is also attributed to the short-time appearance of temperature peaks above 1,300 K [253].

A modification transformation with negative affinity caused by shearing processes is schematically represented in Fig. 4.67. by the example of graphite which is transferred from the hexagonal to the rhombic modification by shifting the lattice planes.

Usually only the transition of one phase to a second is observed in tribochemical modification transformations. But there is also the possibility that several phase transformations proceed parallel or successive during the treatment. As example in Fig. 4.68. the tribochemical transformation of the anatas-modification is represented leading firstly to the formation of an amorphous phase. The latter changes into the brookite phase. As final phase, with increasing duration of milling rutile comes into existence.

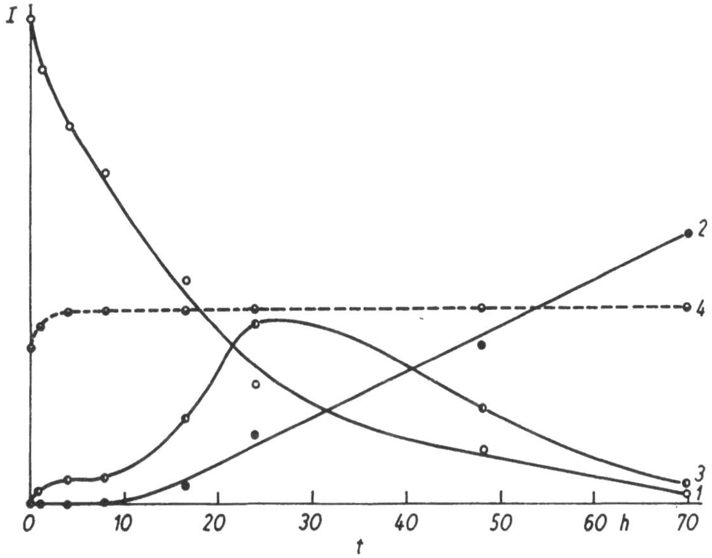

Fig. 4.68. Tribochemical transformation of anastase in dependence on the treatment time [237].
I — X-ray intensity corresponds to the proportion of the modification in the grinding material phase; 1 — anastase content, 2 — rutile content, 3 — brookite content, 4 — course of the background.

In all it can be concluded that a thermodynamic treatment is also possible in modification transformations, that however, the structural specialities of the individual phases determine the kinetics and the mechanism of the tribochemical processes to a larger degree than in gas-solid reactions, and that the development of unified points of view is only at the beginning.

Altogether, the application of the thermodynamics of irreversible processes has led to a deeper understanding of the nature of tribochemical reactions.

References

[1] Bowden, F. P., Tabor, D.: „Reibung u. Schmierung fester Körper", Berlin 1959
[2] Peters, K.: „Symposium Zerkleinern", p. 79, Weinheim/Bergstr. 1962
[3] Carey-Lea, M.: Amer. J. Sci. **46** (1893), 413
[4] Boldyrew, V. V., Avvakumov, E. G., Harenz, H., Heinicke, G., Strugova, L. I.: Z. anorg. allg. Chem. **393** (1972), 152
[5] Fink, M., Hofmann, U.: Z. anorg. allg. Chem. **210** (1933), 100
[6] Bock, N.: Dissertation, Berlin (1973), 137
[7] Heinicke, G.: chapt. 1., [1], p. 101

[8] Heinicke, G.: chapt. 1., [1], p. 141
[9] Thiessen, P. A., Heinicke, G., Schober, E.: Z. anorg. allg. Chem. **377** (1970), 20
[10] Hennig, H.-P.: Dissertation, Berlin 1970
[11] Heinicke, G.: Schmiertechnik **13** (1966), 88
[12] Heinicke, G., Sigrist, K.: Z. Chem. **11** (1971), 226
[13] Thiessen, P. A.: chapt. 1., [1], p. 11
[14] Thiessen, P. A., Heinicke, G., Meyer, K.: chapt. 1., [4] p. 497
[15] Thiessen, P. A.: Z. Chem. **5** (1965), 168
[16] Hertz, H.: „Ges. Werke", Bd. 1, p. 155, Leipzig 1895
[17] Meyer, K., Gragert, E.: phys. stat. sol. **6** (1964), 803
[18] Spangenberg, H. J.: Mber. Dt. Akad. Wiss. **13** (1971), 130
[19] Thiessen, P. A., Franke, E., Florek, H.: Mber. Dt. Akad. Wiss. **7** (1965), 371
[20] Thiessen, K. P., Sieber, K.: Z. phys. Chem. **260** (1979), 410
[21] Prigogine, J., Defay, R.: „Chemische Thermodynamik", Leipzig 1962
[22] Torka, K.: "Reactivity of solids", p. 400, Amsterdam 1961
[23] Timoshenko, S., Goodier, J. N.: "Theorie of Elastivity", p. 372, New York 1951
[24] Meyer, K., Zittlau, W.: Z. phys. Chem. **234** (1967), 44
[25] Wollbrandt, J., Linke, E.: "III. Polish Seminar on Exoelectron Emission and related phenomena", Karpacz 1978
[26] Wollbrandt, J., Linke, E.: "Proceedings of the 5th International Symposium on Exoelectron Emission and Dosimetry" p. 274, Zvikov 1976
[27] Meyer, K., Obrikat, D., Rossberg, H.: Kristall u. Technik **5** (1970), 181
[28] Pick, H.: "Optical Properties of Solids", Amsterdam/London 1972
[29] Fox, P. G.: J. mat. Sci. **10** (1975), 346
[30] Stumpf, H.: „Quantentheorie der Ionenkristalle", Berlin 1961
[31] Schönert, K., Weichert, R.: Chem. Ing. Techn. **41** (1969), 295
[32] Diels, K., Jaeckel, R.: „Leybold Vakuum-Taschenbuch", Berlin/Göttingen/Heidelberg 1962
[33] Spangenberg, H. J.: Mitteilungsblatt d. Chem. Gesellsch. in der DDR **22** (1975), 10
[34] Butjagin, P. Ju., Bystrikov, A. V.: chapt. 1., [9], I. p. 70
[35] Meyer, K.: „Physikalisch-chemische Kristallographie", p. 302, Leipzig 1968
[36] Heinicke, G.: chapt. 1., [1], p. 114
[37] Thiessen, P. A., Heinicke, G., Hennig, H.-P.: Z. Chem. **11** (1971), 195
[38] Heinicke, G.: chapt. 1., [1], p. 117
[39] Heinicke, G., Hoffmann, H.: Surface Science **24** (1971), 484
[40] Thiessen, P. A., Heinicke, G., Meyer, K.: chapt. 3., [1], p. 138
[41] Schober, E.: Diplomarbeit, Berlin (1967), 103
[42] Heikicne, G.: Kolloidn. z. **31** (1969), 596
[43] Heinicke, G., Harenz, H., Sigrist, K.: Z. anorg. allg. Chem. **352** (1967), 178
[44] Heinicke, G., Harenz, H.: Z. phys. Chem. **240** (1969), 327
[45] Thiessen, P. A., Heinicke, G., Meyer, K.: chapt. 1., [4], p. 513
[46] Heinicke, G., Hennig, H.-P.: Schmiertechnik **14** (1967), 86
[47] Heinicke, G.: Z. anorg. allg. Chem. **324** (1963), 177
[48] Heinicke, G.: chapt. 1., [1], p. 108
[49] Thiessen, P. A.: Abh. dtsch. Akad. Wiss., Kl. Chem. Geolog. u. Biol. Nr. 5, p. 44, Berlin 1966
[50] Heinicke, G., Lischke, I.: Z. Chem. **11** (1971), 335
[51] Schöne, R.: Chem. Ing. Techn. **41** (1969), 287
[52] Heinicke, G.: Dissertation, Berlin (1960), 29
[53] Ref. [34], p. 74
[54] Ref. [34], p. 64
[55] Thiessen, P. A., Heinicke, G., Meyer, K.: „III. Intern. Kongreß für grenzflächenaktive Stoffe", Bd. II, p. 518, Köln 1961

[56] AVVAKUMOV, E. G., KOSOBUDSKIJ, I. D.: Izv. Sib. otd. AN SSSR, ser. chim. nauk 5, **46** (1973), 135
[57] HEINICKE, G., RIEDEL, R., HARENZ, H.: Z. phys. Chem. **227** (1964), 62
[58] HEINICKE, G., JECH, Č., WINKLER, K.: Z. phys. Chem. **225** (1964), 161
[59] HEINICKE, G.: chapt. 1., [1], p. 147
[60] PAUDERT, R., HARENZ, H., HEINICKE, G., DÜNKEL, L., PÖTHIG, R., KOLOSOV, A. S., BOLDYREV, V. V., TSCHAIKINA, M. W., TARANZOVA, M. J., GORDEJEWA, G. I.: Z. anorg. allg. Chem. **452** (1979), 141
[61] HEINICKE, G., SCHOBER, E.: Z. Chem. **11** (1971), 219
[62] HEINICKE, G.: ref. [7], p. 146
[63] CHODAKOV, G. S., AVVAKUMOV, E. G.: chapt. 1.,[4], p. 546
[64] PAUDERT, R., HARENZ, H., PÖTHIG, R., HEINICKE, G., DÜNKEL, L., SĆHUMANN, H.: Chem. Techn. **30** (1978), 470
[65] SIGRIST, K., HEINICKE, G.: Chem. Techn. **21** (1969), 288; **22** (1970), 31
[66] AVVAKUMOV, E. G., GIMAUTDINOV, Ju. V., BOLDYREV, V. V.: chapt. 1., [9], II. p. 54
[67] HEINICKE, G.: Habilitationsschrift, Berlin (1965), 136
[68] HEINICKE, G.: chapt. 1., [1], p. 146
[69] HUKKI, R. T., REDDY, I. G.: „2. Europ. Symp. Zerkleinerung", p. 249, Amsterdam 1966
[70] SCHRADER, R.: chapt. 1., [4], p. 528
[71] Ref. [64], p. 474
[72] Ref. [12], p. 233
[73] KRUG, D.: Thermochim. Acta **20** (1977), 53
[74] HEINICKE, G.: Z. Chem. **2** (1962), 284
[75] GROHN, H., PAUDERT, R.: Z. Chem. **3** (1963), 89
[76] MITTASCH, A.: Z. phys. Chem. **40** (1902), 1
[77] ROBERTS, M. W.: "1. Int. Congr. on Metallic Corrosion", p. 84, London 1961
[78] BELCHETZ, L., RIDEAL, E. K.: J. Amer. chem. Soc. **57** (1935), 1169
[79] DOHSE, H.: Z. phys. Chem. **8** (1930), 159
[80] GREENHALGH, R. K., POLANYI, M.: Trans. Faraday Soc. **35** (1939), 520
[81] STAUDE, H.: „Physikalisch-chemisches Taschenbuch", II, p. 2009, Leipzig 1949
[82] RIENÄCKER, G., BADE, H.: Z. anorg. allg. Chem. **248** (1941), 45
[83] TEMKIN, M., RISEV, K.: z. fiz. chim. **13** (1939), 851
[84] BARRER, R. M.: J. chem. Soc., London, 8 (1936), 1256
[85] BARRER, R. M., RIDEAL, E. K.: Proc. Roy. Soc., London A **149** (1935), 253
[86] BLACKWOOD, J. D.: Austral. J. Chem. **12** (1959), 14
[87] SCHRADER, R., STÄDTER, W., OETTEL, H.: Z. phys. Chem. **249** (1972), 87
[88] PÖTHIG, R.: Dissertation, Berlin 1979
[89] SENZKY, U.: Ingenieurarbeit, Berlin 1960
[90] HEINICKE, G., SIGRIST, K.: Z. anorg. allg. Chem. **350** (1967), 157
[91] HEINICKE, G.: chapt. 1., [1], p. 153
[92] HENNIG, H.-P.: Diplomarbeit, Berlin 1963
[93] HEINICKE, G., MEYER, K., SENZKY, U.: Z. anorg. allg. Chem. **312** (1961), 180
[94] BOLDYREV, V. V., AVVAKUMOV, E. G., SCHMIDT, I. V., STRUGOVA, L. M.: chapt. 1., [6], p. 43
[95] HENNIG, H.-P.: HEINICKE, G.: unpublished
[96] LISCHKE, I.: Dissertation, Berlin (1967), 81
[97] SCHRADER, R., TETZNER, G., GRUND, H.: Z. anorg. allg. Chem. **342** (1966), 207
[98] SCHRADER, R., GRUND, H., TETZNER, G.: Chem. Ing. Techn. **39** (1967), 807
[99] Ref. [43], p. 174
[100] TRAPNEL, B.: „Chemisorpcija", p. 176, Moskva 1958
[101] HEINICKE, G., HARENZ, H., RICHTER-MENDAU, J.: Kristall u. Technik **4** (1969), 78
[102] ŽURKOV, S. N.: Vestnik **11** (1957) 78
[103] BUTJAGIN, P. J.: Chem. Techn. **25** (1973), 259

[104] Heinicke, G.: chapt. 1., [1], p. 120
[105] Boldyrev, V. V., Heinicke, G.: Z. Chem. **19** (1979), 353
[106] Heinicke, G.: chapt. 1., [1], p. 141
[107] Hoffmann, H.: Z. phys. Chem. **245** (1970), 37
[108] Heinicke, G., Hoffmann, H.: unpublished
[109] Hoffmann, H., Funke, G.: Z. phys. Chem. **247** (1971), 234
[110] Heinicke, G., Sigrist, K.: Mber. Dt. Akad. Wiss. **8** (1966), 654
[111] Heinicke, G., Sigrist, K.: Z. Chem. **6** (1966), 291
[112] Heinicke, G., Sigrist, K.: Chem. Techn. **26** (1974), 70
[113] Ref. [10], p. A 45
[114] Spangenberg, H. J.: Mber. Dt. Akad. Wiss. **13** (1971), 135
[115] Ref. [43], p. 174
[116] Heinicke, G.: Schmierungstechnik **1** (1970), 69
[117] Heinicke, G., Schober, E.: Mber. Dt. Akad. Wiss. **10** (1968), 445
[118] Heinicke, G., Jech, Č., Friedrich, M.: Mber. Dt. Akad. Wiss. **8** (1966), 344
[119] Heinicke, G., Friedrich, M., Jech, Č.: Z. phys. Chem. **236** (1967), 191
[120] Jech, Č., Heinicke, G.: Wissenschaft u. Fortschritt **10** (1966), 435
[121] Heinicke, G.: chapt. 1., [9], I. p. 39
[122] Heinicke, G., Hennig, H.-P.: chapt. 1., [6], p. 50
[123] Friedrich, M.: unpublished
[124] Schober, E.: Dissertation, Berlin 1970
[125] Lohrisch, H. J., Wagner, W., Wagner, K.: Schmierungstechnik **9** (1978), 198
[126] Fricke, R.: „Handbuch der Katalyse“, IV, p. 1, Wien 1943
[127] Hüttig, G. F.: „Handbuch der Katalyse“, VI, 3, p. 318, Wien 1943
[128] Ref. [101], p. 112
[129] Fricke, R., Schweckendick, W.: Z. Elektrochem. angew. physik. Chem. **46** (1940), 93
[130] Heinikce, G., Harenz, H.: Technik **24** (1969), 313
[131] Eckell, J.: Z. Elektrochem. angew. physik. Chem. **39** (1933), 435
[132] Titchener, A. L., Bever, M. B.: Progr. Metall Phys. **7** (1958), 247
[133] Quinney, H., Taylor, G. I.: Proc. Roy. Soc., London A **143** (1934), 307
[134] Michell, D., Haig, F. D.: Phil. Mag. **2** (1957), 15
[135] Wagendvistel, A., Polak, R.: Aluminium **47** (1971), 195
[136] Chapt. 3., [82], p. 1102
[137] Paudert, R.: Metalloberfläche **21** (1967), 142
[138] Hüttig, G. F., Herrmann, E.: Z. anorg. allg. Chem. **247** (1941), 235
[139] Williams, R. O.: Rev. Sci. Instr. **34** (1963), 639
[140] Fricke, R., Ackermann, P.: Z. Elektrochem. angew. physik. Chem. **40** (1934), 630
[141] Naeser, G., Scholz, W.: Kolloid-Z. **156** (1958), 7
[142] Heegn, H., Tkačova, K., Huhn, H. J., Ludwig, G., Bernhardt, C., Husemann, K.: Chem. Techn. **30** (1978), 348
[143] Schrader, R.: Freiberger Forschungsh. A **392** (1966), 81
[144] Schrader, R.: Technik **24** (1969), 91
[145] Schrader, R., Hoffmann, Br.: Z. anorg. allg. Chem. **369** (1969), 45
[146] Hüttig, G. F., Lewinter, M.: Z. angew. Chem. **41** (1928), 1034
[147] Fuchs, O.: “Reactivity of solids”, p. 407, Amsterdam 1961
[148] Chodakov, G. S., Edelman, L. I.: Kolloidn. z. **29** (1967), 728
[149] Schrader, R., Schumann, H.: Silikattechnik **22** (1971), 344
[150] Dialer, K., Küssner, K.: Kolloid-Z. 251 (1973), 711
[151] Pichogov, Ju., Semenenko, I. N., Babkina, L. A.: Neorganičesk. Mat. **13** (1977), 81
[152] Chapt. 3., [83], p. 67
[153] Tkáčová, K.: Silikaty **4** (1976), 321
[154] Tkáčová, K.: Erzmetall **29** (1976), 207

[155] KRETZSCHMAR, U.: Dissertation, Berlin 1980
[156] HEINICKE, G., SIGRIST, K.: Mber. Dt. Akad. Wiss. 8 (1966), 661
[157] KIHLSTEDT, P. G., FORSSBERG, E.: Silicates industr. **36** (1971), 45
[158] HEINICKE, G.: chapt. 1., [1], p. 123
[159] HEINICKE, G.: chapt. 1., [1], p. 156
[160] HEINICKE, G., SIGRIST, K.: Mber. Dt. Akad. Wiss **11** (1969), 44
[161] HEINICKE, G., HARENZ, H.: Z. Chem. 8 (1968), 34
[162] SIGRIST, K.: Dissertation, Berlin 1967
[163] GUTMAN, E. M.: fiz. chim. Mechanika tv. tel: **5** (1967), 264, 404, 548
[164] THIESSEN, K. P., SIEBER, K.: Z. phys. Chem., **260** (1979), 410, 417
[165] SIGRIST, K., HEINICKE, G.: Mber. Dt. Akad. Wiss. **11** (1969), 48
[166] HAUFFE, K.: „Oxydation von Metallen", Berlin/Göttingen/Heidelberg 1956
[167] GROHN, H., PAUDERT, R., FRIEDRICH, H.: Z. anorg. allg. Chem. **348** (1966), 21
[168] SCHRADER, R., WEIGELT, D.: Z. anorg. allg. Chem. **372** (1970), 228
[169] SCHRADER, R., HOFFMANN, Br.: Z. Chem. **6** (1966), 388
[170] Ref. [145], p. 41
[171] SCHRADER, R., HOFFMANN, B.: chapt. 1., [4], p. 529
[172] SCHLEDDE, A., GANTZKOW, H.: Z. Physik **15** (1923), 184
[173] MATHIEU, M., PAIC, M.: C. R. hebd. Seanč. Acad. Sci. **192** (1931), 147
[174] FREY, F.: Amer. Physics **2** (1948), 147
[175] SMITH, F. G., HILL, V. G.: Acta crystallogr. **9** (1956), 821
[176] TEKIZ, Y., LEGRAND, C.: C. R. hebd. Séanc. Akad. Sci. **261** (1965), 2335
[177] SCHORT, M. A., STEWARET, E. G.: Z. phys. Chem. **13** (1957), 298
[178] SIGRIST, K., HEINICKE, G., STEINIKE, U.: Kristall u. Technik 8 (9173), 393
[179] ADAMS, L. H., DAVIS, B. L.: Amer. J. Sci. **263** (1965), 359
[180] KOLKMEIJER, N. H., VAN HENGEL, J. W. A.: Z. Krist. **88** (1934), 317
[181] BURNS, J. H., BREDIG, M. A.: J. Chem. Phys. **25** (1956), 1281
[182] JAMIESON, J. C., GOLDSMITH, J. R.: Am. Mineralogist **45** (1960), 818
[183] GAMMAGE, R. B., GLASSON, D. R.: Chem. and Ind. (1963), 1466
[184] STERZEL, W.: Z. angew. Chem. **77** (1965), 16
[185] SCHNEIDER, U.: Verfahrenstechnik **1** (1967), 20
[186] NEWTON, R. C., GOLDSMITH, J. R., SMITH, J. V.: Contr. Mineral Petrol **22** (1969), 335
[187] DANDURAND, J. L.: C. R. hebd. Séanc. Acad. Sci., Ser. D **271** (1970), 193
[188] NORTHWOOD, D. O., LEWIS, D.: Can. Mineralogist **10** (1970), 216
[189] FRIES, E., MARHIC, C.: Bull. Soc. Fr. Mineral Crystallogr. **96** (1973), 91
[190] MC. PHERSON, R.: Wear **23** (1973), 83
[191] CRIADO, J. M., TRILLO, J. M.: J. chem. Soc. Faraday Trans. I **7** (1975), 961
[192] SCHRADER, R., HOFFMANN, B.: Z. Chem. **10** (1966), 388
[193] LIN, I. J., SOMASUNDARAN, P.: Powder Techn. **6** (1972), 171
[194] LEISEROWITZ, L., SCHMIDT, G. M. J., SHAMGAR, A.: J. physics chem. Solids **27** (1966), 1453
[195] GAMMAGE, R. B., GLASSON, D. R.: J. Colloid Interface Sci. **55** (1976), 396
[196] NORTHWOOD, D. O., LEWIS, D.: Am. Mineralogist **53** (1968), 2089
[197] CLARK, L. G., ROWAN, R.: J. Amer. Chem. Soc. **63** (1941), 1302
[198] DACHILE, F., ROY, R.: Nature **186** (1960), 34, 71
[199] BURTON, Y. J.: Trans. Inst. Chem. Engrs. **44** (1966), 59
[200] LEWIS, D., NORTHWOOD, D. O., REEVE, R. C.: J. appl. Cryst. **2** (1969), 156
[201] WHITE, W. B.: J. Am. ceram. Soc. **44** (1961), 170
[202] SENNA, M., KUNO, H.: J. Am. ceram. Soc. **54** (1971), 259
[203] LIN, I. J.: Isr. J. Earth. Sci. **20** (1971), 41
[204] LIN, I. J., NIEDZWIEDZ, S.: J. Am. ceram. Soc. **56** (1953), 62
[205] SODERQUIST, R., DICKENS, B.: J. Phys. Chem. Solids **28** (1967), 823

[206] MAMORU, S., HIROSHI, K.: J. Am. ceram. Soc. **34** (1971), 259
[207] REB'KINA, N. I., CHODAKOV, G. S.: Kolloidn. ž. **38** (1976), 3
[208] AZZARIA, L., DACHILLE, F.: J. Phys. Chem. **65** (1961), 889
[209] ZASLAVSKIJ, A. I., KONDRASOV, Ju. D.: Dokl. AN SSSR **75** (1950), 599
[210] BODE, H., VOSS, E.: Z. Elektrochem. **60** (1956), 1053
[211] GUSEV, G. M., NIKOLAEV, A. V.: Dokl. AN SSSR **182** (1968), 343
[212] BREDIG, M. A.: J. Phys. Chem. **46** (1942), 801
[213] Chapt. 3., [66], p. 570
[214] HARRIS, L. A., TAYLOR, S. J.: J. Am. ceram. Soc. **45** (1962), 25
[215] SCHRADER, R., TETZNER, G., GRUND, H.: Z. anorg. allg. Chem. **342** (1966), 205
[216] KOCHANOWSKA, A., SATAVA, V.: Abh. dtsch. Akad. Wiss. (1957) 100
[217] FRICKE, R., MÜLLER, H.: Naturwiss. **30** (1942), 439
[218] HINNÜBER, J.: Powder Metallurgy **8** (1961), 1
[219] BALLEY, J. E., LEWIS, D., LIBRANT, Z. M., PORTER, L. J.: J. Br. ceram. Soc. **71** (1972), 25
[220] MITSUHASHI, T., ICHIHARA, M., TATSUHE, U.: J. Am. ceram. Soc. **57** (1974), 97
[221] BAUDISCH, O., WELO, L. A.: Naturwiss. **21** (1933), 593
[222] SENNA, M., TOJO, S., KUNO, H.: Nippon Kagaku Zasshi **92** (1971), 780
[223] LIN, I. J., GRODZIAN, D. J. M.: "Technion Research and Development Foundation", MERC, No. 016—045, p. 176, Haifa 1974
[224] GRODZIAN, D. J. M., LIN, I. J., NADIV, S.: "104 th AIME Meeting", No 75-B-1, p. 32, New York 1975
[225] ROSS, W. A., GIBBY, R. L.: J. Am. ceram. Soc. **57** (1974), 46
[226] TSUNASHIMA, M., et al.: Kogyo Kagaku Zasshi **74** (1971), 1733, 1740
[227] ZIELINSKI, E.: Roczniki Chem. **27** (1953), 207
[228] YOGANARASIMHAN, S. R.: J. Sci. Ind. Res. Sed. B **21** (1962), 88
[229] KENNEDY, J. D., BENEDICK, W. B.: J. Phys. chem. Solids **27** (1966), 125
[230] HASHIMOTO, K., TODA, Y.: Kogyo Kagaku Zasshi **71** (1968), 1402
[231] PARIS, A.: C. R. hebd. Séanc. Acad. Sci. **271** (1973), 1057
[232] UEHARA, Y.: Zairyo **22** (1973), 702
[233] UEHARA, Y.: J. Soc. Materials Sci. Jap. **21** (1972), 548
[234] KUBO, T., KATO, M., MITARAI, Y., TAKASHI, J., OHKURA, K.: Kogyo Kagaku Zasshi **66** (1963), 318
[235] TEKIZ, Y., LEGRAND, C.: C. R. hebr. Seanc. Acad. Sci. **261** (1965), 3619
[236] KNOLL, H.: Naturwiss. **50** (1963), 546
[237] WANKOVÁ, J., KOCHANSOVKÁ, A.: Kristall u. Technik **1** (1966), 319
[238] HAYAKAWA, S., NAGAVKA, T., ABE, K.: J. Soc. Materials Sci. Jap. **23** (1974), 531
[239] MOTVOKA, T., HASHIZUME, G., KOBAYASHI, M.: J. chem. Soc. Jap. **71** (1968), 1412
[240] GOYAL, Q. R., DESHPANDE, V. V., KARKHANAVALA, M. D.: Indian J. Chem. **9** (1971), 1006
[241] SINJAKOV, E. V., FLEROVA, S. A., ARCHINOV, V. N.: fiz. tv. tela **15** (1973), 1239
[242] SCHOIJET, M.: J. appl. Phys. **15** (1964), 719
[243] KUDRJAVTSEVA, N. L., CHODAKOV, G. S.: Kolloidn. ž. **28** (1966), 524
[244] SCHRADER, R., SCHUMANN, H.: Z. Chem. **7** (1967), 322
[245] ALBRECHT, R., HÄUSLER, H., MÖBIUS, R.: Z. anorg. allg. Chem. **377** (1970), 310
[246] BOEHM, H. P., HOFFMANN, U.: Z. anorg. allg. Chem. **278** (1955), 58
[247] AVVAKUMOV, E. G., RAZVOROTNEVA, L. I.: izv. Sib. otd. AN SSSR Ser. chim. nauk. **9** (1977), 19
[248] DACHILLE, F., ROY, R.: "Reactivity of Solids", p. 505, New York 1961
[249] DAVIS, B. L., ADAMS, L. H.: J. geophys. Res. **70** (1965), 433
[250] LIN, I. J., NADIV, S., GROGZIAN, D. J. M.: Minerals Sci. Engng. **7** (1975), 313
[251] Ref. [29], p. 345
[252] Ref. [145], p. 47
[253] ZENER, C., HOLLOMON, J. H.: J. appl. Phys. **15** (1944), 22

5. On the causes of reactions in tribochemistry

5.1. Chemical effects of triboplasma

The special characteristics of the chemical effect of triboplasma are its short life ($<10^{-7}$ s, Tab. 4.2), its restriction to smallest lattice regions, and the occurrence of high excitation states which lead to the formation of charged and uncharged particles released from the lattice as well as to electrons. The so-called "kinetic temperatures" are indicated with $> 10^4$ K [1]. Since the zone of percussion is not a MAXWELL-space and the particle interactions are of a stochastic nature on principle such reactions may be expected which are impossible according to the relations in thermodynamics (Chapt. 4., [15]). The special characteristics of triboplasmatic reactions also consist in the fact that reaction products built up in the triboplasma are stabilized by rapid quenching and in this way are prevented from further reaction to secondary products.

On account of the stochastic character of the processes it is also understandable that the reaction in triboplasma is largely influenced by the material composition of solid reaction components in the plasmatic region, such as during the establishment of tribochemical equilibrium in the system Cu/CO_2 (see 4.8.3.). Due to the short lifetime and the local limitation of triboplasma only small reaction yields can be expected. Since the post-plasma processes run with a relatively large reaction cross-section the reactions induced by the plasma are overlapped in most cases by the reactions occurring in the subsequent phase. This also applies to reactions with negative affinity which may occur as coupled reactions in the post-plasma.

If the course of tribochemical reactions is connected with an extremely high negative affinity of approximately > 200 kJ/mole, then the reaction cannot be accounted for by thermodynamic processes for under the given conditions, a coupling is very unlikely with a process simultaneously proceeding in the same system but having a many times higher positive affinity (Chapt. 4., [20]). Reactions such as the formation of Au_2O_3 from Au and CO_2 are attributed to the effect of plasma. As Tab. 5.1. shows, the conditions of an extremely negative affinity are given for this reaction (Chapt. 4., [9]).

Table 5.1. Affinity A of the reaction:
$Au + 3/4\,CO_2 \rightarrow 1/2\,Au_2O_3 + 3/4\,C$ at various temperatures T [2].

T, K	A, kJ/mole
298	− 377
1,000	− 477
2,000	− 615

In respect of the short-time and stochastic character there is a certain analogy between the triboplasma and a high-frequency spark discharge. This can also be seen in the chemical effects. For instance ammonia develops tribochemically from the elements (Chapt. 4., [93]). The formation of the same compound is also observed through the initiation of a mixture of hydrogen and nitrogen in an electric spark plasma [3, 4].

Because of this, the tribosublimation of lattice components occurring in the triboplasma, e.g. in NaCl, frequently cannot, or only with great difficulty, be proved, because a spontaneous recombination of the fragments takes place and basically NaCl is left. If, however, substances are introduced forming stable compounds with the one or the other component, e.g. Hg, then a part of the formed Cl atoms may be bound and the remaining elementary Na may be solved in excess Hg and protected from further chemical attack [5].

5.2. The role of elastic stresses in tribochemical reactions

5.2.1. *Tribochemical reactions influenced by constant stresses*

The force constants of the interatomic bonds change by the effect of mechanical stress on solids through elastic deformation, thereby shifting their natural frequency. A change of the interatomic distance and of the valence angles occurs [6]. If a constant stress acts on the solid then the MAXWELL-BOLTZMANN distribution of energy continues to exist in all degrees of freedom so that thermodynamic relations can be applied to tribochemical reactions.

By stretching the chemical bond its energy is reduced. The deformation energy U_{def} of the bond is under equilibrium conditions (Chapt. 4., [103]):

$$U_{\text{def}} = \int_{r_0}^{r} f_u \, \mathrm{d}r \quad \text{or} \quad U_{\text{def}} = \int_{V_0}^{V^*} \sigma_u \, \mathrm{d}V^* \tag{1}$$

(f_u, σ_u — force or stress acting onto the bond, r_0, r — length of the initial and deformed bond, respectively, V_0, V^* — activation volume before and after deformation.)

In order to calculate deformation energy and activation energy U for the breaking of the bonds during the action of a force f_u the equation of the potential-energy curve after MORSE is usually used [7]:

$$\frac{U}{U_0} = \sqrt{1 - \frac{f_u}{F_0}} - \frac{f_u}{2F_0} \ln \frac{1 + \sqrt{1 - \frac{f_u}{F_0}}}{1 - \sqrt{1 - \frac{f_u}{F_0}}} \tag{2}$$

(U_0 — binding energy, F_0 — limit of strength of the bond.)

The curve of function $U(f_u)$ is represented in Fig. 5.1. It indicates that the activation energy of the decomposition of bond is U_0 without the influence of external stresses and that the ARRHENIUS equation is written in the following way:

$$K = K_0 \cdot \exp\left(-\frac{U_0}{RT}\right) \tag{3}$$

(K — constant of the decay rate, s^{-1}, K_0 — pre-exponential factor having the order of 10^{12} bis 10^{13} s^{-1}.)

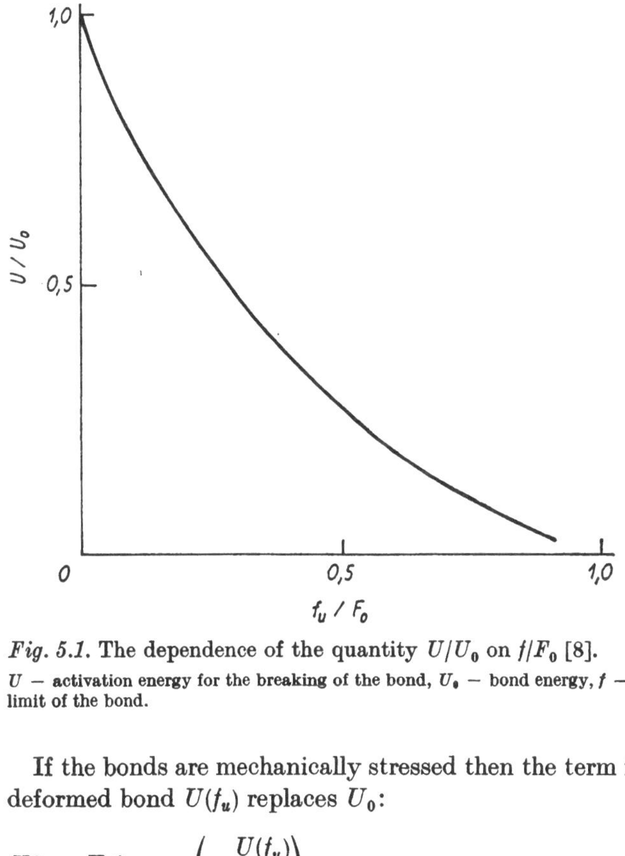

Fig. 5.1. The dependence of the quantity U/U_0 on f/F_0 [8].
U — activation energy for the breaking of the bond, U_0 — bond energy, f — force acting upon the bond, F_0 — strength limit of the bond.

If the bonds are mechanically stressed then the term for the strength of an elastically deformed bond $U(f_u)$ replaces U_0:

$$K^* = K_0^* \cdot \exp\left(-\frac{U(f_u)}{RT}\right) \tag{4}$$

(K^*, K_0^* — constants according to equ. (3) with mechanical stress.)

Since $U(f_u) \leqq U_0$ is valid according to Fig. 5.1. the temperature dependence of the tribochemical reaction is lowered under the influence of mechanical stress in comparison with the mechanically uninfluenced reaction, as is shown in Fig. 5.2. in the example of the hydrolytic reaction of a polyamide (see 4.5.) [9].

If $f_u \geqq F_0$ then the activation energy U for the breaking of the bonds vanishes, as can be seen in the absence of a temperature dependence of the decomposition rate constants. Under these conditions, elastic energy E_{def} is accumulated which is above the strength of the bond ($E_{def} \geqq U_0$). As an example in Fig. 5.3. the formation velocity of the radicals $\frac{d(R^\cdot)}{dt}$ which is proportional to the breaking velocity of supercriti-

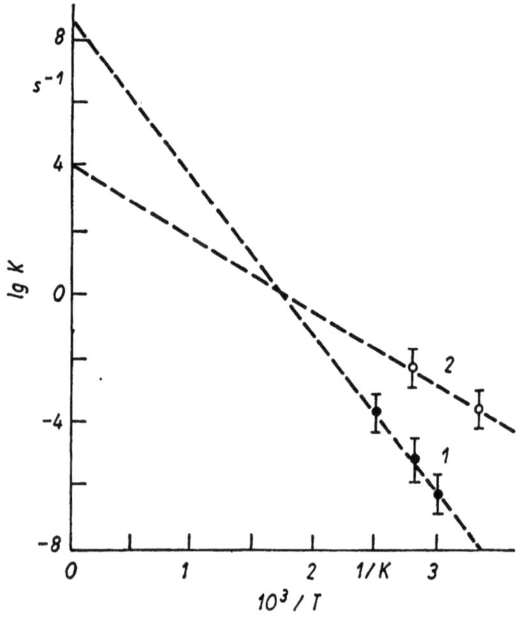

Fig. 5.2. Temperature dependence of the velocity constant K for the hydrolysis of polyamide [9].
(1) — without stresses, (2) — with stresses = $3 \cdot 10^8$ Pa, C_{H_2O} — 3 percent.

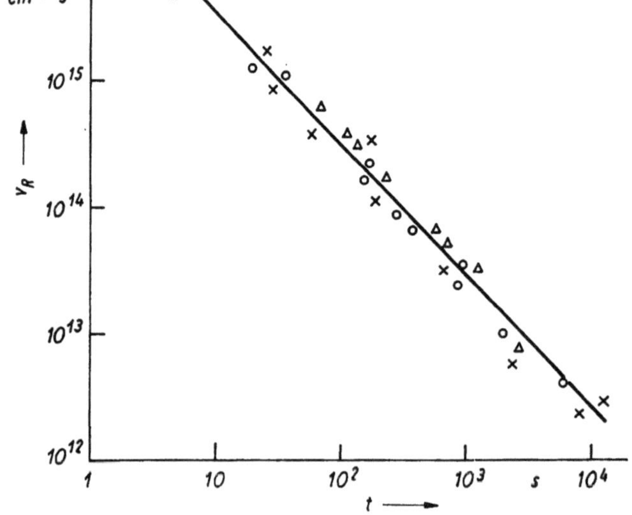

Fig. 5.3. Relations between the formation velocity v_R of the radicals and the time t up to the breaking of the sample for the strain of polyamide with different temperatures (Chapt. 4., [103]).
x — 223 K, o — 291 K, △ — 323 K.

cally stressed bonds is plotted against lifetime of the sample in wide stress and temperature ranges. The values measured at different temperatures are distributed on a common straight line so that the formation velocity of the radicals depends exclusively on the applied stress. The following equation applies for this [10]:

$$\frac{\mathrm{d}\,(R^{\cdot})}{\mathrm{d}t} = B \cdot \exp\,(\beta, \sigma) \tag{5}$$

(B, β — constants.)

As an example the dependence of the concentration of the hydrolytically broken-up bonds of a polyamide on the applied tension stress is represented in Fig. 5.4. The curve shows an exponential course in the measuring range investigated. The same dependence was also measured for the degradation of polypropylen [11].

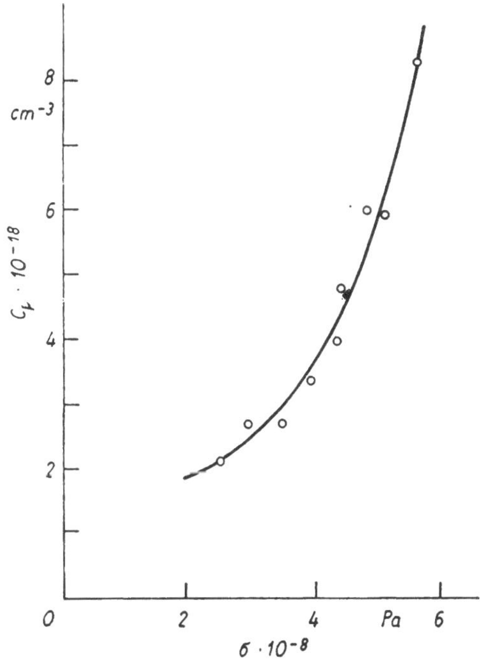

Fig. 5.4. Dependence of the concentration C_p of the hydrolytically broken bonds in polyamide on the acting stress [9].
T — 293 K, t — 11 min, C_{H_2O} — 3 percent.

Since the natural frequencly of the bonds changes under mechanical stress, the stressed bonds upon mechanical deformation can be proved by means of IR-spectroscopy. Fig. 5.5. shows as an example the absorption bands for polypropylene at 975 cm^{-1} during different stresses. When a constant external stress acts, the individual bonds are irregularly stressed in the volume of the solid. The approximate form of the stress distribution function can be established by the analysis of the absorption bands [12]. As Fig. 5.6. shows by way of example, a certain number of bonds are strongly overstressed and stress values are reached on a portion of the bonds which account for more than 10 times the applied stress. At these points with a sufficiently high external stress a breaking of bonds and subsequently a redistribution of the stress comes into existence between the remaining chains. The active primary radicals developing as a result of the breaking of the bonds can be directly proved in the stressed samples in the resonator of the ESR-spectrometer [13].

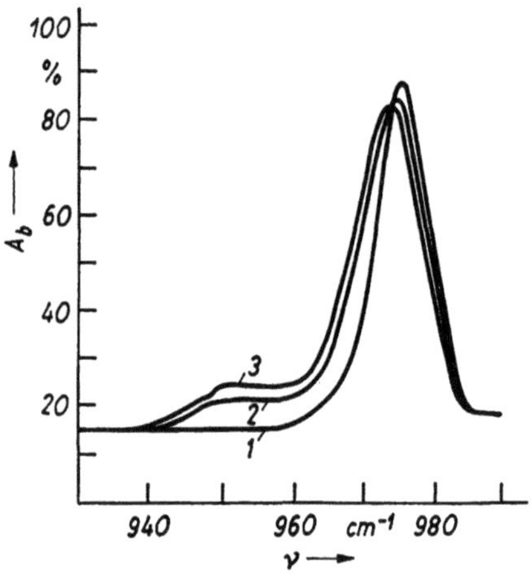

Fig. 5.5. Changes in the absorption band at 975 cm^{-1} in orientated polypropylene under stress (Chapt. 4., [103]).

1 — $\sigma = 0$, 2 — $\sigma = 6.5 \cdot 10^8$ Pa, 3 — $\sigma = 8 \cdot 10^8$ Pa, A_b — absorption, ν = vibration frequency.

If the quantity U_0 is calculated for the mechanical degradation of several polymers according to equ. (1.35) and the activation energy E_A for the thermal degradation according to the ARRHENIUS equation, then as Tab. 5.2. shows, both the quantities are

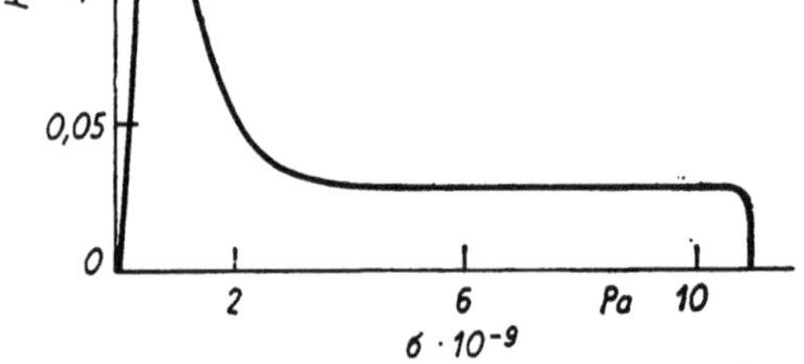

Fig. 5.6. Stress distribution at a polyamide sample stressed by $8.4 \cdot 10^8$ Pa (Chapt. 4., [103]).

H — frequency, σ — stress.

identical for the different polymers [14]. This result also confirms the view that the role of mechanical stresses ($E_{def} = \gamma \cdot \sigma$) in equilibrium processes consists in the decrease of the energy barrier of the thermal reaction. In exceptional cases an inactivation of the reaction can also take place by applying a stress. This is for instance the case in the cyclization of polydienes undergoing a concentration of volume in the course of reaction. By applying a stress the apparent activation energy can be increased and the degree of cyclization can decrease in dependence on the applied stress to the value zero. Evidently steric factors play a dominating part in the present example [15].

The correlations with respect to the influence of stresses on the solid reactivity mainly derived from polymers can also be used for metallic and non-metallic inorganic materials (see Fig. 4.24.). Here above all a part is also played by the natural stresses which for instance are caused by the occurrence of dislocations. As is well known, a dislocation

Table 5.2. Comparison of the activation energy E_A of the thermal polymer degradation with the energy U_0 resulting from the mechanical degradation.

Polymer	Mechanical degradation U_0, kJ/mole	Thermal degradation E_A, kJ/mole
Polymethylmethacrylate	31	30
Polystyrene	33	35
Polyethylene	26	25

represents a crystal defect in which a number of atoms are in distorted lattice positions as against their neighbours, thus accumulating additional energy. As could be shown by the example of the mechanically activated nickel carbonyl formation, a reduction of the activation energy of the reaction will come into being under the influence of these defects (Tab. 4.12.), (Chapt. 4., [101]). The effect of acting stresses on the thermodynamics and kinetics of the dissolution of metals will be discussed in 6.6.

5.2.2. *Tribochemical reactions influenced by altering mechanical stresses*

A further possibility for influencing tribochemical reactions is the application of varying mechanical stresses. These conditions occur very frequently in practice, e.g. in the periodic stresses of solids within the framework of fatigue tests, in the solid stress in mills, in the destruction of surface layers by friction as well as in the action of shock waves. In cyclic actions the material takes up only a portion of the mechanical energy and a mean level of absorbed energy E_{def} establishes itself after a certain time. In addition an energy radistribution having the characteristic duration $\frac{1}{A_t}$ takes place in the system, which is dependent both on the frequency of action and on the real relaxation time of the material.

If the kinetics of the tribochemical process are determined by the redistribution of mechanical energy, the velocity constant must be proportional to the redistribution velocity of the deformation energy between the bonds (A_t in s^{-1}) and the probability of the accumulation of a certain critical deformation energy E^*_{def} at the respective bond being sufficient for the release of a tribochemical reaction. Then is Ref. 4., [103]):

$$K_w \sim A_t \cdot \exp\left(-\frac{E^*_{def}}{\alpha \cdot t_R \cdot N}\right) \tag{6}$$

(K_w — velocity constant in the varying mechanical field, N — applied mechanical power, t_R — relaxation time of the mechanical stress, α — coefficient, the value of which approximates to the tribomechanical efficiency of the process.)

As an example the dependence of the velocity of the tribochemical polymethylmethacrylate degradation in a vibration mill on the intensity of mechanical energy is represented in Fig. 5.7. The value of A_t calculable from this curve was determined to be $10^{-4}\,s^{-1}$ [16]. This value lies within the order of magnitude of redistribution of elastic energy in the solid. Since it thus depends on the physicomechanical properties of the material these results are regarded as confirmation of the assumption that the velocity of the tribochemical process in varying mechanical fields depends on the redistribution velocity of elastic energy in the system.

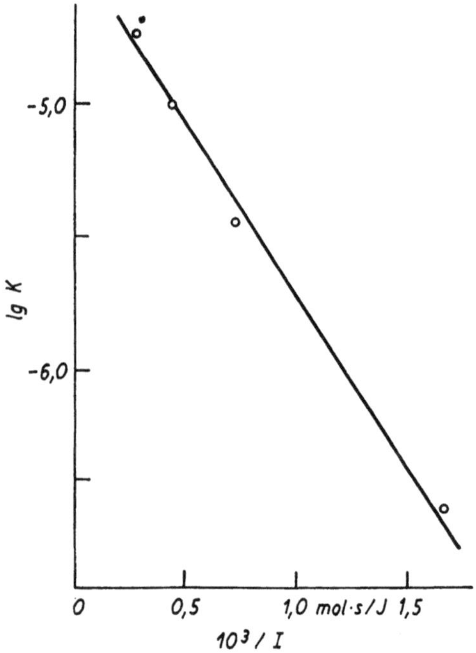

Fig. 5.7. Dependence of the velocity constant K of the mechanical degradation on the intensity I of the mechanical stress of polymethylmethacrylate in a vibration mill (Chapt. 4., [103]).

During the mechanical stress of polymers by means of a vibration mill, e.g. of polymethylmethacrylate, polystyrene, polyvinyl chloride, polyvinyl alcohol, gelatine, cellulose etc., no dependence on T is measured below a certain limiting temperature T_g (Chapt. 1., [4]), (see Fig. 4.5.). In this case it can be assumed that during the vibration milling the limiting value of the elastic energy E^*_{def} is exceeded and that in this way a breaking of the bonds takes place without any additional thermal influence.

In order to guarantee that the bond really breaks up during the fracture stress influence, the stress, whose order exceeds the strength of the bond ($E^*_{def} > U_0$), has to be applied over a lesser time than the vibration period. The velocity w of the stress propagation has therefore to account for:

$$w \approx \frac{10^{-10}}{10^{-12} \dots 10^{-13}} \approx 10^2 \dots 10^3\ \mathrm{m \cdot s^{-1}} \tag{7}$$

with a bond length in the order of 10^{-10} m, which means it has to be comparable with the sound velocity. As the described experiments show this borderline case of purely mechanical degradation is reached with an intensive altering stress, for instance in vibration mills and during the action of shock waves.

Since the primary action of the impact stress discussed in 4.1. similarly consists of an elastic deformation of the surface zone stressed by impacts, the processes running under the influence of elastic stresses, and which are mainly observed on polymers, might also represent one of the important causes of tribochemical initiation in those reactions released by impact in crystalline solids. If for instance the plasma model derived phenomenologically starts inter alia from components being removed from the lattice structure for a very short period of time during impact, and consequently the course of chemical reactions is influenced, then one of the mechanisms responsible for this might certainly be that in local regions and during a correspondingly short time the accumulated stress energy considerably exceeds the strength of the bond ($E_{def} \gg U_0$). The influence of binding energy and of temperature on the course of triboplasma reactions is thus lost. The triboplasma reaction is decisively influenced by factors such as the respective stochasticity of stress and energy accumulations. It has already been possible to prove a broad spectrum of stresses experimentally under equilibrium conditions (see Fig. 5.6.). The preconditions are even more favourable for a much stronger subdivision of the stress spectrum and for the formation of locally high stress peaks under the non-equilibrium conditions of an altering mechanical stress field, as they exist for instance in the impact stress.

Whereas in effect primary valence chains are broken up accompanied by the formation of C-radials under the influence of high stress peaks in polymers, a number of non-thermal secondary processes come into existence in crystalline anorganic solids, such as plastic deformation and fracture (see 3.3.). The changes in solid reactivity resulting from this will be discussed in 5.5. to 5.9.

Furthermore there the possibility exists that the elastic stresses relax and lead to the formation of excited vibrational states which for their part initiate a chemical reaction or contribute to its release. This mechanisms will be discussed in connection with the effect of temperature on tribochemical reactions in 5.3.

5.3. The role of temperature in tribochemical processes

As already shown in 4.1., the opinion was held for a long time that the temperature peaks appearing during the transmission of mechanical energy to solids is the only reason for the release of tribochemical reactions [17]. An important factor in this has been the Bowden-Tabor theory of hot spots, which has facilitated the interpretation of the tribochemical ignition of explosion reactions on the basis of extensive experimental material [18]. Thus it has been found that during solid friction temperatures of more than 1,000 K can appear in small surface areas of 10^{-1}—10 μm for a duration of 10^{-4} to 10^{-3} s. As quantitative calculations show, temperature peaks in an order of magnitude of approximately 1,000 K are also expected to appear with mechanical impacts, e.g. during the steel ball impingement on a NaCl plate [19].

An important assertion of the hot-spots theory is that the temperature peaks do not exceed the melting temperature of the mechanically stressed solid. Proceeding on the basis that certain minimum temperatures are required for the ignition of explosive substances, the most varying substances can be classified as explosion igniters according to their melting points and with respect to their tribochemical effect. As Tab. 5.3. shows, substances having a melting point below approximately 700 K are not able to ignite pentaerythritol tetranitrate by friction, whereas substances having melting points above approximately 700 K are generally able to initiate the explosion by friction. Other properties, such as the hardness of solids, have a secondary influence on the explosion ignition.

Table 5.3. The tribochemical ignition of pentaerythriol tetranitrate in the presence of foreign particles (Chapt. 4., [1]).

Foreign matter	MOHS' hardness	Melting point T, K	Frequency of explosion at friction, per cent
No particles	—	—	0
Ammonium nitrate	2 ··· 3	443	0
Potassium hydrogen-sulphate	3	483	0
Silver nitrate	2 ··· 3	485	0
Sodium dichromate	2 ··· 3	593	0
Sodium acetate	1.5	597	0
Potassium nitrate	2 ··· 3	607	0
Potassium dichromate	2 ··· 3	671	0
Silver bromide	2 ··· 3	707	50
Lead chloride	2 ··· 3	774	60
Silver iodide	2 ··· 3	823	100
Borax	3 ··· 4	833	100
Bismuth glance	2 ··· 2.5	958	100
Glass	7	1,073	100
Copper glance	2.5 ··· 2.7	1,377	100
Lead glance	3 ··· 3.5	1,373	100
Calcite	3	1,612	100

In following the propagation of the explosion it has been found by means of a high-frequency camera that usually a relatively slow chemical process precedes the real detonation. Similar phenomena have been observed for the tribochemical ignition of metallic powders during their milling in pure O_2-atmospheres [20]. As Fig. 5.8. shows, a longer milling process is required in this system before the real fire reaction comes into existence, characterized by a steep increase in the reaction velocity by orders of magnitude which is limited by the complete consumption of oxygen in the reaction vessel.

After a new input of O_2 the initially slow reaction is re-established. This represents the true tribochemical process and in the case of the oxidation reaction can no longer be interpreted by the assumption of hot spots as the cause of the reaction as shown in 5.6. Further results obtained in tribochemical reactions are summarized in 4.1. according to which in the mentioned system at least, the chemical process is not initiated by hot spots. Only during recent years have attempts been made to evaluate the role of temperature in the release of tribochemical reactions (Chapt. 4., [2, 18, 29], [21, 22]).

A comprehensive representation of the physical processes by which mechanical energy is transformed into thermal energy is given in 3.3. An important energy transformation process for the release of tribochemical reactions consists in the relaxation of elastic energy E_{def} during altering mechanical stress of the solid accompanied by

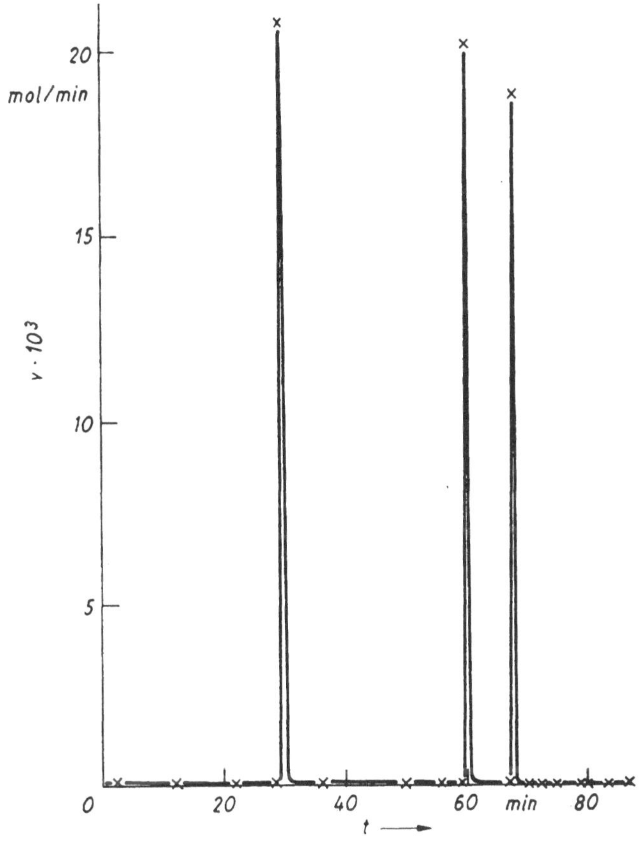

Fig. 5.8. Occurrence of reaction jumps at the mechanical treatment of carbonyl iron powder in an oxygen atmosphere [20].
v – reaction velocity, t – treatment time.

heat development. This process can be depicted by the following scheme:

$$E_{\text{def}} \text{ —— } U^{*}_{\text{osc}} \text{ —— } Q_{\text{heat}} \tag{8}$$

The nature of this relaxation is characterized by the appearance of short-term vibrational states U^{*}_{osc} situated far above the equilibrium states of the thermal vibration spectrum U_{osc}. For these excited vibrational states is valid:

$$\Delta U_{\text{osc}} \gg RT, \tag{9}$$

where

$$\Delta U_{\text{osc}} = U^{*}_{\text{osc}} - U_{\text{osc}}. \tag{10}$$

These are non-equilibrium states coupled to the relaxation process [16]. In a polymer at first the complete flux of the diffusing elastic energy passes through the bonds of the main chain. Therefore bonds in a larger section of chain molecules (some dozen or some 100 links) are set in a vibrationally excited state upon relaxation. On the basis of the excited state fluctuation, the breaking of the polymer chain may appear using the vibrational energy of a complete section of molecules, whereby the following relations apply [13]:

$$E^*_{def} < E_{fr} < U_0 \tag{11}$$

(E_{fr} — energy releasing a fracture.)

If t_{chem} means the characteristic reaction time and t^* the time for which the bond is in a excited vibration state, then the ratio t^*/t_{chem} determines the probability of a reaction. With

$$\frac{t^*}{t_{chem}} > 1 \tag{12}$$

the excitation will most probably lead to reaction. For the decay reaction of the vibrationally excited bond ($\Delta U_{osc} \gg RT$), holds also (Chapt. 4., [103]).

$$t_{chem} \sim t_0 \cdot \exp\left(\frac{U_0}{\Delta U_{osc}}\right). \tag{13}$$

Here a vibration temperature T^* is assigned to the excited vibration states and in a very rough approximation as $RT^* \simeq \Delta U_{osc} \cdot T^*$ does not represent an equilibrium distribution and is therefore designated as equivalent temperature [22]. The lifetime of such excited states is given as 10^{-11}—10^{-10} s in solids, that is to say, it corresponds to 10^2—10^3 vibration periods.

The critical energy ΔU^*_{osc} leading with high probability to the breaking of a bond in the time t^* is considerably smaller than the energy of the bond U_0.

With $t^* = 10^2 \cdot t_0$ is $\Delta U^*_{osc} \approx 0.2 U_0$.

Macroradicals develop by breaking the chain sections which are in an excited vibration state. If thermal reactions are inhibited these macroradials are in a position to react further in low temperatures on the basis of their increased energy content. The most probable reaction of these excited radicals is further degradation by the development of low-molecular compounds. As example Fig. 5.9. shows the mechanical degradation of polyformaldehyde in a mill at 195 K leading to monomers. Comparing the resulting concentration of radicals and the simultaneously developing quantity of low-molecular degradation products, particularly of monomers, the latter is many times larger than the concentration of radicals. It is concluded from this that the excess of vibration energy is sufficient to repel several monomer links.

Highly excited states also occur in solid fracture processes. There is a high concentration of energy particularly at the front of a running fracture which for the most part is transformed into thermal energy. In the first stage this local energy surplus does not represent a temperature and is designated as equivalent temperature in analogy with the above statements. With increasing fracture propagation velocity, increasingly higher equivalent temperatures appear which may exceed the melting temperature by orders of magnitude. By way of example in Fig. 5.10. the equivalent temperature is given for quartz glass in dependence on the distance from the fracture front and on the fracture velocity. Equivalent temperatures of some 1,000 K could be proved by reactivity measurements and with the help of a highly sensitive spectrometer through the analysis of the emergent light (see Tab. 5.4.).

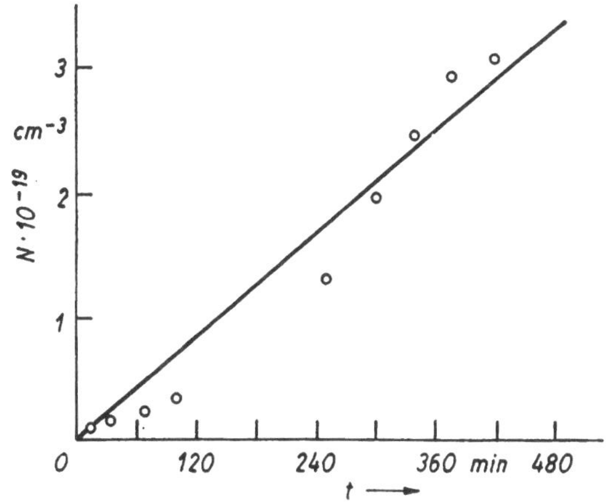

Fig. 5.9. Formation of low-molecular substances during the mechanical degradation of polyformaldehyde at 195 K (Chapt. 4., [103]).

N – molecules/cm³, t – treatment time.

The appearance of these equivalent temperatures is limited to extremely small ranges around the fracture tip and to very short times in the order of magnitude of 10^{-11} s [24].

Regarding the characteristic times for molecular processes (see Tab. 5.5.), the relaxation time for vibration processes lies by orders of magnitude above the lifetime of the

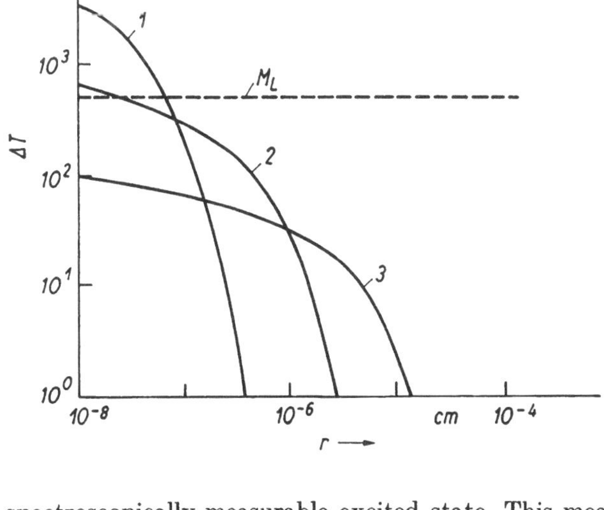

Fig. 5.10. Equivalent temperature T depending on the distance to the fracture front in fracture propagation direction r at several fracture velocities v_B in quartz glass (Chapt. 4., [2]).

1 – $v_B = 10^5$ cm/s, 2 – $v_B = 10^4$ cm/s, 3 – $v_B = 10^3$ cm/s, M_L – softening range.

spectroscopically measurable excited state. This means that there is an extreme non-equilibrium state as is characteristic for a non-thermal system (see Tab. 5.6.). Probably such states are triboplasma components and correspond to the non-thermal plasma of gaseous discharges (Chapt. 4., [33]). This is also the reason why the decomposition of certain salts in the course of a fracture occurs according to another mechanism than for thermal decomposition. Thus there are for instance the following possibilities for

Table 5.4. Equivalent temperature T^* at crack tips of various solids.

Solid	T^*, K	Method of measurement	Ref.
Quartz	4,500	Spectroscopical measurement	[24]
Glass	3,000		[24]
Sugar	2,500		[24]
Calcite	1,250	Reactivity measurement	[23]
Magnesium oxide	1,900		[23]
Magnesium carbonate	7,600		[23]

natrium nitrate decomposition:

$$NaNO_3 \rightarrow NaNO_2 + 1/2O_2 \quad (14)$$

$$NaNO_3 \rightarrow 1/2Na_2O + 1/2H_2 + \frac{5}{4}O_2 \quad (15)$$

$$NaNO_3 \rightarrow 1/2Na_2O + NO + \frac{3}{4}O_2 \quad (16)$$

Table 5.5. Characteristic times t of molecular processes (Chapt. 4., [33]).

Molecular processes	t, s
Average time of molecular interaction (elastic impact)	$\approx 10^{-14}$
Average time of the molecular free path	$\approx 10^{-9}$
Relaxation time for molecular rotation	$\approx 10^{-9}$
Relaxation time for vibration processes	$\approx 10^{-7}$
Relaxation time for O_2 dissociation	$\approx 10^{-7}$
Average effective impact time at surface-barrier energy $E = 250$ kJ/mole	$\approx 10^{-5}$
Life time of excited metastable states	$\approx 10^{-3}$

Table 5.6. Non-thermal systems (Chapt. 4., [33]).

1. No uniform temperature T

   1) 2) 3) 4) 5)

   $T_N \neq T_e;\ T_{tr} \neq T_{red} \neq T_{osz}$
2. No Maxwell-Boltzmann distribution
3. Invalidity of the law of mass action
4. Reaction velocity constant: $\ln k = f\left(\frac{1}{T}\right)$ non linear

a) neutral-particle temperature, b) electron temperature, c) translation temperature, d) rotation temperature, e) vibration temperature.

The thermal decomposition according to equ. (14) takes its course in temperature intervals of 570 ··· 770 K [25].

Above 770 K reaction (15) dominates [26] whereas reaction (16) only becomes possible above 1,000 K [27].

In the initial stage of the fracture, when the propagation velocity of the cracks is still relatively small, the decomposition mainly appears according to equ. (14) and (15). To the degree that the fracture proceeds with higher velocity the decomposition appears according to equ. (16) [28–30].

The conditions are similar in the decomposition of bromates, see equ. (4.1) and (4.2). The reaction shown in equ. (4.2) is only probable (see Fig. 5.11.) with very high temperatures and appears as the dominating reaction during milling (Chapt. 4., [4]).

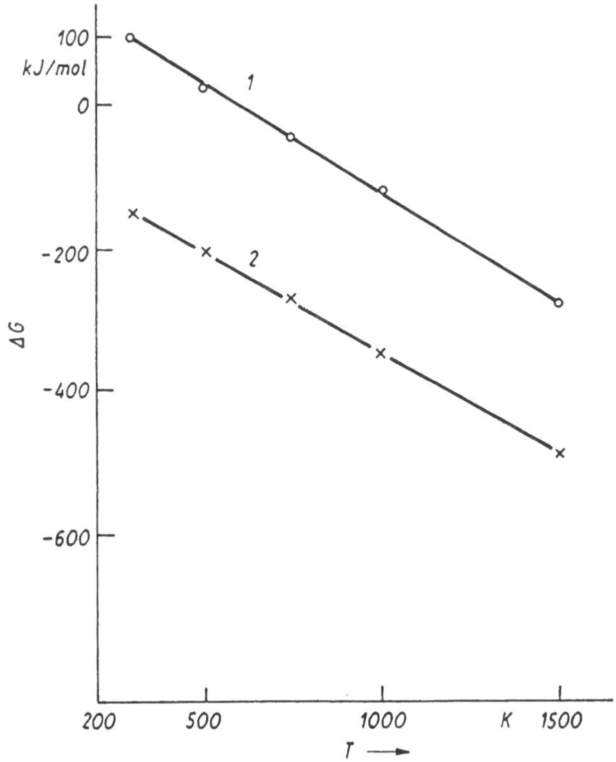

Fig 5.11. Dependence of the free reaction enthalpy ΔG on the temperature T for the reactions: 1 – $KBrO_3 \rightarrow 1/2 K_2O + 1/2 Br_2 + 5/4 O_2$, 2 – $KBrO_3 \rightarrow KBr + 3/2 O_2$, (Chapt. 4., [4]).

Particularly highly excited states in an extreme non-equilibrium state may be assumed to be induced initially during the relaxation of the elastic stresses concentrated on the fracture tip. In analogy to the model given in 4.1. these states turn into low energy ones through the distribution of energy to other vibrational degrees of freedom. If through this the MAXWELL-BOLTZMANN distribution has become established in smaller ranges of the solid these states may be characterized as hot spots.

The decomposition of nitrates and bromates definitely takes place within the network of these dissipation processes. However, due to the short lifetime of the extremely highly excited states the decomposition has a small reaction probability only. It grows with increasing dissipation and decreases again when the remaining excited states are no longer in the position to release the reaction, due to power balance. Thus the reaction probability has a maximum in the dissipation. Whether the decomposition mechanism dominating here is to a greater extent of a thermal of nonthermal nature, can be characterized in detail for instance by comparing the thermal, tribochemical, radiation-chemical, and photochemical decomposition mechanisms of bromates and nitrates.

From the kinetic point of view the tribochemical decomposition of a chemical bond represents a sequence of two processes, that is activation accompanied by formation of the excited state (A^*) and the real decomposition [31], (see equ. (4.6):

$$A \underset{k_2}{\overset{k_1}{\rightleftharpoons}} A^* \xrightarrow{k_3} B \tag{17}$$

(A, B — initial and decomposition product, k_1, k_2, k_3 — velocity constants of the activation, inactivation and decomposition processes, respectively, or of the chemical transformation.)

Provided that A^* has a steady-state concentration the velocity of such a process may be given in the following way [32]:

$$v = \frac{\bar{k}_1 \cdot \bar{k}_3}{\bar{k}_2 + \bar{k}_3} \tag{18}$$

Two borderline cases result from it:

$$1.\ \bar{k}_2 \ll \bar{k}_3 \tag{19}$$

$$\text{Then is } v = \bar{k}_1 \tag{20}$$

$$2.\ \bar{k}_2 \gg \bar{k}_3. \tag{21}$$

$$\text{Then is } v = \frac{\bar{k}_1}{k_2}\, k_3 = \bar{k}' \cdot \bar{k}_3. \tag{22}$$

Equ. (20) is valid, if the initiation process determines the velocity, such as in thermal decomposition. If equ. (22) is valid the chemical reaction is the step determining the velocity.

Photolysis and radiolysis belong to this group of reactions. This subdivision results from the following considerations:

The thermal process of breaking a chemical bond in the solid is achieved by the accumulation of the vibration energy at one bond which is thereby broken up. The vibration energy of the molecules is always much smaller than their binding energy so that the concentration of A^* is limited. Since the activation process has a collective character the activation stage has a relatively large lifetime and therefore a great reaction probability as well, so that on the whole, equ. (19) is valid.

In the case of photolysis or radiolysis the energy of the photon or of the γ-quantum is greater or equal in comparison to the binding energy so that the activation process is not the limiting factor. The highly excited states, however, relax extremely quickly so that theses processes have a small reaction probability and consequently equ. (21) and (22) apply.

In order to determine experimentally the difference in the mechanism of thermal and non-thermal initiated solid reactions the free volume V_f has proved a suitable quantity, for it influences the two types of reaction in different ways.

In thermal reactions the probability of the activation process limiting the reaction as a whole may be represented in first approximation by the probability function K_A for the appearance of a mean frequency ν_a of the normal vibration in the crystal lattice [33]:

$$K_A = \nu_a \cdot \exp(-E_A/kT) \tag{23}$$

(E_A — activation energy of the process.)

The following inequation holds for the alteration of the frequencies of the normal vibrations in the crystal in dependence on the free volume [34]:

$$\frac{\partial \nu_a}{\partial V_f} < 0. \tag{24}$$

According to this the vibration frequency ν_a in the lattice is decreased with the increasing free volume. According to equ. (23) this means a smaller probability of the activation process. Consequently the velocity of the thermal decomposition must increase if a changeover from a large to a smaller free volume appears in a number of chemical components of the same kind.

In the case of radiation-chemical decomposition the following equation is valid [35]:

$$K_s = \exp\left(-\frac{\gamma \cdot V^*}{V_f}\right) \tag{25}$$

(K_s — probability of the radiation-chemical decomposition, γ — constant, $\gamma \cdot V^*$ corresponds in first approximation to the dimensions of the decomposition product developing during the radiolysis in the primary process.)

According to equ. (25) the radiation-chemical decomposition appears with high probability in such crystal regions where the free volume exceeds a certain value. The distance of the fragments produced during the breaking of bonds increases with increasing free volume and through this the re-combination leading to a regeneration of the bond is rendered more difficult and k_2 decreased in equ. (17), respectively. Consequently the yield of radiation G_0 increases, in contrast to thermal decomposition, with growing free volume, whereby $K_s \sim G_0$. Comparing the radiation-chemical decomposition in a series of analogous compounds, e.g. in the bromates and nitrates, a decomposition of groups of the same kind, such as of the NO_3^--groups occurs in them, so that the value $\gamma \cdot V^*$ is equal for all materials of this series. As Fig. 5.12. shows the relation

$$\ln G_0 \sim -\frac{1}{V_f} \tag{26}$$

is valid according to equ. (25) in the series of compounds of the same kind.

Since the radiation-chemical and the thermal decompositions act contrary to each other in dependence on the free volume, the possibility exists to draw conclusions in respect of the character of the mechanism of tribochemical processes by classification of the tribochemical decomposition series. As an example Fig. 5.13. gives the degree

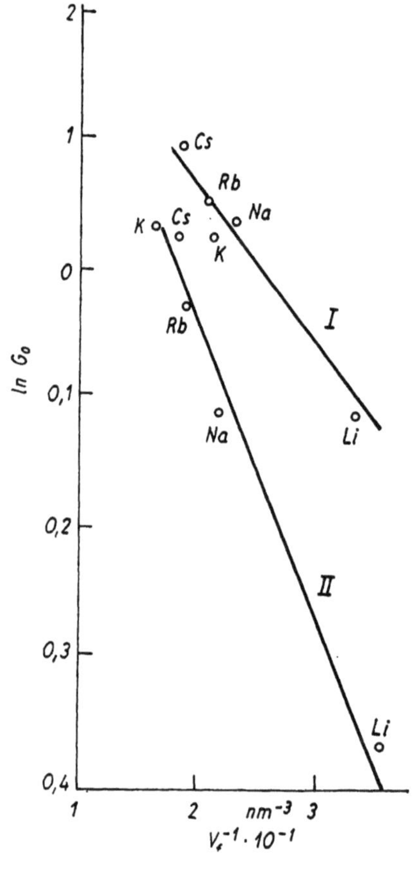

Fig. 5.12. Dependence of the X-ray output G_0 on the size of the free volume V_f for alkali bromates (I) and alkali nitrates (II) [31].

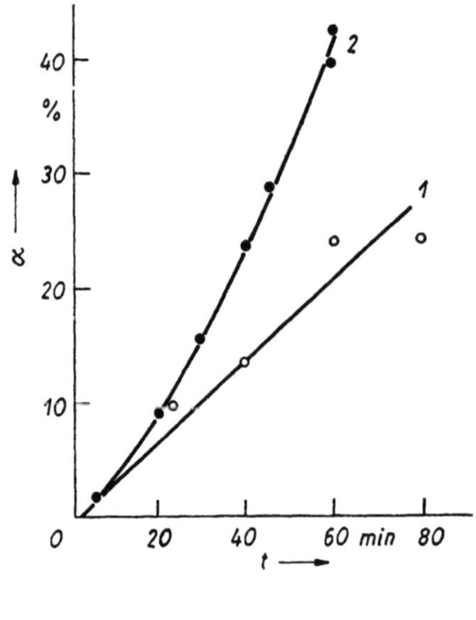

Fig. 5.13. Degree of decomposition α of alkali bromates in dependence on the treatment time *t* (Chapt. 4., [4]).

1 — $NaBrO_3$, 2 — $CsBrO_3$.

of decomposition of Na and Cs bromates in dependence on the treatment time. As in radiation-chemical decomposition, the stability of bromates increases under the influence of mechanical energy, with decreasing atomic weight of the alkali metals. The decomposition series for nitrates and bromates in dependence on the nature of energy influence are summaryed in Tab. 5.7.

Table 5.7. Comparison of thermal, radiation-chemical, photochemical, and tribochemical stability of alkali bromates and alkali nitrates [22, 32].

Substance	Free volume $V_f \cdot 10^2$, nm^3	Temperature T of the start of the thermal decomposition K	Yield			
			Radiolysis molecules/ 100 eV	Photolysis quantum yield, $\eta_F \cdot 10^3$	Tribolysis	
					degree of decomposition η_{tr} at milling, per cent	η_{tr} in the impact wave per cent
$NaBrO_3$	4.3	616	1.6		10	
$KBrO_3$	4.6	663	1.8		14	
$RbBrO_3$	4.8	679	2.4		70	
$CsBrO_3$	5.2	639	3.3		51	
$NaNO_3$	4.4	743	0.37	0.44	0.006	0.43
KNO_3	5.5	766	1.96	3.00	0.616	0.46
$RbNO_3$	5.1	713	0.80	4.00	2.500	1.00
$CsNO_3$	5.3	800	1.37	6.00	0.093	0.55

From this may be drawn the following conclusions:

1. The radiation-chemical, photochemical, and tribochemical decomposition of bromates and nitrates varies almost similarly to each other in the individual homologous series so that by character the tribochemical decomposition comes close to such reactions not limited by the initial stage. The photochemical decomposition also belongs to them. Fundamental differences in character of reaction do not appear under the influence of milling and impact wave stresses. Consequently in both the cases the high-energy vibration states are dominant for the release of the reaction.

2. Accordingly the series of tribochemical and thermal stability do not coincide.
   An analogous behaviour between radiation-chemical and tribochemical decomposition also appears in hydrocarbons such as hexane (tribochemical) and n-heptane (radiation-chemical), respectively. As Fig. 5.14. shows the relation of the decomposition products developing in both cases is nearly identical (Chapt. 4., [10]).

The formation of hydrocarbons from the elements also does not occur according to a simple hot-spot mechanism, as could be observed on the basis of investigations of isotope effect [36]. As shown in Tab. 5.8., a fractionation factor of 1.009 results for the tribochemical reaction:

$$C + 2H_2 \rightarrow CH_4 \quad (27)$$

Table 5.8. Appearance of an isotopic effect in the tribochemical formation of methane from the elements [36].

Substance	Isotopic relation* C^{13}, ‰
Graphite (raw material)	— 25.7
Methane (reaction product)	— 34.5

*) Values related to PDB-Standard.

On the basis of the isotopic effect found [37], the appearance of temperatures higher than 700···800 K can be excluded as decisive factors for the initiation of the reaction. In the given range of temperature, the formation of methane is negligibly small without any influence of mechanical energy. Very probably the breaking of bonds pro-

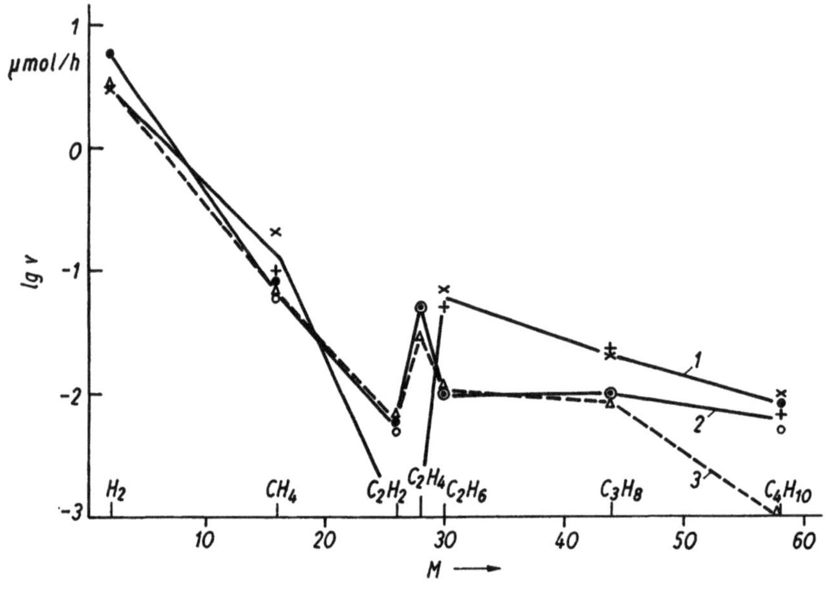

Fig. 5.14. Velocity v of the formation of low hydrocarbons in dependence on the molecular weight M in the tribochemical (1), (2) and radiation-chemical (3) decomposition of paraffins (Chapt. 4., [10]).
1 — n-C_6H_{14}/SiC, 2 — n-C_6H_{14} + O_2/SiC, 3 — n-C_7H_{16} (relative values).

ceeds through the action of highly excited non-equilibrium states produced in the stress relaxation. The bonds of lighter isotopes to which higher frequency vibrations are attributed, are most probably broken up under the influence of these highly excited non-equilibrium states.

Another possibility for obtaining information on the character of the decomposition mechanism consists in studying the effect of additives on the decomposition process. Thus the thermal, radiation-chemical, and tribochemical decomposition of silver oxalate can be variously influenced by the addition of copper ions [38]. The thermal decomposition of silver oxalate takes place via the breaking of the C—C bond in the oxalate molecule accompanied by the formation of CO_2^- radicals, and via the release of the electron to an adjacent interstitial Ag^+ ion. Thus gaseous CO_2 and an uncharged Ag atom develop. Accordingly the probability of thermal decomposition depends on the concentration of the interstitial Ag^+ ions in the oxalate lattice. By the addition of Cu^{2+} ions the concentration of the interstitial Ag^+ ions is reduced and consequently the thermal stability of silver oxalate is increased [39, 40].

However, during photochemical and radiation-chemical decomposition, respectively, because of the higher energy supply, a transport of electrons takes place in the lattice of the solid over a distance which is much larger than the interatomic distances. Consequently a delocalization of the electron transfer comes into being with the Cu^{2+} acting as acceptor of the electrons and the photochemical and radiation-chemical decomposition being accelerated with an increasing addition of Cu^{2+} ions [32, 41].

As Tab. 5.9. shows, the addition of an increasing amount of Cu^{2+} ions in the silver oxalate brings about an increase of the tribochemical decomposition, which is also indicative of a decomposition mechanism analogous to the radiation-chemical and photochemical decomposition.

Table 5.9. Effect of the Cu^{2+} doping on the tribochemical decomposition of silver oxalate [38].

Sample	Degree of decomposition, per cent after comparable treatment time)
$Ag_2C_2O_4$	14
$Ag_2C_2O_4$ + 0.2 mol-% Cu^{2+}	23.2
$Ag_2C_2O_4$ + 0.85 mol-% Cu^{2+}	25.5
$Ag_2C_2O_4$ + 2.5 mol-% Cu^{2+}	25.6
$Ag_2C_2O_4$ + 4.6 mol-% Cu^{2+}	26.9

The non-thermal decomposition mechanism has also been confirmed for tribochemical nitrate decomposition in respect of the dependence of the decomposition velocity on the addition of certain substances which act as electron acceptors. Thus it has been shown that there is respectively a linear correlation between the electronic work function of the additive and the variation of the decomposition velocity or the latter increases with a growing acceptor effect of the additives (see Fig. 5.15.). Donors, however, inhibit the process. The decomposition process can be represented by the following equation:

$$NO_3^- \underset{k_2}{\overset{k_1}{\rightleftharpoons}} NO_3^\circ + e^- \xrightarrow{k_3} NO + O_2 \tag{27}$$

The transition into a highly excited state and the delocalization of the electron are the dominating mechanisms for the tribochemical decomposition, and this also accords with the triboplasma model.

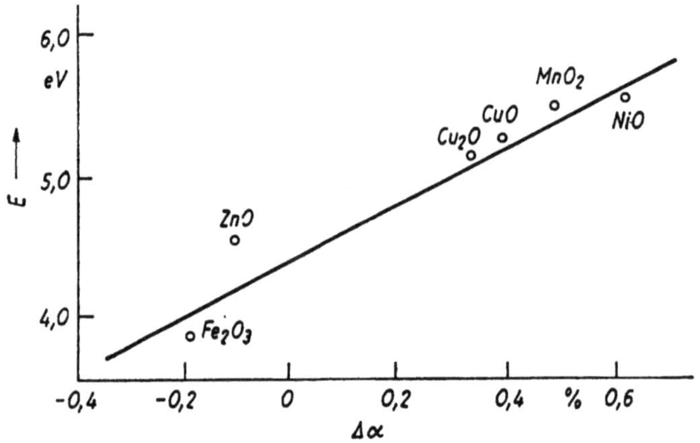

Fig. 5.15. Changes in the degree of tribochemical decomposition of doped sodium nitrate in dependence on the work function E of oxidic dopings [32].

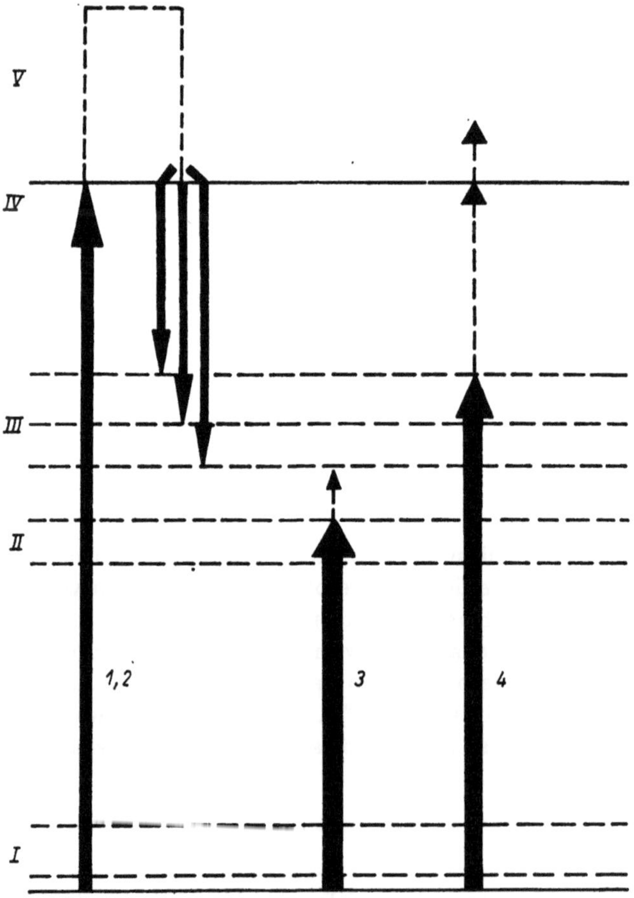

Fig. 5.16. Schematic drawing of several energy effects upon a solid in an energetic diagram [29].
1 — photolysis, 2 — radiolysis, 3 — thermolysis, 4 — tribolysis. (I) — translation energy level, (II) — rotation energy level, (III) — vibration energy level, (IV) — excited electron level, (V) — free electron.

There is no correlation between electronic properties of oxidic additives and the nature of their effect during thermal decomposition. This correlation cannot be expected since the excitation of the electrons only appears with thermal stress in low-energy electron states and consequently no wide-reaching effect can occur with respect to the heterophase additives.

In the radiation-chemical decomposition of nitrates the dependence of the reaction velocity on the work function behaves conversely in comparison to that observed during tribolysis. Here donors accelerate the reaction velocity. This behaviour, which deviates from tribolysis is induced by the fact that in the given case the radiation-chemical decomposition takes place according to another mechanism, namely according to the dissociative attachment of the electron. Here the stages of electron excitation and detachment of nitrate ion are not decisive, rather the secondary processes following the attachment of an electron to the intermediate product [32]. This example shows that radiation-chemical and tribochemical processes cannot be attributed to the same mechanism in even case and that specific differences exist in the nature of initiating reactions. This also follows from the scheme in Fig. 5.16.

The excitation of the high-energy electron levels up to the detachment of free electrons takes place first in radiolysis and photolysis. Only after this does the excitation of the vibration states and the breaking of bonds take place. In thermal decomposition the excitation appears via vibration states by continuously and increasingly initiating the rotation and vibration degrees of freedom.

In tribochemical decomposition as well the process mainly occurs through the excitation of vibration states, but unlike the thermal breaking here, those high energy vibration states are excited which lead at once up to delocalisation and under certain conditions as shown in 3.3., to the release of free electrons.

On the one hand according to 5.2., a spontaneous decomposition of bonds takes place influenced by high stress peaks. On the other hand the initiation of tribochemical reactions according to the considerations at the beginning of this section is decisively caused by thermal states as they exist in the hot spots. Whereas in the latter case the concept of temperature (e.g. lightning temperature) can be applied, the term temperature makes no sense in very short-term excitation states ($< 10^{-7}$ s). Therefore the term "equivalent temperature" is used for these states. Consequently, unlike the other non-thermal and thermal types of initiation, the mechanically induced initiation of chemical reactions is of complex nature. Depending on the specific mechanical initiation and the respective reaction in question, several mechanisms are possible in the network of the excitation states, which control the tribochemical reactions either individually, alongside each other, or successively. As a whole the main characteristic of tribochemical reactions consists in the fact that they run through highly-excited extreme non-equilibrium processes.

5.4. Chemical effects of triboelectric processes

5.4.1. On supercharging disperse substances in milling equipment

As has been stated in 3.3., the effect of mechanical energy on solids also changes their electric and electronic properties. Thus, contact, frictional stress, cohesion fractures as well as the breaking up of adhesion bonds, result in charging, and also, dependent on

Table 5.10. Electrostatic charging Q and charge sign at different solids due to the activation in the disintegrator [42].

Solid	SiO_2	TiO_2	Fe_2O_3	BaO	$CaCO_3$	Na_2CO_3	NaCl	LiF	Polyethylene
Charging $Q \cdot 10^7$, C/g	38	28	64	7	23	10	50	203	21
Charge sign	−	−	−	+	−	+	+	+	−

external conditions, in discharge processes and in the emission of electrons. In milling equipment such as in vibration mills and in disintegrators, the charging of the grain material can also be attributed to several causes overlapping in different ways according to the treatment conditions.

As Tab. 5.10. shows that in practice a charging of the milling products must always be taken into account if the latter have a low conductivity. The kind of sign and the

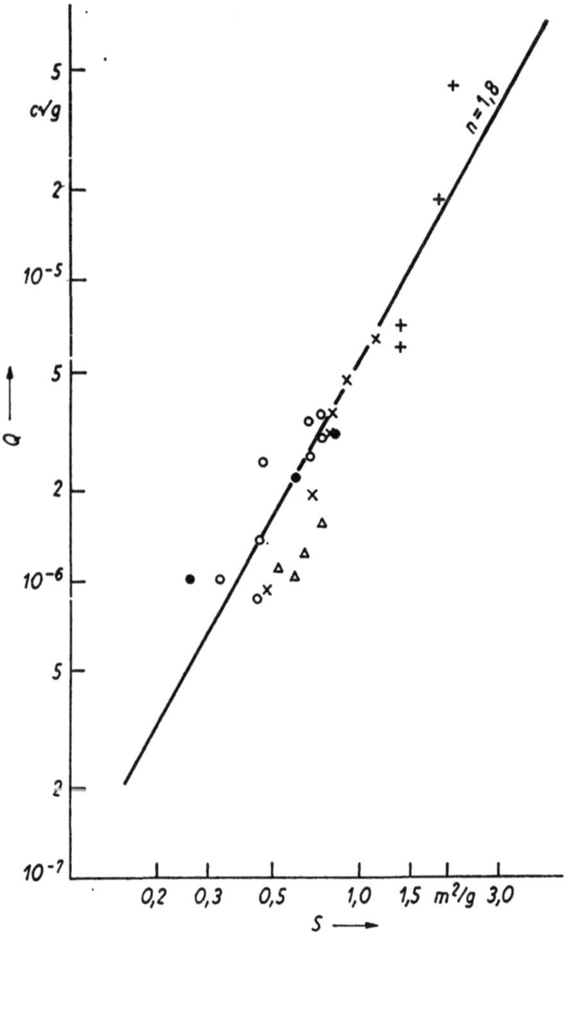

Fig. 5.17. Dependence of the electrostatic charging Q on the specific surface area S at SiO_2 samples activated in a disintegrator [44].

yield of charging are material-specific quantities under comparable stress conditions [43]. However, the charging is also determined by the treatment intensity and other parameters [44]. Thus, for instance, an increase of both the surface area and charging appears with an increasing rotational speed of the disintegrator. As given in Fig. 5.17. the following relationship exists between these two quantities:

$$Q \sim S^{1.8} \tag{28}$$

(Q — charging, S — specific surface area.)

The charging per unit surface area of the powder particles increases with increasing dispersity of the grain material, and consequently becomes a quantity influencing decisively the disperse material properties, e.g. agglomeration [45], bulk material properties [46], adhesion [47], and so on. Furthermore discharge processes arise which may act as initiators in the release of fires and explosions whilst milling inflammable substances such as coal, grain and plant-protecting agents [48]. The influence of mechanically induced charges, discharges ,as well as the emission of electrons on solid reactivity will be dealt with separately in the following.

5.4.2. *The influence of tribo-induced electrostatic charges on chemical processes*

Particularly polymerization processes on solid surfaces have been used for the evaluation of the influence of tribo-induced electric charges on chemical reactions [49 to 53]. It has been found that polymerization reactions are initiated under the influence of extremely high electric fields having field strengths from 10^7 up to some 10^8 V/cm. These processes, known as field polymerization, proceed mainly via a cationic mechanism. If radical binding states are initiated by field ionization a radical mechanism is also observed.

In order to obtain the described field strengths the application of an external electric field is required [54]. Mechanically induced charge concentrations at the solid surface leading to such high electric fields have not been ascertained by means of mainly macroscopic methods. Investigations of the tribo-induced formation of ERS-centres have led to the estimate that electric fields up to 10^7 V/cm occur in the atomic range [55].

A larger number of solids such as LiF, SiO_2, Fe_2O_3, ZnO, graphite, $BaSO_4$, $CaCO_3$, CaF_2, NaCl [56], slate [57], Al_2O_3 [49], BaO, BaS, ZnS [53], etc. possess the capacity under the influence of mechanical stress to initiate the polymerization of styrene, acrylonitrile, methylmethacrylate, and other monomers, whereas no graft process could be observed at the mentioned unstressed solids. Researches on the mechanism of graft polymerization have been carried out in particular for SiO_2 [52] and for LiF and NaCl cleavage surfaces [50, 51, 58]. As an example an LiF cleavage surface treated with styrene is shown in Fig. 5.18.

The graft polymer segregates in the form of certain growth structures on the cleavage surface with a molecular weight of $\approx$ 40,000 [59]. Predominantly a disarranged mosaic structure develops, but also a growth orientation by cleavage steps [42]. In anisotropic crystals such as slate, oriented structures are obtained as well [57]. In the case of SiO_2 a cationic polymerization at positively charged defects [52] and in the case of alkali halides in earlier publications an anionic polymerization at negatively charged surface regions [58] are discussed as mechanisms. Since in both cases scavengers have no influence on the polymerization processes a radical polymerization dominating the reaction is excluded. If in the case of alcali halides monomers like α-methyl styrene which do not polymerize anionically, are applied, then only small exchanges are obtained in comparison to the reaction with styrene [49]. Important differences in the

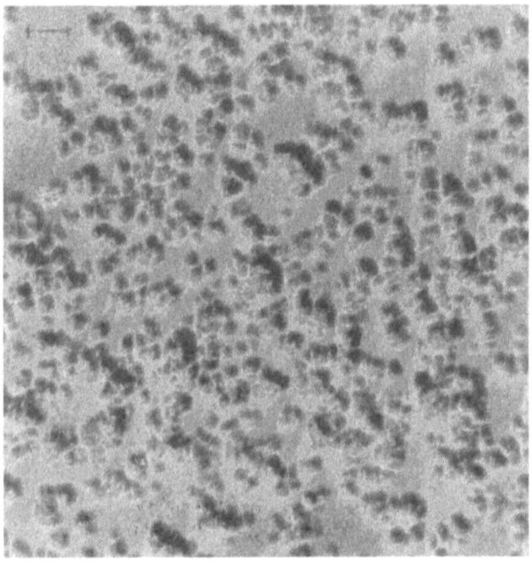

Fig. 5.18. LiF cleavage surface grafted by styrene [42]. ├─┤ – 1 μm.

degree of grafting are measured at the mirror-inverted LiF cleavage surfaces if they differ in the integral charging value. Thus the degree of grafting is sometimes larger at the overall more negatively charged cleavage surface than the one measured at the

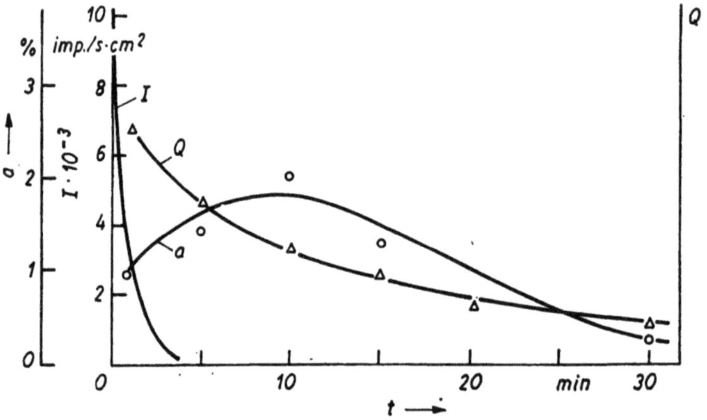

Fig. 5.19. Degree of grafting a, electron emission *I*, and electrostatic charging *Q* on a LiF cleavage surface in dependence on the time *t* after the cleavage [50].

corresponding positively charged cleavage surface [58]. However, recently by more detailed measurements of the distribution of electrostatic charge at LiF cleavage surfaces it was shown that graft polymerization of styrene takes place predominantly at positively charged surface areas. From this is derived a cationic polymerization mechanism [42].

As Fig. 5.19. shows there is a connection in terms of relaxation behaviour between the charging of the cleavage surface and the degree of grafting. However, the mechano-emission of free electrons frequently discussed as the cause of grafting [60] fades out far more rapidly. Since grafting also takes place after the decay of the electron emission this most probably has to be attributed to the electric states having a long lifetime.

Table 5.11. Surface, graft reaction, electrostatic charging, and electron emission of two kinds of slate after mechanical treatment [57].

Kind of slate	Spec. surface area S, m_2/g		Homo-polymer	Grafted polymer	Electro-static charging	Electron emission
	Raw sample	sample milled for 30 min	per cent	per cent	C/cm^2	Imp/s
Green slate	5.5	22	6.6	1.05	$7 \cdot 10^{-7}$	$8 \cdot 10^2$
Gray slate	1.8	16	4.6	0.09	$4 \cdot 10^{-7}$	$8 \cdot 10^1$

In Tab. 5.11. the emission of electrons, the chargings, and the degree of grafting are represented for 2 different samples of slate. In first approximation the different degree of grafting can be related to the varying electric behaviour of the two samples.

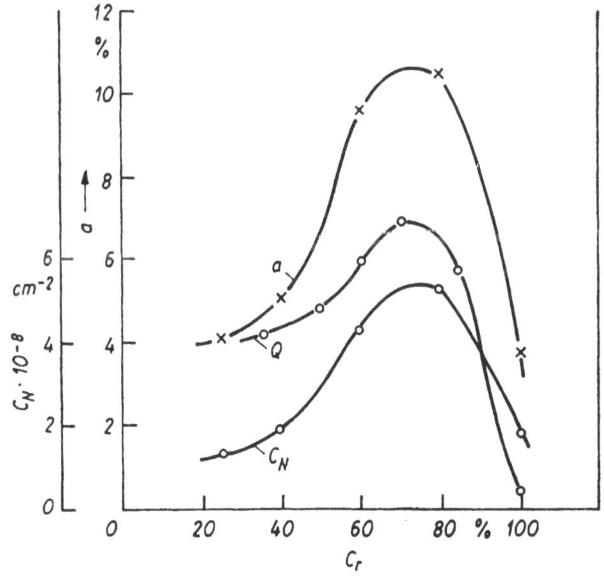

Fig. 5.20. Dependence of the concentration of the grafted centres τ_N, of the degree of grafting a (2), and of the electric charging Q (3) at a NaCl cleavage surface on the relative vapour pressure τ_r of argon [51].

On the basis of the time behaviour of the chemical and electric quantities shown in Fig. 5.19. it can be assumed that in this system the chargings similarly cause the grafting and also in the same way control the emission of electrons.

A close correlation between the charging and grafting of alkali halide cleavage surfaces was also found to exist whilst investigating the influence of water on the degree of grafting. As Fig. 5.20. indicates the mentioned quantities show an analogous behaviour compared to the change in the water concentration. After a longer milling stress, e.g. of slate powder in a vibration mill in the presence of butadiene as monomer, the above processes add up through the continuous formation of fresh surfaces so that polymer contents up to 35 per cent appear in the grinding material (Fig. 5.21)., [61].

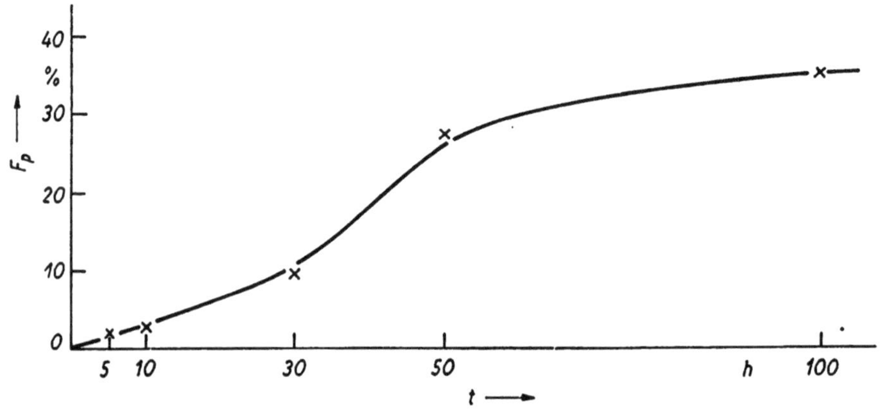

Fig. 5.21. Content of polymers F_p in a slate sample in dependence on the time t of the vibration milling in a butadiene atmosphere [61].

The charging of solids may effect their reactivity in other chemical processes as well, e.g. in dissolution processes of radiation-damaged solids. Thus various defects may occur in the solid under the influence of radiation leading to the development of polarized molecule dipoles and space charges, respectively. Through this additional work is required to overcome the electrostatic interactions during the transition of the reactant from the charged sample into the liquid phase. Altogether this causes a slowing down of the dissolution process [62].

In two phase systems such as inorganic-organic bonds, the strength of the structure relative to compression or to tension is to a high degree determined by the interation forces between the different phases and the adhesion forces at the interfaces, respectively. It has been shown that above all the electric nature of the surface determine the adhesion in a great number of cases as compared to the effect of homopolar bonds and van der Waals forces. By virtue of the various electric properties of substances, an electric double layer having the strength F_A develops in the formation of an adhesion bond, which is calculable from the following equation [47]:

$$F_A = 2\pi \cdot \sigma^2 \cdot S_k \qquad (29)$$

(σ — specific charge density in the double layer, S_k — contact area.)

The donor-acceptor properties of the individual substances are mainly determined by the functional groups at the surface [63]. If their concentration is altered by mechanical treatment, or by discharge, then a change of strength very similar in course arises from it (see Fig. 5.22.). Research in this field has been considerably advanced by the development of an electro-adhesion theory [47].

In summary the investigations carried out on charged solids indicate that certain chemical reactions are caused by the charge state of the solid, in which graft polymerization processes at charged surfaces play a dominating role.

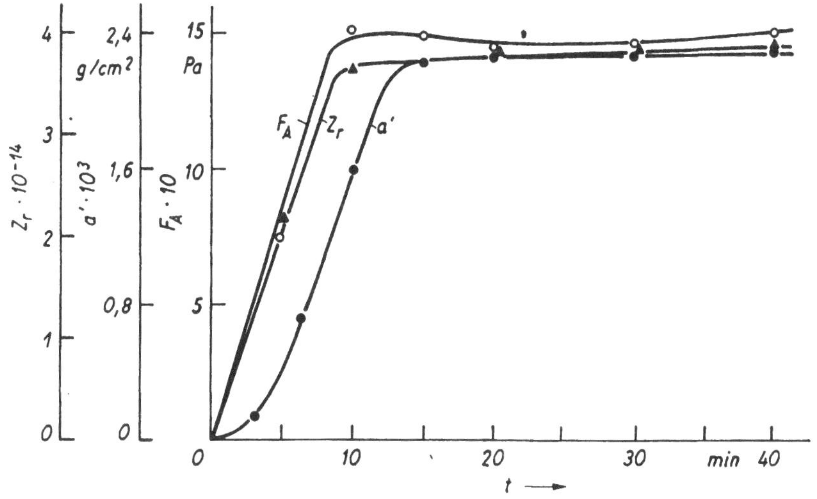

Fig. 5.22. Increase in the adhesion strength F_A, in the specific grafted amount a', and in the number of radicals Z_R as function of the time t of the action of an electrical discharge upon a PTFE surface [64].

5.4.3. *The release of chemical reactions by tribo-induced discharges*

Discharges influenced by mechanical treatment have their origin in the tribo-induced charging of solid surfaces, whereby within certain limits the intensity of the discharge increases very similarly to the field strength of the system [65]. Usually the treated solid particles with different chargings form the electrodes between which an ignition of the gaseous discharge takes place (see 3.3.).

Through the discharge, initiated particles develop which are for the most part in an ionized or radical state, as well as electrons and photons, which lead to the known tribolumescence. By this chemical reactions can also be released in the immediate vicinity of the solid surface, which require a high activation energy for their running. Since the micro-discharge time is $10^{-7} \cdots 10^{-8}$ s [66] the micro-discharges have characteristics very similar to those of the non-thermal plasma of macroscopic gaseous discharges in terms of the initiation of chemical reactions. They are characterized by the criteria given in Tab. 5.6., with the free impact electrons having a decisive importance for the release of chemical reactions. The great similarity of the processes in triboplasma and in non-thermal discharge plasma can be seen for instance by the reaction of the wetted atmospheric nitrogen with magneticbasalt mixtures treated mechanically and stressed by electric spark discharge, resulting in nitrous acid, nitric acid, ammonia, hydrocyanic acid, and hydroxylamine. The common feature of both processes consists mainly in their stochastic nature [67—69].

The influence of tribo-induced gaseous discharges on the release of chemical reactions has been thoroughly studied in the intermittent contact of mercury and glass surfaces [70]. The high charge densities develop by electron transfer from mercury to glass and subsequently a discharge appears having particle activation energies up to 20 eV. Under these conditions higher hydrocarbons are formed in analogy to the milling of corundum in a methane atmosphere [71].

Table 5.12. Relative yields of ethane, ethylene, and acethylene by the decomposition of methane with different energy effects.

Process	Gas pressure, Pa		Relative yield			Ref.
	Methane	Noble gas	Ethane*) per cent	Ethylene per cent	Acetylene per cent	
Triboelectric discharge	$2.4 \cdot 10^4$	Ar $2.6 \cdot 10^3$	100	34	36	[70]
	$2.4 \cdot 10^4$	Kr $2.6 \cdot 10^3$	100	39	38	[70]
	$1.3 \cdot 10^4$	Ar $1.3 \cdot 10^4$	100	42	59	[70]
	$1.3 \cdot 10^4$	Kr $1.3 \cdot 10^4$	100	37	31	[70]
600 V/cm Glow discharge	$9 \cdot 10^2$	— —	100	83	76	[72]
electrons Effect of 15-eV	1.3	— —	100	49	46	[73]

*) Ethane = 100 per cent.

As Tab. 5.12. shows both saturated and unsaturated hydrocarbons come into existence, whose percentage varies slightly in dependence on the gas pressure and the nature of the applied inert gas. On one hand the discharge energy increases up to a range of $10^3 \cdots 10^4$ Pa with falling pressure resulting in a higher reaction yield. On the other hand the velocity of the reaction increases continuously with growing pressure. Both effects overlap and lead to a marked maximum in the pressure range of 3×10^4 Pa (see Fig. 5.23.).

As can also be seen in Tab. 5.12., the percentages of the developing C_2-hydrocarbons are similar to those which can be obtained in a glow discharge [72] and during the bombardement of methane by 15 eV electrons [73], respectively, indicating the uniformity of the excitation mechanisms.

The formation of a CH_2 radical as an elementary step is discussed pursuant to the following equation [70]:

$$CH_4 \xrightarrow{e} CH_4^* \rightarrow CH_2 + H_2 \quad (30)$$

$$CH_4 + e \rightarrow CH_4^* \rightarrow CH_2 + H_2 \quad (31)$$

The CH_2 radical then reacts further, for instance with CH_4, and forms C_2-hydrocarbons:

$$CH_2 + CH_4 \rightarrow C_2H_6^* \begin{cases} \rightarrow C_2H_4 + H_2 \\ \xrightarrow{M} C_2H_6 \end{cases} \quad (32)$$

(M — 3rd participant in the impact.)

The excited elementary particles alter the imperfect structure of the solid and consequently its reactivity. Reactive excited centres of radical and ionic nature develop within the interface layers of SiO_2 [74, 75]. In organic substances, such as PTFE, mainly reactive macro-radicals are formed which develop into stable peroxidic radicals in the

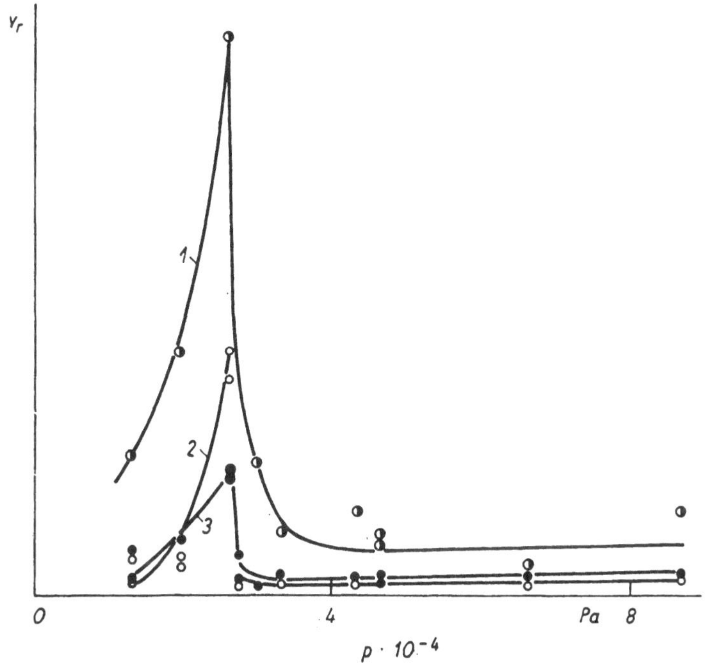

Fig. 5.23. Relative velocity v_r of the formation of ethane (1), ethylene (2), and acetylen (3) from methane by the effect of triboelectrical discharges in dependence on the methane pressure [70].

presence of oxygen and have the capacity for grafting [64]. As Fig. 5.22. shows the number of radicals (3) and the degree of grafting (2) increase very similarly as a function of the discharge time, up to a treshold value. According to quantitative analysis one graft chain is initiated per degrading radical.

Subsequent chemical reactions may also be accelerated through the interfacial distortion by means of discharges such as the reduction of oxidic surface layers on copper. This leads to a respective reduction of the initiation temperature and activation energy of the reaction [76].

The importance of tribo-induced discharge processes for the ignition of explosions in the most varied engineering installations was already pointed out at the beginning (4.8.).

5.4.4. *The release of chemical reactions by fast electrons*

If a charge seperation appears during the breaking or destruction of adhesion bonds in a vacuum with $< 10^{-3}$ Pa, electrons having high energy are then emitted-the so-called mechano-electrons. Here too, the reason for emission are the high electric fields

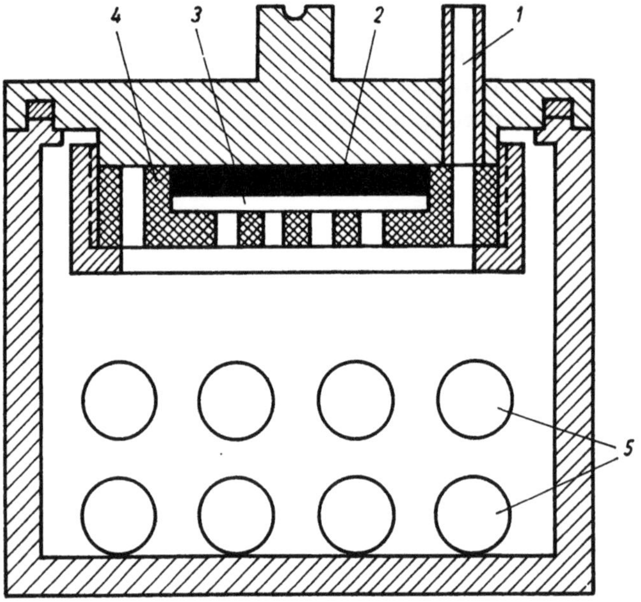

Fig. 5.24. Schematic view of a triboreactor for the detection of the chemical effect of fast electrons [77].

1 — connection, 2 — light-sensitive plate or film made of acrylic acid, 3 — protective filter, 4 — teflon screen, 5 — milling balls.

which occur as a result of stress on the newly built surfaces and which accelerate the electrons up to energies of 100 KeV (see 3.3.). Thus, for instance, the appearance of fast electrons may be proved by blackening a film plate in a vibration mill during the milling of quartz, mica, and polymers in high vacuum (Fig. 5.24.), [77, 78]. The emission of fast electrons may be always reckoned with in the milling of polymers, if the destruc-

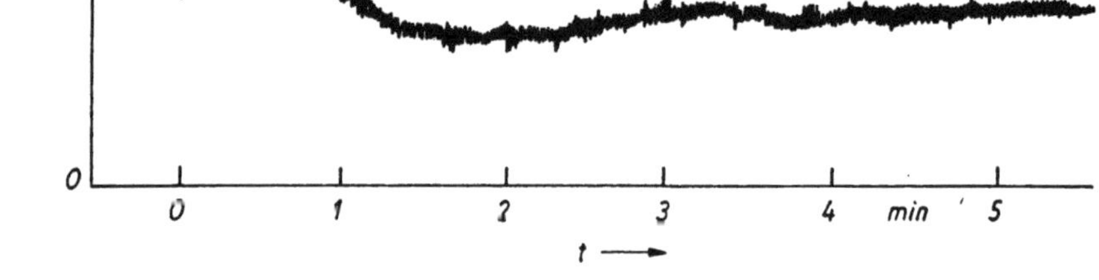

Fig. 5.25. Intensity I of the emission of mechanoelectrons during the destruction of the adhesive bonds existing between the PTFE film treated in a glow discharge and epoxy resin [63].

T — time after the destruction of the adhesive bond.

tion is accompanied by the separation of adhesion bonds, e.g. in the milling of filled or mixed polymers [79—81]. The chemical effect of fast electrons has been investigated particularly in polymers. If, for instance, the film in Fig. 24 is replaced by an acrylic acid layer the free electrons emitted during the milling then release the polymerization to polyacrylic acid [77]. A correlation between intensity of emission and graft yield could be proved [82]. A newly-pulled off adhesion film, e.g. a PTFE film on epoxide resin, also shows a constant emission according to Fig. 5.25. for a certain time, causing graft-active radicals at the surface of an unstressed inactive PTFE sample [63]. The scheme of the process thus taking place is represented in the following equations [66]:

$$\sim CF_2-CF_2-CF_3 \xrightarrow{e} \sim CF_2-\dot{C}F-CF_3 + F \qquad (33a)$$

$$\sim CF_2-CF_2-CF_3 \xrightarrow{e} \sim CF_2-CF_2-\dot{C}F_2 + F \qquad (33b)$$

In the presence of methylcrylate the following graft reaction takes place:

$$\sim CF_2-CF_2-\dot{C}F_2 + \left[\begin{array}{l} CH_2{=}CH \\ \quad\;\; | \\ \quad\;\; COOCH_3 \end{array}\right]_n \longrightarrow$$

$$\sim CF_2-CF_2-CF_2-\left[\begin{array}{l} -CH_2-CH- \\ \qquad\quad | \\ \qquad\quad COOCH_3 \end{array}\right]_{n-1} \begin{array}{l} -CH_2-\dot{C}H \\ \qquad\quad | \\ \qquad\quad COOCH_3 \end{array} \qquad (34)$$

The effect of fast electrons has also been investigated in other systems, e.g. in grafting acrylonitrile to cellulose ester, which runs pursuant to the following equation [83]:

$$\begin{array}{l} \vdash O^{\cdot} \\ \vdash OH \\ \vdash OH \end{array} \begin{array}{c} CH_2{=}CH \\ + \quad\;\; | \quad \\ \quad\;\; CN \end{array} \longrightarrow \begin{array}{l} \vdash O-CH_2-\dot{C}H \\ \vdash OH \qquad\quad CN \\ \vdash OH \end{array} \qquad (35)$$

By the bombardement of compact silver at 77 K with electrons having an energy of $1 \cdots 10^4$ eV, the bombarded silver surface has proved to be an intensive oxygen adsorber reaching an efficiency of 0.1 molecule/electron [84].

Under the influence of fast electrons, chemical reactions usually run in analogy to radiation-chemical reactions. However, the energy of fast electrons is smaller than that of particles used in radiation-chemistry. Therefore mechano-electrons penetrate only the solid top layer and are especially suitable for the activation of surface-chemical reactions.

A polymerization also occurs by the vibration milling of metals, e.g. of Ca, Mg, Al, Fe, Cr, Bi, in the presence of monomers such as methyl methacrylate and styrene. It has been shown that the polymerization activity of the vibration-milled metals grows proportionally to their electronic work function. The initiation of the reaction has been interpreted by an electron transmission [85]. Since the polymerization runs according to an anionic mechanism the electronically founded excitation process is closer to the reaction for initiating polymerization discussed in 5.4.2. than to the processes running by the effect of high-energy particles. Evidently the electron release is the process limiting the reaction velocity.

In summary, the studies discussed in 5.4. show that triboelectrically induced chemical reactions are founded on various mechanisms. Whereas the charge carriers bound to the solid act as a reaction release via a field effect, with free charge carriers, the activation energy required for the run of chemical reactions is mainly provided through impact processes leading to the excitation and breaking-up, respectively, of chemical bonds.

5.5. The dependence of tribochemical reactions on the surface area

5.5.1. The importance of the surface area for the solid reactivity

Already more than 200 years ago the conclusion was formulated [86] that in heterogeneous systems with reactions between a solid reactant and another solid, liquid, or gaseous reactant the velocity of the reaction is not proportional to the amount of solids, but proportional to the surface area of these substances. Therefore solid reactions are generally performed using disperse substances.

The influence of the surface size on the solid reactivity can be seen particularly clearly in a series of metals such as iron and nickel, which ignite themselves or take on pyrophoric properties under the influence of air when exceeding a systemdependent critical degree of dispersion [87].

Especially high degrees of solid dispersion are above all obtained chemically, e.g. by decomposition or precipitation (Chapt. 4., [126]). The main method of obtaining solids with large surface areas is to release fracture and deformation processes by the action of mechanical energy. As discussed in 3.1. this process step plays an important part in technology, because only the comminution of the applied raw materials enables a sufficiently high reaction from the economic point of view in nearly all chemical preparation techniques such as in the processing of ores to metals, in the production and working-up of binding agents for the building industry, in mining of raw phosphates and their further processing to fertilizers (see 7.2.2.2.4.), and so on. Because of its great technological importance the field of mechanical comminution science has developed to a branch of research in its own right. Since this field also has a great importance for tribochemistry, some important aspects of the surface alteration caused by milling and the changes in reactivity resulting from it, will be discussed below.

5.5.2. Several kinds of solid surfaces

The surface size of a perfect monocrystal results from the geometric dimensions of the external boundary surfaces. Imperfect solids, e.g. those disturbed by treatment, however consist of a variety of primary crystallites, that is of coherently diverging regions, and are described as secondary particles. These are often supplied with pores. Three different kinds of surfaces are distinguished for such an imperfect crystal, and these are schematically shown in Fig. 5.26., [88]:

1. The exterior surface area O_A as the outer boundary surface of a solid secondary particle.
2. The inner surface area O_I formed of open pores.
3. The surface of primary crystallites O_P [89].

The BET surface O_{BET} frequently used in literature covers the sum of outer and inner surfaces $O_I + O_A = O_{BET}$ [88].

Therefore generally valid is:

$$O_A \leqq O_{BET} \leqq O_P \tag{36}$$

If fractures and plastic deformation processes are released additionally to the elastic deformation of the solid with the effect of mechanical energy, simultaneously the surface of the stressed solid is permanently changed. Direction and course of the alteration

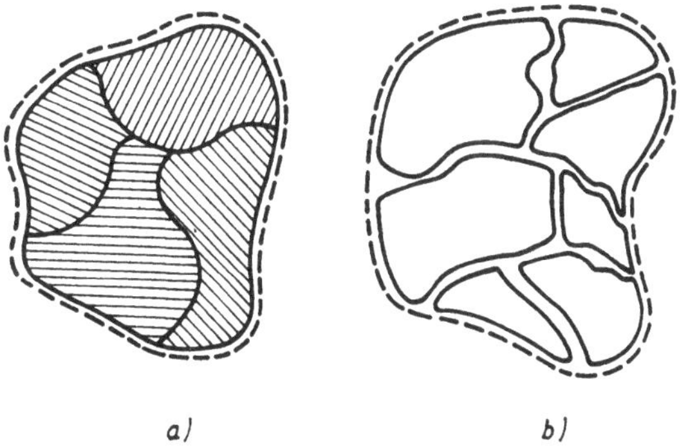

Fig. 5.26. Scheme of various kinds of surfaces.

a – non-porous grain, —— – inner grain boundary, - - - – outer grain boundary, surface along the inner and outer grain bondaries = total surface of the primary crystallites O_p; b – porous grain, —— – inner surface O_I, - - - – outer surface O_A.

depend essentially on the nature of the solid, its initial state, the kind, intensity, and duration of stress, and the surrounding atmosphere. The variation of the 3 mentioned types of surfaces may run according to its own laws in this case.

5.5.3. *Change in primary crystallite sizes by mechanical treatment and its importance for the solid reactivity*

Particularly comprehensive conditions exist in the triboinduced alteration of the primary crystallite size and surface area. As Tab. 5.13. shows, a decrease of primary crystallites takes place with the mechanical stress of the most different substances, e.g. in vibration mills.

After a longer period of milling a final state comes into being which depends both on the initial state of the substance and on the effect of exterior factors, such as kind of treatment, atmosphere, etc. The typical course of changes in primary crystallites is represented for different substances in Fig. 5.27.

With the exception of the NaCl curve the curves of the remaining substances drop continuously and run asymptotically towards a straight line. The following equation was developed to describe the course of the curve (Chapt. 4., [213]):

$$\frac{dr}{dt} = -k_0 \cdot e^{-\frac{U_B}{U_M}} \cdot (r - r_g) + k_r \cdot e^{-\frac{U_A}{\sigma_G \cdot S + \frac{E}{2}\left(\frac{\Delta a}{a}\right)^2 + U_M}} \tag{37}$$

Table 5.13. Changes in the primary crystallite size of several substances by vibration milling. ν_A — initial primary crystallite size, ν_g — final primary crystallite size.

Substance	SiO_2	SiC	Kaolinite	Cement	Chromite
ν_A, nm	800	215	10 ··· 22	115	34
ν_g, nm	150	35	5 ··· 7	48	20
Ref.	[90]	[91]	[92]	[93]	[94]

Substance	MgO	Fe_2O_3 (I)	Fe_2O_3 (II)	(Fe_2O_3 III)	Fe_3O_4	$CaCO_3$ (I)
ν_A, nm	83	200	160	96	63	42
ν_y, nm	26	65	34	32	30	12
Ref.	[95]	[96]	[96]	[95]	[95]	Chapt. 4., [170]

Substance	$CaCO_3$ (II)	$MgCO_3$	Ni	Fe	Zn	CaF_2	NaCl
ν_A, nm	120	73	36	100	24	180	2,200
ν_g, nm	20	18	22	40	18	40	300*)
Ref.	[89]	Chapt. 4., [142]	[97]	[98]	[99]	[100]	[101]

*) In the minimum.

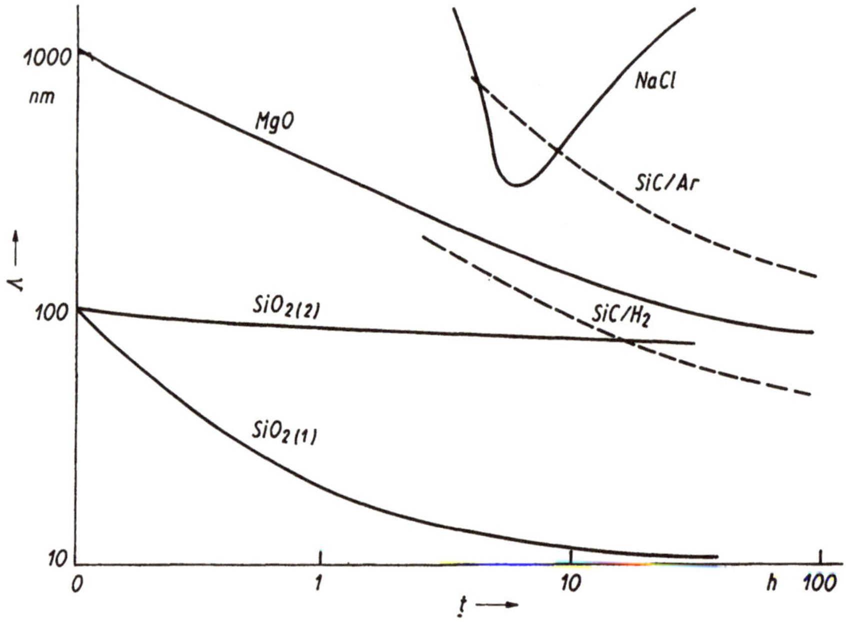

Fig. 5.27. Changes in the primary crystallite size Λ depending on the treatment time *t* for the substances: SiO_2 [90], SiC [91], MgO [101], NaCl (Chapt. 4., [213]).

(r — primary crystallite size, r_g — possible final primary crystallite size, k_0 — constant of velocity for the decrease of primary crystallites, k_r — constant of velocity for the recrystallization, U_A — mean specific activation energy for the recrystallization, U_B — mean specific fracturing energy, U_M — specific mechanical stress power applied in an average time, δ_G — mean specific interfacial surface energy, S — specific surface area of the primary crystallite, E — modulus of elasticity, $\left(\frac{\Delta a}{a}\right)$ — lattice distortion.)

The right hand side augendum of equ. (37) describes the comminution process of primary crystallites and the second addend the recrystallization process. The development of equ. (37) starts out from the presupposition that a recrystallization of

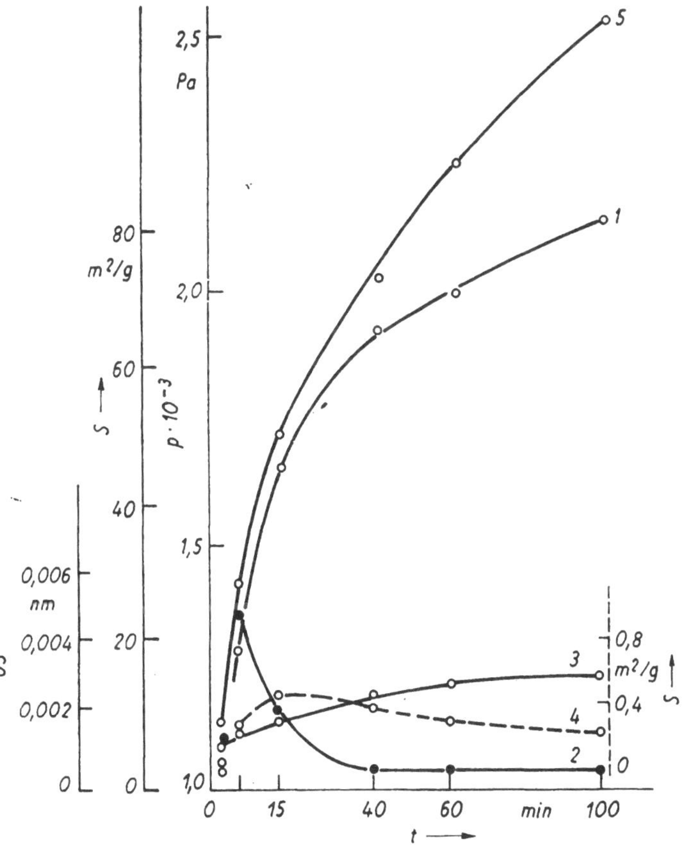

Fig. 5.28. Changes ing the physico-chemical properties of milled calcite [102].

1 — specific surface area O_p of the primary crystallites, 2 — lattice imperfections GS, 3 — specific surface area S_{BET}, 4 — specific surface area of the secondary particles S_A determined by the sedimentation technique, 5 — equilibrium pressure p of carbon dioxide at 973 K.

primary crystallites takes place during the treatment as well. As Fig. 5.27. shows, such a process appears in the measurable range of very simple structured ionic crystals only, such as NaCl. In the remaining substances, especially in compounds having a considerable proportion of homopolar bonds, such as SiO_2, the activation energy for the recrystallization is clearly so high that the second addend in equ. (37) may be neglected and the total process is only determined by the decrease of primary crystallites. The final primary crystallite size is strongly influenced by the stress intensity. As results from Fig. 27 the quartz sample shows a different behaviour with respect to the primary crystallite decrease, if the stress intensity is varied by the alteration of the degree of filling in a vibration mill [90]. The effect of various types of mills on the alteration of solid properties is represented in 3.1. and 7.2.1.1.

The mechanical stress in different gaseous atmospheres may also lead to a different primary crystallite size. As Fig. 5.27. shows a smaller primary crystallite size is obtained by milling SiC in hydrogen than in an argon atmosphere. Tribochemical reactions and sorption processes in the $SiC-H_2$ system (see 6.1.1.) result in increased defects in the solid, in additional stresses, and consequently in stronger primary crystallite decrease [91].

There are only few examples in literature relating the surface of primary crystallites to the reactivity of mechanically stressed solids. As example (Fig. 5.28.) the specific values of exterior surface areas, BET-surface, primary crystallite surface as well as of the lattice imperfections are faced with the CO_2 equilibrium pressure over mechanically activated $CaCO_3$. The best correlation was obtained between CO_2 pressure and specific primary crystallite surface in this system [102].

According to 5.7. the tribodiffusion of impurity particles takes place through a mechanically activated grain boundary diffusion in the lattice of a mechanically stressed solid. The kinetics of these processes are also expected to be influenced by the primary crystallite size.

5.5.4. *Change in the inner and outer surface through mechanical stress*

The surface $(O_A + O_I)$ accessible by the ambient fluid phase has generally a greater importance for the kinetics of heterogeneous reactions at solids as compared to the primary crystallite surface. Thus it nearly always influences the velocity of heterogenous tribochemical reactions. In the following some characteristic surface area changes caused by mechanical treatment will be discussed in dependence on the system of substances and the exterior conditions.

5.5.4.1. *Surface changes in dependence on the system of substances*

The course of the surface area changes during milling can be shown by the following equation for a series of substances [103—105]:

$$O = O_g \cdot (1 - e^{-k_z t}) \tag{38}$$

(O — surface area, O_g — possible final surface area, k_z — constant, t — time of milling.)

According to equ. (38) a constant surface comes into existence after a longer milling The surface values achievable by vibration milling are listed for a larger number of substances in Tab. 5.14.

According to this, surface areas are produced during the milling of pure metals and alkali halides whose specific size only exceeds the value of 1 m²/g under special conditions. For oxides, sulphides, carbonates etc., the achievable surface values are generally in the range of 1—10 m²/g. Substances having a complicated lattice structure such as apatites, SiO_2 having a great proportion of homopolar bonds, and silicate compounds such as cement, can be milled up to surface areas of more than 10 m²/g, whereas the specific surface areas of certain silicates especially phyllosilicates such as kaolin and talcum, can increase to more than 100 m²/g. The course of the surface area in depend-

Table 5.14. BET surface S_{BET} at several substances produced by milling t = treatment time using a vibration mill.

Substance	Ni	Fe	Zn	Ni	NaCl	PbS	Sb_2O_3	Fe_3O_4
t, h	14	14	15	14	*)	150	150	5
S_{BET}, m²/g	0.1	0.3	1	1.5	0.4	1.8	1.9	3.3
Ref.	[106]	[98]	[99]	[106]	[107]	[108]	[108]	[95]

Substance	Fe_2O_3	WO_3	CaF_2	$CaCO_3$	$MgCO_3$	$V_2O\lambda$	Cement
t, h	1	50	2	3	2	1	2.7
S_{BET}, m²/g	6	7	7	6	11	10	17
Ref.	[96]	[109]	[103]	Chapt. 4., [170]	Chapt. 4., [142]	[110]	[93]

Substance	MgO	Apatite	Kaoline	SiO_2	Talcum
t, h	100	1	68	400	**)
S_{BET}, m²/g	18	18	115	40	122
Ref.	[101]	[111]	[113]	[112]	[113]

*) Measured in a continuous-type mill.
**) No values given.

ence on the time of milling is shown for selected substances in Fig. 5.29. If substances having small surface areas are used, e.g. coarse-crystalline quartz, frequently a course according to the curve ($SiO_{2(1)}$) is observed, which means a more rapid surface area increase at the beginning of the treatment and the adjustment of a final steady-state value which does not change further after a longer period of milling (e.g. [93, 106, 114, 115]). Proceeding with products very rich in surface areas such as Aerosil, the milling leads to a continuous decrease of the surface area (see curve $SiO_{2(2)}$), [96, 106, 114].

5.5.4.2. *The* Hüttig *milling equilibrium*

Independently of whether one proceeds from substances with very small or very large surface areas, an average value in terms of surface area comes into being in the course of the milling process which may be very different for the individual substances. This state representing an n-order steady state in analogy to tribochemical equilibrium (see 4.8.3.) is called Hüttig milling equilibrium. It has been possible to show for a large number of substances such as Zn, Al, Fe, glass, and marble, that after disturbing this equilibrium, for instance by the separation of a grain fraction, if the milling continues the same final state is repeatedly reached [116]. It is characterized by the stability of the granulometric composition and the outer surface of the particles, respectively. If no inner surface is produced during the milling then the BET surface remains constant in the milling equilibrium. However, as Fig. 5.30. shows with the example of the carbonates, conditions appear where the BET surface grows continuously with

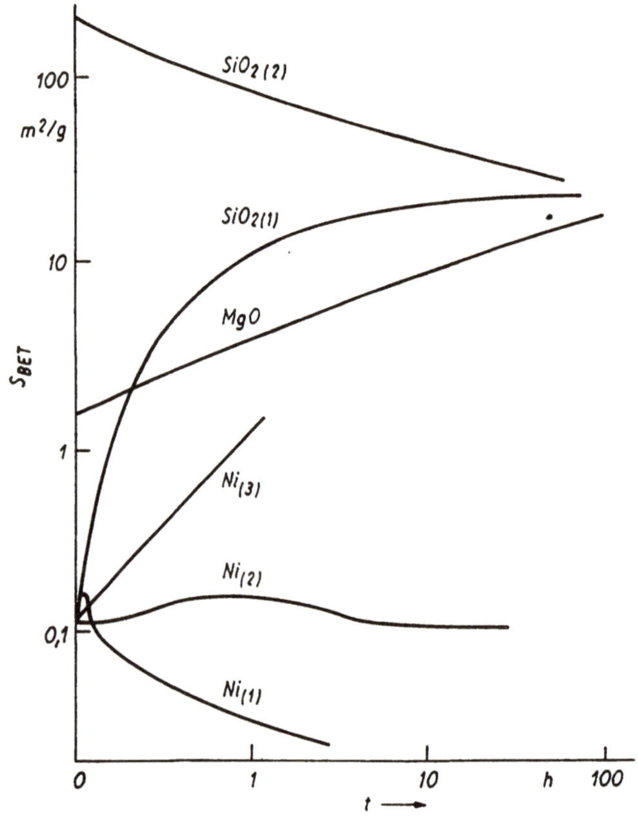

Fig. 5.29. Changes in the specific surface area S_{BET} depending on the treatment time t for the substances: SiO_2 [90], MgO [101], Ni [106].

ongoing milling after the exterior surface becomes constant. Evidently this is brought about by the occurrence of an inner surface area due to the tribochemical decomposition of carbonates [117].

The inner and outer surface areas consequently show different behaviours during the adjustment of a steady state, and this has to be taken into consideration in the analysis of the surface area influence on tribochemical reactions. But a steady state also appears with respect to the inner surface area with a longer period of milling. Samples having different inner initial surfaces also have an approximately similar value of the inner surface area after a milling period depending on the system, as Fig. 5.31. shows.

In the example shown, this inner surface is characterized by the pore volume. As Fig. 5.32. shows, the following main tendencies apply for the changes of the inner sur-

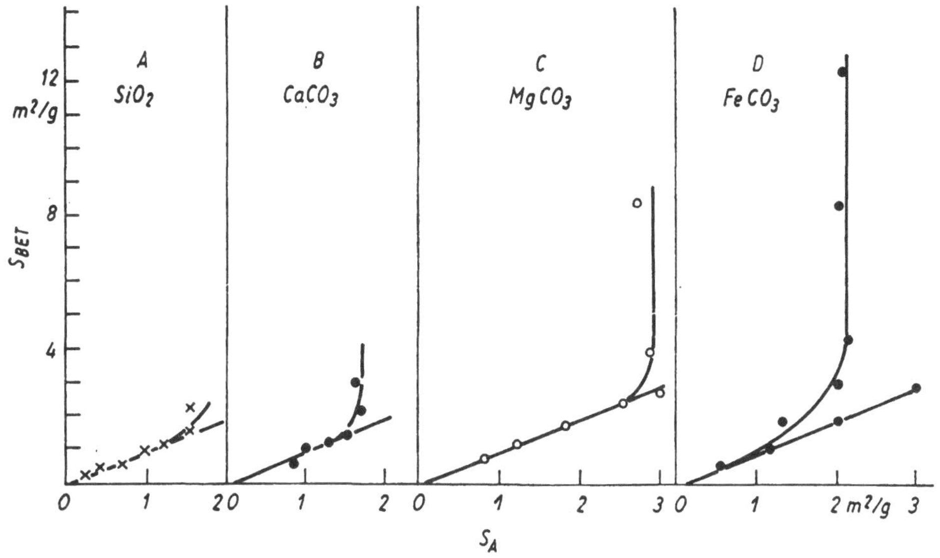

Fig. 5.30. Changes in the specific surface area S_{BET} and in the specific outer surface S_A due to the milling of SiO_2 and some carbonates [117].

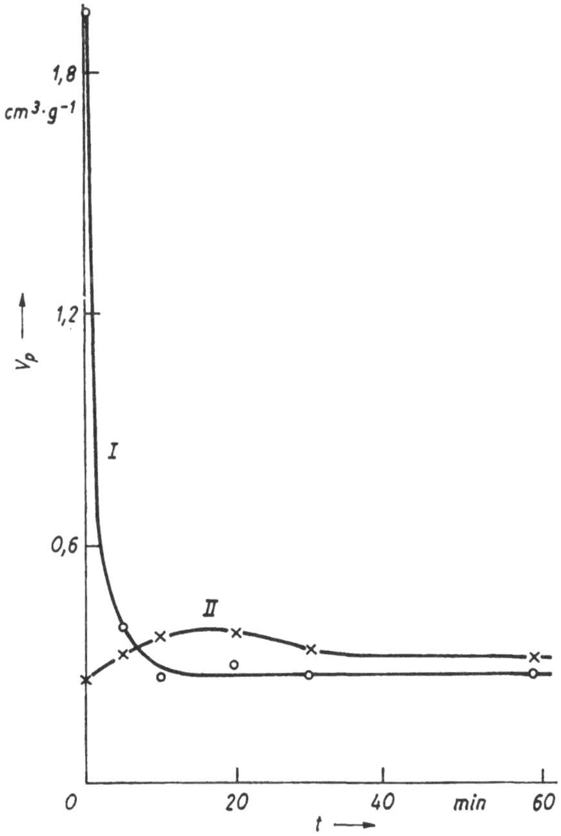

Fig. 5.31. Changes in the pore volume V_p of various α-Fe_2O_3 samples by milling [96].

t — treatment time, sample I — S_{BET} = 41 m²/g, sample II — S_{BET} = 2 m²/g, S_{BET} in the milling equilibrium = 6 m²/g.

face and the pore volume, respectively, by milling:

— Changes in the pore volume caused by milling depend on the nature of mechanical stress. Thus the treatment in impact mills such as in a jet mill, generally leads to a larger pore volume than the stress in vibration mills where a greater compaction of the grinding material results through the treatment of the powder between the milling tools impacting upon one another.
— The milling can lead both to an increase and a decrease of the pore volume. This depends on whether one proceeds with substances with very small or large inner surface area. If the milling continues, steady states come into being depending on the system and on the manner of treatment.

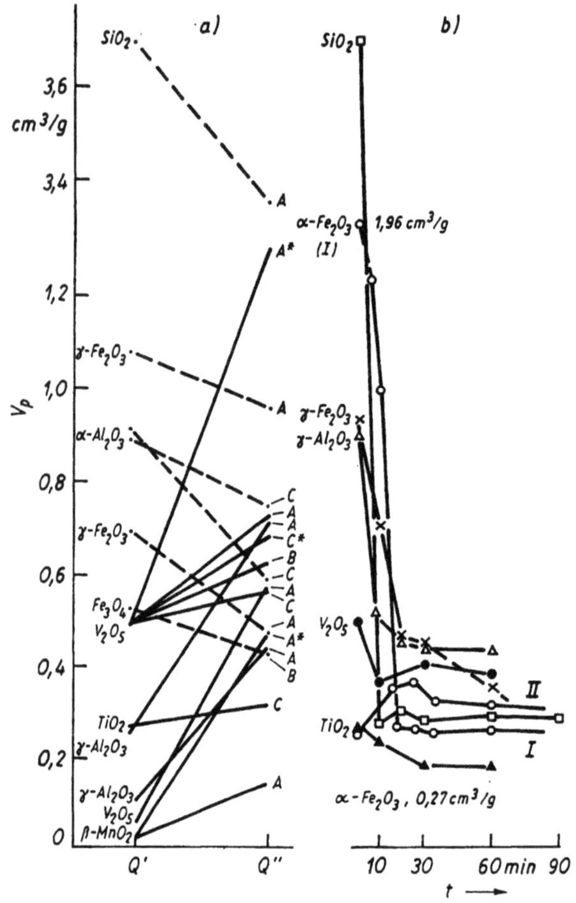

Fig. 5.32. Changes of the pore volume V_p of different substances depending on the treatment method (118).

a) jet milling: *A* — micronizer-jet mill, *B* — spiral-type jet mill after RUMPF, *C* — jet-o-mizer jet mill, * filter bag samples. b) vibration milling: *t* — treatment time, *Q* — initial sample, *Q''* — milled in the jet mill for 1 time.

With a greater number of substances the adjustment of the milling equilibrium runs through a maximum (e.g. curve $Ni_{(2)}$ in Fig. 5.29.), [92, 94, 102, 106, 119]. The occurrence of this maximum can be explained by the fact that the surface area variation during the milling is determined by two processes running in opposite directions: the surface area increasing through fracture and plastic deformation, and the surface area reduction by agglomeration. Evidently the comminution process dominates at the beginning of the stress. In the course of milling the proportion of fine milling material particles increases, the concentration of lattice defects rises, free valences develop (see 5.9.), the charge density grows (see Fig. 5.17.), abrasion develops from the milling tools, and so on, so that altogether compared to the conditions at the beginning of the milling the tendency towards agglomeration increases and a reduction of the surface area is thereby caused. After having reached a maximum the reduction of the surface area depends on the kind of the substance and on the intensity of treatment. Generally the occurrence of a maximum is found in substances of lesser hardness, e.g. in Ni, $CaCO_3$, apatite, etc. The tendency towards agglomeration increases with growing impact intensity. Whereas, for instance, the surface area of apatite changes according to equ. (38) during milling in a vibration mill, a maximum and a steep reduction appear with a higher intensity of treatment by using a planetary mill, that is to say a strong agglomeration already after a short period of milling (Chapt. 4., [60]).

5.5.4.3. *The influence of the outer medium on the milling equilibrium*

Adjustment and position of the milling equilibrium are strongly influenced by the ambient atmosphere, for this affects the comminution and agglomeration processes in different ways. As an example the course of SiC milling in H_2 and Ar atmospheres is given in Fig. 5.33. In both the atmospheres an identical particle size distribution is initially obtained, see Fig. 5.34.a). Whereas the particle size distribution remains constant in an H_2 atmosphere in the course of milling, an agglomeration and a reduction of the BET-surface takes place in argon, see Fig. 5.34.b). If the Ar atmosphere is replaced by an H_2 atmosphere after having reached the changed milling equilibrium then the particle size distribution measured at the beginning is re-established, see Fig. 5.34.c). Apparently the free valences developing during the fracture of SiC are saturated by available reactive gases, like hydrogen in this case, in the phase boundary region and conditions come into existence through tribosorption of hydrogen in the SiC boundary layer which restrain the tendency towards agglomeration.

Similarly if plastically deformable substances such as metals, are milled in Ar, no reaction of the fresh metallic surfaces caused by the treatment occurs, which results in their welding together. Under these conditions, for instance, a fine-grained pure nickel powder is completely cladded to the milling tools (see Fig. 5.29., curve $Ni_{(1)}$), [106]. The tendency towards welding is strongly reduced by chemosorption of the gas on the resulting fresh metallic surface in CO atmospheres so that the disperse distribution of the powder survives during milling and a milling equilibrium comes into existence (see Fig. 5.29., curve $Ni_{(2)}$). As Tab. 5.15. shows, during the milling of SiO_2 in heavily tribosorbing acetylene, after a short period of milling a considerably larger surface area already develops than in an inert argon atmosphere. The effect of liquid additives is even stronger, e.g. in the case of water which leads to a surface area increased by more than one order of magnitude compared to air. As results from Tab. 5.16. water shows an equally strong surface-increasing effect in comparison to apatite. Independently of whether the mentioned substances have mainly proportions of homopolar bonds (SiC), mixed homopolar and heteropolar bonds (SiO_2) or have mainly proportions of heteropolar bonds (apatite), or whether substances with metallic bonds are involved,

Table 5.15. The effect of gases and liquids on the alteration of the specific surface area of quartz (S_{BET}) by milling (treatment time 32 min).

Ambient medium	Kind of additive	S_{BET} m²/g	Ref.
Air	10^5 Pa	6	[90]
Acetylene	10^5 Pa	10	[114]
Benzene	80 mass per cent*)	9	[115]
Acetone	80 mass per cent*)	13	[115]
Water	80 mass per cent*)	73	[115]

*) Related to the solid.

additives leading to a sorption or reaction with the solid during milling frequently cause a considerably stronger increase in surface area than inert additives. The reasons for this are manifold.

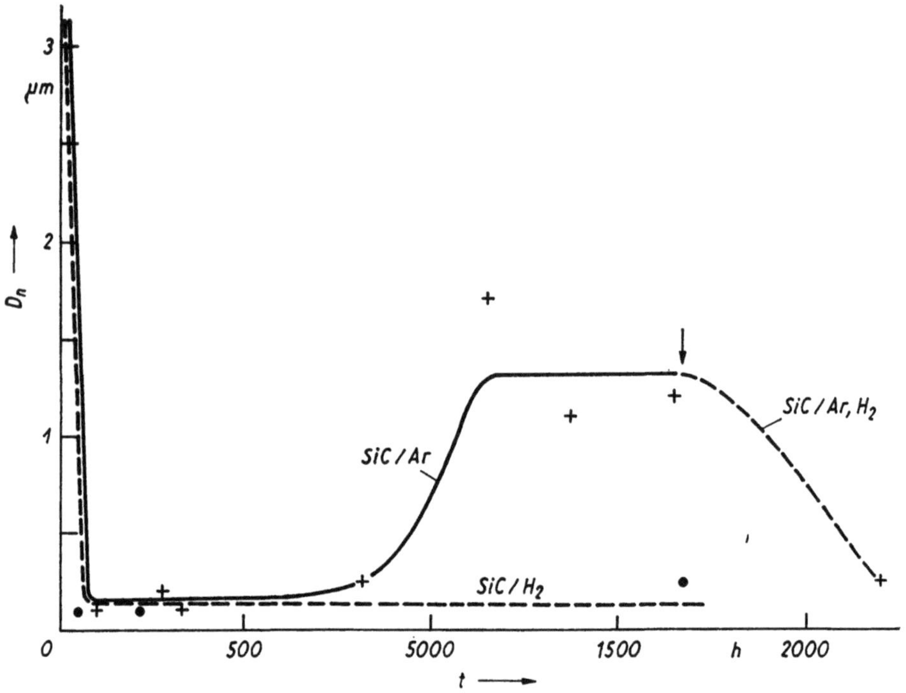

Fig. 5.33. Influence of treatment atmosphere and treatment time t on the most frequently occurring secondary particle diameter D_n of SiC grinding material (Chapt. 4., [10]).
↓ – Ar replaced by H_2.

Thus the mobility of the dislocations is assumed to be increased by adsorption [120]. Furthermore the surface energy of the solid is assumed to be reduced by the influence of the exterior medium and some of its mechanical properties, such as strength and plasticity, are assumed to be changed (REHBINDER effect) [121].

This effect is particularly marked, if solid and ambient medium are related chemically to one another. Thus a great reduction of the surface energy is observed for metals by an application of certain liquid melts of metals, for ionic crystals by the addition of melts of salts as well as polar substances like water, for organic molecule crystals by liquid substances or solutes with homopolar bonds. For instance, Dioxan only minimally reduces the KCl surface tension whereas water reduces it very strongly [122]. An increased plasticity of the solid has been found during the action of the Rehbinder effect, caused by the reduction of the surface potential barrier in the detachment of screw dislocations [123]. Moreover an increased formation of defects appears [124], in particular incipient cracks which are important for comminution, and which are stabilized by sorption and the reaction of the ambient medium with the crack surface [125].

Table 5.16. Relation between the specific surface area S_{BET} of a synthetic fluoroapatite milled for 2 hours and the dipole moment μ of the ambient medium related to the molar volume V [119]

Substance	μ/V (relative values)	S_{BET}, m^2/g
Water	1.0	40
Ethanol	0.4	15
Ethylene glycol	0.4	15
n-Octyl alcohol	0.1	8
Air	—	2

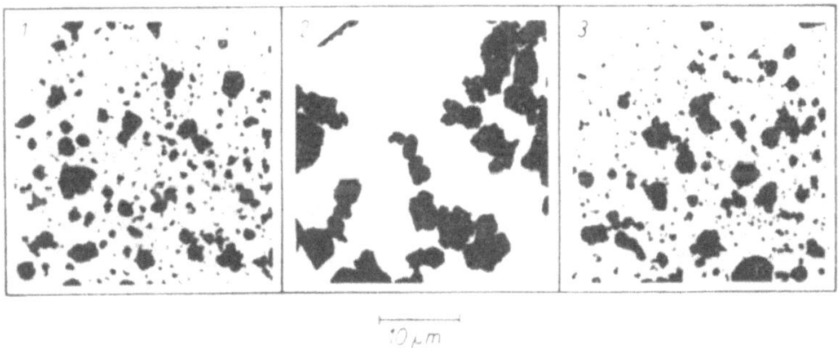

Fig. 5.34. Electron-microscopical exposures of SiC grinding samples stressed differently [91]
1 — SiC 280 h milled in Ar, 2 — SiC 1150 h and 1655 h, respectively, milled in Ar, 3 — SiC 1655 h and 555 h milled in Ar and H_2, respectively.

Since according to the Griffith relation the strength F_s is proportional to the square root of the surface energy σ_G as compared to a brittle fracture [126]:

$$F_s \sim \sqrt{\sigma_G} \tag{39}$$

the Rehbinder effect also gives relief for the brittle fracture. Zinc particles, for instance, are comminuted 200 times faster in a ball mill when 1 per cent gallium is applied as a surface active substance. With an extremely strong surface energy reduction, for instance, to values of 10^{-9} J/cm² a spontaneous dispersion of the solid may even occur; in this particles having diameters in the order of $10^{-5} \cdots 10^{-6}$ cm and smaller coming into existence [120].

The tendency towards agglomeration is also lowered by the reduction of surface energy and consequently of the adhesive forces.

The adhesion of fine particles [127] based mainly on the van der Waals and electrostatic forces is very sensitively influenced by adsorption layers [128, 129]. But films adsorbed to powder surfaces must also have a certain resistance in respect of the mechanical stress so that a formation of fresh surfaces and a locally strong adhesion, or cold welding in the case of metals, do not develop at these areas [130], (see 7.4.4.1.2.). Certain chemical or sorption interactions of the ambient medium with the surface of the grinding material particles are required in order to check the agglomeration arising on account of the two mentioned reasons. In certain cases, for instance with milling solids, the largest surface area was obtained by the application of substances which have the greatest adsorption heat during interaction with the solid [131]. Tab. 5.16. shows that the comminution effect during the milling of apatite in the presence of various liquids increases with the growth of the dipole moment related to the molar volume.

How the ambient medium influences the preceding mentioned processes in detail depends very finely on the special effect fo the medium, on the nature of the solid, on the time, type, and intensity of the stress, and on other factors. As the above-mentioned examples show in the case of the milling of SiC, SiO_2, and Ni and other works [115, 119, 124] generally the drift of the Hüttig milling equilibrium to large surface areas, is however mostly caused by checking the agglomeration effect.

5.5.5. *The importance of the outer and inner surfaces for solid reactivity*

The surface layer accessible by the surrounding liquid phase $(O_A + O_I)$ is generally a contributary factor in the course of heterogeneous reactions on solids, since the degree of the mass transfer depends on its size. This also applies for example to oxidation reactions, to catalytic processes, and to solution reactions. As an example of the fact that the course of the reaction is extensively controlled by the surface change during mechanical stress, the triboxidation of various metals in vibration mills may be mentioned. Thus at the beginning of milling the reaction velocity v may be described for the studied metals by the following equation in good approximation, as can be seen in Fig. 5.35., (Chapt. 4., [57]):

$$\ln v = k \cdot t + C \tag{40}$$

(k, C — constants, t — time of treatment.)

An analogous function has been found for the comminution processes in ball mills (Chapt. 4., [57]):

$$\ln H = -k_0 t + C' \tag{41}$$

(H — mass per cent frequency of a grain-size class, k_0 — velocity constant for the comminution of crystals, C' — constant, t — time of treatment.)

The reaction velocity varies with the time of treatment according to the same relation as the quantity H characteristic for the kinetics of comminution. If the surface area is determined according to BET in the course of milling, then a direct proportionality arises between the surface O_{BET} and the oxidation velocity (see Fig. 5.36.). The lattice defects measured, however, showed no essential variation in the course of milling (Chapt. 4., [57]).

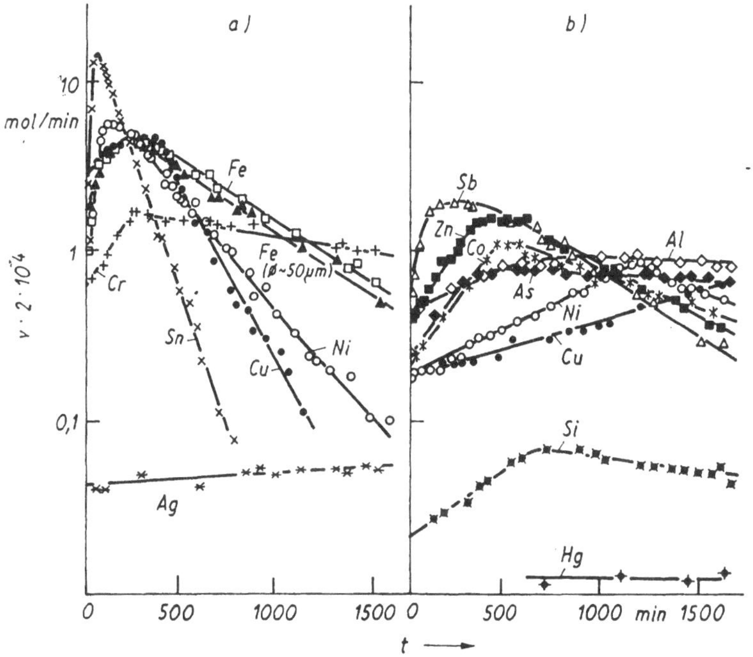

Fig. 5.35. Course of the reaction of the mechanically activated oxidation of various metals (Chapt. 4., [57].)

v – reaction velocity, t – treatment time, a – treatment in a vibration mill (type LS 60) using corundum balls b – treat- ment in a vibration mill (type Vibratom) using porcelain balls.

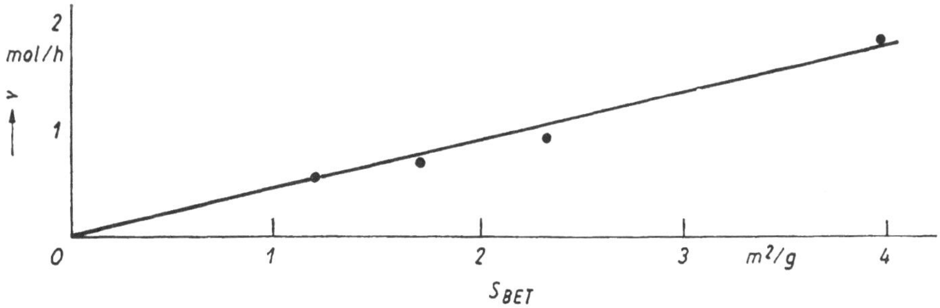

Fig. 5.36. Velocity v of the tribo-oxidation of Ni in dependence on the specific surface area S_{BET} (Chapt. 4., [57]).

The curves in Fig. 5.35. pass through a maximum according to the time of milling dependent on the system and on the intensity of treatment, and then drop linearly in good approximation. The reason for the appearance of maxima is explained by the fact that the surface area tends towards a steady-state limiting value and the milling equilibrium develops, in accordance with the time of treatment depending on the material system and conditions of milling.

The falling branch of the curve in Fig. 5.35. can also be expressed for numerous metals by a logarithmic relation (Chapt. 4., [57]):

$$\ln v = -\varkappa t + C'' \tag{42}$$

($\varkappa$, C'' — constants.)

or differentiated:

$$\frac{\partial c_{\mathrm{Me}}}{\partial t} \sim -c_{\mathrm{Me}} \tag{43}$$

($\delta c_{\mathrm{Me}}/\partial t = v$, c_{Me} — concentration of the still unchanged metal in the solid.)

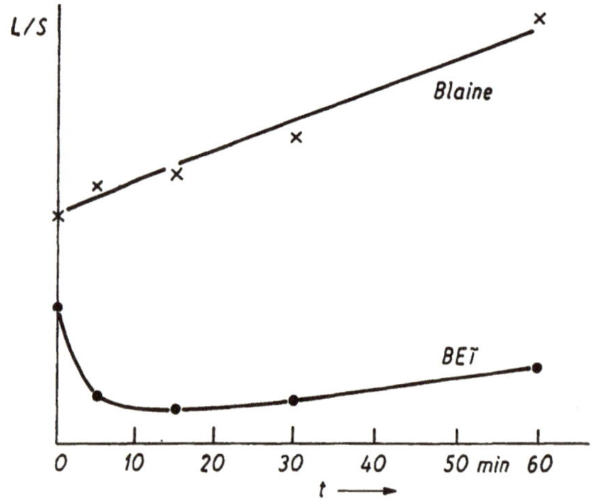

Fig. 5.37. Changes in the specific solubility L/s of apatite after different activation times t by taking the total specific surface area S_{BET} and the outer specific surface S_A as a basis [132].

Thus for the section of the falling branch of the curve the oxidation velocity is directly proportional to the metal concentration in the milling vessel. This simple relation results from the fact that the formation of so-called fresh surfaces steadily continues with the stability of the surface area in milling equilibrium (see 5.6.) and that the proportion of the still unchanged metal on these fresh surfaces decreases with continuing reaction to a corresponding extent. Altogether this example shows that changes in surface area can decisively determine the course of tribochemical reactions. Only when the surface areas are becoming constant will other activation processes determine the further course of the curve, as in the fresh surface formation presented in this example.

However, in other heterogeneous solid reactions particularly in dissolution reactions after the treatment not the whole surface area but the outer one only, influences the course of reaction. Thus an increase in lattice defects is obtained for the specific dissolution velocity of apatite following different treatment times, only if the outer surface area is taken into account in the calculation of the specific reactivity (see Fig. 5.37.). If, however, the specific reactivity is calculated from the solubility and the measured BET surfaces, the changes in reactivity are then not correctly reflected. The reason for this may be explained by the fact diffusing the solutes out of pores filled with solvent runs so slowly that the mass transfer on the inner surface area is only of minor importance to the total process.

Altogether these considerations show that in the analysis of tribochemical reaction kinetics, the variation of solid surface areas caused by the action of mechanical energy frequently decisively determine the course of reaction. This represents a fundamental difference to the radiation-chemical and photochemical initiation of solids, where in most cases no significant variations of surface areas appear.

Conditioned by the variety of the surface area alterations occurring during the mechanical treatment, the course of numerous tribochemical reactions is often decisively determined by the specific milling conditions. The studies of the correlations between milling conditions and surface area alterations on one hand and surface area alterations and reactivity on the other hand are a common objective in the field of comminution science and tribochemistry.

5.6. Fresh surface and transport effects as factors accelerating tribochemical reactions

Chemical reactions involving several solid phases frequently run via a diffusion process of the reacting particles through the lattice of the solids participating in the reaction. In many cases this mass transfer determines the velocity of the whole reaction. This applies for instance to the large group of tarnishing and scale reactions as well as solid reactions such as the formation of cement, the sintering of several solid phases, and so on [133]. Solid reactions controlled by diffusion have been shown to be particularly strongly accelerated mostly by many orders of magnitude, by mechanical treatment or when running with the same velocity at markedly decreased temperatures the diffusion processes controlling the reaction run inmeasurably slowly without the influence of mechanical energy (see example concerning tarnishing reaction for rolling, see 4.1.). Two factors may be named as reasons for the acceleration of reaction, which alter the dominating mechanism in purely thermal energy initiation, and consequently also the velocity of reaction:

1. So-called "fresh surfaces" are continuously developed by deformation and fracture processes during permanent treatment. These have no adsorbate and reaction layers upon coming into existence and interact directly with the ambient reactant. In this way the diffusion step of reacting particles dominating the reaction velocity through the solid matrix and reaction layer, respectively, is obviated, and the progress of the reaction is determined by the chemical reaction on the solid surface area and by the mechanical development of new fresh surfaces [134].

2. The diffusion in the solid is the velocity-determining step of the reaction. However in contrast to the exclusively thermal initiation, according to 4.7., under the influence of the mechanical treatment, a strongly accelerated diffusion of reacting particles through the disturbed solid matrix occurs, as a consequence of increasing the diffusion constants and decreasing the activation energy of diffusion.

Below the aspects mentioned under 1. will be discussed, whilst the mechanism of tribo-diffusion processes will be dealt with in 5.7.

5.6.1. *Sorption and reaction of gases on mechanically developed fresh surfaces*

As is known, chemical reactions of gases with metals are inhibited by oxidic top layers. For instance, the formation of nickel carbonyl from carbon oxide and nickel only occurs if such oxide layers are removed from the nickel surface, e.g. by reduction with hydrogen. But such top layers are also penetrated and metallic fresh surfaces are exposed with a sufficiently intensive mechanical treatment. As shown in Fig. 5.38. the formation of nickel carbonyl only takes place with a measurable velocity after a short-term me-

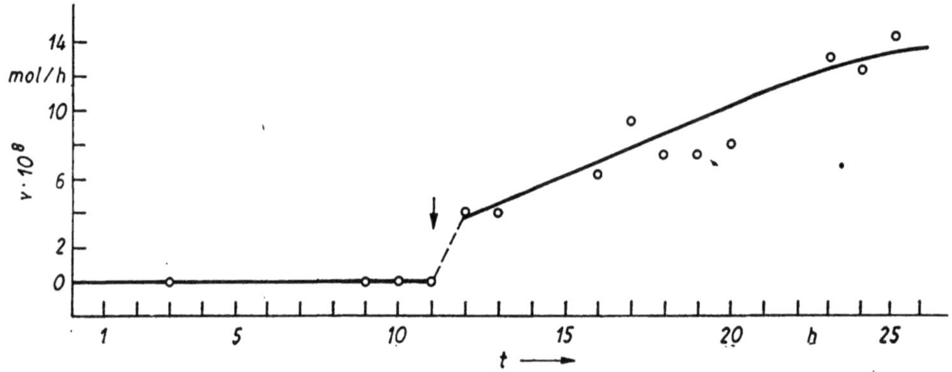

Fig. 5.38. Initiation of the nickel carbonyl formation by shortterm treatment of the nickel surface covered by oxide [134].
v – reaction velocity, t – time; ↓ – mechanical treatment for 2 min.

chanical treatment of nickel surface occupied by oxide, and also accelerates with increasing reaction time after interrupting the treatment.

Fig. 5.39. shows a mechanically stressed zone following the effect of CO. It indicates that not only the stress zone discharged from the oxide layer represents the position of reaction but also etching traces show the course of the nickel carbonyl formation

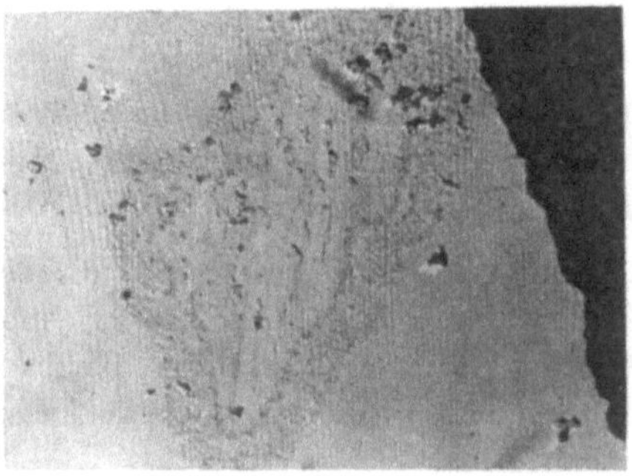

Fig. 5.39. Formation of etching traces in the vicinity of a treatment groove after the action of CO upon a nickel surface [134].

in the environment of the hit position covered by oxide. This is evidently caused by the disturbing of this top layer as a result of the deformation processes (see Fig. 3.3.). Such reaction sequences continuously increase in a chemical way the proportion of the surfaces able to react, albeit over a long period. On the other hand the top layer is already removed after a relatively short period of mechanical treatment when using an apparatus represented in Fig. 4.3. [135].

In contrast to the preceding example, if solid reaction products develop by the reaction of the metallic fresh surfaces with the surrounding gas this very quickly leads to a decline or standstill of the reaction, through the repeated development of a reaction-inhibiting layer. As results from Fig. 5.40., the uptake of oxygen, for instance, by reduced copper above 10^3 Pa with the development of about 5 oxygen atom layers at

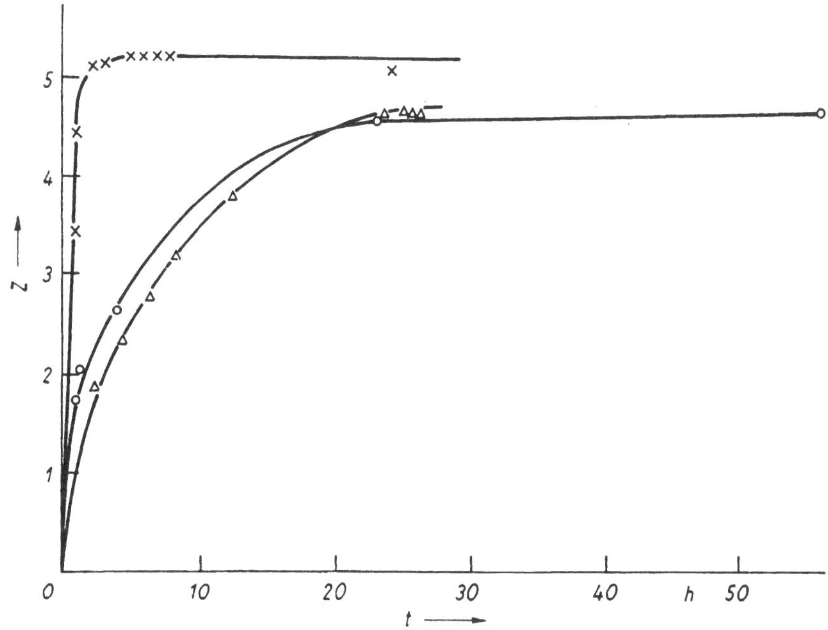

Fig. 5.40. Oxygen intakte at reduced copper with different oxygen pressures [136].
z — number of the oxygen atom layers, t — time, T — 295 K, p_{O_2}: x — $1.6 \cdot 10^3$ Pa, O — 60 Pa, Δ — 30 Pa.

the copper interface will already by completed after some hours, whereas the development of the tarnishing layer runs considerably slower with gas pressures lower than 10^3 Pa. The true fresh surface reaction however consists only in the formation of the first oxygen atom layer on the metal and runs inmeasurably rapidly with the given pressures (see 4.6.). The development of the next atom layer already comes about as a diffusion-dependent reaction with strongly decreased velocity [136].

The proof of metallic fresh surfaces without top layers can be obtained by measuring the work function in the ultra-high vacuum. For instance, an aluminium surface covered with oxide has a value of about 2.7 eV, whereas a value of 4.25 eV was found for single-crystal Al surfaces and evaporated Al films without top layers. The influence of mechanical treatment by means of needle-shaped or cutting edge-shaped tools on the variation of the work function is represented for aluminium in Fig. 5.41. After appropriate treatment the latter increases to the value measured for the pure uncovered surface and can be maintained constantly in the ultra-high vacuum for several hours. Only by adding small amounts of air does the work function drop to a value which is characteristic for a surface layer covered by oxide.

If the treatment of the pure fresh surface or the treatment intensity, are changed, the value for the work function then remains constant. Consequently a changed work function which could be attributed to a different disorder caused by treatment was not proved. Since the TAMMANN temperature of Al is very low and lies in the range of 280⋯370 K a rapid annealing of the produced disorder can already be ecpected to appear at room temperature. The value of 4.25 eV characteristic for pure surfaces is not reached for the work function by using ball-shaped tools (see Tab. 5.17.).

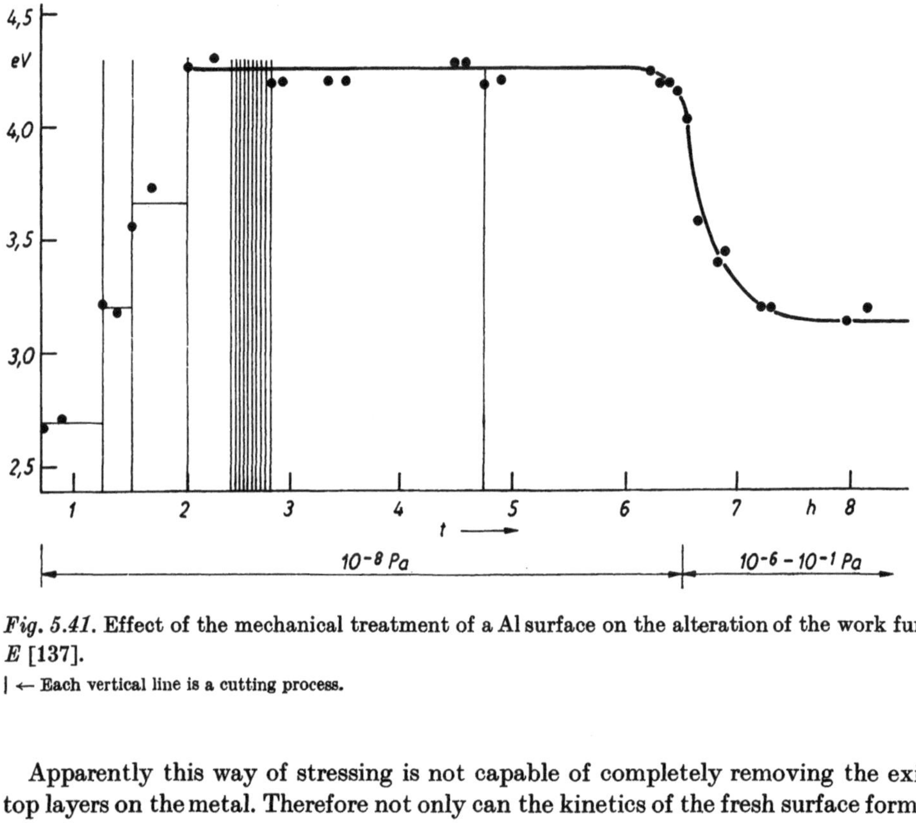

Fig. 5.41. Effect of the mechanical treatment of a Al surface on the alteration of the work function *E* [137].
| ← Each vertical line is a cutting process.

Apparently this way of stressing is not capable of completely removing the existing top layers on the metal. Therefore not only can the kinetics of the fresh surface formation be observed by measuring the work function, but also the purity of metallic fresh surface with respect to the content of sorbates or oxidic additives may be checked.

Table 5.17. The effect of the way of treatment on the work function *E* [137].

Kind of the mechanical effect	*E*, eV
Treatment with cutter shaped and needleshaped tools	4.25
Treatment with ball-shaped tools	3.7⋯4.1

The different treatment effects of cutting edge-shaped and ballshaped tools is the essential reason for the fact that significantly higher reaction velocities are measured in the presence of sharp-edged hard materials such as corundum and carborundum, than by milling the pure solid phase, such as of metal powder (see Fig. 4.17.). As Fig.

5.42. shows, the course of tribo-catalytic reactions also strongly depends on whether the metallic nickel catalyst is treated by itself, comparable with a ball-shaped stress, or in the presence of corundum particles where a more marked cutting treatment appears [139].

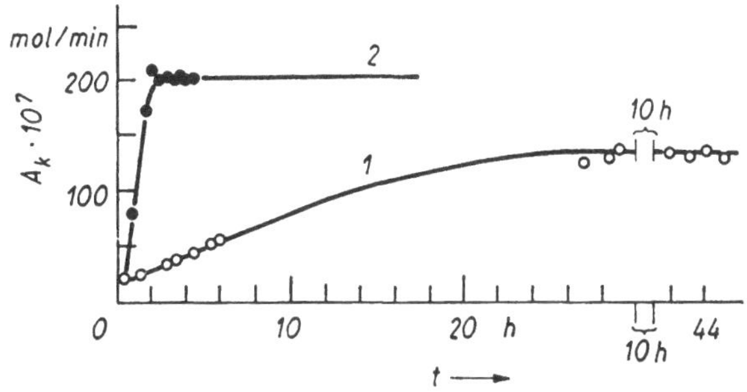

Fig. 5.42. Effect of the addition of SiC as working tool on the benzene hydrogenation at the nickel contact [138].
A_k — catalytic activity, t — treatment time, 1 — treatment without SiC, 2 — treatment with SiC.

With permanent treatment of metallic solids, such as Ni wafers, by using the apparatus shown in Fig. 4.3. a permanent new formation of fresh surfaces occurs being constant for a longer period. If there are gases in the ambient atmosphere capable of interacting with these fresh surfaces by sorption and reaction, such as H_2, O_2, H_2O, or CO

Table 5.18. Sorption of individual gases and of gas mixtures at tribomechanically produced fresh surfaces of nickel.
v_s — sorption velocity, v_F — velocity of the fresh surface formation.

Gas	Sorption per surface site	$v \cdot 10^6$, $Pa \cdot m^3/s$	v_F, cm^2/s***)	Ref.
CO	CO	6.31	1.16	Chapt. 4., [107]
H_2	$1/2\,H_2$	3.12	1.12	Chapt. 4., [107]
O_2	O_2	6.50	1.18	Chapt. 4., [107]
H_2O	$1/2\,H_2O$	2.86	1.14	Chapt. 4., [39]
CO/O_2*)	—	CO: 3.51	—	Chapt. 4., [39]
		O_2: 4.94	—	
CO/O_2**)	—	CO: 2.73	—	Chapt. 4., [39]
		O_2: 6.11		

*) O_2 content 60 per cent,
**) O_2 content 70 per cent,
***) calculated from the sorption velocity.

with nickel surfaces, then a monomolecular sorption layer immediately develops on them. Since according to Fig. 5.40. the development of further layers takes place substantially slower, in practice the consumption of gas is exclusively controlled by the formation of fresh surfaces and their covering with gases from the surrounding atmosphere which are able to sorb. In this the velocity of sorption is not so much determined by the type of these gases, than far more by the formation rate of fresh surfaces. Oy way of example the sorption velocities of four gases on nickel surface areas and the resulting formation velocities of the mechanically produced fresh surfaces are presented in Tab. 5.18. The latter determine the gas consumption in all the four gases, whereas differences in sorption properties of these gases are of minor importance. The kinetics of sorption are solely determined by the velocity of the diffusion of gases to the mechanically produced fresh surface (see Fig. 4.27.). Only if the sorption does not lead to a complete covering of the fresh surfaces such as the sorption of inert gases, H_2 or CO to silver [140], is the simple relation between sorption velocity and formation of fresh surfaces lost (Chapt. 4., [107]).

With pressures over 10^4 Pa the consumption of gas is also determined by the formation of fresh surfaces. The oxidation of iron H_2O and O_2 is given as example in Fig. 5.43. Here the reaction of both gases also takes place according to the same kinetics with constant mechanical treatment [141]. In analogy to the studies discussed here the dependence of the oxidation velocity on the intensity of treatment of some metals discussed in 4.4. (see Fig. 4.12.) can also be attributed to the fact that the production of fresh surfaces is increased with growing intensity of treatment and consequently the consumption of gas dependent on it.

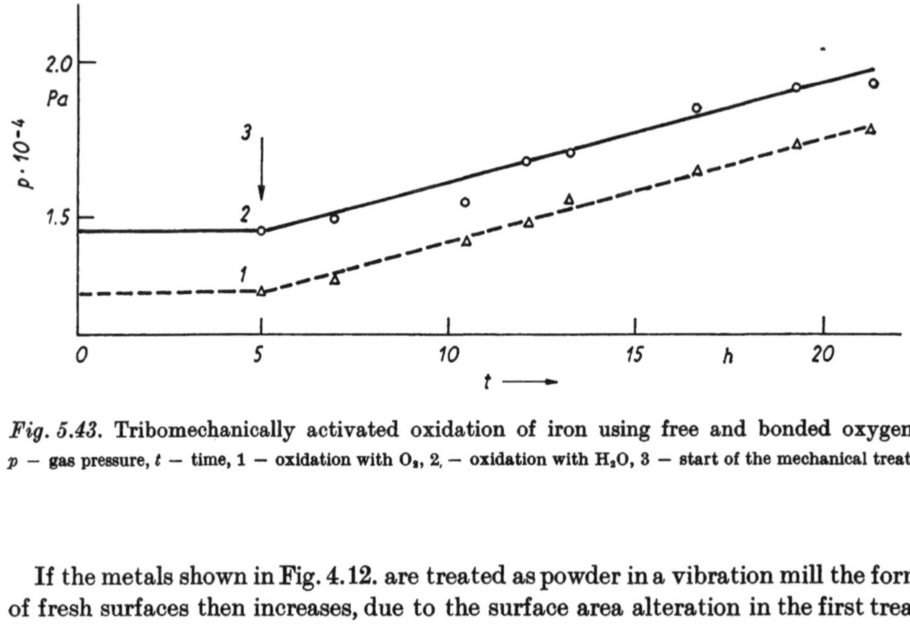

Fig. 5.43. Tribomechanically activated oxidation of iron using free and bonded oxygen [141].
p – gas pressure, t – time, 1 – oxidation with O_2, 2, – oxidation with H_2O, 3 – start of the mechanical treatment.

If the metals shown in Fig. 4.12. are treated as powder in a vibration mill the formation of fresh surfaces then increases, due to the surface area alteration in the first treatment stage explained in 5.5.5., with the milling time (see Fig. 5.35.). Since there is a proportional connection between changes in the surface area and reactivity this must evidently also exist between changes in surface area and fresh surface, see equ. (47). As follows from the comparison of the curves in Fig. 5.35.a) and 5.35.b), by using vibration mills, differences in intensity of the treatment also act upon the velocity of the fresh surface formation.

At the same time the oxidation reaction given in Fig. 5.35. is an example for the possibility to obtain a quantitative reaction of the substance in the mill and consequently to use the tribochemical fresh surface effect for preparation purposes (Chapt. 3., [45]).

5.6.2. *The sorption of gaseous mixtures to mechanically developed fresh surfaces*

As is known, the formation of nickel carbonyl is inhibited by trace amounts of oxygen (Chapt. 4., [47]). The effect of oxygen can be attributed to the phenomenon that O_2 reacts with nickel by the formation of an oxide layer and in this way prevents the sorption and further reaction of CO to nickel carbonyl, as the initial course of the curve in Fig. 5.38. shows.

However, as Fig. 4.6. shows, the formation of nickel carbonyl is also observed during treatment in the presence of oxygen. The reason for this is that new fresh surfaces are constantly produced during the treatment. Both O_2 and CO sorb in a kind of competition sorption at these surfaces. As Tab. 5.18. shows this sorption takes place in first approximation in the order of its partial pressure proportions. An increase of the O_2 partial pressure leads to a corresponding increased tribosorption at the produced fresh surface. The ratio of the sorbed gases is therefore determined by the kinetics of the sorption process and not by a sorption equilibrium appearing after a longer action of gas after interrupting the treatment. In accordance with this the increase of the proportion of air in the CO-air mixture during treatment causes a continuous decrease in the velocity of the nickel carbonyl formation (see Fig. 5.44.). However after interrupting the treatment, according to Fig. 4.6. a standstill of the reaction immediately takes place, for oxygen successively blocks the total surface area due to its greater affinity to nickel (Chapt. 4., [47]). Therefore non-equilibrium states occur at fresh surfaces during the tribosorption of gaseous mixtures, which influence substantially the reaction at such a surface.

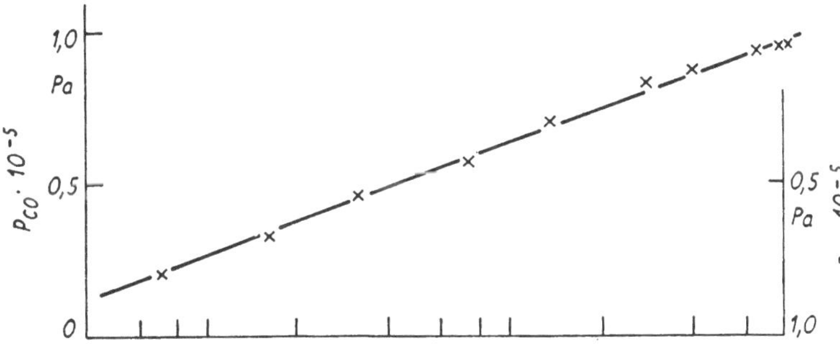

Fig. 5.44: Velocity v of the formation of nickel carbonyl with various partial pressures of oxygen p_{O_2} and carbon oxide p_{CO} (Chapt. 4., [47]).

The course of catalytic reactions is also influenced by these factors. As an example in Fig. 5.45. the course of the hydrogenation of ethylene during the treatment of the catalyst in a hydrogen-ethylene mixture is given, once with and once without, pretreatment of the contact in the vacuum. With further treatment, immediately after the addition of the substrate the pretreated catalyst has a catalytic activity higher by one order of magnitude than the not penetreated one. With further treatment however, this activity, reduces to the value measured at the non-pretreated contact.

After interrupting the treatment a further drop of activity takes place ([139], Chapt. 4., [96]).

The reason for this behaviour lies in the fact that ethylene as component of the substrate blocks the surface area to a significant extent [142]. This behaviour is shown in Fig. 5.46. Whereas a gaseous pretreatment of the catalyst by hydrogen has no influence on its activity, the activity of the catalyst decreases by orders of magnitude with the duration of gaseous pretreatment in ethylene. With continuous treatment

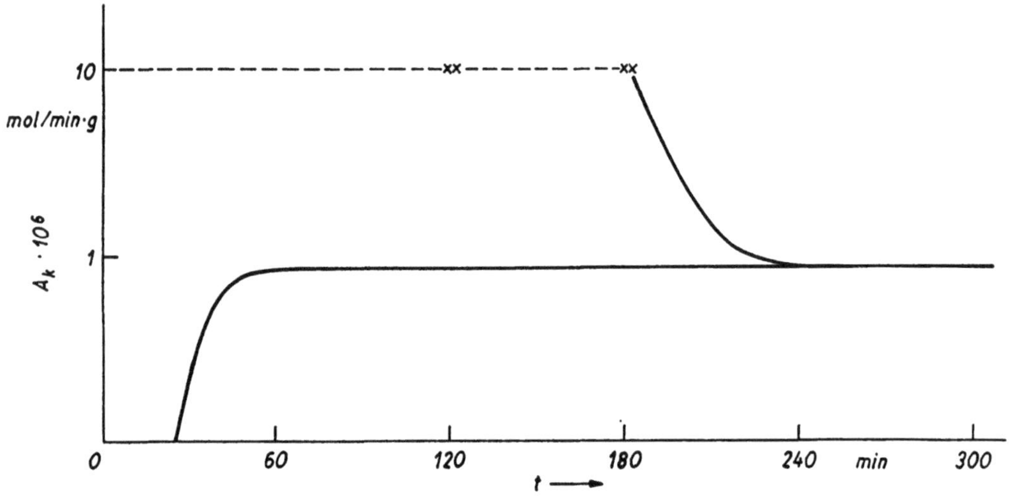

Fig. 5.45. Catalytic activity of an Armco iron catalyst depending on different treatment conditions (Chapt. 4., [50]).

Reaction: hydrogenation of ethylene. A_k – catalytic activity, t – treatment time, – – – treatment in the vacuum, —— treatment in an ethylene hydrogen mixture.

the blocking by ethylene is reduced at the catalyst surface due to a constant formation of fresh surfaces. If the pretreatment takes place in the vacuum there the blocked surface regions are regenerated to an even greater extent and a after higher activity is obtained after the addition of the substrate during running treatment. Here too, the changed sorption ratio of the substrate components can be seen as reason for this increase in activity upon the spontaneous covering of the fresh surface, as compared to the ratio appearing without treatment or after a longer action of the substrate (Chapt. 4., [50]).

The fresh surface effect facilitater the hydrogenation of benzene over in the presence of thiophene with a constant if decreased velocity, whereas without treatment small amounts of thiophene already contaminate the catalyst (Chapt. 4., [50]).

During the tribosorption of two and more gases at metallic fresh surfaces reactions between these gases have also been observed. Thus O_2 and CO from CO_2 which tribodesorbs during the treatment as shown in Fig. 5.47. After the consumption of O_2 and CO by tribosorption at the solid, the CO_2 formed by reaction is once again tribosorbed. The reaction runs according to the following equation (4.39):

$$CO_{ads} + O^-_{ads} + Ni^{3+} \rightarrow \uparrow CO_2 + Ni^{2+} \qquad (44)$$

(CO_{ads}, O^-_{ads} – surface groups developing during the adsorption of CO and O_2 at Ni.)

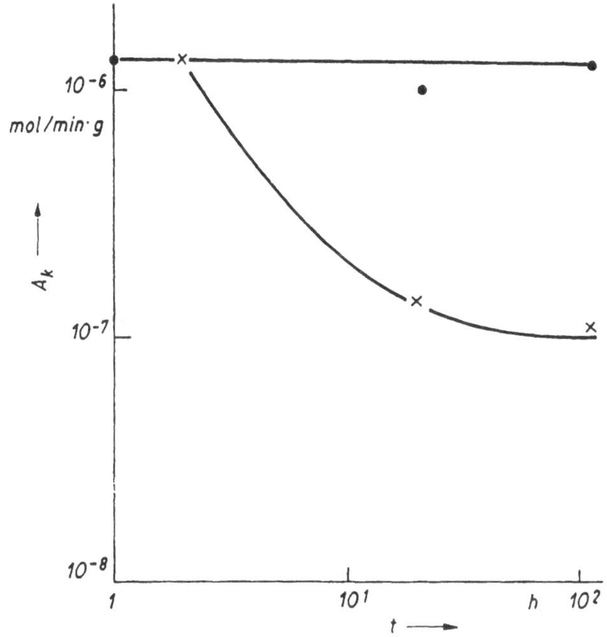

Fig. 5.46. Catalytic activity A_k of an Armco iron catalyst pretreated mechanically for 2 hours in vacuum in dependence on the time t of the gas pretreatment (Chapt. 4., [50]).
• — hydrogen, x — ethylene.

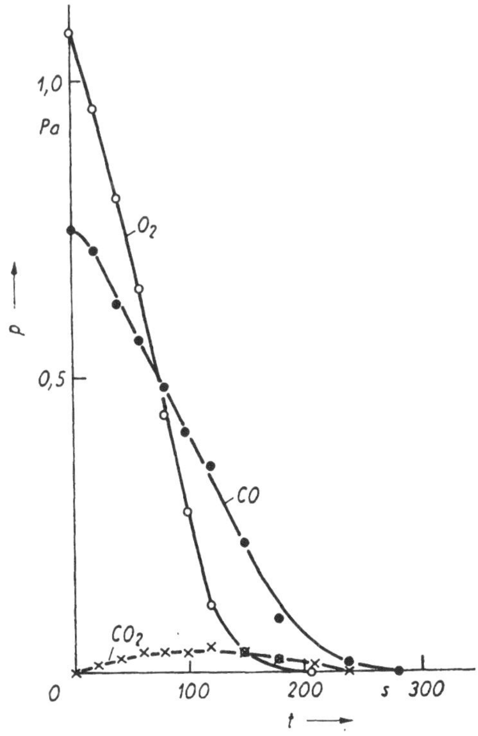

Fig. 5.47. Sorption of an O_2—CO mixture at mechanically treated nickel (Chapt. 4., [39]).
p — partial pressure, t — treatment time.

During the tribosorption of CO and O_2 to silver, CO_2 is developed, whereas during the tribosorption of O_2 and CO_2 to silver, CO is formed [140].

The results of tribosorption to fresh surfaces mentioned here are identical for instance, with the results obtained for evaporated metal films [143]. However, the kinetics of sorption processes and the further chemical reactions of gases may be more comprehensively investigated even without using a high-vacuum on account of a reproducable and adjustable velocity of the fresh surface formation.

5.6.3. *Tribomechanically induced displacement sorption*

A decomposition of H_2O molecules into atomic hydrogen and OH radicals takes place during the tribosorption of water to nickel [144]. A part of the hydrogen recombines and is released as molecular hydrogen to the gaseous phase again. If in addition to water, oxygen also takes part, then the latter causes a displacement of the sorbed hydro-

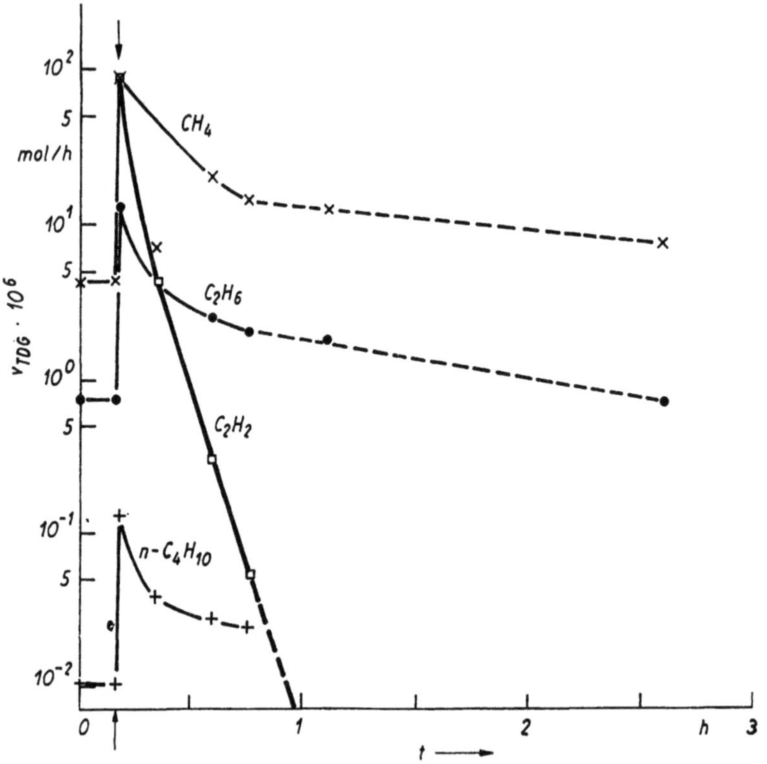

Fig. 5.48. Effect of air on the tribodesorption velocity V_{TDG} of hydrocarbons at the SiC—H_2 system [145].

t — treatment time, ↓ — addition of air, temperature: —— 710 K, – – – 840 K.

gen during the sorption to the nickel surface. Consequently an additional tribodesorption of molecular hydrogen is realized by the effect of oxygen (Chapt. 4., [39]).

Numerous hydrocarbons such as methane, ethane, acethylene, and so on develop during the mechanical treatment of SiC in a hydrogen atmosphere (see 6.1.1.). These gases partially remain sorbed at the solid, and are partially tribodesorbed. Strongly chemosorbing gases such as acetylene are however not released to the gaseous phase. Here too, the velocity of tribodesorption could be increased by more than one order of magnitude by the application of air to a number of hydrocarbons (Fig. 5.48.). As Fig. 5.48. also shows, acetylene could also be tribodesorbed in a relatively large quantity under these conditions. In this the displacement sorption of acetylene is bound to the treatment process [145].

As already has been shown by the example of CO_2 in Fig. 5.47., with treatment the simultaneous course of tribosorption and tribodesorption processes can always be expected (see tribosorption equilibrium in 4.7. and 4.8.3.). Thus with a large supply of sorbable gases tribodesorbed reaction products can be proved in the gas chamber. Conversely with a large supply of fresh surfaces the gases are once again tribosorbed according to the course of the CO_2 curve in Fig. 5.47. Reaction products newly developing during the treatment remain sorbed to the strongly unsaturated solid.

To sum up, these investigations show the treated solid acts more and more as a getter (tribogetter) with a growing fresh surface supply [145]. Reaction products developing under these conditions similarly remain sorbed at the solid. Conversely a large supply of sorbable gas, particularly of O_2 which strongly interacts both with SiC and with metals, may release a desorption of previously sorbed gases. Evidently the adjustment of the sorption ratio of various gases sorbed to the solid seems to be a dynamic process running reversibly in dependence on pressure, type, and composition of the outer gaseous atmosphere during treatment.

5.6.4. *Fresh surface reactions in liquid phase*

In analogy to the fresh surface reaction occurring in the gaseous phase, during the treatment of solids fresh surface effects also become effective in terms of the release or influencing of chemical reactions in the liquid phase. Thus GRIGNARD reactions frequently can be accelerated if the Mg-chips required for them are produced directly by a cutting process in the liquid medium [146]. GRIGNARD reactions have also been released, which till now had not been observed without a simultaneous treatment of the metal [147]. As expected the fresh surface effect generally plays a leading part in the synthesis of organometallic compounds. The reaction of tin and benzyl chloride resulting in the formation of stannanes can be put forward by way of example in Fig. 4.20., (Chapt. 4., [75]). A hexameric complex having a complicated structure $(Ni(SC_9H_{19})_2)_6$ develops in the same way during the treatment of nickel in the presence of organic disulphide, e.g. *n*-dinonyl disulphide [148].

5.6.5. *Fresh surface reactions caused by the transport effect between solid phases*

Whereas in the fresh surface interaction of a solid with the fluid phase surrounding it the delivery of the gaseous and liquid reactants takes place by a mostly rapid diffusion, this possibility does not exist for the interaction between two solid reactants. The latter only appears, if the particles are brought into an atomic contact, e.g. during the press-working of powder mixtures for the making of sintered products. If the reaction is carried out thermally, the reaction phase coming into existence in the chemical reaction is mostly formed between the solid reaction components, thereby inhibiting the further progress of the reaction. Therefore thermally induced reactions between solid phases nearly always depend on diffusion and generally run at higher temperatures only, if the lattice components have sufficiently large diffusive mobility.

In contrast to all the other methods of initiation the effect of mechanical energy may lead to a fundamentally different reaction mechanism, for in a triboreactor the reactants are brought intermittently into an atomic contact (transport effect) and the reaction products thereby coming into existence are removed from the surface (fresh surface effect). The preconditions for "diffusionless" kinetics are given by the joint acting of both the effects. Thus, for instance, nickeliodide already develops in a

Table 5.19. Tribochemical reactions between two solids.

No.	Reaction*)	Ref.
1	$Ni + J_2 \rightarrow NiJ_2$	Chapt. 4., [47]
2	$3\,Ni + KClO_3 \rightarrow 3\,NiO + KCl$	Chapt. 4., [47]
3–8	analogous to BaO_2, $K_2S_2O_8$, PbO_2, $KMnO_4$, $K_2Cr_2O_7$, KJO_3	Chapt. 4., [47]
9	$SrO + WO_3 \rightarrow SrWO_4$	[149]
10, 11	analogous to CaO, BaO	[149]
12	$3\,Fe + C \rightarrow Fe_3C$	Chapt. 4., [92]
13	analogous to Ni	Chapt. 4., [110]
14, 15	analogous to Al, Ca	[150]
16	$Fe_3O_4 + 4\,C \rightarrow 4\,CO_2 + 3\,Fe$	[151, 152]
17–19	analogous to CuO, Co_3O_4, MgO	[150]
20	$MeCN + S \rightarrow MeSCN$	[153]
21	$Mg + S \rightarrow MgS$	[150]
22–28	analogous to Cd, Zn, Sn, Pb, P, Sb, Bi	[150]
29	$2\,Fe + MoS_2 \rightarrow 2FeS + Mo$	[154, 155]
30	$Fe + 2\,S \rightarrow FeS_2$	[156]
31	$FeS_2 + Fe \rightarrow 2\,FeS$	[157]
32	$HgCl_2 + 2\,KJ \rightarrow HgJ_2 + 2\,KCl$	[158]
33	$Si + C \rightarrow SiC$	Chapt. 4., [10]
34	$Ag_2SO_4 + Na_2SO_3 \rightarrow Ag_2SO_3 + Na_2SO_4$	Chapt. 4., [10]
35, 36	analogous to Na_2CO_3, $K_2C_2O_4$	
37	$Na_2CO_3 + SiO_2 \rightarrow Na_2SiO_3 + CO_2$	
38	analogous to $CaCO_3$	Chapt. 4., [2]
39	$Na_2SO_3 + S \rightarrow Na_2S_2O_3$	[160]
40	$NiO + Al_2O_3 \rightarrow NiAl_2O_4$	[161]
41	$FeO + Cr_2O_3 \rightarrow FeCr_2O_4$	[161]
42	$SnO_2 + Si \rightarrow SiO_2 + Sn$	[162]
43–46	analogous to C, Al, W, CaC_2	[163, 164]
47	$SiO_2 + CaO \rightarrow CaSiO_3$	[165]
48	$SiO_2 + CaO + H_2O \rightarrow$ CSH phases	[166]
49	Double reactions of alkali halides	[167, 168]
50	$NaBH_4 + NH_4Cl \rightarrow NaCl + NH_4BH_4$	[169]
51	$2\,KJ + PbCl_2 \rightarrow PbJ_2 + 2\,KCl$	[170]
52–60	analogous to $PbBr_2$, $PbNO_2$, Pb(CNS), $Pb(CH_3COO)_2$, $PbSO_4$, PbB_4O_7, $PbSiO_3$, $PbCO_3$, $Pb_3(PO_4)_2$	[170, 171]

*) Substances which are exchanged in the reaction equation are underlined.

relatively short time through milling nickel powder and iodine using a ball mill at room temperature (Chapt. 4., [47]). By milling a powder mixture consisting of SrO and WO_3 the quantitative exchange to strontium wolframate similarly takes place [149]. Other examples are given in Tab. 5.19.

In order to derive a kinetic equation the reaction velocity between two solids is assumed to be proportional to the surface of contact F_K of the mechanically produced fresh surface [162, 170]:

$$\frac{d\alpha}{dt} = k_F F_K \tag{45}$$

(α — degree of exchange, k_F — velocity constant of the reaction.)

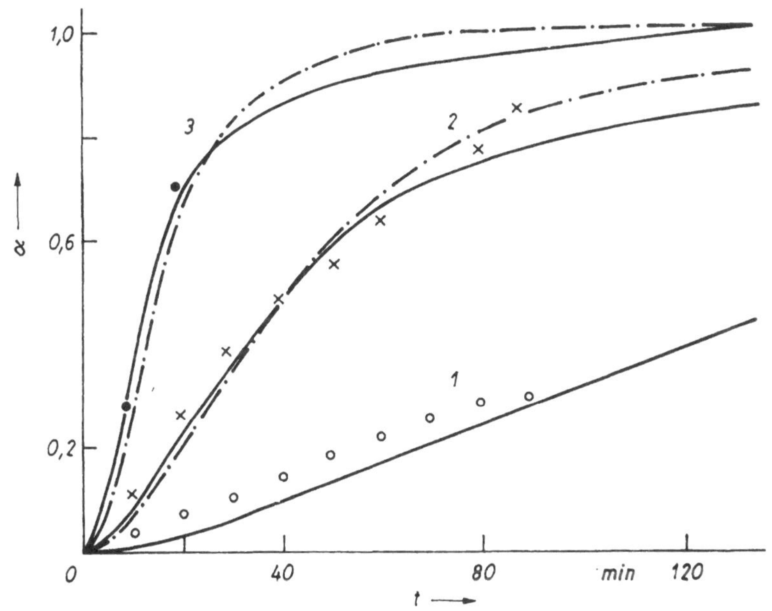

Fig. 5.49. Degree of transition α for the reaction: $SnO_2 + Si \rightarrow Sn + SiO_2$ in dependence on the treatment time t and the power input N of the mill [162].

—•— — curve calculated according to equ. (48), ——— — curve calculated according to equ. (49).
Exerpimental value: o — curve 1 for N = 470 J/min, x — curve 2 for N = 930 J/min, • — curve 3 for N = 2,620 J/min.

With increasing exchange the surface of contact of the initial components is more and more reduced due to the chemical reaction. These conditions may be taken into consideration by the following equation:

$$\frac{d\alpha}{dt} = k \cdot F_K(1 - \alpha) \cdot (1 - \varepsilon\alpha) \tag{46}$$

(α, $\varepsilon\alpha$ — degree of the substances A and B, $\varepsilon = n/m$ — mixture ratio coefficient.)

Further, the ratio of the surface of contact to the total surface area 0 is assumed to be constant at milling:

$$F_K = \gamma \cdot O \tag{47}$$

(γ — constant, O — outer total surface area.)

According to equ. (38) O is:

$$O = O_g(1 - e^{-k_z t}) \tag{38}$$

The following expression is obtained by the integration of equ. (46) taking into account equ. (47) and equ. (38):

$$\frac{1}{\varepsilon - 1} \ln \frac{1 - \alpha}{1 \dot{-} \varepsilon\alpha} = K_F \left[t - \frac{1}{k_z} (1 - e^{-k_z t})\right] \tag{48}$$

$(K_F - k_F \cdot O_g \cdot \gamma)$

For a stoichiometric mixture results the equation:

$$\frac{\alpha}{1 - \alpha} = K \left[t - \frac{1}{k_z} (1 - e^{-k_z t})\right] \tag{49}$$

The validity of these relations has been proved by means of the chemical reaction of SnO_2 with Si. As Fig. 5.49. shows, there is a good conformity between experimental and calculated values.

In summary it may by noted that the fresh surface effect has an important influence on tribochemical reactions. It leads to a mechanism which is independent of the solid diffusion and consequently to a strong acceleration of chemical reactions.

5.7. The tribodiffusion mechanism of gases

Studies of the mechanism of tribodiffusion have been carried out mainly for the diffusion of gases in impact-stressed metals, particularly of inert gases (Chapt. 4., [58, 121, 123, 124].

As is known, the solubility of inert gases is very small in metals and generally not provable by experiments.

Only if there are very high pressures (> 3 GPa) could a permeation of argon through steel walls be observed [172]. However, inert gases can be introduced non-thermally into the metal matrix, for instance, by bombardement with inert gas ions (Chapt. 4., [120, 173]) or may be produced in the solid during the activation with neutrons [147]. Whereas in the latter case a homogeneous solution of the inert gas develops in an atomic-dispersal form or in stills, with bombardement by ions a certain penetration depth of varying geometry comes into being. One-dimensional and multidimensional lattice defects have been found to be the active points for capture for heavy inert gases these are mostly vacancies and conglomerations of vacancies [175—177].

The mobility of inert gases only becomes measurable with the mobility of lattice defects, that is at average temperatures of some 100 K, in the stage of the so-called "damage diffusion", or in the lattice self-diffusion only [178]. Since the presence of inert gases represents an extreme non-equilibrium state in the solid under normal pressure the thermally activated diffusion leads to a practically complete desorption.

According to the considerations in 4.7., inert gases already penetrate the impact-stressed solid at very low temperature, and are conveyed further inside the solid. The tribo-induced uptake of gas is schematically given in Fig. 5.50.

For the nickel-argon system the argon distribution in the treated nickel boundary layer was determined by a successive etching of the nickel surface by CO (Fig. 5.51.).

Pursuant to this about 90 per cent of the inert gas is contained in a 6 nm layer thickness. The inert gas concentration accounts for $\tau_0 = (n_{Ar}/n_{Ni}) \cdot 100 = 2.5\%$ in the lattice top layers.

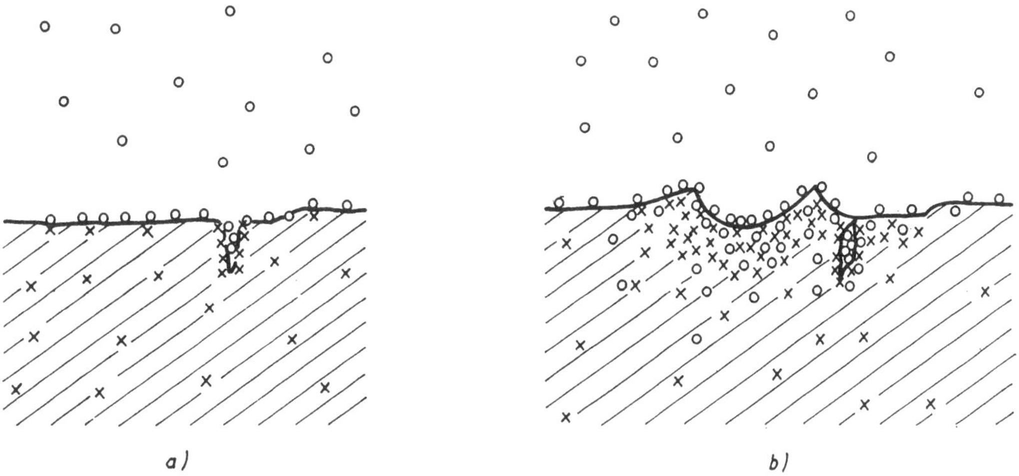

Fig. 5.50. Tribosorption model [179].

/// – solid, o – gas molecule, x – lattice imperfection, ↓ – impact place, a – before the treatment, b – after the treatment.

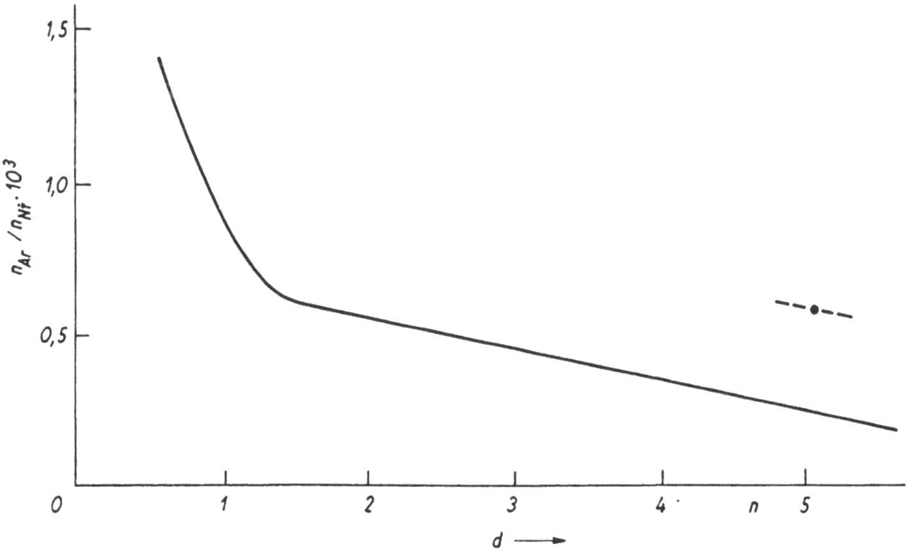

Fig. 5.51. Depth distribution of the argon tribosorbed in nickel (Chapt. 4., [123]).

n_{Ar}/n_{Ni} – ratio of the argon atoms to the nickel atoms in the lattice, d – penetration depth; x – $4 \cdot 10^3$ s treatment time, • – $4 \cdot 10^5$ s treatment time.

The kinetics of the tribosorption process are shown in Fig. 4.37. Depending on the period of treatment, the sorption process may be divided in 3 stages for a number of systems, such as Ar/Ni, He/Ni, Kr/Ni, etc., characterized by the following properties (Chapt. 4., [123]):

1st stage: Strong non-linear decrease of the pressure p according to the relation:

$$p \sim t^{\frac{1}{2}} \tag{50}$$

2nd stage: Linear decrease of the pressure in the course of a longer period of treatment
3rd stage: constancy of pressure.

Equ. (50) is typical for diffusion processes. Thus the following relation is valid for the diffusion velocity in a plate of semi-infinite expansion [180]:

$$\frac{(M/F)}{\Delta t^{\frac{1}{2}}} = 2 \cdot \tau_0 \left(\frac{\omega}{\pi}\right)^{1/2} \tag{51}$$

(M — amount of sorbed gas, F — surface, τ_0 — concentration at the surface area, ω — coefficient of diffusion.)

Since equ. (50) correlates with equ. (51) the latter was used for the calculation of the diffusion coefficient of tribosorption ω_T. This is given for two different temperatures and compared with constants of diffusion measured in other systems in Tab. 5.20.

Table 5.20. Diffusion coefficients D of noble gases and nitrogen in several solids.

System	D, m²/s				Energy effect	Ref.
	Found by experiment		Extrapolated			
	T_1, K	D_1	T_2, K	D_2		
Ar/Ni	83	$2 \cdot 10^{-21}$	293	$3 \cdot 10^{-10}$	mechanical	Chapt. 4., [121]
	96	$1 \cdot 10^{-19}$			mechanical	Chapt. 4., [123]
Ar/CaF_2	473	10^{-17}	83	10^{-35}		[174]
Rn/Fe_2O_3	600	$5 \cdot 10^{-19}$				—
	293	$7 \cdot 10^{-20}$			thermal	[181]
N/Fe	293	$9 \cdot 10^{-21}$				[182]
	323	$2 \cdot 10^{-19}$				[183]
H/SiC	293	$2 \cdot 10^{-13}$			mechanical	[184]

Comparative considerations taking into account the different conditions of diffusion in the systems in Tab. 5.20. permit the conclusion that the tribodiffusion runs faster by orders of magnitude than the corresponding thermally activated diffusion.

In order to determine the activation energy of the process, the dependence of the constant of diffusion on temperature was measured in the range of 78 ··· 90 K, and from this the apparent activation enthalpy E_A was calculated. As Fig. 5.52. shows the applicability of the ARRHENIUS relations is possible for the tribodiffusion process studied:

$$D_T = D_0 \cdot e^{-\frac{E_A}{RT}} \tag{52}$$

Since D_T and, according to equ. 52, E_A may be determined experimentally, the frequency factor can be calculated for the tribodiffusion via the ARRHENIUS equation (4.121):

$$D_T = \left(9 \pm \frac{21}{10}\right) \cdot 10^{-6} \frac{\mathrm{m}^2}{\mathrm{s}} \cdot \exp\left(-\frac{(25.4 \pm 1) \cdot 10^3 \ \mathrm{J/mole}}{RT}\right) \tag{53}$$

Ideas as to the character of the diffusion mechanism were obtained by comparisons with values in literature. The frequency factor of tribodiffusion was shown to agree with the factor of self-diffusion along grain boundaries of nickel [182].

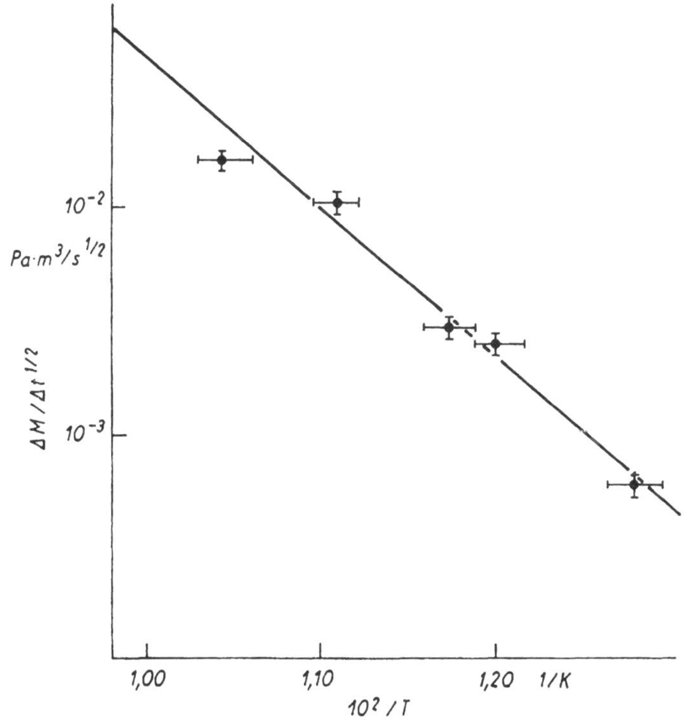

Fig. 5.52. Temperature-dependence of the triboabsorption of argon in nickel controlled by diffusion (Chapt. 4., [123]).
M – quantity of triboabsorbed gas, t – treatment time.

As is known, grain boundaries are sites with a higher degree of disturbance in comparison to the bulk phase. Therefore diffusion processes run faster by orders of magnitude at grain boundaries as compared to the volume diffusion. A vacancy migration is preferred as the mechanism in discussions of this [182].

Therefore it is most likely that the incorporation of Ar in Ni takes place at vacancies and the transport of the inert gas atoms via vacancies. This mechanism is supported by the desorption studies in the Ar/Ni system represented in 4.7.

The activation enthalpy hereby determined is compared with the values of the various diffusion processes being possible in nickel in Tab. 5.21. The activation enthalpy of desorption having a value of 136 kJ/mole agrees with the vacancy migration for mono-vacancies. Consequently the tribosorbed Ar seems to be collected at vacancies from the beginning. The tribodiffusion process may also be characterized by the position change frequency Γ which may be calculated by the following equation [181]:

$$\Gamma = \frac{6D}{f \cdot \lambda^2} \tag{54}$$

(λ — average jump distance, for Ni = 0.25 nm, f — correlation factor, for kfz-lattice is $f = 0.78$.)

Table 5.21. Activation enthalpies E_A of different processes in nickel.

Process	E_A, kJ/mole	Ref.
Thermal desorption of		Chapt. 4., [123]
tribosorbed Ar from Ni	136	
Tribodiffusion in the Ar/Ni system	25	Chapt. 4., [123]
Self-diffusion Ni^{63}/Ni	276	[185]
Grain-boundary self-diffusion	118	[185]
Surface self-diffusion	90	[186]
	75	[187]
Recrystallization (in pure nickel	120	[188]
Monovacancy migration in nickel	135	[189]

A value of 0.2 s^{-1} results for the tribodiffusion. This is a considerable position change frequency which lies within the order of magnitude of the recrystallization-interchange frequency ($10^{-2} \cdots 1\ s^{-1}$), (Chapt. 4., [123]). By virtue of the mechanical treatment the diffusing argon has a mobility corresponding to the mobility at large-angle grain boundaries at recrystallization temperature.

Comparing the activation enthalpy E_A calculated according to equ. (52) for the tribodiffusion process with the activation enthalpy found for the grain boundary diffusion (see Tab. 5.21.), then in conformity with the laws discussed in 4.5. the E_A value is considerably below the grain boundary diffusion value found for purely thermal influence.

Apparently a considerable part of the energy required for tribodiffusion is supplied directly through the treatment process. This is also supported by the result that the tribodiffusion process discussed here is only running during the action of mechanical energy. Since the validity of the ARRHENIUS relation has been proved, the tribodiffusion ought to be a typical example for a coupling process in the sense of the thermodynamics of irreversible processes. Contrary to the examples considered in 4.8. there is a system here where vectorial fluxes and forces are coupled. The concentration gradient and the high shear stresses occurring during the treatment are considered as the coupling components for the generalized forces, influencing the mass flux together and simultaneously. Thus a shear stress of 1.2 GPa is calculated for the example mentioned above from the experimental data (Chapt. 4., [123]).

In the 2nd stage of tribosorption a pressure course linear in time is observed (see Fig. 4.37.). The tribosorption velocity is smaller by a narrow one order of magnitude in this range than in the 1st stage. However, it remains constant for a greater period. According to the studies given in 5.6.1. the treatment of nickel leads for a certain period of time to a continuous production of fresh surfaces coupled with a proportional increase of the surface area. In this way with running treatment the preconditions are

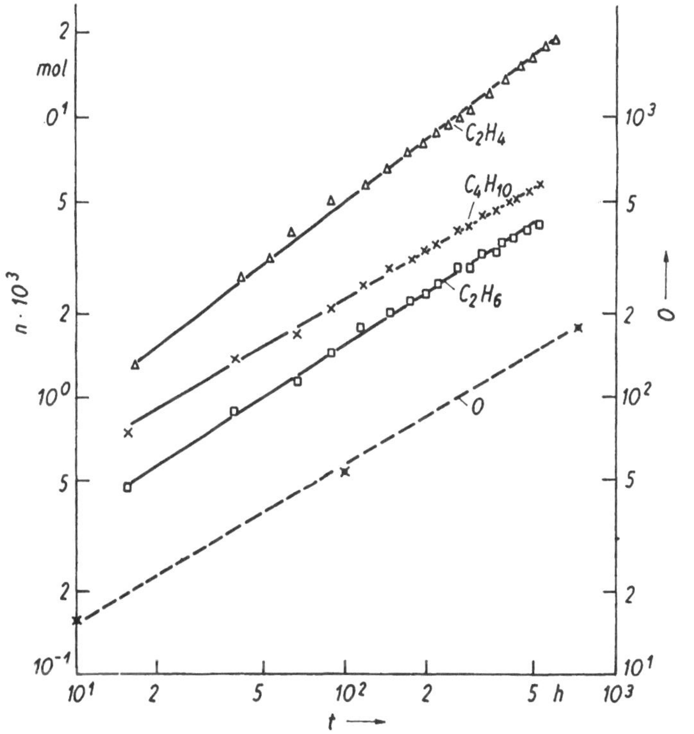

Fig. 5.53. Dependence of the tribosorbate *n* of different hydrocarbons and of the surface *O* on the treatment time *t* at SiC (Chapt. 4., [10]).

provided for a further tribosorption as a steady-state process. This stage is also observed with the tribosorption of hydrogen and hydrocarbons at SiC if the treatment process takes place over a longer period. As Fig. 5.53. shows, the amount of tribosorbed gas increased in first approximation proportionally with the surface development (Chapt. 4., [10]).

If corundum is added as a tool in the tribosorption of argon at nickel the tribosorption velocity is then raised in terms of the intensified increase of the surface area (Chapt. 4., [123]).

After a sufficiently long treatment time, the increase in surface area drops to zero without the use of hard materials accompanied by the establishement of the Hüttig milling equilibrium. Under these conditions stage III is observed where measurable tribosorption no longer takes place (see Fig. 4.37.). A stationary state comes into existence, the so-called tribosorption equilibrium, where absorption and desorption of the inert gas run with the same velocity (see 4.8.3.). However in the presence of corundum as a working tool, the surface area increase lowers considerably more slowly during running treatment, so that the stage III is not observed. Consequently stages II and III are essentially determined by the surface area variation caused by dependence on the type of treatment.

On the basis of a vacancy mechanism, mathematical model conceptions have been developed for the tribodiffusion process which also take into consideration the simultaneous course of tribochemical reactions. It has been possible to obtain a good conformity between the experimentally measured and the theoretically calculated course of the curve for Ar/Ni/corundum, Kr/Cu, CO_2/Cu systems (see Fig. 4.47.), [190], Chapt. 4., [125]). A diffusion mechanism has also been postulated for the H_2 uptake in SiC and a relatively high diffusion coefficient calculated by use of the FICKIAN diffusion law (see Tab. 5.20.), [184].

A dislocation transport is also discussed as a mechanism for the penetration of the gas [191]. This mechanism is coupled with the plastic deformation of the solid, where not only gases but also clusters of submicroscopic and microscopic size are transported inside the solid surface layer stressed by impact (Chapt. 3., [15]), Fig. 3.4.). Which of the mentioned mechanisms dominates the tribodiffusion process depends above all on the elastic and plastic properties of the stressed solid.

As investigations on the tribosorption of H_2 in SiO_2 showed, the purely physical diffusion process is overlapped by simultaneously appearing tribochemical reactions with formation of SiH and SiOH bonds. In analogy to the reaction considered in equ. (4.44) these not only occur at the phase boundary also but on the basis of the tribodiffusion inside the solid matrix to a considerable proportion too [192]. This is also basic reason for the acceleration of the chemical reaction of solids during the action of mechanical energy.

Tribosorption and tribodiffusion of hydrogen in steel have a great importance in practice, because the so-called hydrogen embrittlement, much-feared in technology is caused by them. Generally it is found through mechanical stress of Fe samples, that these possess an absorptive capacity for hydrogen, caused by deformation and the formation of adsorption-active and absorption-active interface regions coupled with it [193]. Both a vacancy mechanism [190] and a dislocation transport [191] are discussed for the diffusion of gas, with the H diffusion representing the limiting process in the embrittlement [194]. It can be understood from this that the embrittlement effect of hydrogen is decreased with falling hydrogen pressure and with increasing deformation velocity in the range between 3 and 300 per cent/min [195]. The decohesion of grain boundaries, the interaction of solved hydrogen atoms with dislocations [196] and the dissociative H_2 adsorption especially in the developing cracks and the advancement of the brittle fracture propagation of small cracks associated with it [197], the formation of highly concentrated interstitial mixed crystals accompanied by the development of internal stresses [198], the appearance of methane according to the following equation:

$$2H_2 + Fe_3C \rightarrow 3Fe + CH_4 \quad (55)$$

and the development of considerable stresses by the formation of hydrocarbon are discussed as causes of hydrogen embrittlement [199].

5.8. The influence of top layers on tribochemical reactions (top-layer effect)

Different processes mechanically stressed surfaces lead to the formation of top layers of varying thickness on the treated solid. As far as metals are concerned these mainly are oxidic reaction layers formed by tribochemical reaction and having a sufficient adhesion strength on the metallic base during the mechanical treatment with low intensity. Very frequently a top layer develops from very fine-grained abrasion coming into existence during the treatment in a mill, both on the reaction material to be treated

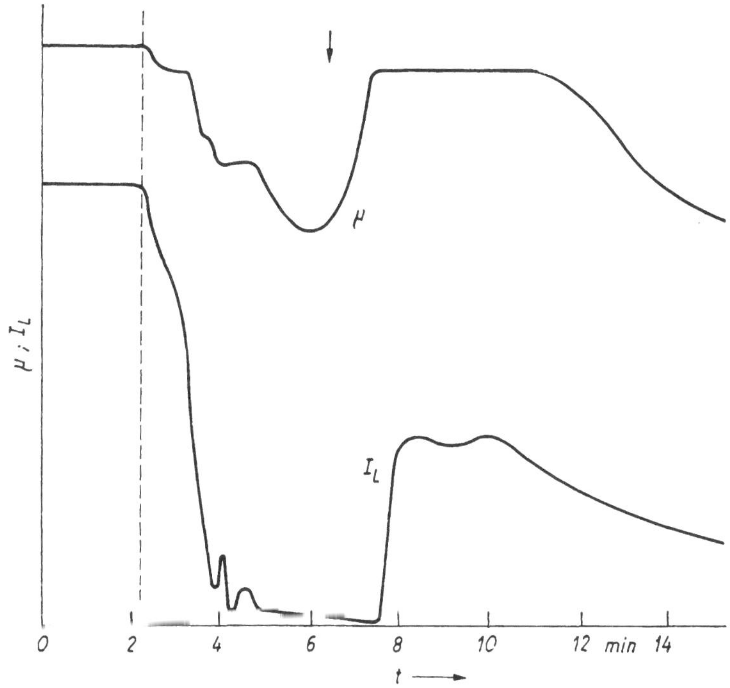

Fig. 5.54. Similar changes of friction strength μ and intensity of luminescence I_L due to the treatment of lithium fluoride by a steel cutter with and without water [201].

t – treatment time, ¦ – addition of H_2O in the form of gas, ↓ – removal of H_2O by evacuation.

and on the surfaces of the milling tools. This top layer may be coupled very strongly with the base and is often able to significantly alter the interface-mechanical behaviour of the solid. Thus, for instance, during the milling of SiC by means of steel balls, a fine-grained SiC layer is formed on the latter which almost completely prevents the further abrasion of the milling tools [200].

Both the interface-mechanical and the physical and chemical behaviour of the solid is changed by these top layers. If, for instance, an alkali halide crystal is treated by means of a steel cutter in a high-vacuum then a triboluminescence is thereby released (see 3.3.). With water vapour acting on the cutter surface, tribooxidation and the formation of oxidic layers will occur. Due to the reduction of the μ-value conditioned by this, the mechanical energy transferred onto the solid will be reduced thus causing reduction of the number of running elementary processes (see Fig. 5.54.).

If the water vapour atmosphere is removed and the treatment continues in the high-vacuum then a regeneration of the oxide layer removed due to the friction is not possible. This leads to an increase of the μ value and luminescence intensity [201].

Investigations using the experimental scheme described in 4.2. have shown that the conditions of the treatment of Ni surfaces are changed by the formation of top layers in the presence of certain substances, such as Na, Ca, MoS_2, FeS_2. In this way the course of tribochemical reactions is influenced as well.

Thus, for instance, the tribochemical formation of nickel carbonyl is completely tied up by calcium. As calcium is a very soft metal the mechanical energy transferred to a large extent dissipates under the release of deformation processes in the top layer, whereas the nickel does not undergo any activation (Chapt. 4., [47]).

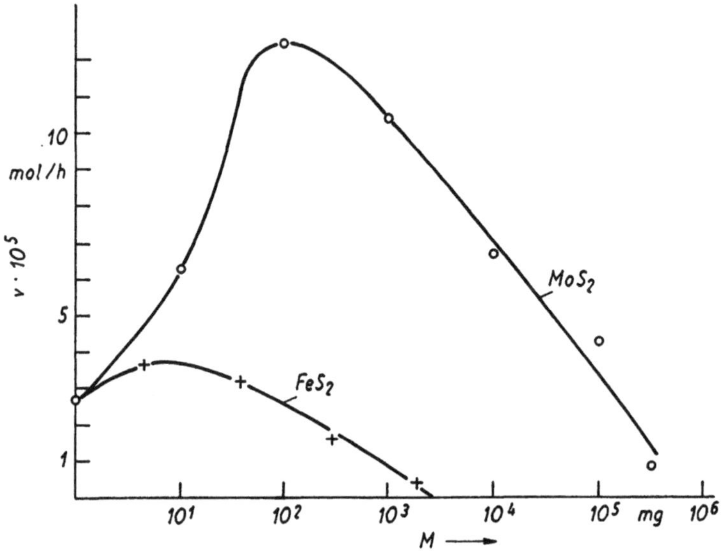

Fig. 5.55. The effect of FeS_2 and MoS_2 of the tribomechanically excited formation of nickel carbonyl [203].
v – reaction velocity, M – added FeS_2 and MoS_2, respectively.

If the formation of nickel carbonyl is investigated with successive addition of increasing amounts of MoS_2 and FeS_2, then the curves given in Fig. 5.55. are obtained. In both cases the velocity of the nickel carbonyl formation increases with the addition of small amounts of sulphide. This effect is attributed to the tribochemical reaction of the applied sulphide with the nickel by formation of a γ-nickel sulphide which the formation of nickel carbonyl sensitizes according to section 6.2. With the addition of small quantities of sulphide, upon treatment the metallic surface is activated and on account of this the solid reaction

$$2\,Ni + MoS_2 \rightarrow Mo + 2\,NiS \tag{56}$$

initiated [202].

The addition of larger amounts of MoS_2 decreases the formation velocity of nickel carbonyl again. The top layer effect gains more and more influence and leads finally to an extensive inhibition of the chemical reaction. An MoS_2 layer with some 1,000 molecular layers already brings about a lowering of the reaction velocity to about 30 per cent in comparison to the value measured on nickel without a top layer [203].

Because of the low shear strength of MoS_2, during the treatment the applied mechanical energy is extensively dissipated in this thin layer and is therewith lost for the activation of nickel.

Altogether these examples show that a considerable inhibition of the tribochemical reaction may appear by the action of the top layer effect. This process is intentionally used for the extension of the lifetime of tribomechanically stressed parts of engines in frictional and wear processes, as shown in section 7.4.

5.9. The effect of tribomechanically caused structural changes on the solid reactivity

5.9.1. *The influence of crystal imperfections on the solid reactivity*

As has been presented in section 3.2., various crystal defects develop in dependence on the properties of solids during the latter's mechanical treatment, whose concentration depends on the type and intensity of the treatment as well as on the duration of stress. Since generally the different crystal defects appear in different concentrations in dependence on the pretreatment until now it has only been managed in a few cases to derive an unambiguous correlation between certain types of defects in the crystal and their reactivity. Researches in this field have only recently begun to be carried out more intensively.

Studies have centred on a characterization of lattice defects and their reaction behaviour leading to a phenomenological coupling of parameters of structure and reactivity and providing an integral conclusion on the disorder of the mechanically treated substance. A discussion of these works therefore should precede the discussion on the influence of chemical reactions through individual defects.

5.9.1.1. *The influence of the degree of disorder on the reactivity*

The parameters of disorder which have been most investigated up until now are lattice imperfections, lattice distortions, and crystallinity and amorphization, respectively (see section 3.2.). These quantities have particularly proved their worth for the determination of disorders in finedisperse milled substances. In Fig. 5.56. the alteration of the mentioned quantities is plotted together with the changes in reactivity for selected solids in dependence on treatment time. Further examples are summarized in Tab. 5.22.

They indicate that the quantities of structure and reactivity in a large number of cases change to each other in a very similar way during the treatment. The defect structure in the solid brought about by the action of mechanical energy consequently belongs to the decisive quantities determining its reaction behaviour.

As will be shown in detail the structure-reactivity relations are system-dependent, that is to say they depend both on the nature of the solid and the lattice defects and

Table 5.22. Changes of chemical and physico-chemical processes by tribomechanically chaused.

No.	Substance	Kind of defect*)	Kind of stress**)	Treatment time t, h	Concentration of defects after the time t	Occurrence of an imperfection maximum	Chemical and physico-chemical processes changing similarly with the concentration of defects	Ref.
1	Hydrargillite	GS	S	3	0.012 nm	–		[205]
2	Boehmite	GS	S	5	0.011 nm	–	Solubility in sodium hydroxide solution	[206]
3		GS	K	25	0.011 nm	–		[206]
4	Corundum	GS	S	25	0.011 nm	–		[207]
5	β-PbO_2	GS	S	200	0.003 nm	+	Decrease of the oxidation number	[208]
6	MgO	GS	S	4	0.006 nm	+	Change of the bulk density	[209]
7	MgO	GS	S	20	0.005 nm	+	Hydration	[210]
8	Fe_3O_4	GS	S	1	0.01 nm	+	SO_2 oxidation	[211]
9	CaO	GS	S	80	0.01 nm	+	Hydration	[212]
10	Fe_3O_4	GS	S	5	0.012 nm	+	H_2S adsorption	Chapt. 4., [143]
11	Ni	GS	S	2.3	0.005 nm	–	Benzene hydrogenation	Chapt. 4., [143]
12	α-Fe_2O_3	GS	S	1.5	0.012 nm	+	SO_2 oxidation	[96]
13	Zn	GS	S	3.3	0.007 nm	–	$Zn + Cd^{2+} \rightleftharpoons Cd + Zn^{2+}$	[213]
14	SiC	GS	S	90	0.015 nm	–	Change of the quantity of H_2 tribosorbate	[91]
15	NaCl	GS	S	***)	0.028 nm	–	Catalytic decomposition of *i*-propyl chloride	[107]
16	CaF_2	GS	S	2	0.013 nm	+	Solubilization by acids	[214]
17	CaF_2	GV	S	2	0.43%	–	Solubilization by acids	[214]
18	Ni	GV	S	4.2	1.2 %	–	$Ni + 4\,CO \rightarrow Ni(CO)_4$	[106]
19	Ni	GV	S	2.3	0.15%	–	Benzene hydrogenation	Chapt. 4., [143]
20	Corundum	GV	S	25	0.3%	–	Solubility in NaOH solution	[207]
21	NaCl	GV	S	***)	0.05%	+	Catalytic decomposition of isopropyl chloride	[107]
22	Kola-apatite	GV	S	3	0.3%	–		[215]
23	Marocco-phosphate	GV	S	4	0.45%	–	Solubility in ammonium citrate solution	[215]
24	Kola-apatite	A	P	1	70%	–		Chapt. 4., [60]
25	SiO_2	A	S		100%	–	Solubility in NaOH solution	[115]
26	SiO_2	A	S	400	100%	–		[216]

Table 5.22. (Continuation)

No.	Substance	Kind of defect*)	Kind of stress**)	Treatment time t, h	Concentration of defects time t	Occurrence of an imperfection maximum	Chemical and physico-chemical processes changing similarly with the concentration of defects	Ref.
27	Hydrargillite	A	S	25	100%	—	Solubility in NaOH solution	[205]
28	Boehmite	A	S	25	100%	—	Solubility in NaOH (Solubility: 90%)	[206]
29	Boehmite	A	K	25	40%	—	Solubility in NaOH (Solubility: 35%)	[206]
30	β-MnO_2	A	S	200	75%	—	Decrease of the oxidation number	[217]
31	Kaolinite	A	S	25	100%	—	Acid solubilization	[92]
32	Chromite	A	S	200	100%	—	Acid solubilization	[218]
33	CaC_2	A	S	1	65%	—	$CaC_2 + N_2 \rightarrow CaCH_2 + C$	[219]
34	CaC_2	A	K	1	15%	—	$CaC_2 + N_2 \rightarrow CaCH_2 + C$	[219]
35	SiC	A	S	90	55%	—	Changes of the quantity of H_2 tribosorbate	[91]

*) GS — lattice imperfection, GV — lattice distortion, A — amorphization.
**) S — vibration mill, K — ball mill, P — planetary mill.
***) Determined in the continuous-type mill.

on the character of the chemical reactions in question. As an example, in Fig. 5.56.a) the formation of methane taking place during the treatment of SiC in hydrogen atmosphere and the tribosorbed amounts of H_2 are related with the lattice imperfections. This shows that the formation of methane takes place independent of the degree of lattice imperfections, and according to 6.1.1. runs via the more excited states not measurable by X-rays, appearing directly during the mechanical treatment. On the other hand the

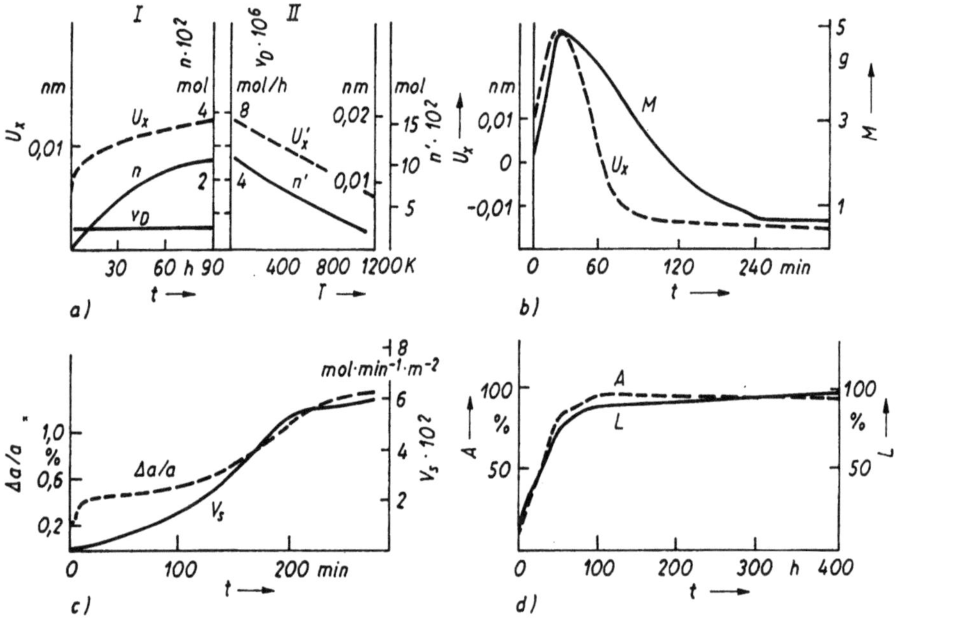

Fig. 5.56. Dependence of roentgenografically detectable crystal defects of various solids and their reactivity on the treatment time *t* using vibration mills and the annealing temperature *T*.

a — I — treatment of SiC in H_2 atmosphere [91], II — annealing of treated SiC [91], u_x, u_x' — lattice imperfections, n,n' — tribosorbate, v_D — methane desorption velocity.

b — H_2S intake at treated magnetic [204], u_x — lattice imperfections, M — H_2S intake.

c — specific velocity of the nickel carbonyl formation at nickel treated mechanically [106], a — lattice distortions, v_s — pecific reactivity.

d — solubility of quartz treated mechanically [97], A — amorphous part, L — solubility.

amount of the H_2 tribosorbate correlates very well with the lattice imperfections, which also appears from the annealing curve of Fig. 5.56.a), a.II. The amount of H_2 tribosorbed in the SiC boundary layer falls with growing temperature, very similarly to the reduction of lattice imperfections. Hence generalizations with regard to the structure-reactivity relation, are possible to a limited extent only and can only be mode if the same types of reactions in solids are studied, such as decomposition reactions with charge transfer (see 5.3.), decomposition reactions without charge transfer (see 5.5.), carbonylization reactions (see 4.4.), hydrogenation reactions (see 6.2.), reactions of paraffins (see 6.1.1.), solution reactions (see 6.1.2.), and so on, the bond character of which is very similar such as in metals like iron, nickel, cobalt, in alkali halides such as NaCl, LiF (see 5.4.) or in substances having proportions of strongly homopolar bonds such as SiC (see 6.1.).

As shown in Fig. 5.56. b) the treatment not only leads to the formation of certain types of lattice imperfections but also accelerates their annealing. In such cases with continuing milling a maximum appears in the concentration of the lattice imperfection, whose position depends on the nature and intensity of the mechanical stress [220]. The appearance of defect maxima is observed particularly frequently with lattice imperfections. However, the removal of lattice distortions by treatment could only be found for NaCl and is connected with the increase of the primary crystallite size (see Fig. 5.27.). In the case of amorphization, (see Fig. 5.56. d) however innocase could the appearance of a maximum be observed or the reduction of the degree of amorphization be found under the influence of mechanical energy. Apparently according to equ. 37 the probability for the recrystallization of the amorphous state is so small and the activation energy for the reforming process so large, that the change over of the amorphous into the crystallized phase during the treatment may be neglected in comparison to the reverse running process.

Corresponding to this a tribomechanically released amorphization comes into existence, especially in such substances whose bonds are of a strongly homopolar nature, such as SIC (Tab. 5.22., No. 25, 26, 31), and which have a complicated lattice structure, such as kaolinite (Tab. 5.22., No. 31), feldspar, chlorite, montmorillonite, and so on [113].

From this the conclusion may be drawn that the reactivity of the solid does not increase in every case with growing treatment time, but passes a maximum in numerous cases. According to Tab. 5.22. this applies above all to such chemical processes which run on metallic oxides and depend on lattice defects.

The final state obtainable with treatment in terms of lattice imperfections and reactivity, depends to a large degree on the nature and intensity of the mechanical stress. As shown by the example of CaC_2 (Tab. 5.22., No. 33, 34) the treatment in a vibration mill leads to a considerably higher degree of amorphization with the same treatment time and consequently also to a higher reaction velocity than by using a ball mill. The same results have been obtained for boehmite, too Tab. 5.22., No. 28, 29).

The increase of the solid imperfection by a more intensive mechanical treatment is one of the essential reasons for the increase of the chemical exchange with growing intensity of treatment (see 4.4.).

As can be shown with the example of the activation of Kola apatite the dominating defect structure is qualitatively changed by increasing the intensity of treatment, e.g. by transfer from vibration mill to planetary mill. Whereas lattice distortions also occur in the vibration mill after a longer treatment time, the same substance becomes vastly amorphous in the planetary mill. This can also be seen by the jump-like increase in solubility (see Tab. 5.22., No. 22—24).

Here a possibility is given to produce solids having selected reactivity properties by means of the type of mechanical stress.

5.9.1.2. The reaction behaviour of solids in dependence on the distribution of crystal imperfections

The concentration of imperfections measured roentgenographically only represents a mean value. As discussed in 3.2. however, with mechanical stress of grain material the imperfections are particularly concentrated in grain boundary layers and grain surfaces. These are also the sites of increased reactivity. As an example a nickel surface etched with atomic hydrogen is shown in Fig. 5.57. a). Some grain boundaries are weakly developed. As Fig. 5.57. b) shows, the formation of nickel carbonyl is accelerated by the influence of CO at these grain boundaries and becomes visible through the occurrence of deep scratch traces (Chapt. 4., [101]). In the vicinity of scratch traces the slip bands generated by the deformation may similarly become visible by etching with CO, Fig. 5.57. c).

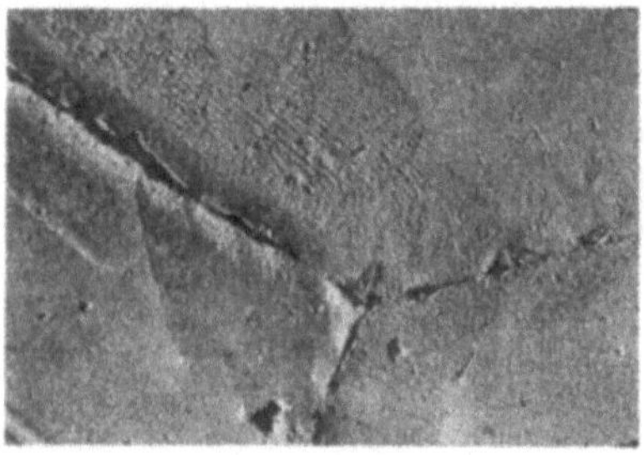

a – nickel surface after the action of atomic hydrogen,

b – nickel surface after etching with CO for 320 hours,

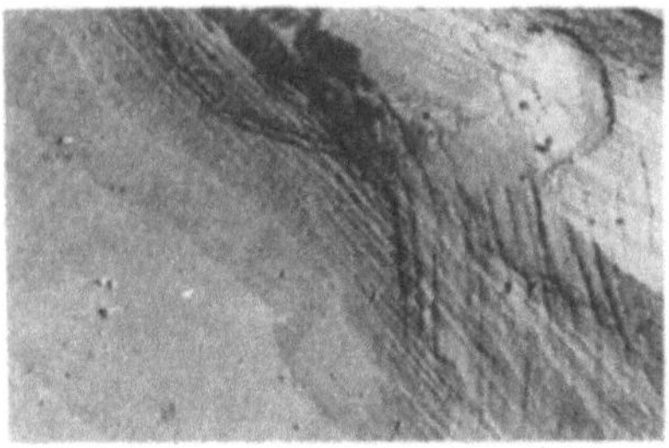

c – treated nickel surface after a treatment with CO for 300 hours.

Fig. 5.57. Etching effect of CO at mechanically treated and untreated nickel surfaces after having been treated by atomic hydrogen (Chapt. 4., [101]).

As shown in Fig. 5.28., the decomposition pressure of $CaCO_3$ grows with increasing surface of the primary crystallites. Since the decomposition of $CaCO_3$ as a reaction without charge transfer is strongly influenced by lattice imperfections [151], the conclusion may be drawn, taking into account the result given in Fig. 5.28., that the imperfections existing with increased concentration in the grain boundaries considerably determine the decomposition of $CaCO_3$.

Detailed studies on the distribution of X-ray amorphous lattice ranges in vibration-milled quartz samples have also shown that the amorphisation of quartz particles mainly takes place in the lattice layer near to the surface whereas the core is only little disordered (see 6.1.2.).

It is possible by use of different types of treatment, to influence to a certain extent the distribution of lattice imperfections. According to 6.1.2. a different distribution of lattice imperfections will also result in changed kinetics of the solution process. As a result of the local accumulation of disturbances, a formation of local elements appears through the impact treatment of metals in an electrolytic solution.

The chemical processes caused by this will be discussed in 6.4.

5.9.1.3. *Structure-reactivity correlations in the presence of several crystal imperfections in the solid*

Frequently several crystal imperfections are produced by the mechanical treatment of solids (see Tab. 5.22., e.g. boehmite, hydrargillite, NaCl, chromite, etc.). In such cases an analysis of the structure- reactivity correlations is made more difficult and enables qualitative evaluations only. Thus, for instance, the katalytic behaviour of NaCl (Tab. 5.22., No. 15, 21) is determined by the combined effect of lattice imperfections and lattice distortions in the decomposition of *i*-propyl chloride [107].

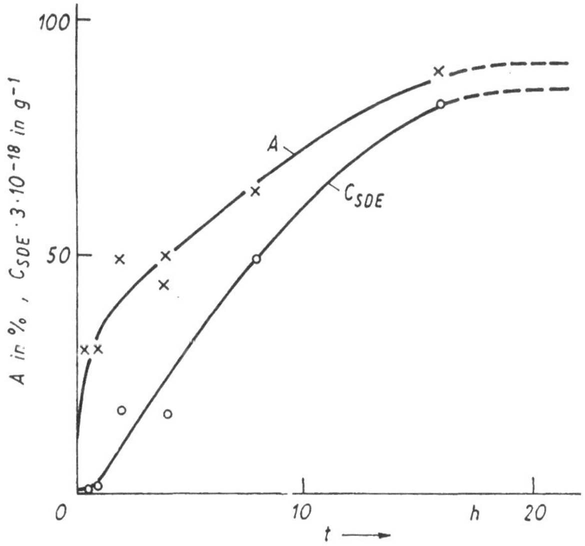

Fig. 5.58. Amorphous part A and oxygen defect electron concentration τ_{SDE} of mechanically stressed SiO_2 depending on the treatment time t [221].

However, one type of crystal imperfections frequently dominates which then becomes the decisive factor for the alteration of the solid reactivity. As an example the alteration of the amorphous phase content and the number of the oxygen defect electron centres through the treatment of quartz in dependence on treatment time, is presented in Fig. 5.58. The centres may be removed by annealing at 700 K without reducing the degree of amorphization. In this way the solubility of quartz can be shown to determine exclusively the amorphization. Since the ratio of the number of the oxygen defect electron centres to the amount of amorphized SiO_4 tetrahedron accounts for 1 : 1,000 no significant influence on the solution process is to be expected either, due to the relatively low concentration of centres [221].

In some cases one type of defect dominates at the beginning of the milling process, e.g. lattice imperfections. After a longer period of treatment other crystal defects are emphasized, e.g. lattice distortions or amorphous ranges. Thus it could be shown by the example of the milling of corundum that lattice imperfections are the decisive factors for the solubility at the beginning of the treatment, and lattice distortions after a longer period of treatment [207].

The analysis of structure-reactivity relations is also made more difficult by the fact that surface alterations appear in addition to structure alterations during mechanical stress (see surface layer effect, 5.5.), and particularly in metals, the formation of fresh surfaces, too (see fresh surface effect 5.6.). Because of this several factors changing in the course of the action of mechanical energy influence the run of a solid reaction in a frequently unclear way. The combined action of these factors can be seen by the example of nickel carbonyl formation in Tab. 5.23.

Table 5.23. Changes in the chemical and physical properties of two different nickelpowders in dependence on the treatment time *t* [106].

t, min	Velocity of reaction $Ni + 4\,CO \rightarrow Ni(CO)_4$ $v \cdot 10^5$, mole/min · g	Spec. surface area S, m^2/g	Spec. reactivity $v_s \cdot 10^5$, mole/min · m^2	Lattice distortion $\frac{\Delta a}{a}$, %	Dislocation density $\rho \cdot 10^{-10}$, cm^{-2}
Pyrophoric Ni powder					
0	39.8	7.38	5.4	—	—
20	84.9	4.89	17.4	0.63	0.2
60	101.1	2.19	45.8	1.06	0.5
120	101.6	1.98	55.5	1.51	3
240	120.4	1.71	70.4	2.20	—
840	100.8	1.49	68.2	1.72	7
Annealed Ni powder					
0	0.05	0.11	0.5	0.27	—
20	0.3	0.15	2.0	0.42	—
60	0.8	0.15	5.3	0.46	0.12
120	2.0	0.14	14.1	0.59	0.2
240	6.5	0.11	59.2	1.30	1
840	6.4	0.11	58.8	2.20	3

In Tab. 5.23. the reactivity of both highly active, pyrophoric Ni powder and low reactive, tempered Ni powders and its dependence on the individual influencing factors are compared. Pyrophoric and tempered nickel powders differ in their reactivity by about 3 orders of magnitude. According to Tab. 5.23. for this can be atributed above all to tho very different surface area of the two nickel samples and to the presence of a top layer inhibiting the reaction on the tempered nickel sample (see 5.6.). The top layer is removed by treatment. It is also the cause of the specific reactivity increasing much more slowly than the concentration of lattice distortions at the beginning of the treatment, (see Fig. 5.56. c).

If the strong influence of the surface area is eliminated by the calculation of the specific reactivity, then the latter only still depends on the concentration of the lattice distortions through a linear relation which is identical for both samples independent of the preparation of the samples (see Fig. 5.59.).

As the last mentioned example shows, in most cases there are several causes of importance for the alteration of the velocity of tribochemical reaction. For this reason, in order to clarify the decisive structure-reactivity correlations, all the influencing factors which change under the treatment have to be included in the process analysis.

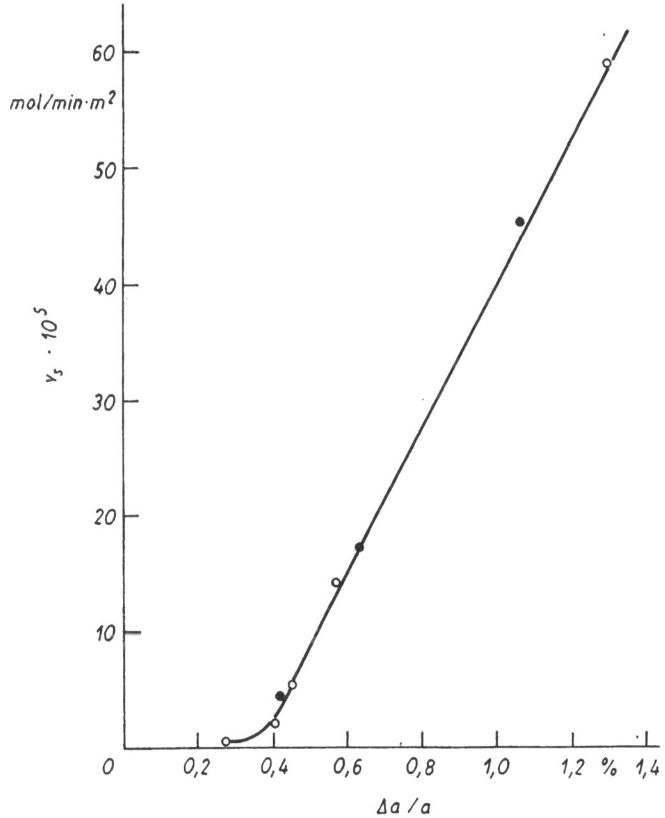

Fig. 5.59. Specific velocity v_s of the nickel carbonyl formation at pyrophoric and annealed nickel powder in dependence on the lattice distortion $\Delta a/a$ [106].

• — pyrophoric nickel powder, ○ — annealed nickel powder.

5.9.2. *Correlations between individual structural imperfections and reactivity of solids*

It is possible in individual cases to attribute the alteration in reactivity to the effect of defined structural defects, e.g. to dislocations, point defects, and so on (see 3.2.). A number of these defects decisively influencing in particular the reactivity of mechanically treated solids will be discussed in detail in the following sections.

5.9.2.1. *The influence of dislocations on the reactivity of solids*

As has been shown in 3.1. the mechanical stress of solid surfaces leads to deformation processes by the production and motion of dislocations. Particularly high dislocation densities occur in the vicinity of impact-stressed surface ranges, see Fig. 5.57. c), Fig. 3.3. The concentration of dislocations also increases with the treatment time during milling, e.g. of metallic powders, and after a period depending on the material reaches a steady-state final value [222]. The calculation of the dislocation densities may be done, for instance, for treated nickel powder from the lattice distortions measured roentgenographically [223]. As Tab. 5.23., column 6 shows, an increase in the velocity of the nickel carbonyl formation related to the unit surface area appears very similar in course to the increase of the dislocation. In tribocatalytic reactions too, the alteration of the specific activity may be attributed to changes in the concentration of dislocations. According to Fig. 5.60. there is a linear correlation between the constant of the specific hydrogenation catalyst activity and dislocation density, dislocation energy, and lattice distortions.

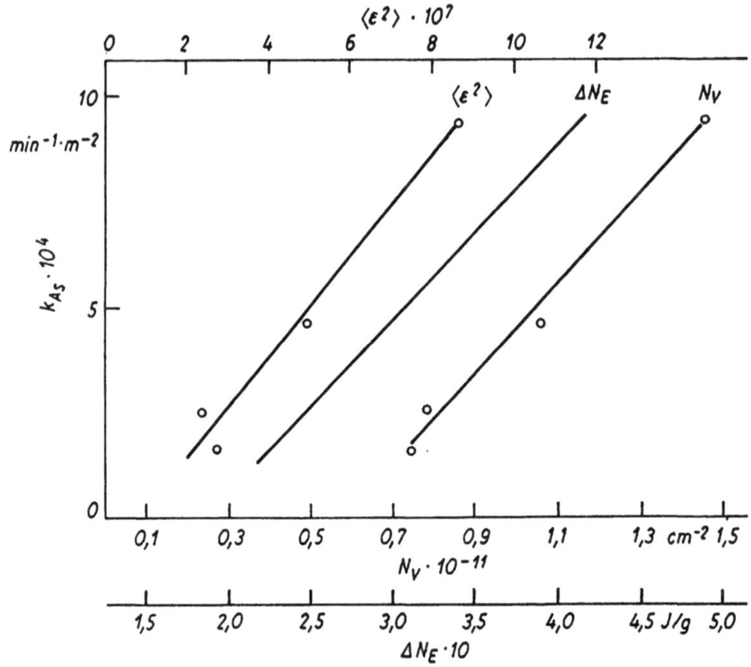

Fig. 5.60. Dependence of the specific activity constant of a mechanically treated hydrogenation catalyst, k_{A_s} on the dislocation density N_v, the dislocation energy ΔN_E, and the lattice distortion $\langle \varepsilon^2 \rangle$ for the fat hardening (Chapt. 4., [87]).

As already discussed in 5.2. the bonds undergo elastic, lattice deformations in the vicinity of a dislocation, which lead to the accumulation of excess energy in the zone of dislocation. According to equ. (4.34) the activation energy of the chemical reaction

$Ni + 4CO \rightarrow Ni(CO)_4$ as well as of the catalytic reaction (hydrogenation) is reduced by this (see Tab. 4.10.).

Fig. 5.61. shows that there is a close correlation between the dislocation energy ΔN_E and the reduction of the activation energy ΔE_A .The product of ΔN_E, and ΔE_A remains constant during the whole activation time, in which the dislocation density changes by more than 200 per cent. The catalytic activity of different mechanically treated metals, such as copper, nickel, platinum, depends on their dislocation densities in the dehydrogenation of ethanol and in the transformation of ortho-para hydrogen as well [224–226].

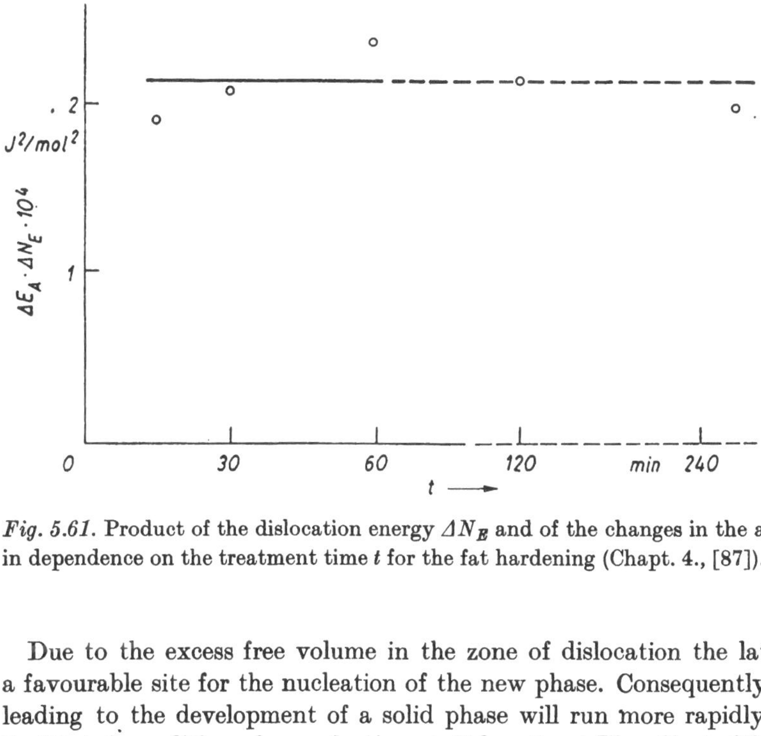

Fig. 5.61. Product of the dislocation energy ΔN_E and of the changes in the activation energy ΔE_A in dependence on the treatment time t for the fat hardening (Chapt. 4., [87]).

Due to the excess free volume in the zone of dislocation the latter also represents a favourable site for the nucleation of the new phase. Consequently chemical reactions leading to the development of a solid phase will run more rapidly on account of the facilitated conditions for nucleation at dislocations. The effect of H_2S on mechanically treated nickel may be given as an example. This leads to the selective development of NiS according to the following equation:

$$Ni + H_2S \rightarrow NiS + H_2 \quad (57)$$

at the slip lines produced by impact stress (see Fig. 5.62.).

As Fig. 5.63. shows, the reaction starts mainly within the slip bands and in the point of intersection of crossing slip bands [227]. Very rapidly compact NiS layer composed of γ-NiS with strong lattice imperfections [228] develops directly in the range of impact stress, (see Fig. 5.62.).

Very strong dislocations also act on decomposition reactions running without electron transfer [151]. This fact could be proved for instance by investigations of mechanically deformed $CaCO_3$ monocrystals. During the decomposition a dependence of the relative dislocation activity on their crystallographic orientation has been observed. Thus the reactions appear mainly at dislocations slipping on 100-areas but not so markedly at dislocations in 111-areas [229]. Similar graduations have been found in the oxidation of graphite, too [230, 231].

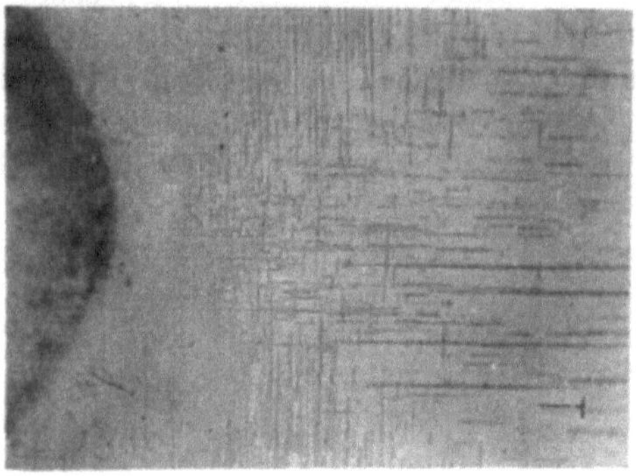

Fig. 5.62. Ball indentation at a [100] nickel surface with surrounding slip system decorated by H_2S [227].
|—| – 20 µm.

Since upon exceeding the limit of elasticity the mechanical treatment leads to a plastic deformation and consequently to the production and migration of dislocations as well as to their multiplication, in contrast to radiation chemistry this type of imperfection has a special importance as a cause of changed reactivity of solids in tribochemistry.

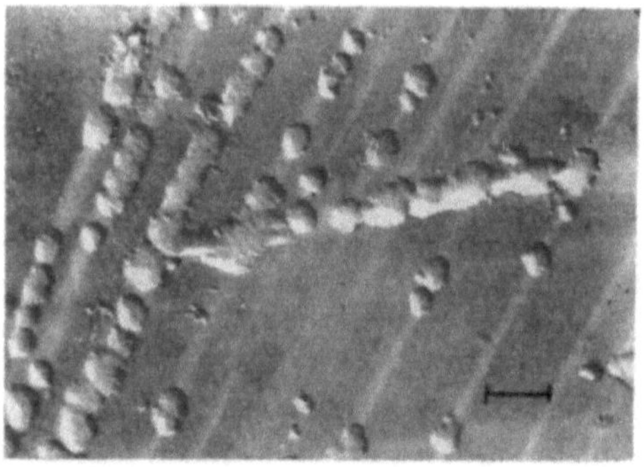

Fig. 5.63. Formation of NiS within the slip line and in the cut region of crossing slip lines due to the treatment of a mechanically stressed Ni surface by H_2S [227].
|—| – 1 µm.

5.9.2.2. *The influence of atomic lattice defects on the course of tribochemical reactions*

Atomic lattice defects influence very strongly the reaction velocity of decomposition reactions whose course is coupled with a charge transmission. For instance the decomposition of AgN_3, NH_4ClO_4, and $KMnO_4$ belong to this group of reactions [151]. As example the decomposition of silver oxalate may be discussed, the chemical decomposition of which takes place according to the following mechanism:

$$C_2O_4^{--} \rightarrow 2 \cdot CO_2^{-} \xrightarrow{-2e} 2CO_2 \qquad (58)$$

Here the electron is released to an interstitial silver ion [232]. The concentration of interstitial ions and consequently the velocity of the thermal decomposition is reduced in accordance with the requirements of the electro-neutrality principle, by the introduction of divalent ions, e.g. copper ions.

The concentration of atomic lattice defects, however, is markedly increased by mechanical treatment, a fact which can be proved for instance by an increased conductivity of the stressed oxalate samples (see Fig. 5.64.). Comprehensive studies have shown that the increase in conductivity must be mainly attributed to the increased concentration of interstitial ions in the solid surface [234]. As Fig. 5.65. shows the decomposition

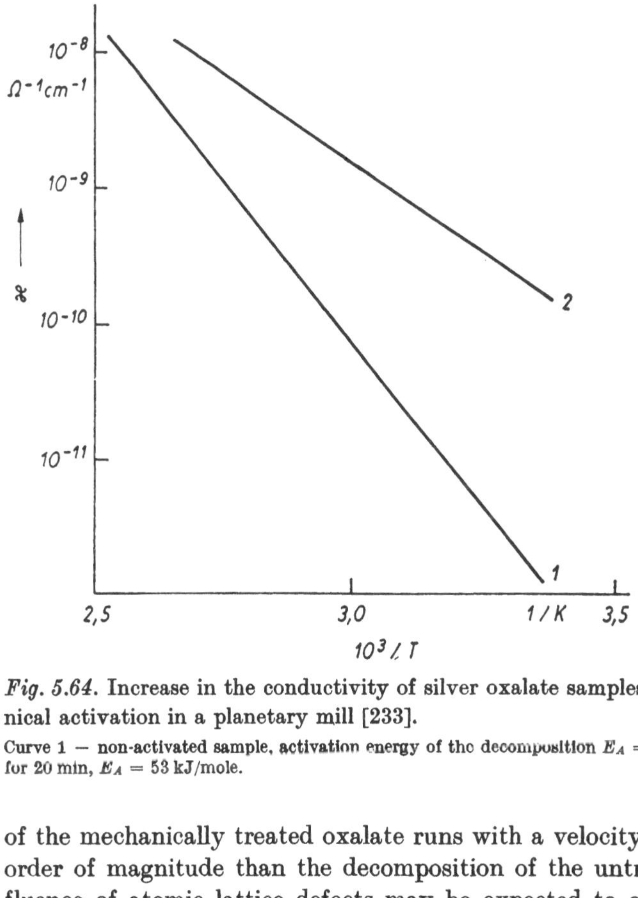

Fig. 5.64. Increase in the conductivity of silver oxalate samples with temperature by the mechanical activation in a planetary mill [233].

Curve 1 — non-activated sample, activation energy of the decomposition $E_A = 85$ kJ/mole, curve 2 — sample activated for 20 min, $E_A = 53$ kJ/mole.

of the mechanically treated oxalate runs with a velocity which is higher by about one order of magnitude than the decomposition of the untreated sample. This strong influence of atomic lattice defects may be expected to occur in all chemical reactions where the electron transition to one of these defects represents the step determining the velocity.

As can be seen the example of tribodiffusion in 5.7. the vacancy concentration may also exert a strong influence on the migration of reactants in the mechanically stressed lattice.

5.9.2.3. *The importance of radical states for the course of tribochemical reactions*

As shown in 5.1.—5.3. due to the effect of mechanical energy a breakdown of chemical bonds comes into existance according to different mechanisms. In principle two cases are possible:

$$\sim A - B \sim \longrightarrow \sim A^{\cdot} + {}^{\cdot}B \sim \qquad (59)$$

$$\sim A - B \sim \longrightarrow \sim A^{+} + {}^{-}B \sim \qquad (60)$$

In the first case the energy input required for the decomposition is equal to the strength of the bond $A - B$ whereas in the second case additional energy has to be applied being equal to the difference of the ionization potential of the fragment $\sim A(I_A)$ and the affinity for the electron of the fragment $\sim B(A_B)$, that is $I_A - A_B$. This difference is not smaller than 4 ⋯ 5 eV in most organic compounds, so that the formation of

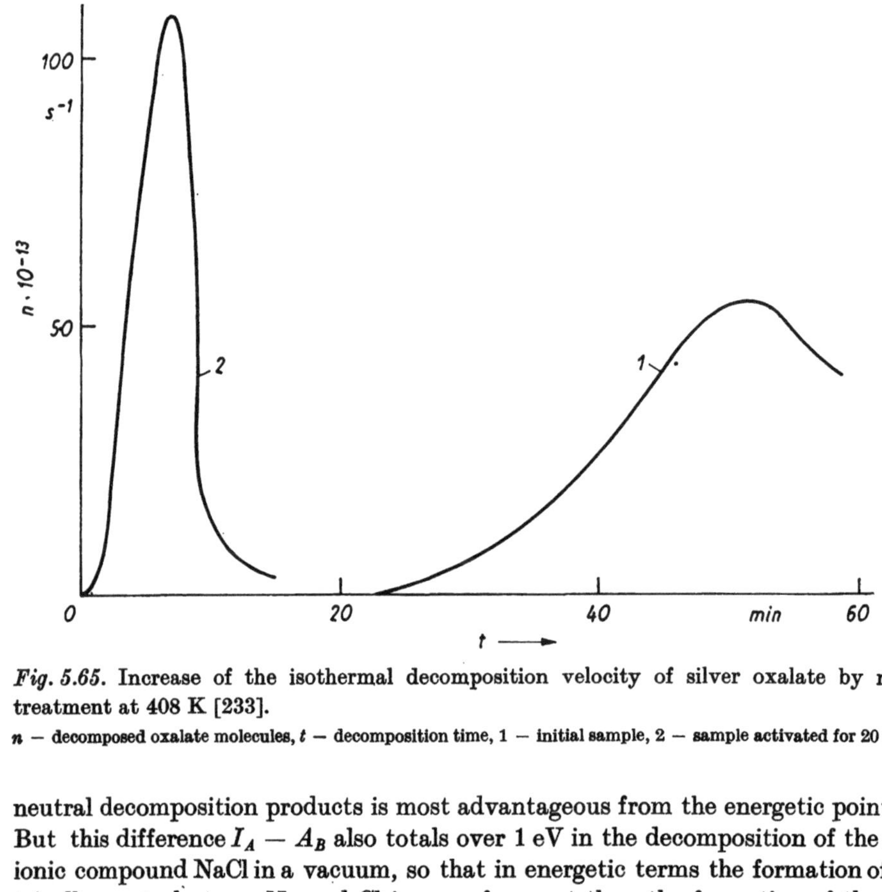

Fig. 5.65. Increase of the isothermal decomposition velocity of silver oxalate by mechanical treatment at 408 K [233].

n — decomposed oxalate molecules, t — decomposition time, 1 — initial sample, 2 — sample activated for 20 min.

neutral decomposition products is most advantageous from the energetic point of view. But this difference $I_A - A_B$ also totals over 1 eV in the decomposition of the typically ionic compound NaCl in a vacuum, so that in energetic terms the formation of the electrically neutral atoms Na and Cl is more frequent than the formation of the ions Na^+ and Cl^-. Generally the breaking-up of bonds by mechanical means, primarily results in the development of radical centres which are capable of entering into further subsequent reactions because of their high reactivity. The proof of these subsequent reactions is furnished ESR-spectroscopically. In this way chain fragments having terminal radicals come into being during the tribochemical degradation of polymers [235]:

$$\sim CH_2{-}CH_2{-}CH_2{-}CH_2 \sim \;\rightarrow\; \sim CH_2{-}H_2C^{\cdot} + {}^{\cdot}CH_2{-}CH_2 \sim \qquad (61)$$

The primary free radicals developing during the degradation of various polymers are summarized in Tab. 5.24.

Pursuant to the following equation chain radicals may also be developed as a result of the elimination and addition of neighbouring molecules [234]:

$$\sim CH_2{-}H_2C^{\cdot} + \sim CH_2{-}CH_2{-}CH_2 \sim \;\rightarrow\; \sim CH_2{-}CH_3 + \sim CH_2{-}\dot{C}H{-}CH_2 \sim \qquad (62)$$

Due to the high activity of the terminal radicals not only the primary but also the more stable secondary chain radicals exist even at 80 K and in the vaccuum after a chainbreaking. Peroxide radicals develop in the presence of oxygen according to the

Table 5.24. Structure of the primary free radicals developing by the mechanical degradation.

Polymer	Structure of radicals	Ref.
Polyethylene	$\sim CH_2{-}H_2O^{\cdot} + .CH_2{-}CH_2 \sim$	[236]
Polytetrafluoroethylene	$\sim CF_2{-}F_2C^{\cdot} + .CF_2{-}CF_2 \sim$	[237]
Polystyrene	$\sim CH(C_6H_5){-}H_2C^{\cdot} + .\,CH(C_6H_5)CH_2 \sim$	[235]
Polyamide	$\sim CH_2{-}CH_2C^{\cdot} + .CH_2{-}(NHCO) \sim$	
Polyethylene oxide	$\sim x{-}H_2C^{\cdot} + .CH_2{-}x \sim$	[238, 239]
Polyethylene sulphide		
($x = 0, S$)		

following equation [240]:

$$\sim CH_2{-}HC^{\cdot} + O_2 \rightarrow \sim CH_2{-}HCOO^{\cdot} \qquad (63)$$

Depending on the property of the radical and the reaction conditions a great number of secondary reactions may appear in analogy to equ. (62), such as substitution reactions, addition reactions, recombination reactions, disproportionations, interactions with available admixtures, or the release of polymerization reactions. This will be discussed in detail in 6.3.

By means of ESR measurements, in the mechanical treatment of inorganic solids, such as of C [221, 241, 242], SiC [243], SiO_2 [221, 244—246] and so on, it was also possible to prove paramagnetic centres. The interpretation of the nature of these centres has proved to be partially much more complicated than for polymers. Even today it is not yet possible in certain cases to assign the centres measurable by ESR spectroscopy unambiguously to a particular structure. The paramagnetic centres obtained by the milling of graphite are mainly arranged at the surface of the prismatic grain boundaries. Their concentration amounts to about 2×10^{14} cm^{-2}, that is approximately 5 ··· 10 per cent of the total number of the surface carbon atoms [247]. It is assumed that a breaking of the graphite lattice due to the treatment will lead to the development of radical carbon atoms having an increased reactivity. Thus the appearance of the ESR spectrum changes in the presence of oxygen whereas the total concentration of the centres remain constant. The following reaction is assumed to appear:

$$>C^{\cdot} + O_2 \rightarrow\ >C{-}OO^{\cdot} \qquad (64)$$

Both the primary and the oxidized centres are stable up to 600 ··· 800 K [248].

The paramagnetic centres at SiO_2 have also been comprehensively studied. The formation of centres is attributed to the mechanically initiated breaking of the $>Si{-}O{-}S<$-bonds:

$$>Si{-}O{-}Si< \ \rightarrow\ >Si^{\cdot} + .\,O{-}Si< \qquad (65)$$

When developing new surfaces, for instance, as a result of a breaking process, those atoms being on the boundary surface have free unsaturated valencies in the stage of their development. However, such a surface having unsaturated valencies is instable from the energetic point of view and leads to a rapid alteration of the bond structure (kinetics, see 4.3.).

But a small concentration of radical centres will be maintained for a longer period. One of these centres is the so-called *E'* centre corresponding to the $\equiv Si^{\cdot}$ radical and coming into existence during the treatment in the vacuum or in inert gases. The ratio of the *E'* centres still ascertainable by ESR spectroscopy after finishing the treatment to the broken-up Si—O—Si bonds, was estimated to be 6 : 1,000 for quartz [249]. Similar results were also obtained with the comminution of diamond [248].

Consequently the number of *E'* centres only represents a small fraction of the really broken-up SiO bonds. The development of *E'* centres could also be observed when SiO_2 surfaces were treated by plasma activation [250] or exposed to neutron or gamma irradiation [252].

The *E'* centre reacts chemically with a series of gases. Thus CO_2^- radical ion develops in the presence of CO_2 [249, 251]:

$$\equiv Si^{\cdot} + CO_2 \rightarrow \equiv Si^{+}CO_2^{-}, \tag{66}$$

that is to say a partial electron transfer takes place from the adsorbens to the adsorbate. The dissociation appears with thermal treatment up to 473 K, whereby the *E'* centre is reformed [249].

When SiO_2 is treated in an oxygen atmosphere, paramagnetic centres called oxygen defect electron centres occur. Until now the interpretation of this centre has not yet been completed. The assignement of the centres to the following structures is under discussion [221]:

1. $\equiv Si{-}O^{+}{-}Si\equiv$,

2. $\equiv Si{-}O^{+}{-}Hi$,

3. $\equiv Si{-}O{-}O^{\cdot}$

The following reaction is assumed to be responsible for the formation of the last mentioned group in analogy to equ. (63) and equ. (64), [244]:

$$\equiv Si^{\cdot} + {}^{\cdot}O{-}O^{\cdot} \rightarrow \equiv Si{-}O{-}O^{\cdot} \tag{67}$$

$$2 \equiv Si{-}O^{\cdot} + {}^{\cdot}O{-}O^{\cdot} \rightarrow 2 \equiv Si{-}O{-}O^{\cdot} \tag{68}$$

Because of the great differences in electronegativity of Si and O, the surface group is regarded as a radical ion of the silicon hyperoxide type with a redox-amphoteric character. The oxidizing and reducing effects of the radical ion could be proved. In analogy to the behaviour of the CO_2^- radical ion, a decomposition of the radical ion appeared by heating up to 453 ··· 473 K combined by the development of oxygen and the re-forming of the $\geqslant$Si$^{\cdot}$-radical [244]. Further statments with respect to the development of tribo-induced centres on SiO_2 can be found in 6.1.2.

In summary we may say that the results represented in 5.9. show that a large number of defects appear at mechanically treated solids, influencing in different ways the reactivity of solids in dependence on the type of reaction and the exterior conditions. In individual cases one has already succeeded in combining structure and reactivity properties with each other and creating the precondition for the clarification of the elementary mechanism in the tribochemical reaction.

However, the investigations in this field are at the beginning only. The research orientation depieled in 5.9.2.3. will also be of importance for the investigation of short-lived centres appearing in the post-plasma phase and which should be attributed to atoms, in a strongly unsaturated state of valency and in an energetically exited state, respectively.

Altogether the results represented in Chapt. 5. show that there is no uniform mechanism for the initiation of tribochemical reactions. On the basis of the existence of different dissipation mechanisms of mechanical energy several causes have to be taken into consideration for the initiation of reactions. In dependence on the reaction cause dominating at the time, tribochemical reactions may run in analogy to thermally-excited reactions or they may be very much like radiation-chemical and photochemical reactions. The characteristic feature of tribochemistry is above all based on the typical excitation processes due to the effect of mechanical energy, such as accnmulation and transformation of elastic energy, which have not been observed with thermal initiation or radiation activation.

References

[1] Thiessen, P. A., Meyer, K., Heinicke, G.: chapt. 1., [1], p. 514
[2] Chapt. 4., [9], p. 28
[3] Thiessen, P. A., Haase, G., Altnau, H., Gutsche, Ch., Florek, D.: Z. Chem. **16** (1976), 8
[4] Thiessen, P. A., Haase, G., Altnau, H., Pusch, Ch., Florek, D.: chapt. 1., [10], p. 45
[5] Thiessen, K. P.: Mber. Dt. Akad. Wiss. **10** (1968), 200
[6] Žurkov, S. N., Vettegren, V. I., Korsukov, V. E., Novak, I. I.: fiz. tv. tela **11** (1969), 290
[7] Kauzman, W., Eyring, H.: J. Amer. Chem. Soc. **62** (1940), 3113
[8] Tomaševskij, Ė. E.: fiz. tv. tela **12** (1970), 3202
[9] Berstein, V. A.: chapt. 1., [9], 2, p. 72
[10] Žurkov, S. N., Tomaševskij, E. E.: "Conf. Physical Basis of Yield and Fracutre", p. 200, Oxford 1966
[11] Vettegren, V. I., Kovsukov, V. E., Novak, I. I.: Plaste u. Kautschuk **19** (1972), 86
[12] Žurkov, S. N., Vettegren, V. I., Novak, I. I.: "II. Intern. Congress of Fracture", p. 545, Brighton 1969
[13] Butjagin, P. Ju., Radzig, V. A.: Plaste u. Kautschuk **19** (1972), 81
[14] Regel, V. A., Pozdnjakov, O. F.: Plaste u. Kautschuk **19** (1972), 100
[15] Sedov, V. V., Kuz'minskij, A. S., Kiršchstein, N. I.: chapt. 1., [9], 2, p. 80
[16] Butjagin, P. Ju.: Vysokomol. soed. **9 A** (1967), 136
[17] Gilbert, L.: chapt. 4., [2], p. 94
[18] Chapt. 4., [1], p. 397
[19] Chapt. 4., [114], p. 133

[20] Heinicke, G., Harenz, H.: Technik 24 (1969), 313
[21] Butjagin, P. Ju.: Vysokomol. soed. 9A (1967), 136
[22] Boldyrev, V. V.: Kinetika i Kataliz 13 (1972), 1414
[23] Fox, P. G., Soria-Ruiz, J.: Proc. Roy. Soc. London A 317 (1970), 79
[24] Weichert, R., Schönert, K.: Chem. Ing. Techn. 49 (1977), 278
[25] Borbjutkova, E. A., Porcekno, P. I., Venerovskaja, L. N.: Ž. neorg. chim. 40 (1967), 1438
[26] Freeman, E. S.: J. Phys. Chem. 60 (1956), 1487
[27] Bartos, R., Margrave, J. L.: J. Phys. Chem. 60 (1956), 256
[28] Boldyrev, V. V., Regel, V. A., Pozdnjakov, O. F., Urakaev, F. Ch., Byl'skij, B. Ja.: Dokl. AN SSSR 221 (1975), 634
[29] Boldyrev, V. V., Heinicke, G.: Z. Chem. 19 (1979), 353
[30] Urakaev, F. Ch.: Dissertacija, Sverdlovsk 1978
[31] Boldyrev, V. V.: Int. Radiation Phys. Chem. 3 (1971), 155
[32] Boldyrev, V. V.: Z. phys. Chem. 256 (1975), 342
[33] Salichov, K. M., Medvinskij, A. A., Boldyrev, V. V.: Chim. vysok. ėnerg. 1 (1967) 381
[34] Born, M., Kuhn, Ch.: „Dimičeskaja teorija kristalličeskich pešėtok", p. 64, Moskva 1958
[35] Cohen, M. H., Turnbull, D.: J. Chem. Phys. 31 (1959), 1164
[36] Maass, I., Nitzsche, H. M., Hübner, H., Hennig, H.-P., Heinicke, G.: Isotopenpraxis 14 (1978), 94
[37] Biegeleisen, J., Wolfsberg, M.: Adv. chem. phys. 1 (1958), 15
[38] Winkler, E., Jost, H., Heinicke, G.: Z. anorg. allg. Chem. 480 (1981), 205
[39] Boldyrev, V. V., Eroškin, V. I., Pismenko, V. T., Ryžak, I. A., Kefeli, L. M.: Kinetika i Kataliz 9 (1968), 260
[40] Boldyrev, V. V., Eroskin, V. I., Ryžak, I. A., Kefeli, L. M.: Dokl. AN SSSR 173 (1967), 117
[41] Belous, V. M., Boldyrev, V. V., Zenkevič, I. G.: Kinetika i Kataliz 11 (1970), 1400
[42] Hennig, H. P., Heinicke, G., Eifler, R. J.: unpublished
[43] Kihlstedt, P. G.: „I. Europäisches Symposium „Partikelmeßtechnik", p. 95, Nürnberg 1975
[44] Hennig, H. P., Heinicke, G., Steinert, K. M.: unpublished
[45] Kottler, W., Krupp, H., Rabenhorst, H.: Z. angew. Physik 24 (1968), 219
[46] Sokolov, V. N., Osipov, V. I., Novik, V. K.: Vor. Inž. geol. gruntoved. 3 (1973), 188
[47] Derjagin, B. V., Krotova, N. A., Smilga, V. P.: Adgezija tverdych tel", Moskva 1973
[48] Rauchfuss, S., Sack, R.: „Elektrostatische Aufladungen", Berlin 1976
[49] Kargin, V. A., Plate, N. A., Litvinov, I. A., Sibaev, V. P., Lure, E. G.: Vysokomol. soed. 3 (1961), 1091
[50] Kuznecov, V. A., Janova, L. P., Tolstaja, S. N.: Plaste u. Kautschuk 26 (1979), 134
[51] Janova, L. P., Tolstaja, S. N.: Plaste u. Kautschuk 26 (1979), 198
[52] Benson, R. E., Castle, J. E.: J. Phys. Chem. 62 (1958), 840
[53] Antonova, L. A., Chrustalev, Ju. A., Barambojm, N. K., Krotova, N. A.: Kolloidn. z. 38 (1976), 535
[54] Beckey, H. D., Röllgen, F. W.: Naturwiss. 58 (1971), 23
[55] Butjagin, P. Ju.: Ž. vsesojuzn. chim. obščestva 18 (1973), 92
[56] Taubman, A. B.: chapt. 1.9., ref. [1], p. 79
[57] Heinicke, G., Janova, L. P., Chrustalev, Ju. A., Krotowa, N. A.: Z. Chem. 19 (1979), 118
[58] Janova, L. P., Kuznecov, V. A., Tolstaja, S. N.: Dokl. AN SSSR 237 (1977), 1418
[59] Janova, L. P., Kuznecov, V. A., Taubman, A. B.: Kolloidn. z. 37 (1975), 1023
[60] Janova, L. P.: Plaste u. Kautschuk 19 (1972), 97
[61] Thiel, S.: Diplomarbeit, Freiberg (1974), 21
[62] Gromov, V. V.: „Konf. Vozdejstvie ionizirujuščego izučenija na geterogennye sistemy", S. 4, Moskva 1976
[63] Krotova, N. A.: Z. Chem. 14 (1974), 141
[64] Stefanowitsch, N. N., Radzig, V. A., Vilenskij, A. I., Vladykina, T. N., Krotova, N. A.: Plaste u. Kautschuk 19 (1972), 94

[65] Knjazeva, N. P., Mambetov, D. M.: chapt. 1., [6], p. 106
[66] Krotova, N. A., Heinicke, G.: Wissenschaft u. Fortschritt **27** (1977), 454
[67] Thiessen, P. A., et al.: Sitzungsber. AdW der DDR, Math.-Naturw.-Techn. (1981), 3N, 65
[68] Thiessen, P. A., Haase, G., Altnau, H., Pusch, Ch., Florek, D.: chapt. 1., [10], p. 45
[69] Thiessen, P. A.: Spektrum **8** (1977), 13
[70] Acoock, W. G., Hayward, E. J., Mile, B., Ward, B.: Can. J. Chem. **50** (1972), 3813
[71] Chapt. 4., [10], p. 45A
[72] Yeddanapalli, L. M.: J. Chem. Phys. **10** (1942), 249
[73] Ausloos, P., Lias, S. G.: J. Chem. Phys. **38** (1963), 2207; **48** (1968), 392
[74] Meyer, K., Tiller, H. J., Welz, E., Kühn, W.: Z. Chem. **14** (1974), 146
[75] Tiller, H. J., Kühn, W., Meyer, K.: Z. Chem. **14** (1974), 450
[76] Momose, Y., Tamai, Y.: J. inst. metals **98** (1970), 110
[77] Simakov, Ju. S., Barambojm, N. K.: Vysokomolek. soed. **8** (1967), 567
[78] Poljakov, A. M., Kurdubov, Ju. F., Barambojm, N. K., Krotova, N. A.: Dokl. AN SSSR **175** (1967), 72
[79] Chrustalev, Ju. A., Calych, A. E., Krotova, N. A.: Plaste u. Kautschuk **26** (1979)
[80] Chrustalev, Ju. A.: Dissertacija, Moskva 1978
[81] Chrustalev, Ju. A., Prochorova, L. I., Barambojm, N. K., Kurdubov, Ju. F., Krotova, N. A.: Kolloidn. ž. **38** (1976), 814
[82] Krotova, N. A., Chrustalev, Ju. A., Stevanovič, N. N., Vladykina, T. N., Toporov, Ju. P.: ref. [62], p. 31
[83] Krotova, N. A., Morozova, L. P.: Dokl. AN SSSR **177** (1967), 639
[84] Bailitis, E.: Z. phys. Chem. N. F. **64** (1969), 302
[85] Grohn, H., Paudert, R., Hösselbarth, B.: Plaste u. Kautschuk **13** (1966), 1
[86] Wenzel, C. F.: „Lehre von der chemischen Verwandtschaft", p. 28, Dresden 1777
[87] Feitknecht, W., Durtschi, A.: Helv. **47** (1964), 174
[88] Koglin, B., Leschonski, K., Alex, W.: Chem. Ing. Techn. **46** (1974), 984
[89] Schrader, R.: Dechema-Monogr. **41** (1961), 303
[90] Steinike, U., Kretzschmar, U., Hennig, H. P.: unpublished
[91] Heinicke, G., Hennig, H. P., Steinike, U.: Kristall u. Technik **8** (1973), 379
[92] Schrader, R., Kutzer, H. J., Hoffmann, B.: Tonind. Ztg. **94** (1970), 4
[93] Chapt. 1., [4], p. 539
[94] Schrader, R., Hoffmann, Kl.: Chem. Ing. Techn. **42** (1970), 38
[95] Schrader, R.: Technik **24** (1969), 88
[96] Paudert, R.: Chem. Techn. **21** (1969), 477
[97] Chapt. 4., [143], p. 94
[98] Schrader, R., Schneider, F.: Neue Hütte **13** (1968), 349
[99] Paudert, R., Liebscher, S., Körlin, P., Trinks, W.: Chem. Techn. **20** (1968), 276
[100] Schrader, R., Oese, W.: Chem. Techn. **23** (1971), 741
[101] Kretzschmar, U., Steinike, U., Hennig, H. P.: unpublished
[102] Chapt. 4., [63], p. 545
[103] Chodakov, G. S.: „Fizika izmel'čenija", Moskva 1972
[104] Jimbo, G.: Kogyo Kakaku Zasshi **71** (1968), 1309
[105] Tanaka, T.: Kagaku Kogaku **18** (1954), 160
[106] Heinicke, G., Bock, N., Steinike, U.: Z. anorg. allg. Chem. **443** (1978), 231
[107] Schrader, R., Balarew, Chr.: Z. anorg. allg. Chem. **376** (1970), 156
[108] Lindner, K. H.: Staub, Reinhaltung der Luft **28** (1968), 467
[109] Schrader, R., Westphal, H.: Neue Hütte **15** (1970), 582
[110] Leithel, E.: Dissertation, Berlin (1971)
[111] Heinicke, G., Paudert, R., Harenz, H.: chapt. 1.9., [3], p. 87
[112] Schrader, R., Wirsing, R., Kubsch, H.: Z. anorg. allg. Chem. **365** (1969), 191
[113] Juhász, Z.: Aufber.-Techn. **10** (1974), 558

[114] HENNIG, H. P., STEINIKE, U., KRETZSCHMAR, U.: unpublished
[115] REHBINDER, P. A., CHODAKOW, G. S.: Silikattechn. **13** (1962), 200
[116] HÜTTIG, G. F.: Proc. Intern. Symp. React. Solids Gothenburg (1952), 980
[117] TKAČOVA, K., HOCHMANOVA, I., MIHALIK, V.: Freiberger Forschungsh. **A 553** (1976), 167
[118] LEITEL, E., HEINICKE, G., PAUDERT, R.: Z. Chem. **10** (1970), 157
[119] KUBO, T.: Kogyo Kagaku Zasshi **71** (1968), 1302
[120] WESTWOOD, A. R. C.: "Surface-Sensitive Mechanical Properties, Chemistry and Physics of Interfaces", p. 161, Washington 1965
[121] HEINICKE, G., BEYER, K.: Wissenschaft u. Fortschritt **27** (1977), 522
LICHTMAN, W. I., REHBINDER, P. A., KARPENKO, G. W.: „Der Einfluß grenzflächenaktiver Stoffe auf die Deformation von Metallen", Berlin 1964
[122] ŠČUKIN, E. D.: Vestnik **11** (1973), 30
[123] SABENKO, V. I., KOČANOVA, L. A., ŠČUKIN, E. D.: Kristallografija, XVII (1972), 995
[124] SCHEIBE, W., HEEGN, H., LUDWIG, G.: chapt. 4., [152], p. 327
[125] TANAKA, T.: Zement—Kalk—Gips **15** (1962), 28
[126] RUMPF, H.: Chem. Ing. Techn. **31** (1959), 697
[127] LOCHER, F. W., V. SEEBACH, H. M.: „Haus der Technik, Vortragsveröffentlichungen", Heft 247, p. 35, Essen 1970
[128] DERJAGIN, B. V., SURKOV, Ju. A., TOPOROV, Ju. P., ŠANDOR, V. V.: Neorgančies. Mat. **11** (1975), 2221
[129] DERJAGIN, B. V., TOPOROV, Ju. P., ŠANDOR, V. V.: Pribory i Technika Eksperimenta AN SSSR **3** (1974), 198
[130] KNAPPWOST, A., GAYKO, M., WÜSTEFELD, B.: Tribologie u. Schmierungstechnik **19** (1972), 63
[131] GÖTTE, A.: Freiberger Forschungsh. **A 281** (1963), 5
[132] PÖTHIG, R.: unpublished
[133] HEDVALL, J. A.: Einführung in die Festkörperchemie, Braunschweig 1952
[134] CHODAKOV, G. S.: chapt. 1., [5], p. 182
[135] HEINICKE, G., HARENZ, H.: Z. phys. Chem. **240** (1969), 325
[136] HARENZ, H., HEINICKE, G.: Schmierungstechnik **2** (1971), 214
[137] BORN, D.: Dissertation, Berlin 1975
[138] HEINICKE, G.: Ropa a Uhlie **4** (1962), 150
[139] HEINICKE, G., LISCHKE, I.: Z. Chem. **3** (1963), 355
[140] HOFFMANN, H., FUNKE, G.: Z. phys. Chem. **247** (1971), 233
[141] HEINICKE, G., HARENZ, H.: Schmierstoffe u. Schmierungstechnik **30** (1968), 4
[142] SCHWAB, G. M.: „Handbuch der Katalyse", Bd. 5, II, p. 246, Wien 1957
[143] KLIER, K., JIRATOVA, M.: "Proc. Third Intern. Congr. on Catalysis, Amsterdam, 1964", p. 763, Amsterdam 1965
[144] SURHMANN, R., HERAS, J. M., DE HERAS, J. V., WEDLER, G.: Ber. Bunsenges. phys. Chem. **68** (1964), 511, 990
[145] Chapt. 4., [10], p. 130
[146] SHAW, M. C.: J. appl. Mechan. **15** (1948), 37
[147] KOLONKO, G.: Dissertation, Halle 1952
[148] MÜLLER, H., WALLBRAUN, O.: Wiss. Ztschr. Friedrich-Schiller-Univ. Jena, Math.-Nat. R. **20** (1971), 205
[149] ALBRECHT, R., HÄUSLER, H., MÖBIUS, R.: Z. anorg. allg. Chem. **382** (1971), 177
[150] GROHN, H., PAUDERT, R., BISINGER, H. J.: Z. Chem. **2** (1962), 88
[151] GROHN, H., PAUDERT, R., FRIEDRICH, H.: Z. anorg. allg. Chem. **348** (1966), 23
[152] BOLDYREV, V. V., MEYER, K.: chapt. 1., [4], p. 38
[153] BALLCZO, H., PETERS, K.: Microchim. Acta (1960), 291
[154] HEINICKE, G., HARENZ, H.: Schmierstoffe u. Schmierungstechnik **21** (1967), 23
[155] KNAPPWOST, A., GRIGUTSCH, F. D.: Wear **9** (1966), 193

[156] GUSEV, G. M.: chapt. 1., [8], p. 225
[157] AVVAKUMOV, E. G., KOSSOBUDSKIJ, I. D., PAVLJUCHIN, Ju. T., BOLDYREV, V. V., TYR ŠKINA, O. G.: Lit. 1.8, p. 224
[158] PARKER, L. H.: J. Chem. Soc., London, **113** (1918), 396
[159] PAVLJUŠČENKO, M. M., GILEVIČ, M. R.: izv. AN SSSR **139** (1961), 648
[160] PETERS, K., PAJAKOFF, S.: Microchim. acta (1962), 314; (1964), 429
[161] YAMAGUCHI, S., SAWAMINA, H.: Werkstoffe u. Korrosion **16** (1965), 465
[162] AVVAKUMOV, E. G., STRUGOVA, L. I.: Izv. Sib. otd. AN SSSR, ser. chim. nauk (1974), 37
[163] MATICIN, A. I., AVVAKUMOV, E. G., D-JAKOV, V. E., STAVER, A. M.: chapt. 1., [9], II. p. 3
[164] AVVAKUMOV, E. G., D-JAKOV, V. E., STRUGOVA, L. I., BOLDYREV, V. V.: chapt. 1., [8], p. 211
[165] SCHLEGEL, E.: chapt. 1., [10], p. 27
[166] TAYLOR, H. F. W.: "Int. Symp. on autoclaved calcium silicate buildings", Vol. B, London 1965
[167] UIBO, L. Ja., PAE, A. Ja., MJUJRSEPP, T. K., KIJSLER, A. Ch.: chapt. 1.9., [2], p. 15
[168] PAE, A. Ja., PALM, U., UIBO, L. Ja.: chapt. 1., [8], p. 250
[169] AVVAKUMOV, E. G., BAGRJANCEV, G. I., VOLKOV, V. V.: chapt. 1., [9], [2], p. 26
[170] AVVAKUMOV, E. G.: chapt. 1., [8], p. 207
[171] BOLDYREV, V. V., EREEMEEVA, Ju. E.: Učenie zap. Tomsk. gos. univ. **29** (1959), 27
[172] BRIDGEMAN, P. W.: Proc. Amer. Acad. Arts Sci. **81** (1952), 165
[173] HEINICKE, G., JECH, Č., PAUDERT, R., HARENZ, H.: to be published
[174] LAGERWAL, T.: Trans. Chalmers Univ. Technol. No 307, 1965
[175] BIERSACK, J.: Nukleonik **10** (1967), 267
[176] RIEMER, E. E., COTTRELL, A. H.: Phil. Mag. **2** (1957), 1345
[177] TOBIN, G. M.: Acta metallurgica **5** (1957), 398; **7** (1959), 701
[178] KELLY, R., JECH, Č.: Transaction British Ceramic Soc. **66** (1967), 259
[179] THIESSEN, P. A., HEINICKE, G., MEYER, K.: chapt. 1., [4], p. 517
[180] BARRER, R. M.: Diffusion in and through Solids, New York 1951
[181] JOST, W.: "Diffusion in Solids, Liquids, Gases", New York 1960
[182] 7. Metalltagung, „Diffusion in metallischen Werkstoffen", Dresden 1970
[183] SEITH, W.: „Diffusion in Metallen", Berlin 1955
[184] ICHIOKA, K., ITO, T., SHIRAHAWA, T., TOI, K., TOKUDA, T.: Bull. Chem. Soc. Jap. **49** (1976), 1199
[185] HÄSSNER, A., LANGE, W.: phys. stat. sol. **8** (1965), 77
[186] MEHNED, A. J.: appl. Phys. **38** (1967), 1885
[187] BLAKELY, J. M., MYKINA, H.: Acta metallurgica **9** (1961), 23
[188] DRESSLER, G.: Dissertation, Berlin 1964
[189] SEEGER, A.: Proc. "Jülich Conf. Vacancis and Interstitials in Metals", Amsterdam 1970
[190] LOHRISCH, H. J., WAGNER, W., WAGNER, K.: Schmierungstechnik 8 (1977), 227, 415
[191] JOHNSON, H. H., HIRTH, J. P.: Metallurgical transactions **A 7** (1976), 1543
[192] MARX, G., HENNIG, H. P., HEINICKE, G.: unpublished
[193] HOFMANN, W., KAULS, W.: Arch. Eisenhüttenwesen **34** (1963), 925
[194] MOROS, L. S., MINGIN, T. E.: Dokl. AN SSSR **160** (1965), 311
[195] HOFMANN, W., KAULS, W., VOGT, J.: Acta metallurgica **10** (1962), 688
[196] SMIALOWSKI, M.: „Symp. Spannungskorrosion u. Wasserstoffversprödung", p. 18, Dresden 1975
[197] PETCH, J., STABLES, P.: Nature (London), **169** (1952), 842
[198] SUHRMANN, R., WEDLER, G., GENTSCH, H.: Z. phys. chem. **17** (1958), 350
[199] KÜNTSCHER, W., KELGER, H., BIEGLER, H.: Technische Baustähle, p. 73, Halle 1958
[200] HENNIG, H. P.: unpublished
[201] HOFFMANK, K., LINKE, L.: Kristall u. Technik **11** (1976), 842
[202] HEINICKE, G., HARENZ, H.: Schmierstoffe u. Schmierungstechnik **21** (1967), 23

[203] HEINICKE, G., HARENZ, H.: Mber. Dtsch. Akad. Wiss. **11** (1969), 59
[204] SCHRADER, R.: Freiberger Forschungsh. **A 292** (1964), 53
[205] SCHRADER, R., RUMP, H., SANNER, G.: Z. anorg. allg. Chem. **350** (1967), 130
[206] SCHRADER, R., ARNHOLD, M., RUMP, H.: Z. anorg. allg. Chem. **350** (1967), 120
[207] SCHRADER, R., RUMP, H.: Z. anorg. allg. Chem. **350** (1967), 137
[208] SCHRADER, R., WEIGELT, D.: Plaste u. Kautschuk **16** (1969), 739
[209] SCHRADER, R., KNESCHKE, G.: Silikattechnik **19** (1968), 172
[210] SCHRADER, R., HORN, G., HOFFMANN, Br.: Kristall u. Technik 8 (1972), 369
[211] PAUDERT, R.: Mber. Dt. Akad. Wiss. **9** (1967), 719
[212] SCHRADER, R., HOFFMANN, B., PLÄNITZ, H., HENNEBERGER, J.: Zement—Kalk—Gips **23** (1970), 194
[213] PAUDERT, R., LIEBSCHER, S., KÖRLIN, P., TRINKS, W.: Chem. Techn. **20** (1968), 276
[214] SCHRADER, R., OESE, W.: Chem. Techn. **23** (1971), 744
[215] PÖTHIG, R., DÜNKEL, L., PAUDERT, R., HARENZ, H., GRIMMER, A. R.: Kristall u. Technik **13** (1978), 13
[216] Chapt. 3., [127], p. 63
[217] SCHRADER, R., PREISSLER, B., BRÜCKEL, J.: Z. Chem. **7** (1967), 322
[218] SCHRADER, R., HOFFMANN, K.: Chem. Ing. Techn. **42** (1970), 38
[219] SCHRADER, R., KRESSNER, R.: Chem. Techn. **25** (1973), 159
[220] SCHNEIDER, U.: Aachener Blätter f. Aufber., Verkoken, Brikett. **18** (1968), 8
[221] HENNIG, H. P., EBERT, I., STEINIKE, U.: unpublished
[222] Chapt. 4., [87], p. 93, 97
[223] WILLIAMSON, G. K., SMALLMAU, R. E.: Phil. Mag. 8 (1956), 34
[224] UHARA, I., YANAGIMOTO, S., TANI, K., ADACHI, G.: Nature (London) **192** (1961), 867
[225] UHARA, I., KISHIMOTO, S., HIKINO, T., KAGEYAMA, Y., HAMADA, H., NUMATA, Y.: J. Phys. Chem. **67** (1963), 996
[226] KISHIMOTO, S.: J. Phys. Chem. **67** (1963), 1161
[227] MEYER, K.: chapt.1., [1], p. 74
[228] HEINICKE, G., HARENZ, H.: Z. anorg. allg. Chem. **324** (1963), 185
[229] THOMAS, J. M., RENSHAW, G. D.: J. chem. Soc. (London) (A) (1967), 2058
[230] HUGHES, G., THOMAS, J. M.: Carbon **1** (1964), 339
[231] THOMAS, J. M., HUGHES, G.: Carbon **1** (1964), 209
[232] BOLDYREW, V. V.: Z. Chem. **8** (1968), 453
[233] CHAIRETDINOV, E. Ju., GALISIN, Ju. G., JOST, Ch.: Izv. Sib. otd. AN SSSR Pečatyj
[234] MICHAILOV, Ju. I.: Dissertacija, Novosibirsk 1971
[235] ZAKREVSKIJ, V. A., TOMAŠVSKIJ, Ė. E., BAPTIZMANSKIJ, V. V.: fiz. tv. tela **9** (1967), 1434
[236] RADCIG, V. A., BUTJAGIN, P. Ju.: Vysokomol. soed. A 9 (1967), 2549
[237] ZAKREVSKIJ, V. A.: Dissertacija, Leningrad 1969
[238] RADCIG, V. A., BUTJAGIN, P. Ju.: Vysokomol. soed. **7** (1965), 922
[239] ZAKREVSKIJ, V. A., TOMAŠEVSKIJ, Ė. E.: Vysokomol. soed. 8 (1966), 1295
[240] SAKREVSKIJ, V. A., KORSUKOV, V. E.: Plaste u. Kautschuk **19** (1972), 92
[241] MROZOWSKI, S.: Carbon **6** (1968), 243
[242] URBANSKI, T.: Nature (London) **216** (1967), 5115
[243] HEINICKE, G., HENNIG, H. P.: chapt. 1., [6], p. 54
[244] SCHRADER, R., WIRSING, R., KUBSCH, H.: Z. anorg. allg. Chem. **365** (1969), 191
[245] VOLAND, U., SCHRADER, R., SCHNEIDER, H.: Z. anorg. allg. Chem. **368** (1969), 317
[246] ARENDS, J., DEKKER, A. J., PERDOK, W. G.: phys. stat. sol. **3** (1963), 2275
[247] KUSELEV, V. F., NIKITINA, O. V.: Dokl. AN SSSR **171** (1966), 374
[248] BOEHM, H. P.: Z. Polymere u. Kolloid-Z. **227** (1968), 17
[249] HOCHSTRASSER, G., ANTONINI, J. F.: Surface Science **32** (1972), 644, 665
[250] MEYER, K., TILLER, H.-J., WELZ, E., KÜHN, W.: Z. Chem. **14** (1974), 146
[251] EBERT, I., HENNIG, H. P., ULBRICHT, K.: Z. Chem. **16** (1976), 10

6. Tribochemical reactions in special systems

6.1. Tribochemical reactions in solids with different bond character

The reactivity of solids is strongly influenced by the character of bonding states in the crystal lattice. The differences in bonding states are an essential reason for the fact that no uniform reaction model can be established for tribochemical reactions in spite of a larger number of common features (see Chapt. 4. and 5.). However, certain reaction models exist for various solid systems having the same, or similar, bonding states. By way of example, in the following sections tribochemical reactions of solids with different bond types will be studied.

Thus the bonds in the SiC crystal lattice are mainly of a homopolar nature (see 6.1.1.). SiO_2 belongs the those solids in which the bonds between lattice components are in nearly equal proportion of a homopolar and heteropolar nature (see 6.1.2.), whereas apatites have a predominantly heteropolar bond structure (see 6.1.3.).

6.1.1. Tribochemistry of silicon carbide

6.1.1.1. Structure and reaction behaviour of silicon carbide

Silicon carbide exists in a cubic and hexagonal modification [1]. These are very similar in their structure. They consist of Si and C layers, respectively, perpendicular to the hexagonal axes with the arrangement of layers to each other being very different in the individual modifications. In these layers the central structural component is the C atom surrounded tetrahedrally by 4 Si atoms. In the same way each Si atom is surrounded by 4 C atoms [2].

Silicon carbide has exclusively Si—C bonds because of this structure. These are mainly homopolar [3]. Since the energy of each individual bond totals $\approx$ 240 kJ/mole [4], an energy of nearly 1,000 kJ/mole is required for the detachment of a centrally bound Si and C atom, respectively. Because of the high bond strength SiC is a stable product up to temperatures of about 3,000 K and belongs to the chemically most inert substances. In order to markedly oxidize silicon carbide temperatures of about 1,300 K are required. However, experiments to obtain methane from SiC and H_2 at this temperature had no success [5]. Since SiC belongs to the hardest substances, upon tribomechanical stress locally high energy excitation states are expected to occur for the release of tribochemical reactions (Chapt. 1., [1]).

6.1.1.2. Changes in structure of SiC by tribomechanical treatment

The alteration of the secondary structure, e.g. particle size and surface area, depends on the type of mechanical stress. If SiC grains (diam. about 2 mm) are treated in a vibration mill then a strong increase in surface area takes place (by more than 2 orders of magnitude) similar to numerous other solids. The secondary particle diameter appearing in the milling equilibrium accounts for about 0.2 μm for the $SiC-H_2$ system for instance (see Fig. 5.33.).

By contrast, using a vibration reactor described in section 4.2. leads hardly to a grain comminution. In comparison to the conditions in the vibration mill the surface development is accordingly significantly smaller (factor 10, see Fig. 4.36.).

For SiC the following equation is valid for the dependence of the specific surface area S on the treatment time t (Chapt. 4., [10]):

$$S = a \cdot t^b \tag{1}$$

(a, b — constants.)

It corresponds to the relationship generally describing the increase in surface area of brittle bodies. In spite of the differences in the surface area increase by using different triboreactors no qualitative differences were found with respect to the reactivity of the treated SiC (Chapt. 4., [10]).

In addition the mechanical treatment leads to changes in the primary structure accompanied by the formation or increase of concentration of structural defects. As Fig. 6.1. shows, the variously measured structural quantities such as lattice imperfections, primary crystallite size, amorphous portion, and radical concentration are very similar in course to each other. Lattice distortions could not be proved by roentgenography [6].

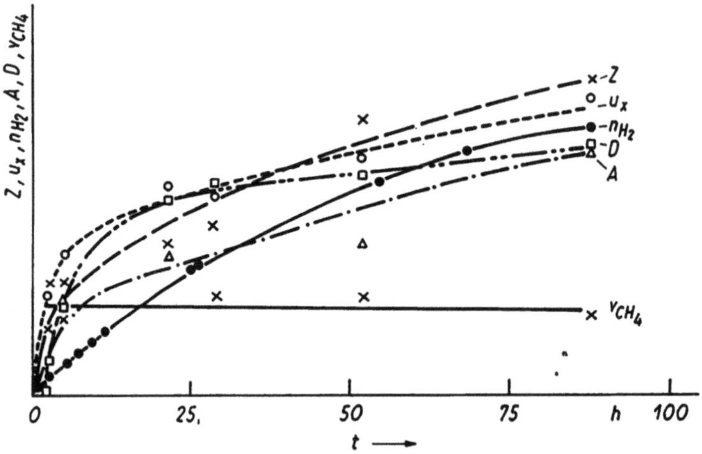

Fig. 6.1. Effect of the treatment time t on the primary structure and the tribochemical reactions at the SiC/H_2 system (Chapt. 4., [10]).

Z — concentration of radicals, u_x — lattice distortions, n_{H_2} — H_2 tribosorbate, D — primary crystallite size, A — amorphous part, v_{CH_4} — CH_4 tribodesorption velocity.

Emphasis should be given to the occurrence of an EPR signal having a g-value of 2.0031 ± 0.0006 and a line width of 5.0 ± 0.5 G, which is not influenced in position and shape by stress atmosphere and duration and may be assigned to a typical solid radical [7, 8]. The g-value approximates very much to that one of the free electron with 2.0023. Amorphous SiC films produced by sputtering yield an analogous signal

[9]. The proportionality between intensity of the signal and thickness of the layer shows that the centres are in the bulk phase.

The conclusion is drawn that chemical bonds are broken by the impact of mechanical energy on SiC and as a result free C and Si valencies, respectively, exist as A_3B complexes (A, B = C, Si).

Comminuted diamond provides an analogous signal to SiC, whereas the values at the comminuted Si ($g = 2.0055$, line width = 7···8 G) deviate from it [8]. Therefore it must be assumed that the $\geqslant C^{\cdot}$ and $\geqslant Si^{\cdot}$ radicals in SiC can not exist isolated side by side, because otherwise two signals would have been expected in the EPR spectrum.

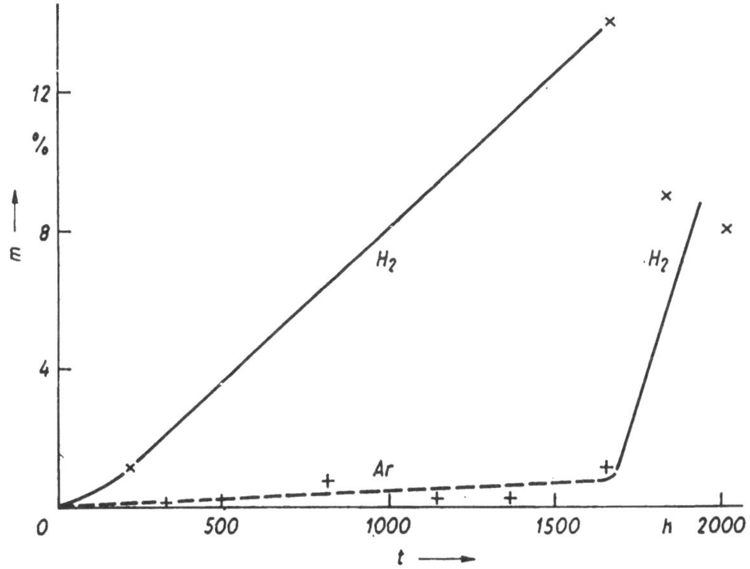

Fig. 6.2. Effect of the treatment atmosphere (Ar, H_2) and treatment time t on the mechanically activated SiC decomposition (Chapt. 4., [10]).
m – elementary Si.

In spite of the radical character, since the EPR centres produced in SiC mainly appear in the bulk phase, they are stable at room temperature and remain constant in their concentration even with longer storage of the samples in air (more than 5×10^3 h) ([8], Chapt. 4., [10]).

The mechanical treatment goes beyond the breaking up of individual SiC bonds and leads to the development of carbon and silicium phases with correspondingly long and intensive treatments. In Fig. 6.2. the development of free silicium is shown in dependence on treatment time and milling atmosphere. Thus during milling in argon only a small decomposition of SiC in the elements takes place. If hydrogen is present a part of the carbon reacts to hydrocarbons so that a considerable amount of silicium accumulates in the ground material due to the prevented reverse reaction. Conversely, the development of Si–C bonds could also be proved IR-spectroscopically during the milling of carbon and silicium (see Fig. 6.3.), [10]. Whether the cubic modification which was rather to be expected in comparison to the hexagonal SiC modification was thus developed, could also not be decided by roentgenography [11]. During the treatment of SiC not only a breaking up of the Si—C bond but also a new development of Si—C bonds take place (see tribochemical equilibrium, 4.8.3.).

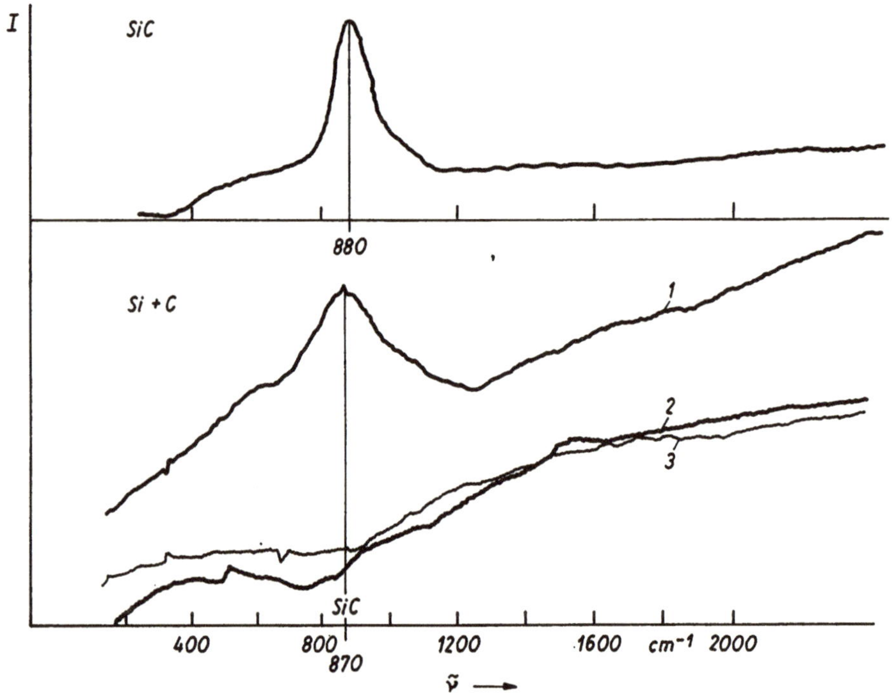

Fig. 6.3. IR spectra of differently pretreated Si + C mixtures as compared with mechanically stressed SiC [10].

I – absorption intensity, $\tilde{\nu}$ – wave number. 1 – Si + C milled together, 2 – Si + C milled separately and mixed, 3 – Si + C unmilled mixed.

6.1.1.3. *Tribosorption at SiC*

During the treatment of SiC gases such as H_2, hydrocarbons, O_2, N_2, NH_3, H_2O, HCN, are tribosorbed (Chapt. 4., [10], [12]) (Tribosorption see 4.7. and 5.7.). As has been proved in the SiC—H_2 system, considerable amounts of gas (10 mole per cent H_2), (Chapt. 4., [10]) may be tribosorbed without already attaining a saturation of the solid. Assuming an exclusive adsorption of the tribosorbed gas molecules and a monomolecular coverage, it follows from a comparison of the area required with the surface area really measured that the tribosorbed gases are situated above all in the phase boundary and only to a small extent on the surface area (see Fig. 5.53.), (Chapt. 4., [10], [12]).

During the vibration milling of SiC in steel vessels, Fe abrasion also comes into the phase boundary (comp. Fig. 3.4.) by which the ground material obtains ferromagnetic properties. That Fe is not only to be found on the surface area of SiC could be proved by the fact that the ferromagnetic properties did not disappear with longer treatment of the grinding material in boiling hydrochloric acid (Chapt. 4., [10]).

As Fig. 6.1. shows the amount of H_2 tribosorbate increases similarly in course to the tribomechanically produced structural disorders. According to Fig. 5.56. a) II the amount of tribosorbate again decreases parallel to the decrease of lattice imperfections during the heating. From this it is concluded that the tribosorbed gas molecules are bound to these defect centres. Thus a solid disorder intensified by pre-activation also leads to an increased tribosorption velocity of the gas (Chapt. 4., [10]). The velocity of the formation of HD from H_2 and D_2 on mechanically stressed SiC also runs faster, with an increasing active state [13].

In analogy to studies on vibration milled quartz samples (see 6.1.2.) the disordered zone should be situated above all in the mechanically stressed peripheral layer of the SiC grains. As a consequence of the treatment, SiC grains also attain a certain porosity (Chapt. 4., [10]).

The dependence of the H_2 tribosorption velocity v_{TS} on the H_2 pressure p_{H_2} can be expressed by a relationship analogously to the LANGMUIR adsorption isotherm (Chapt. 4., [10]):

$$v_{TS} = \frac{K_1 \cdot p_{H_2}}{K_2 + p_{H_2}} \tag{2}$$

(K_1, K_2 — constants.)

The tribosorption velocity runs up to 40 Pa according to first order and above 5×10^4 Pa according to zero order. Generally the tribosorbate depends on the amount of adsorbed gas. The more gas molecules there are on the surface area the more are tribosorbed (see 4.7.2.).

The amount of tribosorbed gas n_x increases with constant gas pressure for e.g. $x = CH_4$, C_2H_2, and C_4H_{10} in dependence on the treatment time t according to the power function:

$$n_x = a \cdot t^b \tag{3}$$

(a, b — constants.), see Fig. 5.53.

With constant treatment conditions b is nearly constant, whereas a differs by more than one order of magnitude in dependence on the type of gas. Consequently the constant a is a function of the chemical and physical properties of the sorbed gases (see Fig. 4.41. and 4.42.). However, the constant b should depend on the treatment conditions analogously to the system shown in Fig. 4.26.

Comparing the changes in structure at SiC during stressing in H_2 and Ar atmospheres under otherwise similar conditions, it has been proved that a heavy gas uptake by the solid leads to an increased disorder in structure. As an example in Fig. 6.4. the influence of the treatment atmosphere on the spin concentration of tribomechanically disordered SiC is shown. Since a part of the produced radicals is definitely used by chemical interaction with hydrogen during the stress in H_2 atmosphere, the increased disorder of the SiC must have even substantially higher values through the H_2 tribosorbate, as can be calculated from the difference of the radical concentrations measured in the two gases.

The increased disorder in structure because of tribosorbed gases can also be proved by means of the primary crystallite size in the mentioned systems [6].

The largest part of the tribosorbed hydrogen exists in physisorbed form as can be concluded from measurements of desorption. It can be seen from Fig. 6.5. that the

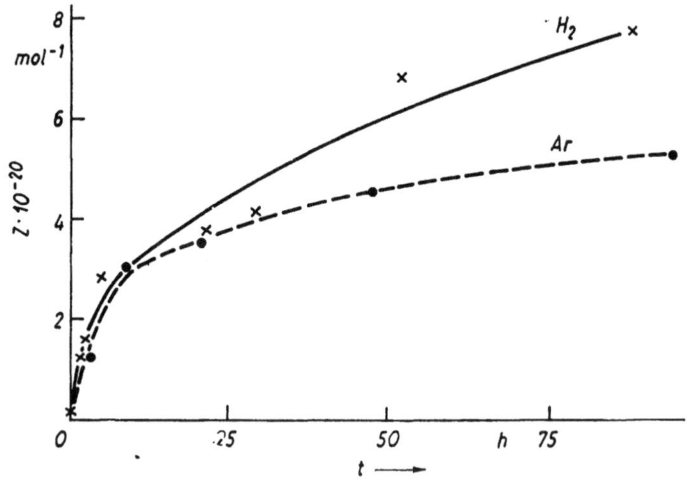

Fig. 6.4. Effect of the treatment atmosphere (H_2, Ar) on the spin concentration Z in SiC [7].
t – treatment time.

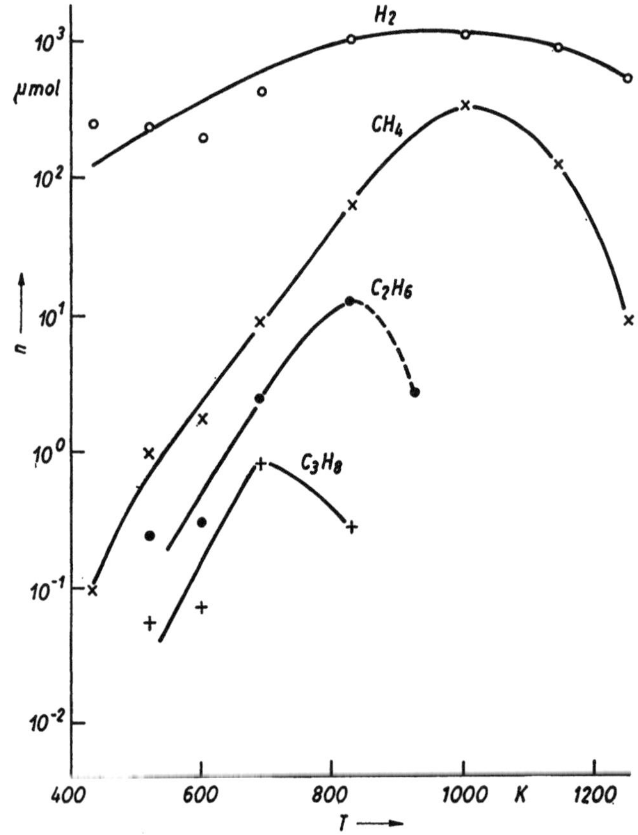

Fig. 6.5. Thermally activated desorption of H_2 and hydrocarbons at pretreated SiC (treatment atmosphere: H_2) (Chapt. 4., [10]).
n – desorbate, T – temperature.

thermally activated desorption already begins at temperatures of 400 K. It runs with an apparent activation energy of 17 kJ/mole (Chapt. 4., [10]).

There are only relatively weak bonding forces between the most tribosorbed H_2 molecules and the solid. In other gases such as CH_4, n-C_6H_{14}, C_2H_2, and HCN, the physisorbed portion is relatively small. These gases are for the most part bound by chemisorption, like a small part of tribosorbed hydrogen. That a proportion of H_2 is bound by chemisorption to SiC is also evidenced by the fact that He and Kr are not tribosorbed to SiC [12]. The formation of C—H, Si—H, Si—O, and C—N bonds during treatment in H_2, O_2, and N_2 atmospheres, respectively, could be proved by means of IR-spectroscopic investigations on SiC [10].

6.1.1.4. *Tribochemical reactions at SiC*

As already mentioned above SiC belongs to the most unreactive substances and shows no reactivity towards gases at room temperature. According to Tab. 6.1. oxidation reactions are thermodynamically possible by the use of oxygen and water but due to strong kinetic inhibitions they only appear with measurable velocity at high temperatures. However, according to Tab. 6.1. reactions of SiC with hydrogen accompanied by the development of hydrocarbons do not appear voluntarily with thermal excitation, and have not been observed until now [5].

Table 6.1. Free enthalpy G_{298K} and equilibrium constant K_p of chemical reactions with silicon carbide (Chapt. 4., [10]).

	Reaction*)	G_{298K} kJ/formula conversation	K_p at T, K		
			300	1,000	2,000
1	$SiC + 2H_2 \rightarrow Si + CH_4$	59	10^{-10}	10^{-6}	10^{-5}
2	$2SiC + 3H_2 \rightarrow 2Si + C_2H_6$	177	10^{-31}	10^{-14}	10^{-10}
3	$2SiC + H_2 \rightarrow 2Si + C_2H_2$	430	10^{-75}	10^{-20}	10^{-8}
4	$SiC \rightarrow Si + C$	110	10^{-19}	10^{-6}	10^{-3}
5	$SiC + 2H_2O \rightarrow SiO_2 + 2H_2 + C$	− 222	10^{38}	10^{9}	10^{2}
6	$SiC + 2H_2O \rightarrow SiO_2 + CH_4$	− 272	10^{47}	10^{9}	10^{0}
7	$SiC + 2O_2 \rightarrow SiO_2 + CO_2$	− 1100	10^{190}	10^{51}	10^{21}
8	$C + 2H_2 \rightarrow CH_4$	− 51	10^{9}	10^{0}	10^{-2}

*) SiO_2 as quartz, C as graphite.

On the other hand during tribomechanical treatment reactions both with N_2, O_2, and oxygenic gases and with H_2 and hydrogenous gases, have already been observed at room temperature (see Tab. 6.2.). The last mentioned reactions have been systematically investigated with respect to kinetics and mechanism.

The quantitative proof of the reacted products is relatively difficult because only the tribodesorbed parts are directly determinable by gas analysis, whereas the reaction products remaining behind in the boundary layer can only be determined by subsequent heating of the sample (see Fig. 6.5.). Here thermal reactions cannot be excluded. Proportionality is assumed to exist in first approximation between that part of the reaction products appearing in the boundary layer and the part delivered by tribodesorption to the gaseous phase.

The correlation between tribosorption and tribochemical reaction is shown in Fig. 6.6. At the beginning of the treatment the tribodesorbate and tribosorption velocity curves rise so that a proportionality arises between the concentration of the tribosorbed gas and reaction velocity, and consequently equation (4.44) can be used for the SiC—H_2 system as well.

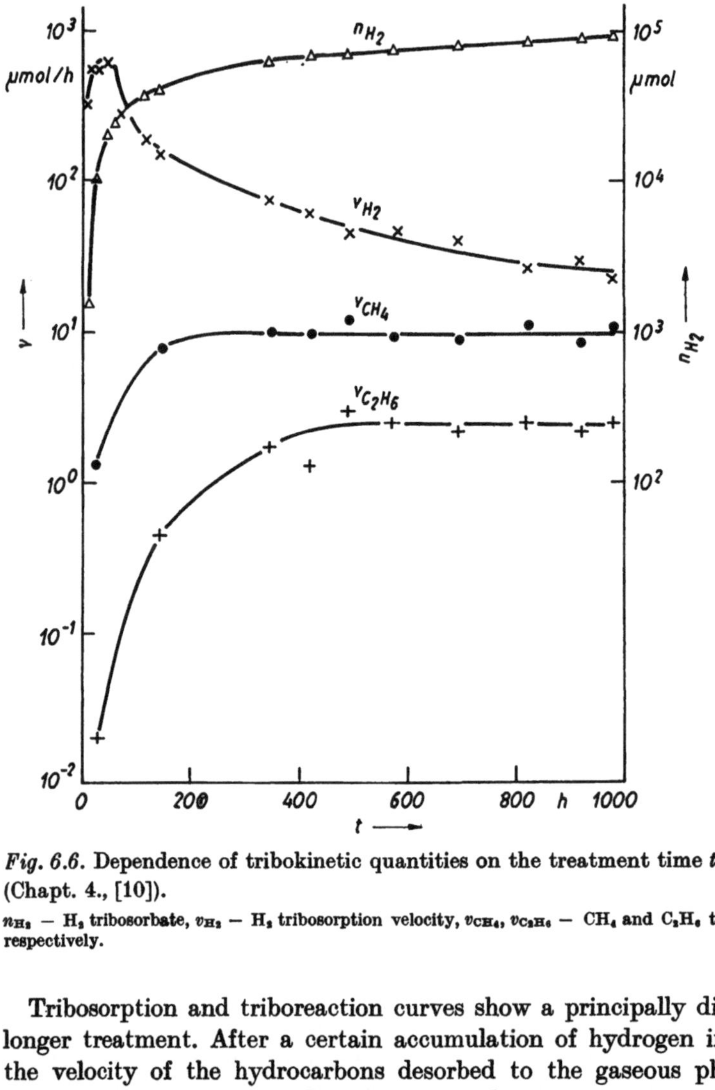

Fig. 6.6. Dependence of tribokinetic quantities on the treatment time *t* at SiC/H_2 system (Chapt. 4., [10]).

n_{H_2} — H_2 tribosorbate, v_{H_2} — H_2 tribosorption velocity, v_{CH_4}, $v_{C_2H_6}$ — CH_4 and C_2H_6 tribodesorption velocity, respectively.

Tribosorption and triboreaction curves show a principally different behaviour with longer treatment. After a certain accumulation of hydrogen in the phase boundary, the velocity of the hydrocarbons desorbed to the gaseous phase remains constant, whereas the amount of tribosorbate caused by the increase of the surface area (see Fig. 5.53.) and the tribodiffusion of H_2 into the phase boundary of SiC grains, continues to grow. Consequently the tribodesorption is independent from the surface area and is in the first instance determined by number and intensity of impacts. As Fig. 5.56. Ia) shows, in contrast to tribosorption, tribodesorption does not depend either on the concentration in lattice imperfections, out of the rise phase. Evidently it is exclusively controlled by the high excitation states appearing briefly in the impact centre.

Table 6.2. Reaction products due to the chemical treatment of SiC in different gaseous atmospheres (Chapt. 4., [10]).

Gas applied	Reaction products detected
—	Si, C
O_2	SiO_2, CO, CO_2
H_2O	CO_2, H_2, CH_4, C_2H_6, C_3H_8
H_2	Si, CH_4, C_2H_6, C_3H_8, n-C_4H_{10}, i-C_4H_{10}, n-C_5H_{12}, i-C_5H_{12}, n-C_6H_{14}, i-C_6H_{14}, C_2H_4, C_2H_2
CH_4, C_2H_6⋯n-C_6H_{14} C_2H_2, C_2H_4	H_2, CH_4, C_2H_6, C_3H_8, n-C_4H_{10}, i-C_4H_{10}, n-C_5H_{12}, i-C_5H_{12} *) **)
$H_2 + O_2$	CO_2, CO, CH_4, C_2H_6
$N_2 + H_2$	NH_3, HCN
H_2, O_2, N_2	CH_3CHO, $CH_2(NH_2)COOH$, $CH_3-CH(NH_2)COOH$, $(CH_3)_2CH-CH(NH_2)COOH$

*) Using the individual gases not all reaction products could always be detected in the gaseous phase, partially the detection was possible in the subsequent thermal desorption only.
**) Using C_2H_2 the formation of C_2H_4 could be detected, too.

As Tab. 6.2. shows, a larger number of saturated and unsaturated hydrocarbons come into being during the reaction of SiC with hydrogen. Methane appears as the main reaction product. The concentration of higher hydrocarbons decreases with growing C numbers in the reaction mixture (see Fig. 6.7.).

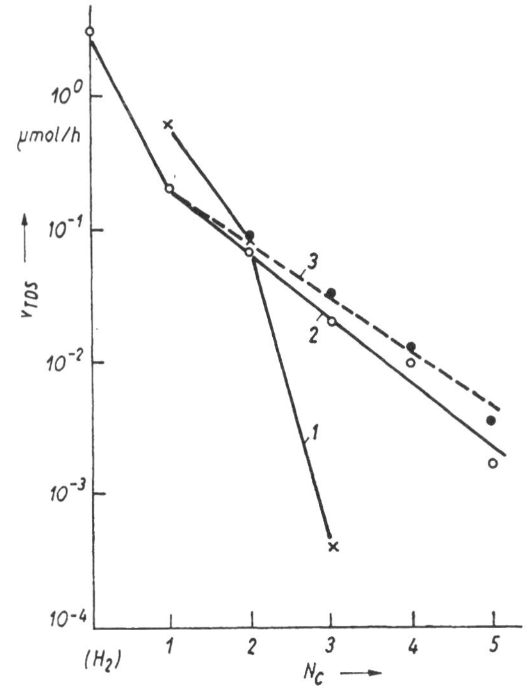

Fig. 6.7. Correlation between tribodesorption velocity v_{TDS} and C-number N_C of the hydrocarbon developed at mechanically stressed SiC/n-C_6H_{14} and SiC/H_2 systems (Chapt. 4., [10]).

1 – SiC/H_2 system, 2, 3 – SiC/n-C_6H_{14} system (2 – measured, 3 – calculated).

A similar distribution of the hydrocarbon concentration can be obtained by the treatment of SiC in a hexane atmosphere. The falling of the tribodesorption velocity v_{TDS} of C_1 to C_5 alkanes is described by the following e-function according to the curve represented in Fig. 6.7.:

$$v_{TDS} = a \cdot e^{-b \cdot m} \qquad (4)$$

(a, b – constants, m – C-number of the tribodesorbed alkane.)

Since H_2 occurs as the main reaction product in the tribochemical decomposition of hexane, ist was drawn in curve 2, too (Fig. 6.7.).

In dependence on the gas applied, higher hydrocarbons are synthesized or decomposed. In addition to H_2, as main decomposition product hydrocarbons are again developed during the decomposition of hydrocarbons, the formation velocities of which fall with increasing C-number. Intensity and duration of the mechanical treatment have no significant influence on this. A more exact analysis shows that these processes run statistically.

There are 1⋯5 C—C break possibilities for hexane. And in turn there are 1 to (m – 1) break possibilities for the decomposition products (m – C-number of the decomposition products.)

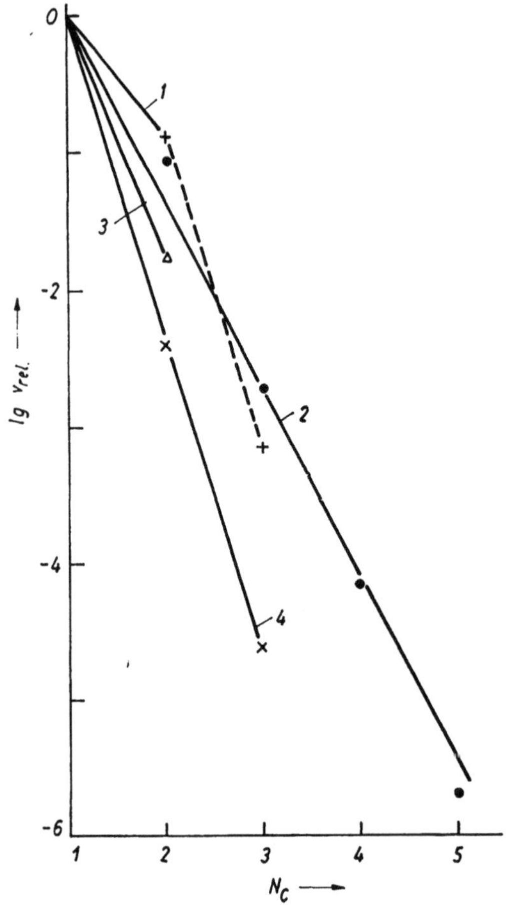

Fig. 6.8. Effect of iron addition on the decrease of the relative alkane tribodesorption velocity v_{rel} at the mechanically stressed SiC/H_2 system (Chapt. 4., [10]).

N_C – C-number of the developed hydrocarbon, Fe additive: 1 – without, 2 – 0.05 mole, 3 – 0.1 mole, 4 – 0.9 mole; (v_{rel} in μ mol/h).

If all the decomposition products having the same number of C atoms are added and standardized to the CH_4 tribodesorption velocity then curve 3 in Fig. 6.7. results.

The experimental and calculated values lie in a semi-logarithmic form on a straight line, showing a similar course.

By applying increasing amounts of iron — which as a catalyst raises the hydrogenating effect of hydrogen — curves are obtained which also comply with equ. (4) and drop increasingly steeply with rising iron addition (see Fig. 6.8.).

If unsaturated instead of saturated hydrocarbons are used, then H_2 and hydrocarbons with growing C-numbers are similarly produced in decreasing amounts. But additionally further mechanisms may also become effective with the use of unsaturated hydrocarbons, as expressed by the corresponding increased formation velocities of individual hydrocarbons (Fig. 6.9.). For acetylene a straight line is obtained conforming with the curves of Fig. 6.7. and 6.8. The sum of the tribodesorption velocities of ethylene and ethane (listed at C = 2 in Fig. 6.9.) is no exception here. The occurrence of an amount of ethylene 5 times larger than the amount of ethane speaks for an additional mechanism becoming effective, the immediate hydrogenation of acetylene. No un-

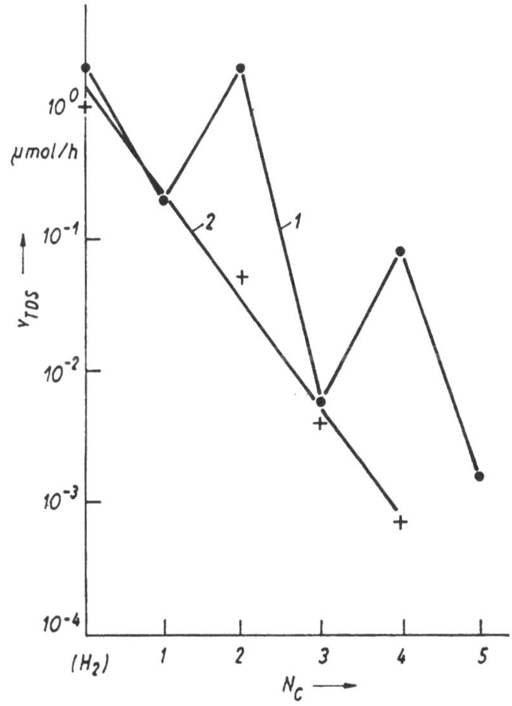

Fig. 6.9. Relationship between tribodesorption velocity v_{TDS} and C-number N_C of the hydrocarbon formed at mechanically stressed SiC/C_2H_4 (1) and SiC/C_2H_2 (2) systems (Chapt. 4., [10]).

saturated hydrocarbons are demonstrated in the tribodesorbate under analogous conditions if saturated hydrocarbons are used. In the $SiC—C_2H_4$ system this 2nd mechanism is directly expressed by the increased tribodesorption velocity of ethane (Fig. 6.9.). The tribodesorption velocity of the C_4 hydrocarbons, which additionally deviates from the linear drop, also speaks for a preferred dimerization.

Altogether the investigations show that the tribochemical decomposition of hydrocarbons takes place purely statistically and is little influenced by the bonding strength of the individual C—H and C—C bonds. The graded tribodesorption, and consequently also the reaction velocities, result exclusively from the different concentration of the hydrocarbons existing in the phase boundary and from the probability of the breakdown of the individual C—C bonds arising from this. The reactions run via C_xH_y groups (x, y = 1, 2, 3, ...) which are bound to the solid. According to IR-spectroscopic and kinetic measurements, CH, CH_2, and CH_3 groups predominate. As investigations of acetylene and particularly on ethylene show, further mechanisms may influence reaction behaviour, such as hydrogenation and dimerization.

Since the development of hydrocarbons could be proved up to butane and pentane in small concentration, C_3H_y, C_4H_y, and higher groups are expected to exist in the phase boundary, via which the synthesis and decomposition of higher hydrocarbons take place.

The pattern of synthesis and decomposition reactions of hydrocarbons during the treatment of SiC is given in Fig. 6.10. The more the steps that are required for the C—C linkage in order to develop a determined molecule, the smaller is the probability of its development and consequently also of its tribodesorption. However, with growing C-number the decomposition of the developed hydrocarbons gains significance. That synthesis and decomposition reactions are closely linked is shown by Fig. 6.7. to 6.9., according to which for both the reaction types the concentration distributions of reaction products are described by an exponential function.

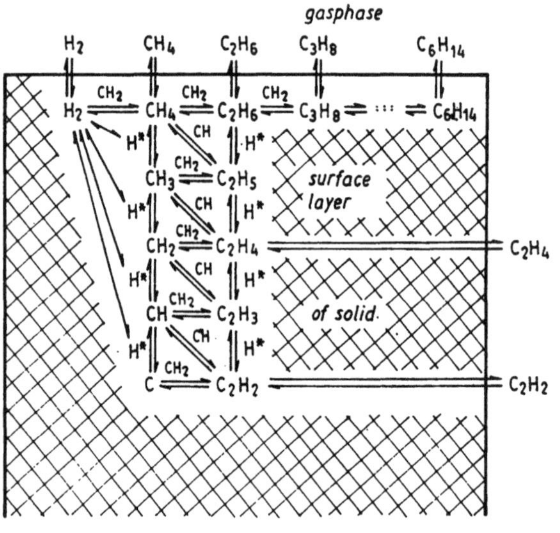

Fig. 6.10. Model conception on the tribochemical formation and transformation of hydrocarbons (Chapt. 4., [10]).

The fact that a statistical breakdown of the bonds takes place speaks for the tribochemical reaction mechanism discussed in 5.3. according to which the breaking up of the bonds takes place stochastically through very short-term highly excited energy states. Hence tribochemical decomposition is very similar to radiation-chemical decay, and it is possible to obtain completely similar concentration distributions in the mixture of the decomposition products, as Fig. 5.14. shows.

The H_2 coming from the more stable C—H bond (in comparison to the C—C bond) appears as the main reaction product in both activation methods.

With very long treatment times both sides aspire to steady states at which synthesis and decomposition reactions run with the same velocity.

That this tribochemical equilibrium differs drastically from the thermodynamic equilibrium is shown by the comparison of the equilibrium constants in Tab. 6.3. These results, too, speak for a non-thermal mechanism of hydrocarbon reactions on SiC.

Table 6.3. Comparison of thermodynamic equilibrium constants K_p with tribochemical equilibrium constants $K_{p(\text{tribochem.})}$ for different reactions at 298 K (Chapt. 4., [10]).

Reaction	K_p	$K_{p(\text{tribochem.})}$*)
$SiC + 2\,H_2 = Si + CH_4$	10^{-10}	$10^{-1} < K_{p(\text{tribochem.})} < 10^2$
$C_{\text{graphite}} + 2\,H_2 = CH_4$	10^9	$10^{-2} < K_{p(\text{tribochem.})} < 10^1$

*) Calculated from the partial pressures obtained after longer treatment times in both the directions of the reaction equations (without taking into consideration the disturbed structure of the mechanically activated solids).

6.1.1.5. *Extension of the statistical reaction model to other solids*

If graphite is treated in a hydrogen atmosphere the tribosorption processes and the tribochemical reactions are similar to the corresponding processes observed with silicon carbide. As Fig. 6.11. shows by the example of the formation of methane, the hydro-

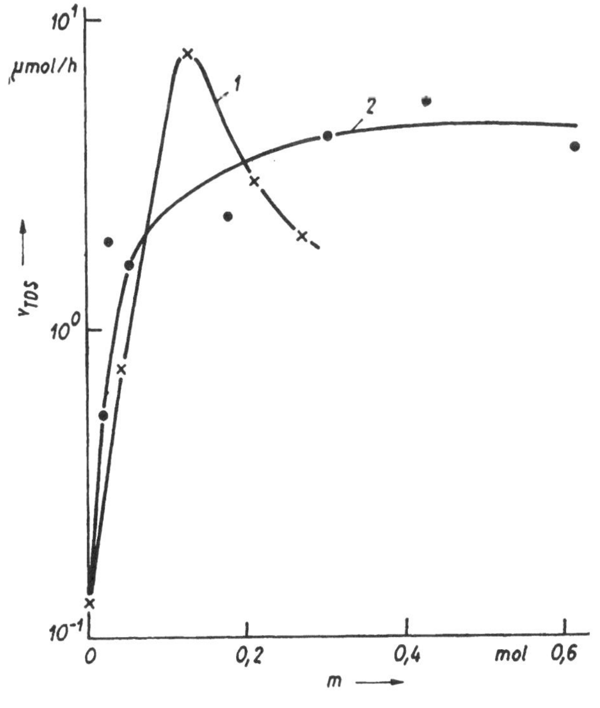

Fig. 6.11. Effect of the used amount of graphite and SiC on the CH_4 tribodesorption velocity v_{TDS} at mechanically stressed graphite/H_2 (1) and SiC/H_2 (2) systems, respectively (Chapt. 4., [10]).

m — carbon applied (as graphite and bound chemically in SiC, respectively).

carbon-tribodesorption velocity is even in the same order of magnitude with the use of small amounts of SiC and C, although the thermodynamic equilibrium positions for the hydrogenation of SiC and graphite differ principally (see Tab. 6.1.).

The dropping of the tribodesorption velocities of hydrocarbons (butane could still be detected) is similarly approximately linear at the graphite-H_2 system if shown semi-logarithmically (Chapt. 5., [36]).

During the tribomechanical treatment of corundum in the presence of various hydrocarbons, the same synthesis and decomposition reactions have been found as during the impact stress of SiC. As Tab. 6.4. shows, the individual processes also run with similar velocities in the various solids. Consequently the development of the same C_xH_y groups as with SiC can be expected to appear. The linkage of these groups to the individual solids need not to be identical on account of different bond conditions in the solids themselves.

Table 6.4. Hydrocarbon-tribodesorption velocities v_{TDS} at mechanically stressed solid-hydrocarbon systems (Chapt. 4., [10]).

System		$v_{TDS} \cdot 10^6$, mole/h				
		H_2	CH_4	C_2H_6	C_2H_4	C_3H_8
SiC/C_2H_2		$1 \cdot 10^0$	*)	$9 \cdot 10^{-3}$	$1 \cdot 10^{-1}$	$\approx 4 \cdot 10^{-3}$
Al_2O_3/C_2H_2	a)	$1 \cdot 10^0$	*)	$1 \cdot 10^{-2}$	$1 \cdot 10^{-1}$	$\approx 10^{-3}$
SiO_2/C_2H_2	b)	$4 \cdot 10^{-1}$	$8 \cdot 10^{-2}$	$3 \cdot 10^{-2}$	$1 \cdot 10^{-1}$	$4 \cdot 10^{-3}$
$SiC/n\text{-}C_6H_{14}$		$3 \cdot 10^0$	$2 \cdot 10^{-1}$	$7 \cdot 10^{-2}$	—	$2 \cdot 10^{-2}$
$Al_2O_3/n\text{-}C_6H_{14}$	a)	$5 \cdot 10^0$	$2 \cdot 10^{-1}$	$9 \cdot 10^{-2}$	$2 \cdot 10^{-3}$	$4 \cdot 10^{-2}$
$Al_2O_3/n\text{-}C_6H_{14}$	c)	$2 \cdot 10^1$	$7 \cdot 10^{-1}$	$3 \cdot 10^{-1}$	$3 \cdot 10^{-3}$	$8 \cdot 10^{-2}$

a) Corundum, b) quartz, c) sapphire.
*) Was not measurable for reasons of measurement technology.

Since Al_2O_3 does not contain carbon, which could participate in the reaction, corundum has only the function of an adsorbent for gases as well as an energy accumulator for the production of the highly excited energy states required for the breakdown of the hydrocarbons. The analogous results with corundum have proved that the reactions are not restricted to solids containing C, such as SiC and graphite.

Also if quartz is used as a solid the same tribochemical synthesis and decomposition reactions can be observed as with treated silicon carbide. As Fig. 6.12. shows, similar tribodesorption-velocity curves result in the mechanically stressed SiO_2—C_2H_2 and SiO_2—C_2H_4 systems as with the use of SiC. The maxima (here with C_2H_2, too) have to be attributed to the hydrogenation mechanism additionally becoming effective — as already discussed for the SiC—C_2H_4 system. More intensive treatment in the first instance only brings about a parallel displacement of the curves to higher values of the tribodesorption velocities.

Altogether the investigations show that the reaction model given in Fig. 6.10. is able to sum up the tribochemical reaction courses from uniform aspects not only with regard to a larger number of gases but also with regard to numerous solids, and to attribute them to reaction mechanisms which are based upon the highly excited non-equilibrium states occurring during the dissipation of mechanical energy in the first stage. These are the same mechanisms form the basis of the degradation of polymers (see 6.3.).

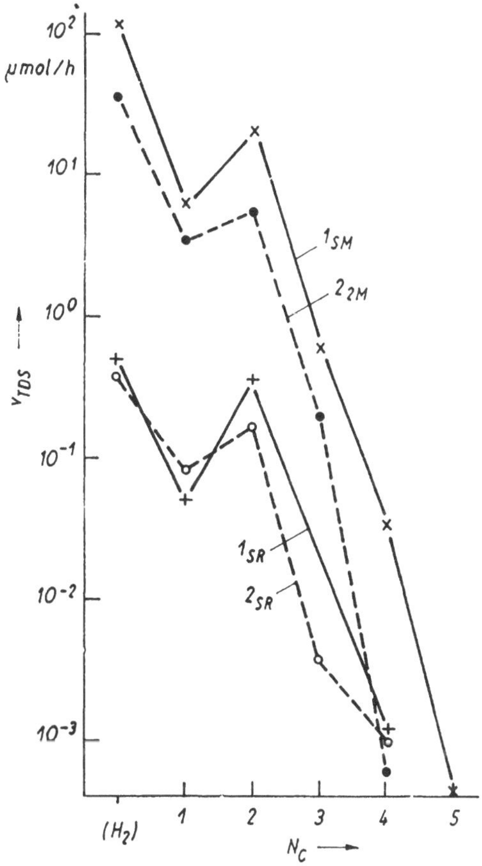

Fig. 6.12. Correlation between tribodesorption velocity v_{TDS} and C-number N_C of the hydrocarbon formed at mechanically stressed quartz/C_2H_4 (1) and quartz/C_2H_2 (2) systems.
SM – treatment in a vibration mill, *SR* – treatment in a screening machine.

6.1.2. *Tribochemistry of quartz*

Quartz belongs to the most frequently used solids in respect of mechanical treatment (e.g. Chapt. 3., [127], Chapt. 4., [250], Chapt. 5., [115], [15–19]. This can already be gathered from the numerous examples mentioned in the preceding chapters. The reason for this lies in its technological importance inter alia for building material (see 7.2.3.), in the glass, foundry, and ceramic industries, as hard material as well as in recent years as a filler in the plastics industry (see 7.2.2.4.3.). The manifold fields of technical application of quartz are based on the presence of natural deposits of quartz of varying quality extending to pure rock crystal. The scientific interest in quartz in terms of mechanical treatment not only derives from the different branches of the processing industry but also from the importance of such investigations for the research on silicosis [20, 21, 22].

6.1.2.1. Structure and properties of quartz

The low temperature form of quartz (α-quartz) is the most important of numerous SiO_2 phases, and crystallizes trigonal-trapezohedrally. Quartz belongs to the skeletal silicates. The characteristic component units of quartz are tetrahedrons connected with one another via the edges. In each case 1 Si atom is arranged inside the tetrahedrons. In line with the chemical composition of the quartz, the O atoms at the edges belong to 2 tetrahedrons in each case. The tetrahedrons are arranged in non-planar six-membered rings in the lattice (see Fig. 6.13.). There are also SiO_4 tetrahedrons as structural elements in amorphous SiO_2, and this latter will also be included in the subsequent discussion. These structural elements form a three-dimensional network. The latter consists of non-planar, distorted four-membered, five-membered ... eight-membered

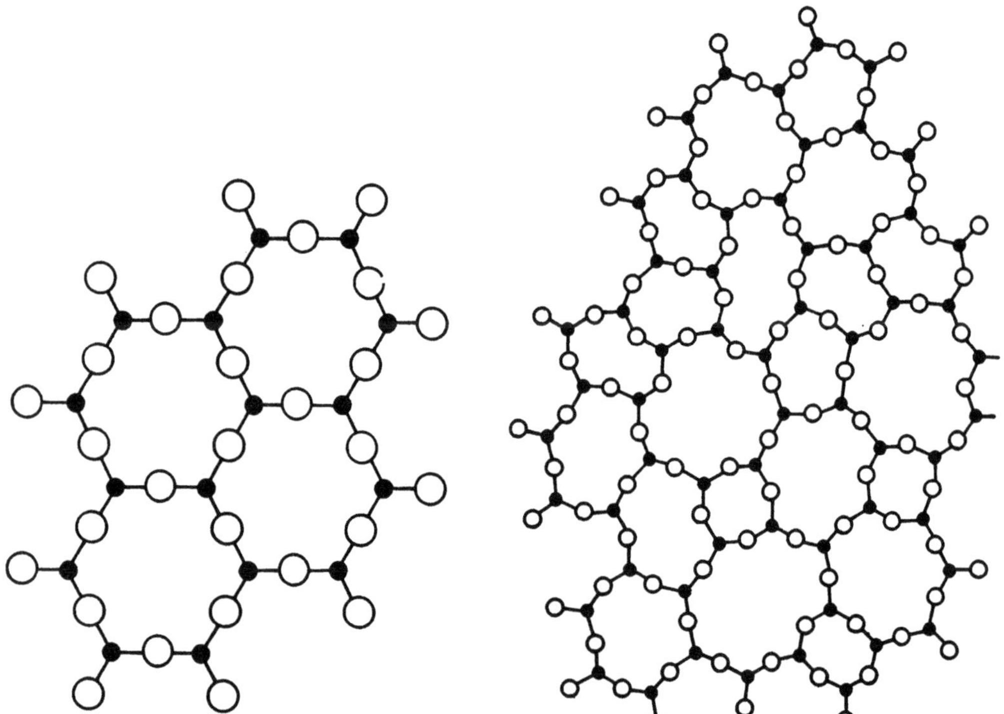

a) b)

Fig. 6.13. Arrangements of atoms in crystalline quartz (a) and amorphous SiO_2 (b) [23].
Two-dimensional model, ● – Si, ○ – O.

rings which are arranged irregularly (see Fig. 6.13.). Worth mentioning and of interest for later remarks (see 6.1.2.2.) is the α-β quartz modification transformation running reversibly at 846 K.

Quartz possesses exclusively Si—O bonds. These are of 50 per cent homopolar and heteropolar nature respectively [3]. Hence the interpretation of the quartz lattice components as Si and O atoms as well as Si^{4+} and O^{2-} ions only represents the borderline cases of the real bond conditions. The latter are the reason for the assumption of a homolytic and heterolytic breaking of Si—O bonds, respectively, during the mechanical treatment (see 6.1.2.2.). The binding energy of the Si—O bond amounts to 444 kJ/mole [24]. Quartz is a hard solid (hardness 7 according to the Mohs hardness scale).

Quartz is chemically resistent to lyes and acids with the exception of HF. In alkaline melts (NaOH, Na_2CO_3) quartz is easily converted into silicate.

The solubility of quartz in water is very small with 6···11 ppm (standard conditions) [15, 25—29]. Temperature (up to approx. 400 K) and pH (pH range 1···9) have a smaller influence on the solubility than, for instance, particle size and crystalline state of order (see 6.1.2.3.1.). The dependence on the crystalline state of order can be seen in Fig. 7.19. where the quartz solubility in water is compared with that of amorphous silicic acid at different temperatures. The solubility of amorphous silicic acid is higher by about one order of magnitude at room temperature than quartz (Chapt. 4., [250], [15, 26, 28—32]).

6.1.2.2. *Changes in structure of quartz by tribomechanical treatment*

Quartz sands occurring in nature already have to one degree or another disordered surface and lattice regions [24]. As will be shown in the following a much stronger lattice disorder may be obtained by the effect of mechanical energy.

If treated by mechanical energy quartz shows a conchoidal fracture. Although quartz shows brittle-elastic behaviour [33] a plastic behaviour cannot be excluded for particles smaller than 1 μm as well as in corresponding surface regions [24] (see 3.1.1.2.).

As has frequently been proved in the preceding chapters the alteration of the secondary structure, such as particle size, specific surface area, and grain shape, and of the primary structure such as primary crystallite size and amorphous portion, depends on the conditions of mechanical treatment, such as mode, intensity, and duration of the mechanical energy effect. In particular, the different stress mechanisms occurring in the various comminution sets (comp. 3.1.2.1.) are of decisive importance here. The treatment atmosphere (reactive gas or inert gas) as well as applications of water or surface-active substances may also influence the changes in structure. If no further precise details are given in the following, then millings in air and without additives (dry milling) are always involved.

Quartz powders activated by impact stress (see 3.1.2.1., e.g. in disintegrators, jet mills) result mainly in sharpedged particles of irregular shape [15, 34], whereas the quartz powders with comparable surface area activated by double impact stress consist chiefly of rounded particles having not very rough surface areas [17] (see 3.1.2.1., e.g. in ball and vibration mills). Consequently the treatment mechanism may be inferred from the particle shape [15, 17, 34]. In the same way conclusions as to their genesis may be drawn from the particle shape of unmilled quartz sands [24].

The secondary particles of quartz produced by impact stress have a particle size down to about 0.2 μm [15]. Particle sizes down to about 0.01 μm appear with double impact stress [17]. Differences in the fineness of grinding manifest themselves in the specific surface areas produced. The specific surface areas of quartz samples ground in the disintegrator do not exceed 5 m^2/g (Chapt. 3., [42], Chapt. 4., [157], [15, 35]), whereas surface areas up to 30 m^2/g are obtained during vibration milling after a longer period of treatment (Chapt. 3., [68, 127], [17]). Generally the fineness of grains and the specific surface area of the samples increase with growing mechanical energy applied up to the attainment of a milling equilibrium characteristic for the respective milling conditions (see Fig. 5.29.).

In addition to the secondary structure which is of special importance to surface reactions, the primary structure chiefly plays a decisive part for classification of structure-reactivity relations in mechanically treated quartz (comp. 6.1.2.3.). Frequently with increasing intensity and duration of mechanical energy, the changes in secondary and primary structure run in a similar way (see Fig. 6.14.). (n and Z will be discussed in detail in 6.1.2.3.2.)

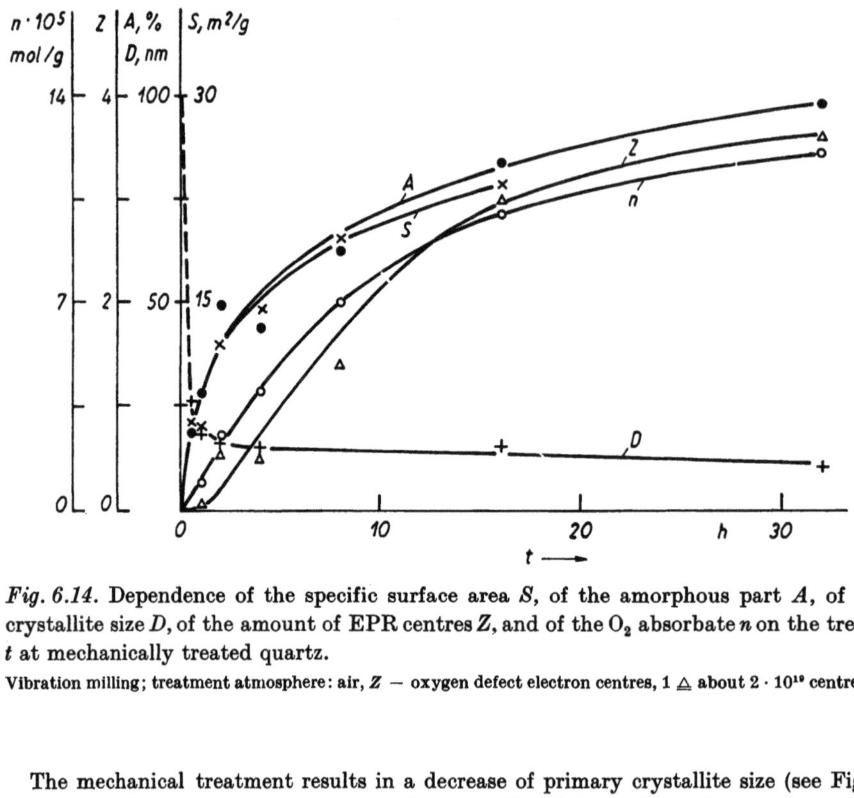

Fig. 6.14. Dependence of the specific surface area S, of the amorphous part A, of the primary crystallite size D, of the amount of EPR centres Z, and of the O_2 absorbate n on the treatment time t at mechanically treated quartz.

Vibration milling; treatment atmosphere: air, Z — oxygen defect electron centres, 1 △ about $2 \cdot 10^{19}$ centres/g.

The mechanical treatment results in a decrease of primary crystallite size (see Fig. 5.27.) and in an amorphization of the quartz powders, see Fig. 5.56.d). A nearly complete amorphization can be attained for the quartz samples activated by double impact stress (Chapt. 3., [127], [38]) and the average primary crystallite size can thus total $\leq$ 10 nm [37]. Usually a good conformity results between the amorphous portions measurable by different methods (Chapt. 3., [127], [15, 17, 38]).

Since lattice distortions primarily appear on crystals tending to plastic behaviour, no larger rates have to be expected with quartz because of its brittle-elastic behaviour. This also accords with the experimental results [17, 35, 38]. However, lattice distortions in quartz are mentioned in some cases (Chapt. 3., [66], [36]).

The changed structure of mechanically activated quartz, described up till now as amorphous portion and amorphous phase, respectively, has long been the object of numerous investigations due to its importance for chemical reactivity. Discussion on the structure of this phase has still not finished to date. It can be taken for granted, that a primary crystallite effect is involved here [19, 38].

It is still a matter of controversy whether the proved primary crystallite effect extends to the real amorphization (see 3.2.5.) or only to the formation of micro-crystallites, which like an amorphous phase no longer provide any characteristic X-ray diffraction diagrams. Therefore the amorphous portion is characterized in a great variaty of ways, according to the respective conceptions of the structure, e.g. as micro-crystalline [19], lattice distorted [39], lattice-destructed (Chapt. 4., [157], [15], disordered [40], or amorphous (Chapt. 3., [127], Chapt. 5., [115]). Also customary is the combination with the method of investigation, e.g. X-ray-amorphous. Not customary but equally justified are IR-amorphous and DTA-amorphous.

According to more recent results from X-ray wide-angle scattering diagrams of mechanically activated quartz powders the developed amorphous phase are partial-crystalline regions (Chapt. 3., [68]).

It could be concluded from radial electron-density distribution diagrams that through mechanical stress in a vibration mill the lattice structure existing in quartz remains constant up to about 50 per cent in the short-range order (< 0.65 nm). This becomes especially apparent in the partially preserved arrangement of the SiO_4 tetrahedrons in the six-membered rings typicial of α-quartz. A complete destruction of this arrangement is not reached under the given conditions (time of treatment up to 70 h). In this way the quartz destroyed mechanically in its structure differs on the one hand from the state of the amorphous silicic acid and also on the other hand from the state of the crystalline quartz in the short-range order. A structure similar to the amorphous silicic acid is reached in the spatial regions outside the SiO_4 tetrahedron rings (long-range order; > 0.65 nm).

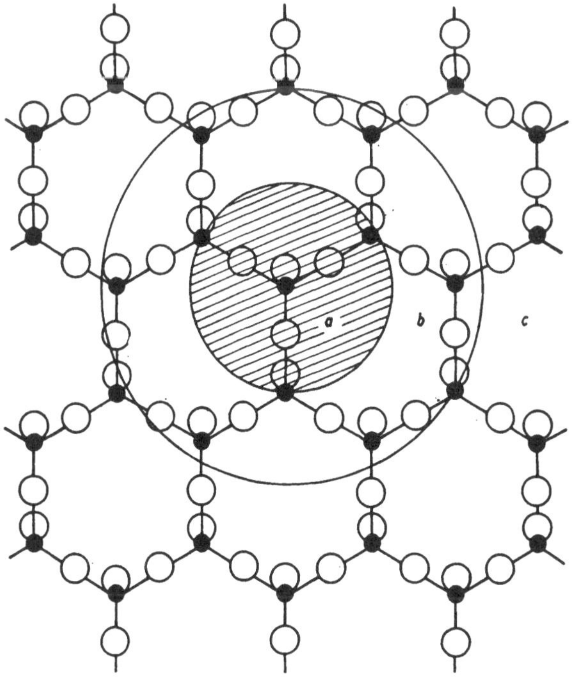

Fig. 6.15. Model of the reach of the quartz structure destruction by the action of mechanical energy [41].

a) little disturbed region of the SiO_4 tetrahedron component (< 0.30 nm), b) region of the partial cancellation of the ring arrangement (< 0.65 nm), c) amorphous region outside the SiO_4 tetrahedron rings (> 0.65 nm).

With respect to the structural order of the mechanically activated quartz, 3 regions are distinguished (Fig, 6.15.). The term "partial-crystalline phase" best expresses in line with current knowledge, the similarity and the differences both to the quartz and to the amorphous SiO_2 phases.

But for the sake of simplification the terms "amorphous phase" and "amorphous portion" will continue to be used in the following, nevertheless taking the preceding into account.

In addition to the structure of the amorphous phase, its distribution and amount in the quartz grain is also of interest. The distribution decisively depends on the type of mechanical energy supply (Chapt. 3., [50], [15, 36]), as already discussed in 3.1.2.3. The amount of amorphous phase formed is also influenced by the intensity and duration of treatment (comp. 5.9.1.). Therefore the data on the thickness of the amorphous layer is very different according to the experimental conditions in each case and lie between 150 and 2 nm, but most frequently about 20···30 nm (Chapt. 4., [250], Chapt. 5., [115], [15, 20, 26, 40, 43—52]).

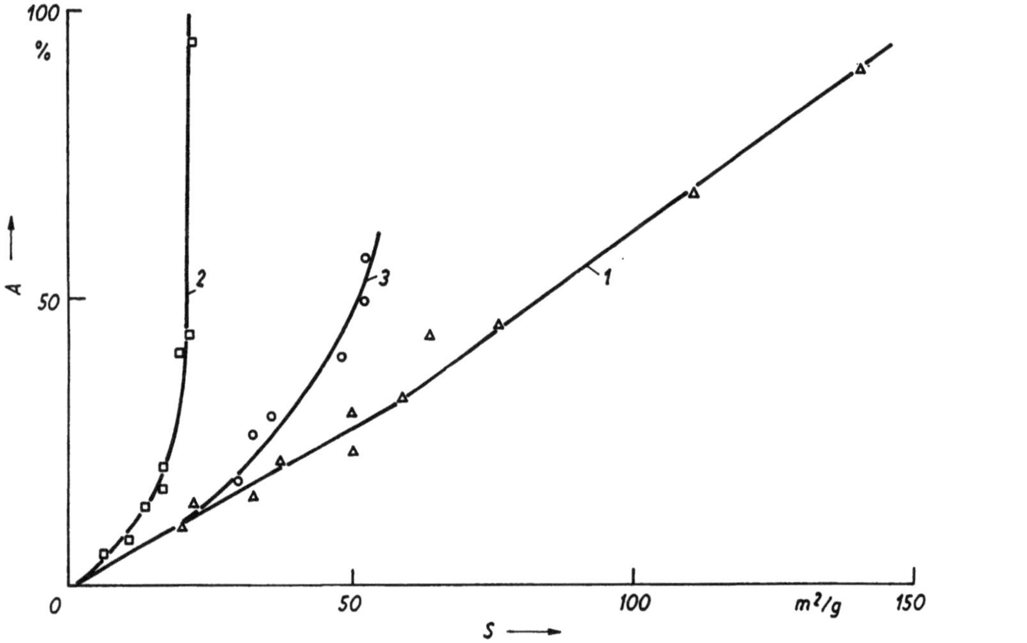

Fig. 6.16. Relationship between amorphous part *A* and specific surface area *S* of mechanically activated quartz Chapt. 5., [115] (vibration milling).
1 — milling with 80 per cent water, 2 — milling in dry air, 3 — milling with 1.2 per cent water.

The presence of water during the mechanical treatment also has a decisive influence on the formation of the amorphous phase and the thickness of layer (Chapt. 5., [115], [15]). Whereas the specific surface area grows much more during wet milling than during dry milling (Chapt. 5., [115], [16, 53]), (see Tab. 5.15.) the amorphous portions formed (see Fig. 6.16.) and the thickness of the amorphous layer are much smaller (about 2 as opposed to 15···16 nm) (Chapt. 5., [115]) than during dry milling.

Altogether the model conception represented in Fig. 6.17. results for the structure of a quartz grain activated by double impact stress, with the transition between core and layer occurring continuously. The thickness of layer is to a very large extent independent from the specific surface area S and from the amorphous portion A ($S = 1.4$ to $12\,m^2/g$; $A = 10-90$ per cent) [44].

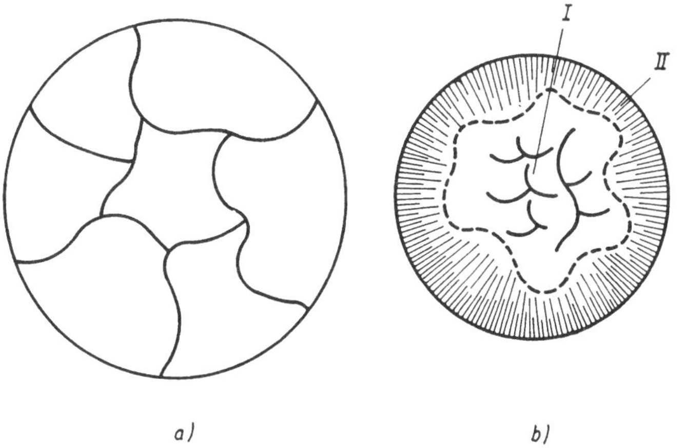

Fig. 6.17. Effect of an impact stress on a quartz grain (model conception) [44].

a – initial grain with regions scattering coherently (primary crystallites); primary crystallite size > 100 nm, crystallinity: 100 per cent, particle size = 0.2–0.4 mm.

b – grain after the stress; crystallinity: 10–90 per cent. I – relatively little disturbed center with regions scattering coherently, primary crystallite size ≈ 90 nm; II – partial-crystalline region, thickness ≈ 20 nm, primary crystallite size ≈ 1–10 nm.

During the comminution of quartz, a rupture of Si—O bonds appears at the cleavage planes (see 5.9.). In addition to the generally assumed homolytic breaking of the Si—O bonds (see equ. 5.65), (Chapt. 5., [112, 249], [54, 34]) the assumption of a heterolytic breaking on account to the bond conditions of the Si—O bonds can also to a certain extent be found in literature (see 6.1.2.1.), [55, 56]. From the $\rangle$Si$^{\cdot}$ and $\rangle$SiO$^{\cdot}$ fragments expected from the homolytic breaking the first mentioned radicals could be detected EPR-spectroscopically as so-called E'-centres after mechanical treatment in vacuum, inert gas atmosphere, or in H_2 (see 5.9. and 6.1.2.3.2.), (Chapt. 5., [246, 249], [37, 57]), whereas until now there has not been any clear success with respect to the latter. However such paramagnetic $\rangle$SiO$^{\cdot}$ centres have been detected in irradiated silica glass [58].

The concentration of paramagnetic centres increases in the same way as the amorphous portion (see Fig. 6.14.). As is evident from the gradual dissolution of the amorphous layer and the CSH-phase formation, the paramagnetic centres are situated in this layer (see Fig. 6.18.). Consequently they can be used as indicators of the amorphous layer (Chapt. 3., [125], [37, 38]).

The changed structure of quartz, particularly the formation of the amorphous phase leads to a reduction of the density (Chapt. 3., [127], Chapt. 4., [250]). Moreover the structural and electronic defects in quartz cause an increase in excess enthalpy (see 4.8.1.). In line with the treatment conditions chosen and the following methods of analysis values up to 33 ([15], DTA), 38 ([35], calorimetric), 48 (Chapt. 4., [148], calorimetric), and 86 kJ/mole (Chapt. 4., [153], calorimetric), respectively, are measured. A simple correlation could be derived between crystallinity, specific surface area, and the difference in heat of solution of milled and unmilled quartz samples as well as of carbonates (comp. equ. (4.59.)). According to it the excess enthalpy increases with the growth of surface area and amorphous portion, a fact which has also been observed by others [35].

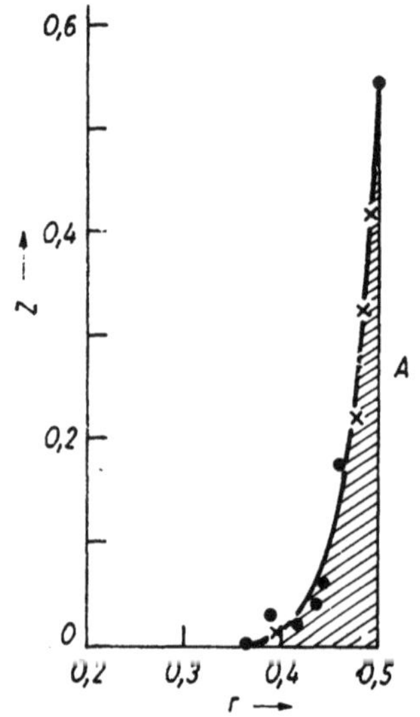

Fig. 6.18. Distribution of the paramagnetic centres in a quartz grain after impact stress [44].

Vibration milling, treatment atmosphere — air, oxygen defect electron centres (Z — relative amount of centres). r — relative radius of the grain, A — edge of the grain, × — stepwise detachment with NaOH, ● — gradual reaction of the amorphous phase with formation of calcium silicate hydrate (comp. 7.2.3.1.).

The exothermic peak of treated quartz samples occurring in DTA covers a temperature range of 100 ··· 600 K and is explained by the recombination of annealing, of radicals (Chapt. 3., [127], Chapt. 4., [157], [15, 59]).

A comparison of the Si—O bonds ruptured during the recombination — according to their concentration of a maximum 0.1 per cent (comp. 6.1.2.3.) — released energy with the differential-thermoanalytically measured energy [15] being higher by 2 orders of magnitude, speaks for further energy-producing processes. Thus the structural order in the amorphous phase may increase, for instance by annealing, without already being detectable by roentgenography. This follows from luminescence studies with quartz samples following mechanical treatment, X-ray radiation, and subsequent annealing (Chapt. 3., [125]).

In summary, it follows that new surface area and an amorphous phase are developed by mechanical treatment, which are connected with the rupture of Si—O bonds. Altogether an increased energy content of the solid results from this.

6.1.2.3. *Structure — reactivity correlations at mechanically treated quartz*

The structural defects dealt with in the preceding section, particularly in the form of the amorphous phase, with the resultant excess free enthalpy, lead to an increased reactivity of the mechanically treated quartz. The latter shows a similar behaviour in its reactivity to amorphous SiO_2 phases, e.g. silicic acid and silicagel. But whereas in amorphous SiO_2 phases no positive gradient exists in respect of the structural order in the SiO_2 grain from the outside to the inside, this is generally the case with the mechanically treated quartz grain (see Fig. 6.17.). The smaller the amorphous portion the more the mechanically treated quartz will resemble the untreated quartz in structure and reactivity.

The increased reactivity of mechanically treated quartz expresses itself, for instance, by a sintering which already beging at lower temperatures and runs faster at the same temperature (see 7.2.3.4.). In the same way as the change in the sintering behaviour (Chapt. 4., [141], [19, 36]) the phase transformation of mechanically treated quartz occurring in cristobalite at temperatures of 1,473 (Chapt. 3., [127]) and 1,800 K [36], respectively, may also be explained by the amorphous phase formed.

The increased reactivity of mechanically treated quartz may also be proved in comparison to a 2nd solid, liquid or gaseous phase. Of the reactions of quartz with a 2nd solid, e.g. HgJ_2 (Chapt. 4., [193]), Fe-compounds [60], MgO [61], and CaO, occurring directly during the treatment or in connection with a subsequent heating, the reactions with CaO are of special interest in terms of the formation of binding agents (see 7.2.3.3.).

In the following section (6.1.2.3.1.) the solution processes of mechanically treated quartz will be dealt with in more detail because of their importance to, for instance, the formation of calciumsilicate-hydrate phases in building materials (see 7.2.3.).

The investigations on the interaction of mechanically treated quartz with gases are also of interest for a variety of reasons, e.g. the detection of elementary reactions at the ruptured Si—O bonds, and surface modifications with respect to the manufacture of inorganic-organic composites (see 7.2.2.4.). Hence, this will be discussed in more detail in 6.1.2.3.2.

6.1.2.3.1. The solubility of mechanically treated quartz

Like the etching methods (at quartz for example HF or hydrothermal treatment [24]), the increase in the dissolution velocity and solubility of mechanically treated quartz in liquid phase, e.g. water and sodium hydroxide solution is founded on the increased energy content of the structural disorder produced. In powders this particularly occurs in the form of the integrally determined amorphous phase, see Fig. 5.56.d).

Table 6.5. Dependence of the calculated solubility L of quartz in water on the particle diameter D at 300 K [29].

D, nm	L, ppm
∞	6
100	7
10	28
5	120
3	930

Like that of silicic acid (see Fig. 6.19.), the solubility of untreated and mechanically treated quartz is relatively pH independent, with pH between 1.0 and 9.0 [15, 31, 32].

It amounts to 100 ··· 140 ppm for intensively mechanically treated quartz powders in water under standard condition (Chapt. 5., [115], [15]). It is comparable with that of silicic acid (see Fig. 6.19.) and is higher by one order of magnitude than the solubility of quartz with 6···11 ppm mentioned in 6.1.2.1.

The solubility is known to depend also on the particle size according to the OSTWALD-THOMSON-FREUNDLICH relation (modified by DUNDON and MACK) [29]. The solubilities calculated for various particle sizes according to this are listed in Tab. 6.5. It follows

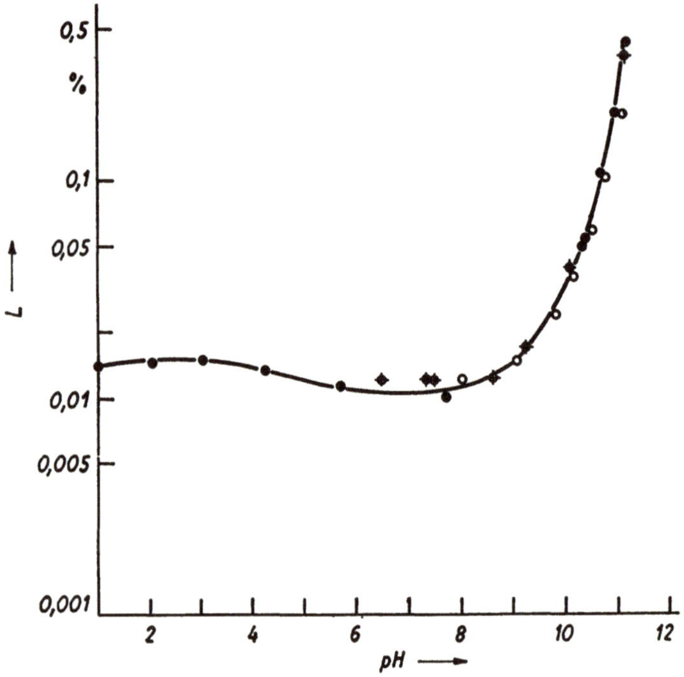

Fig. 6.19. Influence of the pH on the solubility L of amorphous silicic acid in water [31].

from this that the mechanically treated quartz powders with a solubility of 100···140 ppm ought to have a medium particle diameter of 5 nm, but this is in fact at the most the case for only a few grains. Therefore the raised solubility can be primarily attributed to the developed amorphous phase and not to a corresponding particle size reduction (Chapt. 3., [127], Chapt. 4., [250], Chapt. 5., [115], [15, 38]). If the amorphous phase is step by step removed by repeated extraction with water, a lowering of the solubility in water occurs [26, 61]. Conversely the value of 100···140 ppm dissolved silicic acid cannot be exceeded by the application of amorphous silicic acid, as is the case if only amorphous silicic acid is used. Polycondensation and precipitation will appear with such an exceeding [62]. This does not exclude, however, that oversaturated solutions may also be durable in certain circumstances for a longer time [25]. The amount of SiO_2 detached from the untreated as well as from the mechanically treated quartz is not in equilibrium with the crystalline or partial-crystalline phase, (for mechanically treated quartz; see 6.1.2.2.). Upon oversaturation only the precipitation of amorphous silicic acid always occurs.

The elementary step of quartz dissolution is discerned in a partial enlargement of the Si—O—Si bond spacings through the defects in the crystal lattice, by which the screening of the Si^{4+} ions is reduced [42]. By this an attack by H^+ and OH^- ions becomes possible. Altogether the dissolution process for SiO_2 is very complex since

1. each detachment of a lattice element is simultaneously associated with a chemical reaction, the hydrolysis of the Si—O bond,
2. the silicic acid gone into solution can change by condensation and hydrolysis, and through this
3. aggregates can be formed of all possible sizes from molecular to coarsely disperse, of which the latter may be already counted as "solid" bottoms again [32].

In analogy to silicic acid [63], the dissolution velocity in mechanically activated quartz should also be directly connected with the concentration of adsorbed OH^- ions on the quartz surface.

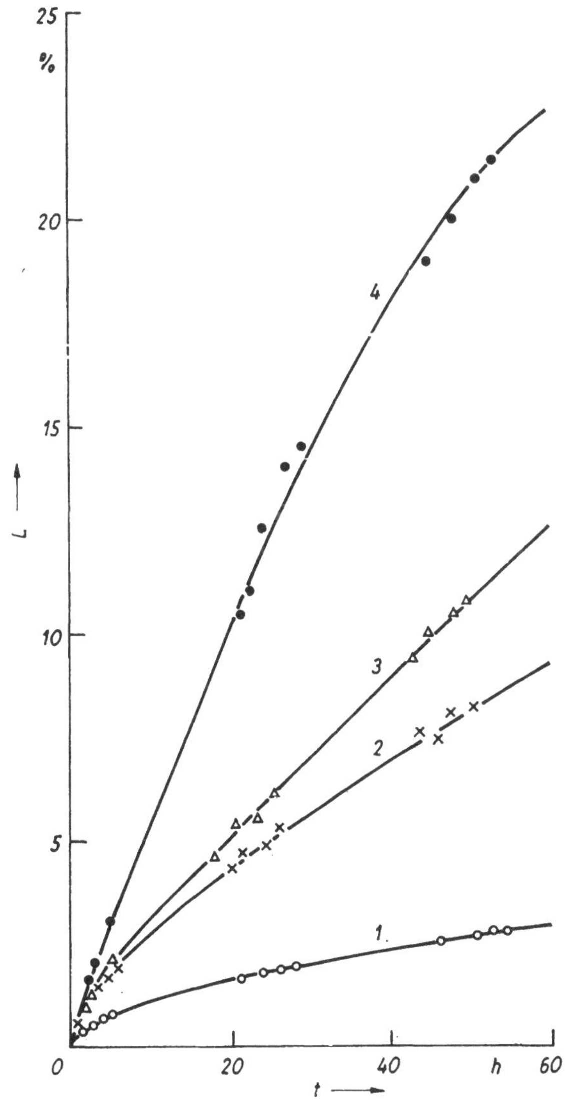

Fig. 6.20. Dissolution of different SiO_2 phases in 0.1 n NaOH at 298 K [64].

L – dissolved SiO_2, *t* – dissolution time, 1 – quartz, 2 – cristobalite, 3 – tridymite, 4 – vitreous glass.

The dissolution velocity of mechanically treated quartz powders in water runs according to a first-order velocity law (Chapt. 5., [115]).

Through this a dependence of the dissolution velocity constant on the used milling set results. An activation energy of 142 kJ/mole was evaluated for the dissolution

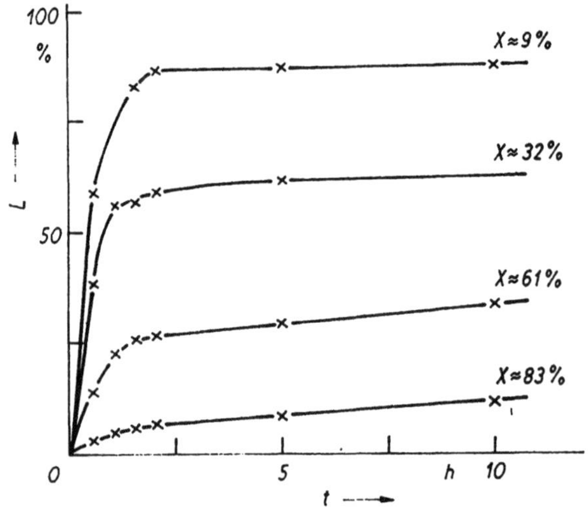

Fig. 6.21. Dissolution of mechanically activated quartz of different crystallinity *X* [38].

Vibration mill, *L* – dissolved SiO_2, *t* – dissolution time.

of quartz in water [27], whereas, for instance, activation energies between 31 and 47 kJ/mole were found for the dissolution of silicic acid in dependence on the pH value (pH between 3 and 10) at temperatures between 293 and 353 K [63].

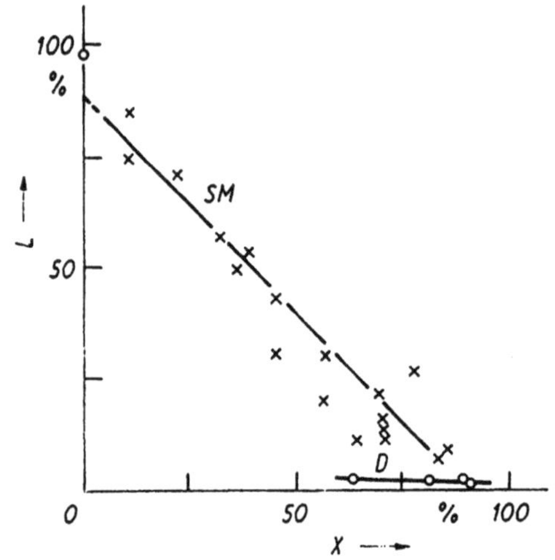

Fig. 6.22. Dependence of the dissolved quantity of SiO_2 *L* of mechanically treated quartz on the crystallinity *X* [41].

1 – quartz milled in a vibration mill, 2 – quartz milled in a disintegrator.

The following relations result for the specific dissolution velocities L_S (pH-range 7···8) of different SiO_2 phases related to the surface area [63]:

$$L_{S(\text{electric arc – silicic acid})} > L_{S(\text{Aerosil})} > L_{S(\text{quartz glass})} > L_{S(\text{cristobalite})} \cong L_{S(\text{quartz})} > L_{S(\text{coesite})}.$$

According to this the phases with the most disordered structures have the highest dissolution velocities. This basically also holds for the alkaline environment, but there is in fact here a larger difference between quartz and cristobalite (see Fig. 6.20.).

Numerous investigations have also shown a correlation between solubility in NaOH and amorphous portion (or surface area) for mechanically treated quartz powders (Chapt. 3., [127], [35, 38]), (see Fig. 6.21.).

Characteristic differences between samples milled in a vibration mill and in a disintegrator [35, 41] become apparent in terms of dependence of the solubility on the amorphous portion. The same also holds for the solubility — surface area dependence (see Fig. 6.22.). The differences may be accounted for by the different distribution of the amorphous phase in the quartz grain resulting from the various stress mechanisms in both milling sets (see 3.1.2.3.).

In summary, it may be stated that dissolution velocity and solubility of mechanically treated quartz are decisively determined by the amount and distribution of the amorphous phase. There are parallels with amorphous SiO_2 phases, such as silicic acid.

6.1.2.3.2. Interaction of mechanically treated quartz with gases

Investigations of the interaction between solid and gas generally provide the opportunity to obtain information on the reactivity of fresh fracture surfaces and the fundamental steps of reactions, on the lifetime of active centres (see 4.3.), and on the excitation states occurring during the mechanical treatment. That is a.o. the case if, as in quartz, the fragments of ruptured chemical bonds survive at least partially or can be detected by reaction with the surrounding gaseous or liquid phase in the form of characteristic groups by means of EPR, IR, or NMR.

There are numerous examples for the investigation of the interaction of mechanically treated quartz with simple gases or liquids. Use has been made of H_2 [18, 37, 57], O_2 (Chapt. 5., [112, 245], [18, 38, 54]), CO ([18, 54, 57, 68]), CO_2 (Chapt. 5., [249], [65, 66, 68]), H_2O in gaseous or liquid phase (Chapt. 5., [112], [15, 16, 53]) as well as of low hydrocarbons (Chapt. 5., [251], [67]). Millings in vacuum or inert gas are frequently used for comparative studies (Chapt. 5., [249], [18, 37, 38, 57, 67, 68]). In the following, a more detailed discussion will be conducted on the relationships between structure and tribosorbate as well as the influence of gas type and gas pressure on the radicals (paramagnetic centres) produced on the quartz during the mechanical treatment.

In addition to the mechanical treatment conditions the tribosorption of gases on the quartz is generally decisively determined by the physical and chemical properties of gases, as has already been shown for SiC (see 6.1.1.3.). The tribosorbed gases are situated in the amorphous SiO_2 phase and have partly reacted with SiO_2. The tribosorbate (n) as well as the chemically transformed portion (Z) increase with the treatment time and consequently with the amorphous portion (A) formed (see Fig. 14 n, Z, A), as has also been proved for different gases with SiC (see 6.1.1.). Hence together with the step by step detachment of the amorphous phase, the remaining sorbed gas volume and the chemically transformed portion decrease [38]. The tribosorbate, which may differ by orders of magnitude for the various gases, has no decisive influence on the alteration of the surface area and of the amorphous portion [37].

Different solid radicals (EPR centres), such as E' centres, oxygen defect electron centres, CO_2^- and CO_3^- radical ions (Chapt. 3., [68], Chapt. 5., [112, 249], [54, 57, 65]) are formed by the mechanical treatment of quartz in the presence or in the absence of specific gases. These solid radicals permit the inference of elementary reaction steps in the interaction of simple gases with quartz, in which a rupture of the Si—O bond is assumed to occur for the time being (see 5.9. and 6.1.2.2.).

One of the most intensively investigated paramagnetic centres on the quartz is the E' centre appearing primarily during the rupture of the Si—O bond. This E' centre is caused by electrons in dangling orbitals, that is to say in non-binding sp^3-hybrid orbitals at Si atoms missing an oxygen neighbour (Chapt. 5., [252]). Put simply, these are Si radicals (see 5.9. and 6.1.2.2.). The E' centre may also be produced by the effect of other non-thermal types of energy on different SiO_2 phases as already mentioned in 5.9. (Chapt. 5., [250], [252], [69]). It does not require any interaction with a gaseous or liquid phase.

The structure of the oxygen defect electron centre has not been clarified until now (see 5.9.). A similar centre develops during the action of other non-thermal types of energy on SiO_2 (Chapt. 3., [68]). The fact that this centre only develops in the presence of O_2 during the mechanical treatment speaks for the direct participation of oxygen and consequently for the meachanism mentioned according to que. (5.67) and (5.68). The redox-amphoteric character of this radical has already been referred to 5.9.

Outside an initial stage, a constant value results between the amount of sorbed gas and the number of EPR centres, which is different for different gases, amounting for instance to 2 with O_2 [37, 54]. This means the formation of 1 oxygen defect electron centre on average during the sorption of 2 O_2 molecules. In comparison to an amorphous portion of more than 90 per cent after a longer period of treatment, the maximum attainable concentration of oxygen defect electrons is about 0.3 per cent only. Therefore this portion is without any importance for dissolution processes.

An average number of 5×10^{18} ruptured Si—O bonds per m^2 newly produced surface area is expected to come into existence during the cleavage of quartz crystals (Chapt. 5., [249], [70]). If the concentration of E' centres and oxygen defect electron centres produced by double impact stress (vibration mill) in quartz powders are related to the surface area (about $1 \times 10^{18}/m^2$ in each case [71]), then the difference in terms of order of magnitude mentioned in 5.9. between the expected and measured E' concentrations does not result (there: glass vibration vessel with a small mechanic activation effect, characterized for instance by the specific surface area of the activated quartz powder: 0.5 m^2/g (Chapt. 5., [249]); here: vibration mill, specific surface area: see Fig. 6.14.). The reason is that during double impact stress in a vibration mill, the centres are not at the surface but mainly in the formed amorphous phase (see Fig. 6.18.). Generally such centres are formed more easily at stressed siloxane bridges [77].

All the E' centres produced during mechanical treatment react with tribosorbed oxygen with the forming of oxygen defect electron centres under standard conditions in air or oxygen atmospheres. If the O_2 partial pressures are low the absorbed O_2 amount is not sufficient to saturate all the ruptured Si—O bonds. In this case E' centres may also be detected in addition to the oxygen defect electron centres [71]. The relationship between the two kinds of centres can be seen in Fig. 6.23.

Using H_2 as treatment atmosphere, at normal pressure only a part of the E' centres reacts with the tribosorbed hydrogen. The remaining E' centres are in their concentration smaller by 1 order of magnitude than with treatment in Ar — as can be detected by EPR-spectroscopy [37]. In the quartz samples treated in H_2 a doublet of protons also appears in addition to the E' centre which may be attributed to the presence of SiH and/or SiOH groups [37, 57]. A similar doublet of protons also appears on irradiated silica glasses [72, 73, 74] and in Aerosil following the action of a low-pressure plasma (Chapt. 5., [250]). The formation of $\geqslant$SiH and $\geqslant$SiOH groups can be detected by IR-spectroscopy [37, 57, 75]. If Aerosil, which is much more reactive in comparison to quartz, is thermally activated in the presence of H_2 a detectable formation of $\geqslant$SiH groups only takes place above 800 K [76].

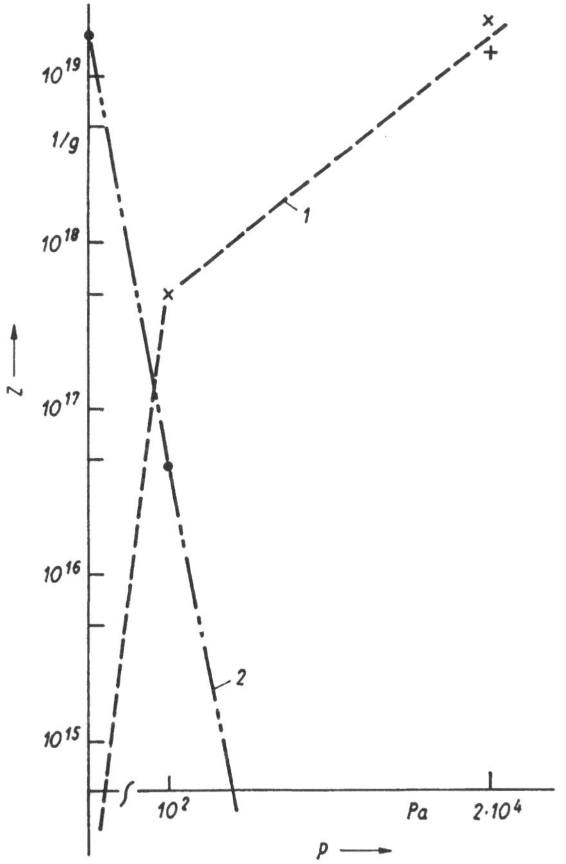

Fig. 6.23. Effect of the O_2 pressure p on kind and number Z of the EPR centres developed in quartz during the mechanical treatment (course on principle) [71].

1 — oxygen defect electron centres, 2 — E' centres.

Comparing the total amount of H_2 (n_{H_2}) absorbed during a double impact stress (vibration mill, normal conditions) with the $\geqslant$SiH and $\geqslant$SiOH groups formed (measured by IR-spectroscopy) as well as with the $\rangle\dot{\mathrm{Si}}$H and $\rangle\dot{\mathrm{Si}}$OH groups (measured by EPR-spectroscopy), the following relation results [71]:

$$n_{H_2} > n_{\geqslant SiH, \geqslant SiOH} > n_{\rangle \dot{Si}H / \rangle \dot{Si}OH} \qquad (5)$$

Since no E' centres can be detected when hydrocarbons are used, as with air (in normal conditions), they must have reacted with the E' centres. The C˙ radicals appearing with the use of hydrocarbons demonstrate the decomposition of hydrocarbons down to elementary carbon (Chapt. 5., [251]), as is also expressed in the model conception developed for tribochemical hydrocarbon transformation (see Fig. 6.10.).

A survey of the spectra, characteristic g values, line widths, and concentrations of some centres produced in quartz during mechanical treatment in different gaseous atmospheres, is given in Fig. 6.24. and Tab. 6.6.

Table 6.6. Summary of some EPR centres produced by mechanical treatment at quartz in various gaseous atmospheres.

Gaseous atmosphere	EPR centre	g-values	Line width H, G	Quantity of EPR centres Z, 1/g	[1])	Ref.
Ar	E'	$g_\perp = 2.0001$ $g_\parallel = 2.0015$		$2 \cdot 10^{19}$	a)	[37]
H_2	E'	$g_\perp = 2.0001$ $g_\parallel = 2.0015$		$2 \cdot 10^{18}$	a)	[37]
	pd[2])		≈ 73[3])	+[4])		
O_2 (air)	SDE[5])	$g_1 = 2.0018 -$ 2.0022 $g_2 = 2.0044 -$ 2.0088 [6]) $g_3 = 2.056 -$ 2.061		$2 \cdot 10^{19}$	a)	(Chapt. 3., [68])
CO_2	CO_2^-	$g_1 = 1.9974$ 1.9975 $g_2 = 2.0027$ 2.0024 [6]) $g_3 = 2.0033$ 2.0031		$1 \cdot 10^{18}$ $8 \cdot 10^{15}$	a) b)	
	CO_3^-	$g_1 = 2.0048$ $g_2 = 2.0063$ [6]) $g_3 = 2.0248$		$4 \cdot 10^{16}$	b)	[65]
	$C^{\cdot}$	$g = 2.0031$	6.5	$< 2 \cdot 10^{14}$	b)	(Chapt. 5., [251])
C_2H_2	$C^{\cdot}$	$g = 2.0026 -$ 2.0034	$\approx 8-9$	$1 \cdot 10^{18}$	a)	(Chapt. 5., [251])
C_2H_4	$C^{\cdot}$	$g = 2.0026 -$ 2.0034	≈ 12	$1 \cdot 10^{17}$	a)	(Chapt. 5., [251])

[1]) Vibration mill, treatment time 8 hours, a) in relation to b) treated more intensively; [2]) pd — proton doublet, [3]) lines distance, [4]) about 25 per cent of the quantity of E' centres, [5]) SDE — oxygen defect electrons, [6]) zero passage.

Altogether a change in the chemical properties (modification) of the surface and, furthermore of the amorphous phase of quartz is effected by the sorption and reaction of gases. This is of direct practical interest for instance for the production of inorganic-organic composites. This will be discussed in more detail in 7.2.2.4.

In summary, it may be stated that the E' centres produced on the mechanically treated quartz ($\geq$ Si˙ radicals) are to be attributed to the rupture of Si—O bonds. The formation of other paramagnetic centres can be explained by the reaction of these fragments with the respective gas.

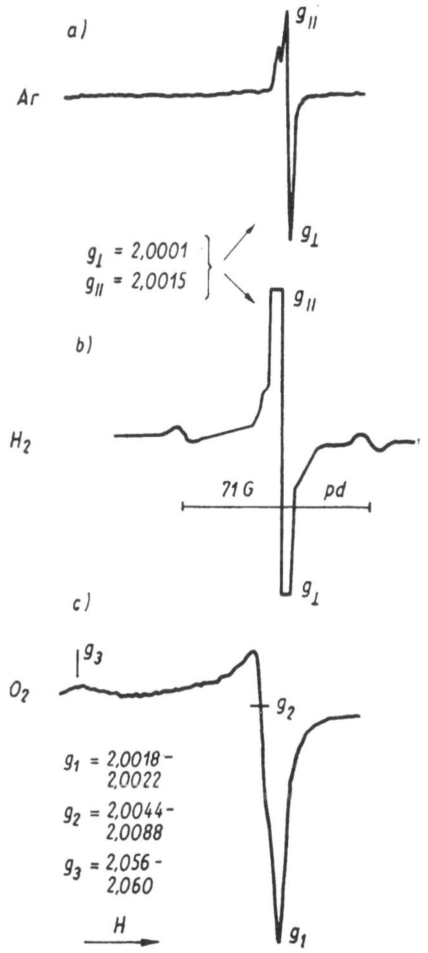

Fig. 6.24. EPR spectra of quartz samples after the treatment in different atmospheres [71].
a — *E'* centre, b — *E'* centrum and proton doublet (pd), c — oxygen defect electron centre.

6.1.3. *Tribochemistry of apatites*

6.1.3.1. *Structure of apatites*

The general formula of apatites is

$$Me_{10}(PO_4)_6X_2, \tag{6}$$

in which Me can equal Ca^{++}, Mg^{++}, Sr^{++}, Ba^{++}, $2Na^+$ und

X can equal F^-, Cl^-, OH^-, $1/2CO_3^{--}$

As a rule apatites crystallize in the hexagonal dipyramidal class and belong to the space-group P $6_{3/m}$. The ions form a primitive lattice with a hexad screw axis parallel to the c-direction and in vertical reflector plane to it (see Fig. 6.25.). There is a canal structure along this screw axis in which the X^--ions are chemically incorporated [81, 82]. A part of the divalent positive ions is arranged in triangular-shaped form a round the screw axis. The size of these ions determines the canal diameter. This arrangement leads to a facilitated diffusion of lattice components along the c-axis over long distances of more than 100 μm [83].

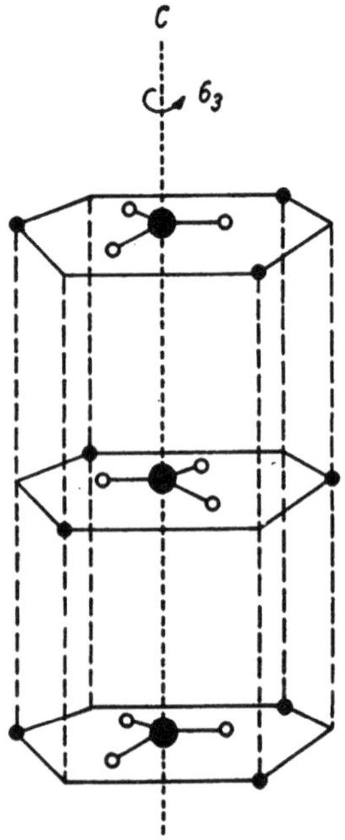

Fig. 6.25. Atomic arrangement in fluoroapatite along a 6-al screw axis *c* [80].
Ca – ○, *F* – ●, P – ●.

Apatites form complex ionic lattices with predominantly heteropolar bond portions in which the bond character is of ionic nature between PO_4^{3-} and Ca^{2+} ions [84]. However, the Ca—X bond between the ions in the canal has a mainly covalent character [85].

The apatite lattice has an extraordinarily high stability expressed in the relatively high melting point of approximately 2,000 K. Corresponding to this, apatites belong to the substances extremely difficult to dissolve in water.

The pure fluoroapatite $Ca_{10}(PO_4)_3F_2$ possesses the least reactivity, because the diffusion process is strongly inhibited due to a relatively small canal diameter. With growing size of the X^--ions an enlargement of the canal diameter, and associated with it an accelerated diffusion in the canal and an increased reactivity R of the apatite, appear. From this results the following graduation of reactivity (Chapt. 4., [88], [85]):

$$R_{fluoroapatite}\ (X^- = F^-) < R_{hydroxylapatite}\ (X^- = OH) < R_{chlorapatite}\ (X^- = Cl^-) \quad (7)$$

There are two possibilities for the chemical incorporation of carbonate:

1. Substitution of PO_4^{3-} groups by CO_3F^{3-} groups and
2. substitution of X^--ions in the canal structure.

The apatites occurring in nature and precipitated apatites belong to the first mentioned type of substitution. A X^--ion substitution by the chemical incorporation of carbonate has been observed only in synthetic apatites through the thermal treatment of hydroxylapatites at 1,200 K [86]. Here an enlargement of the canal diameter and an increase in reactivity appear.

Natural apatites from different deposits show a great variability in the composition, caused by Me^{++}, X^-, and PO_4^{3+} ion substitution. The composition of sedimentary apatites may be expressed in first approximation by the formula [87]:

$$(Ca, Na, Mg)_{10}(PO_4)_{6-x}(CO_3)_x F_y(F, OH)_2 \quad (8)$$

Two typical examples of this class of substances are the magmatic Kola-apatite, which is a relatively undisturbed fluoroapatite, and the sedimentary Marocco-apatite, which is a carbonate-fluoroapatite having a relatively strongly disturbed structure and an increased reactivity in comparison to Kola-apatite. The most important parameters of these two phosphates are listed in Tab. 6.7.

Table 6.7. Comparison of important parameters of Kola-apatite and Marocco-phosphate (Chapt. 4., [88]).

		Kola-apatite	Marocco-phosphate
Composition			
(essential components)	P_2O_5	38.7	32.9%
	CaO	52.4%	51.9%
	CO_3^{2-}	0.2%	8.0%
	F^-	3.2%	4.1%
Primary crystallite size		250 nm	30 nm
Lattice distortion		(standard)	0.35%
Solubility in ammonium citrate		1.4%	2.0%
Solubility in citric acid		7.8%	25.0%

6.1.3.2. *Tribomechanically caused changes in structure and reactivity of apatites*

Under the influence of mechanical activation, for instance by vibration milling, the reactivity of apatites is strongly increased. Their efficiency is mainly characterized by their solubility in citric acid and ammonium citrate solution, respectively, for the solubility also represents a measure of their plant-physiological efficiency with the use of phosphates as fertilizers under specific boundary conditions. As Fig. 6.26. shows by the example of several natural apatites having different compositions, a strong increase of solubility in citric acid solution takes place particularly at the beginning of the activation. Soft-earth sedimentary phosphates like the African phosphates reach a higher solubility than the hard-earth magmatic Kola-apatite. In some cases the activation leads to a total solubility in citric acid solution under specific conditions [88].

The solubility of apatites originating from Soviet deposits could be raised by the factor 1.5 ··· 19 during their activation (Chapt. 5., [29], [89]). For Florida-phosphate an increase in solubility has been observed from 34 to 92 per cent [90].

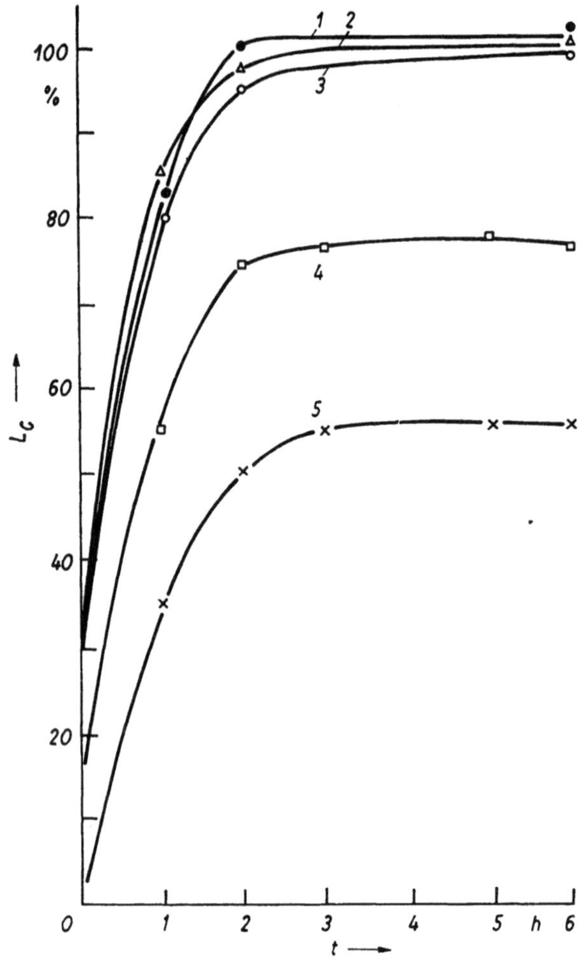

Fig. 6.26. Dependence of the solubility in citric acid L_C on the treatment time t for apatites of different deposits [88].

1 — Annophos, 2 — Tunesian phosphate, 3 — Egyptian phosphate, 4 — Marocco-phosphate, 5 — Kola-apatite.

Table 6.8. Effect of particle size and activation of Kola-apatite on its specific initial solution velocity v_s and equilibrium solubility L_G (Chapt. 4., [88]).

Treatment time of the activated samples t, min	Particle fraction of the non-activated screening material, μm	v*), mg/min	S**), m^2/g	v_s, $g/min \cdot m^2$	$L_{G,3}$ g/m^3
	160···180	0.2	0.09	2.22	33.5
	45··· 71	0.57	0.26	2.23	38.0
	40··· 45	0.80	0.36	2.22	40.0
60		7.35	1.11	6.62	51.5
120		9.20	1.06	8.68	62.0

*) v — initial solution velocity.

**) S — specific surface area after Blaine.

The correlation between structure and reactivity has been particularly investigated for Kola-apatite and Marocco-phosphate. The changes in some structure and reactivity quantities brougth about by vibration milling are given for these materials in Fig. 6.27. in comparitive form, in dependence on the treatment time [91].

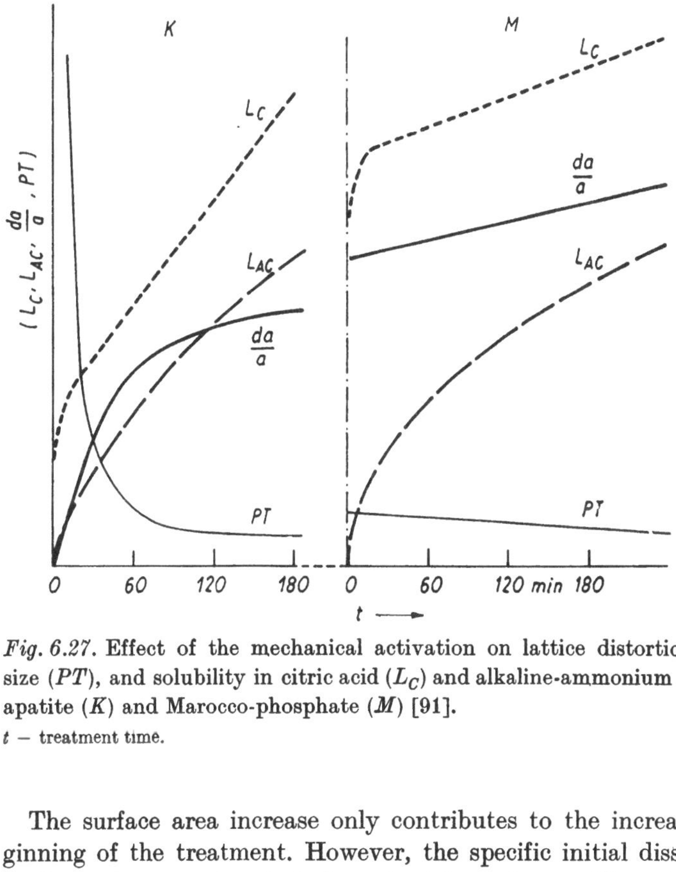

Fig. 6.27. Effect of the mechanical activation on lattice distortion ($\Delta a/a$), primary crystallite size (PT), and solubility in citric acid (L_C) and alkaline-ammonium citrate solution (L_{AC}) at Kola-apatite (K) and Marocco-phosphate (M) [91].
t – treatment time.

The surface area increase only contributes to the increase in solubility at the beginning of the treatment. However, the specific initial dissolution velocity v_s related to the surface area is only independent from the specific surface area in a non-activated apatite, as Tab. 6.8. shows. v_s however, strongly increases with the treatment, indicating that the tribomechanically produced lattice defects are the fundamental reason for increases in reactivity.

Table 6.9. Changes in the specific surface area S and in the solubility L of Marocco-phosphate in citric acid depending on the treatment time t in a planetary mill (Chapt. 4., [60]).

t, min	S, m²/g	L, per cent
0	8.9	27
1	18.0	33
5	8.0	38
10	3.9	53
60	3.8	75

Whereas the surface area of Marocco-phosphate increases to a constant value during the activation in a vibration mill, using a planetary mill it drops even below the value of unmilled Marocco-phosphate. In contrast to this, according to Tab. 6.9. the solubility continuously increases (Chapt. 4., [60]).

The specific surface area of apatites may be raised by the factor 5···10 if milled with water as compared with a dry milling (Chapt. 5., [119]). By contrast, the structure of apatites is significantly less disturbed under the influence of water than with dry milling [92, 93]. The reactivity of the products thus obtained is smaller [94]. Altogether these observations show that the surface size has a secondary influence on the solubility of activated apatites.

According to Fig. 6.27. the changes in citric acid solubility nearly correlate with the primary crystallite size and the degree of lattice distortion measurable by roentgenography. With Kola-apatite, after an activation time of 180 min, values are attained for the mentioned quantities as nearly shown by non-activated Marocco-phosphate.

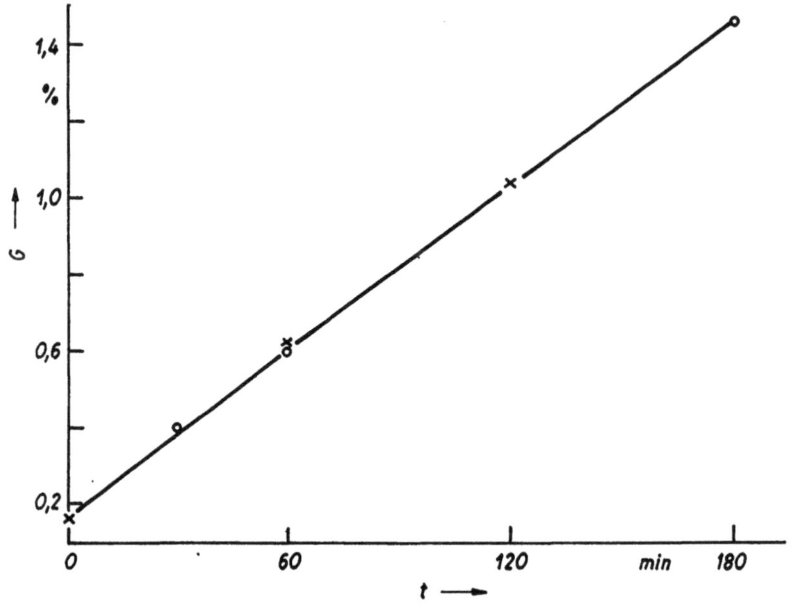

Fig. 6.28. Dependence of the CO_2 content G in Kola-apatite on the treatment time t (Chapt. 4., [88]).

x — chemically determined, o — determined by thermal decomposition in the range of 1,000 to 1,300 K.

One of the reasons for the above-mentioned behaviour can be seen in the fact that Kola-apatite takes up CO_2 from the surrounding atmosphere (see Fig. 6.28.). It is concluded from IR-spectroscopical measurements that CO_2 in the form of CO_3F^{3-}-groups is chemically incorporated onto the lattice sites of phosphate groups (Chapt. 4., [88]). This effect is essentially increased if the activation does not take place in air but in CO_2. The chemical incorporation of carbonate is associated with a reduction of the primary crystallite size and with an increase of the degree of lattice distortion (see Fig. 6.27.).

According to Tab. 6.7. in Marocco-phosphate a considerable portion of CO_2 is already contained in the parent material which is expressed by a relatively high degree of lattice distortion and a high citric acid solubility.

As further results from Fig. 6.27., the solubility measured in alkaline ammonium citrate solution is very low, both with untreated Kola-apatite and untreated Marocco-phosphate. Evidently the treatment process produces a roentgenographically measurable defect state exceeding the lattice distortions which above all leads to an increase of the apatite solubility strongly reduced in the alkaline range [95].

Thus no additional increase in the carbonate content takes place during the activation of Marocco-phosphate but a displacement of lattice carbonate from the sites of PO_4^{3-}-groups to canal sites (Chapt. 4., [88]). This leads to an enlargement of the canal diameter and to a distortion of the ion bonds attached in the canal.

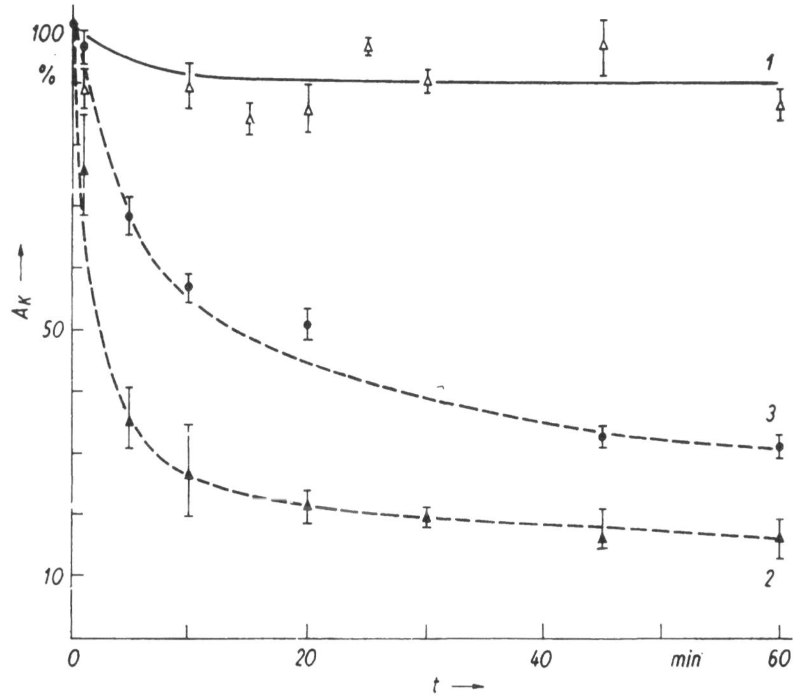

Fig. 6.29. The crystalline portion A_K of Kola-apatite and Marocco-phosphate in dependence on the treatment time *t* using different triboreactors (Chapt. 4., [60]).
1 – Marocco-phosphate, vibration mill, 2 – Marocco-phosphate, planetary mill, 3 – Kola-apatite, planetary mill.

It is concluded from IR-spectroscopic measurements that the disturbance induced by mechanical activation consists in particular of a distortion of the CaO_6F-complex existing in the canal [96] which also includes the change in the symmetry of the PO_4^{3-}-ion [91]. As was shown by ^{31}P-NMR measurements, the order of magnitude of distortion of the PO_4^{3-}-tetrahedron can be compared to that caused by P—OH groups [91]. This degree of distortion is sufficient for the dissolution process, for the first step of the dissolution consists of a hydrolytic breaking of the Ca—OP bond and in the formation of $(Ca—OH)^+$ and $(O_3P—OH)^{2-}$ ions (Chapt. 4., [88]). By this an approximation takes place to the symmetry of the diphosphate being soluble in alkaline ammonium citrate solution. The formation of $(O_3P—OH)^{2-}$ ions on the surface of activated Kola-apatite, could be detected by the ^1H—NMR-spectra [61].

The increased degree of disturbance in apatite has to be attributed to the formation of X-ray amorphous regions. While this process plays a secondary role for activation in a vibration mill, by using a planetary mill it is possible to transform to a considerable extent both Kola-apatite and Marocco-phosphate into the X-ray amorphous state (see Fig. 6.29.). The lowering of crystallinity leads to a loss in symmetry in the apatite lattice which is detectable by IR-spectroscopy. In accordance with this the values for the ammonium citrate solubility L and the X-ray crystalline portion change reciprocally to

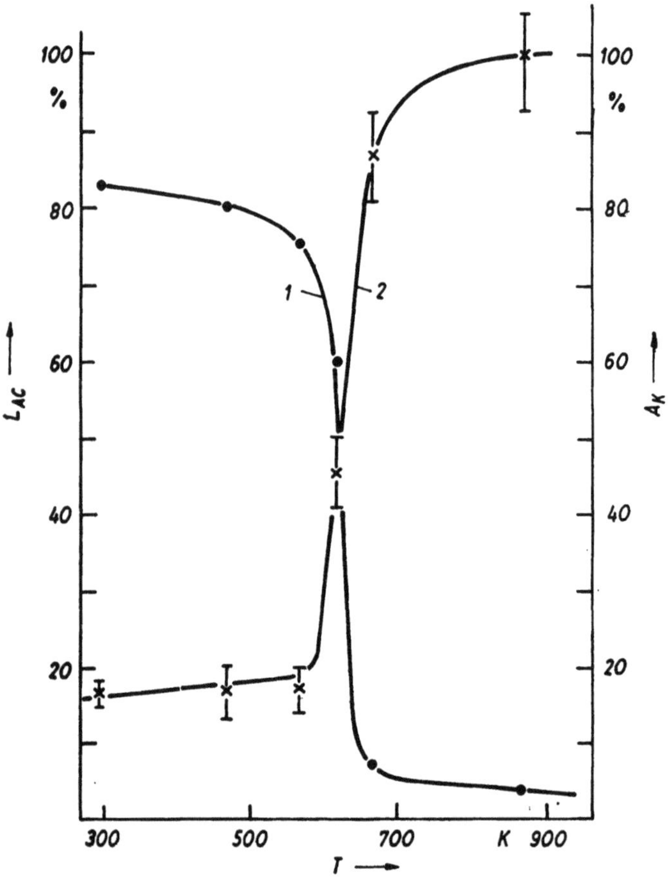

Fig. 6.30. The relationship between solubility in ammonium-citrate L_{AC} (1) of Marocco-phosphate and crystalline portion A_K (2) at annealing (Chapt. 4., [60]).

each other. This relationship also results for the annealing process, particularly in the range of 600···700 K (see Fig. 6.30.), (Chapt. 4., [60]).

The above-mentioned structure-reactivity correlations become more pronounced if the changes coming into existence during the annealing process are comprehensively considered. As Fig. 6.31. shows, a reduction of the ammonium citrate solubility already takes place from 400 K, and the release of the water bound to the apatite surface takes place in a way similar to this course. The hydrolytic breaking of the Ca—OP bond is annuled by the retrogression of the strongly disturbed regions.

A decrease of the specific surface, of the lattice distortion measurable by roentgenography, and of the citric acid solubility correlated with the latter only takes place at much higher temperatures of 600 ··· 800 K.

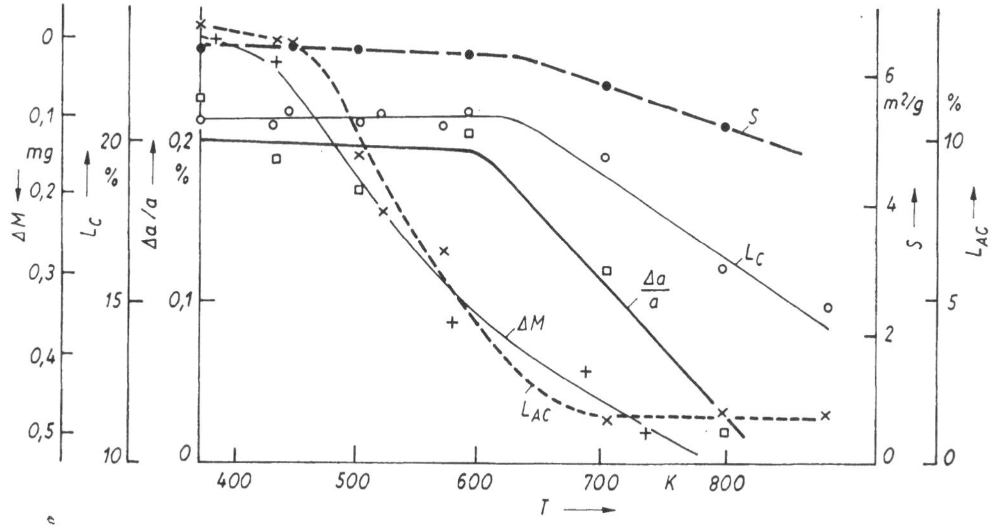

Fig. 6.31. The effect of the annealing lattice distortions ($\Delta a/a$), specific surface area (S), solubility in citric acid (L_C) and alkaline ammonium citrate solution (L_{AC}), and on the mass loss (ΔM) of mechanically activated Kola-apatite [91].
T – temperature.

Consequently there are different structure-reactivity correlations for citric acid solubility and ammonium citrate solubility. In this the ammonium citrate solubility can evidently be attributed to more highly activated lattice regions. An additional increase in the solubility of ground fluoroapatites has been observed in the milling of fluoroapatites if the milling is conducted in the presence of SiO_2 [94]. From this a tribochemical reaction with the fluoroapatite results, leading to the formation of the more reactive hydroxylapatite via the defluorination of the initial material [90].

Under appropriate activation and dissolution conditions a complete solubility may be attained of certain apatites, e.g. in citric acid solution (see Fig. 6.26.). According to the common methods used for the evaluation of apatitic phosphates only a portion of the activated phosphate is dissolved however.

It has now been found that these products may be completely dissolved if they are subjected to a multiple extraction. For Marocco-phosphate activated in vibration and planetary mills, the multiple solvent treatment is given in comparison with a non-activated sample in Fig. 6.32. Whereas only 7 per cent of the non-activated product is dissolved after 4 extraction steps, a solubility of 100 per cent is obtained for the product activated in a planetary mill (Chapt. 4., [60]).

Even greater differences result with the multiple extraction of activated and non-

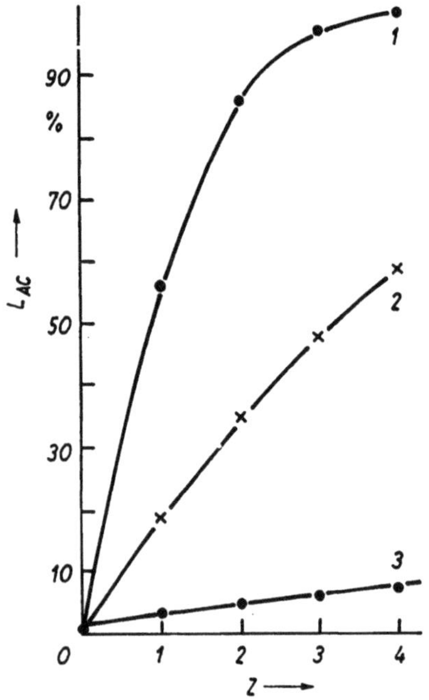

Fig. 6.32. Solubilities in ammonium citrate solution L_{AC} of differently mechanically activated Marocco-phosphate following repeated treatment by solvents (Chapt. 4., [60]).

1 – planetary mill (activated for 10 min), 2 – vibration mill (activated for 60 min), 3 – non-activated raw material, z – extraction steps.

activated Kola-apatites [7]. At the same time the investigations show that not only the regions near the surface of an apatite are activated, as is the case with quartz (see 6.1.2.), but the total particle volume is throughly disturbed, albeit graduated.

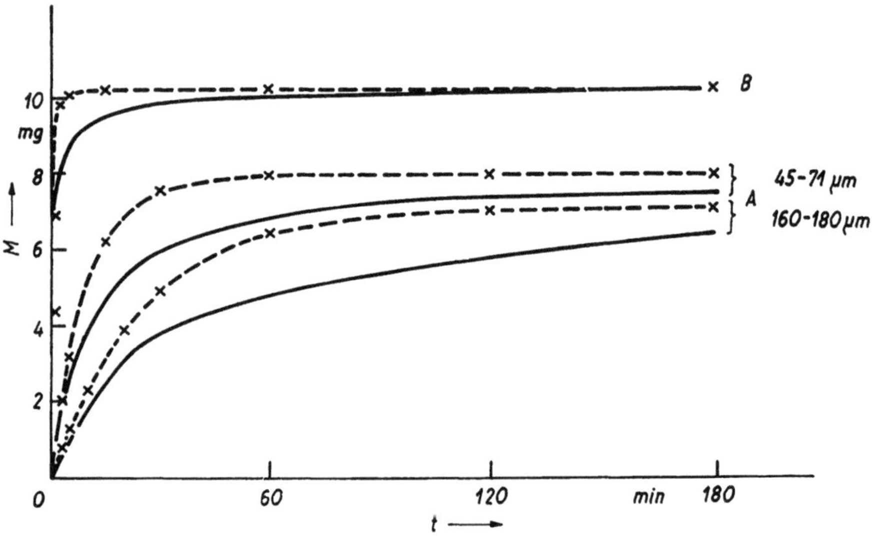

Fig. 6.33. Comparison of the observed (——) and calculated (– – –) concentration-time curves of non-activated Kola-apatite samples (A) and Kola-apatite activated for 60 min (B) (Chapt. 4.,[88]).

M – dissolved quantity, *t* – time.

6.1.3.3. *Dissolution kinetics of activated apatites*

The dissolution of apatite may be followed as dominating process only at the beginning. The velocity of the dissolution process measured in this phase is described as the initial dissolution velocity (see Tab. 6.8.). The dissolution velocity decreases with increas-

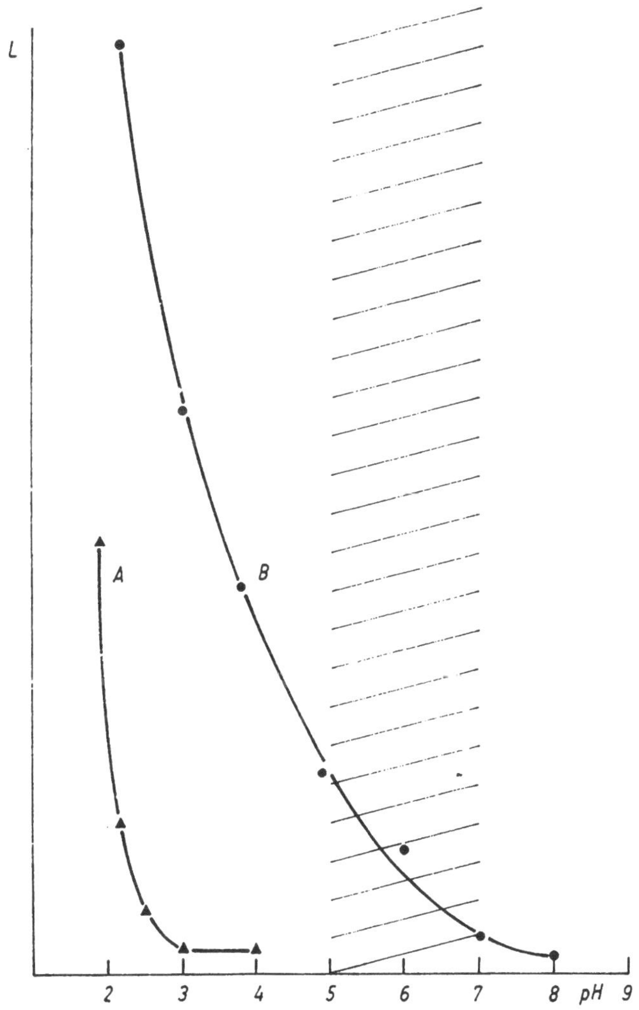

Fig. 6.34. The effect of the mechanical activation on the solubility *L* of Kola-apatite in sodium acetate buffer solution at different pH [91].
A – unmilled Kola-apatite, *B* – Kola-apatite milled for 60 min.

ing reaction time through the growing influence of the opposing reaction, the reprecipitation of the dissolved ions. Provided there is a stoichiometric dissolution the following equation could be derived for the dissolution velocity of the apatites

(Chapt. 4., [88], [97]):

$$\frac{1}{c} \approx \frac{1}{k_1 M_0} \cdot \frac{1}{t} + \frac{1}{\bar{c}} \tag{9}$$

(c — concentration of the $H_2PO_4^-$-ions, k_1 — constant, M_0 — initial quantity of apatite, $\bar{c}$ — equilibrium concentration, t — time).

Fig. 6.33. shows the concentration-time curves of non-activated and activated Kola-apatite measured experimentally and calculated according to equ. (9). Calculated and experimentally determined curves show a good conformity in the initial stage of the dissolution and in the equilibrium state. Consequently equ. (9) permits the calculation of the initial dissolution velocity and of the equilibrium concentration. Furthermore it can be seen from Fig. 6.33. that the mechanical activation leads both to an important nicrease of the initial dissolution velocity and to an increased equilibrium concentration. The strong increase of the initial dissolution velocity during the transition from the

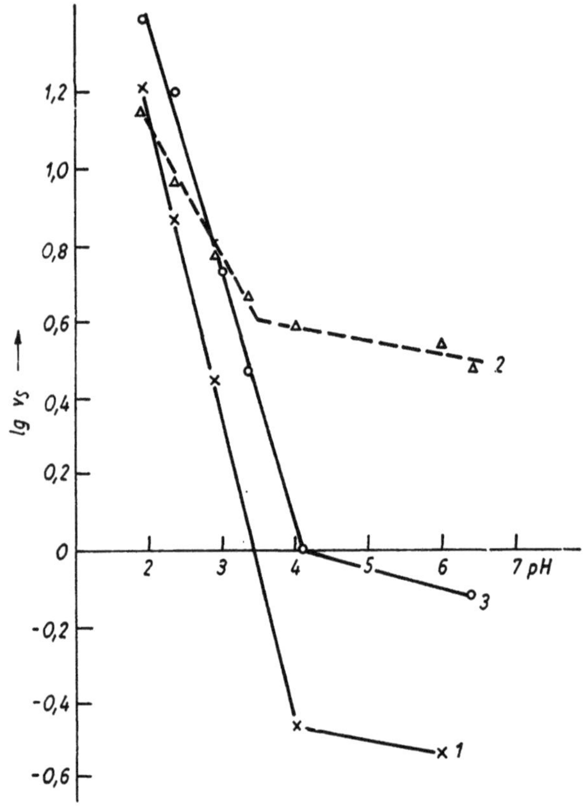

Fig. 6.35. The dependence of the specific initial dissolution velocity v_s of non-activated and activated Marocco-phosphate and Gafsa-phosphate, respectively, in citrate buffer solution on pH (Chapt. 4., [88]).
1 — Marocco-phosphate non-activated, 2 — Marocco-phosphate activated, 3 — Gafsa-phosphate non-activated.

non-activated to the activated Kola-apatite has to be attributed to the reduction of the activation energy of the reaction by approximately one order (29.8 kJ/mole → 3.4 kJ/mole) (see Tab. 4.10.) due to the storage of excess enthalpy ($\Delta H \approx 25$ kJ/mole) (4.88) in the activated Kola-apatite (see 4.5.).

Both the dissolution velocity and the solubility of apatite are strongly influenced by the pH of the solvent. The initial dissolution velocity in the acid range has a linear dependence on the concentration of protons (Chapt. 4., [88]). The activated samples show a smaller dependence on pH as compared to non-activated ones so that the activated apatites still have an increased solubility in a pH-range existing in cultivated soils (Fig. 6.34., hachured area).

If complexing agents, e.g. a citrate buffer solution, are used for the dissolution, then the linear dependence between the initial dissolution velocity and the proton concentration is not maintained. In the pH-range from 3.5···4.2, as Fig. 6.35. shows, points of discontinuity appear at all the curves in which a smaller dependence of the initial dissolution velocity on the proton concentration exists above such a point of discontinuity of a curve. The sharp dropping of the curves is attributed to the change of the dissolution mechanism.

Whereas the proton transfer to the solid has to be understood as the dominating reaction step in the strongly acid range, the dissolution beneath the point of discontinuity of the curve is explained by the complexation of citrate ions with Ca^{2+} cations and the secondary processes subsequent to this (Chapt. 4., [88]).

As has been shown above there are different structure-reactivity correlations for the dissolution of tribomechanically activated apatites in citric acid and alkaline ammonium citrate solutions. According to the latterly discussed kinetic investigation, these differences arise from a different reaction mechanism.

The tribochemistry of apatites is of great practical importance for the processing of rock phosphates and their use as fertilizers in agriculture [98]. The problems resulting from this will be dealt with in 7.2.2.2.4.

6.2. Tribocatalysis

Tribocatalysis is that subsection of tribochemistry dealing with the changes of catalytic properties of solids due to the effect of mechanical energy [99].

The changes in the catalytic activity of solids are primarily caused by the mode of mechanical stress [56]. The treatment of catalysts has been mainly conducted in activation sets, particularly in vibration mills [100]. Other treatment practices have also been used in tribocatalytic investigations such as rolling [101, 102], pressing [103], twisting [104], cutting [105], splitting [106, 107], and high pressures [108, 109].

A summary of the tribocatalytic reactions running on metallic and oxidic catalysts is given in Tab. 6.10.

Generally the catalytic activity of solids is considerably increased by treatment, in particular cases by some orders (Chapt. 4., [96]). But a decrease in catalytic activity as result of mechanical treatment has also been observed in specific catalysts, [138, 147]. The reasons for this are manifold and have been investigated in detail only in a few examples. Some essential relations between mechanical treatment of the catalyst and the changed catalytic activity will be discussed in the following sections.

Table 6.10. Summary of tribocatalytic reactions at metallic and oxidic catalysts.

Catalyst	Reaction	Ref.
Fe	o-p H_2 transformation	Chapt. 5., [98]
Fe	Benzene hydrogenation	Chapt. 4., [50], [96]
Fe	Decomposition of methanol	Chapt. 1., [1]
Fe	Formation of methane	Chapt. 4., [74], [110]
Fe	Ethylene hydrogenation	Chapt. 4., [50], [96]
Fe	Formation of NH_3	[111–115, 116]
Ni/SiO_2	Benzene hydrogenation	Chapt. 4., [50], [96], [98], Chapt. 5., [138], Chapt. 4., [139], [12, 13, 112, 115, 117–120]
Ni	Benzene hydrogenation	
Ni	Phenol hydrogenation	[121, 122]
Ni	Ocenol hydrogenation	[121, 122]
Ni	o-p H_2 transformation	Chapt. 5., [225], [121, 122]
Ni	Fat hydrogenation	Chapt. 4., [87], [123]
Ni	Cyclohexane hydrogenation	[124]
Ni	2-propanol dehydrogenation	[124]
Ni	Cinnamic acid hydrogenation	Chapt. 5., [225]
Ni	Ethanol dehydrogenation	Chapt. 5., [224, 225], [104]
Ni	Decomposition of H_2O_2	Chapt. 5., [225]
Ni	Cinnamic acid ethyl ester hydrogenation	[115]
Ni	Formic acid dehydrogenation	[102, 125, 126]
Ni	Formation of water	[112]
Ni	Ethylene hydrogenation	Chapt. 4., [50], [101]
Co	Benzene hydrogenation	Chapt. 4., [97], [127, 128]
Cu	GATTERMANN reaction	[104]
Cu	Ethanol dehydrogenation	[104, 129]
Pt	Decomposition of H_2O_2	[118]
Ge	Decomposition of N_2O	[106]
Ge	Decomposition of H_2O	[107]
Ge	Ethanol dehydrogenation	[130]
Fe_2O_3	Oxidation of H_2S with formation of S	[131, 132]
α-Fe_2O_3	CO oxidation	[133]
α-Fe_2O_3	SO_2 oxidation	Chapt. 5., [96], [211]
Fe_3O_4	CO oxidation	[134]
Fe_3O_4	SO_2 oxidation	Chapt. 5., [211]
V_2O_5	SO_2 oxidation	Chapt. 5., [110], [135, 136]
ZnO	Decomposition of H_2O_2	[137, 138, 139]
CaO	Decomposition of N_2O	[140]
CuO	Decomposition of N_2O	[141]
CuO	o-p H_2 transformation	[142]
NiO	Decomposition of H_2O_2	[139]
ZnO, Cr_2O_3, CuO	2-propanol decomposition	[143]
MoO_3, V_2O_5, α-Al_2O_3	Acrolein oxidation	[144]
Kaolinite	Ethanol dehydratation	[145]
Mordenite	Conversion of toluene and cumene	[146]

6.2.1. *Catalysts by means of tribochemical reactions*

The influence of tribomechanical treatment on the catalytic activity of a solid may consist in the fact that new chemical compounds or changed phases are tribochemically formed due to the treatment having an increased catalytic activity as compared with the initial substances. The formation of metallic carbonyls belongs to this group of tribocatalytic reactions, e.g. Ni- (Chapt. 4., [44], Chapt. 5., [228], [148, 149]) Fe- (Chapt. 5., [228], [150]), and Mo-carbonyls [150] from carbon oxide and metals in the presence of sulphur or sulphureous substances.

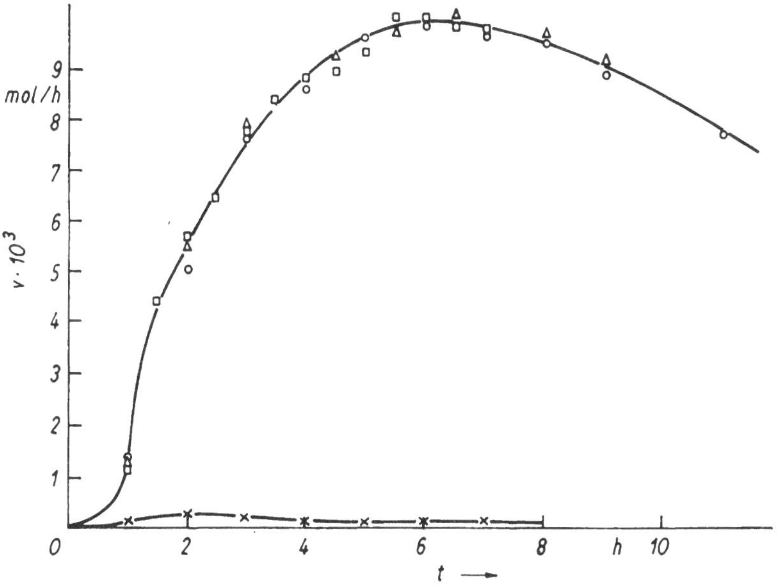

Fig. 6.36. The velocity v of the iron carbonyl formation at the treatment of pure iron and of iron treated by sulphur in CO atmosphere [150].
x — pure iron, △, o, □ — 3 series of measurement iron treated by sulphur, t — treatment time.

The mechanism of this reaction may be represented by the following scheme using the example of the formation of iron carbonyl from iron and carbon oxide in the presence of sulphur [150]:

$$\text{iron} + \text{sulphur (or sulphureous compound)} \rightarrow \text{iron sulphide} \rightarrow Fe_3S_2(CO)_9$$
$$\xrightarrow{\text{disproportionation}} \text{iron sulphide} + Fe(CO)_5 \qquad (10)$$

Through mechanical treatment iron and sulphur react tribochemically to iron sulphide, which in the presence of carbon oxide forms carbonyl sulphide. This compound disproportionates to carbonyl and iron sulphide or iron sulphide compound, respectively. As Fig. 6.36. shows, the sulphur induces a reaction velocity being higher by the factor of about 100. Since 6···8 per cent of sulphur in the input material is sufficient for the accelerated reaction of the used iron the sulphur has a catalytic effect on the formation of carbonyl. The real catalyst, the iron sulphide, represents an intermediate product re-forming again during the disproportionation of carbonyl sulphide accompanying the formation of the final product.

γ-NiS, which for example develops during the treatment of nickel in the presence of H_2S and leads to the formation of compact sulphide layers [148], is the catalyst in the nickel carbonyl formation sensitized by sulphur (see Fig. 5.62.). Since the concentration of the γ-NiS developing on the nickel surface depends on the amount of applied hydrogen sulphide there is a direct proportionality between the velocity of nickel carbonyl formation and the amount of the applied moles of H_2S (see Fig. 6.37.). In this the molare carbonyl yield is higher by nearly ten times than the applied molare amount of H_2S. Consequently it represents a mechanically activated intermediate substance catalysis.

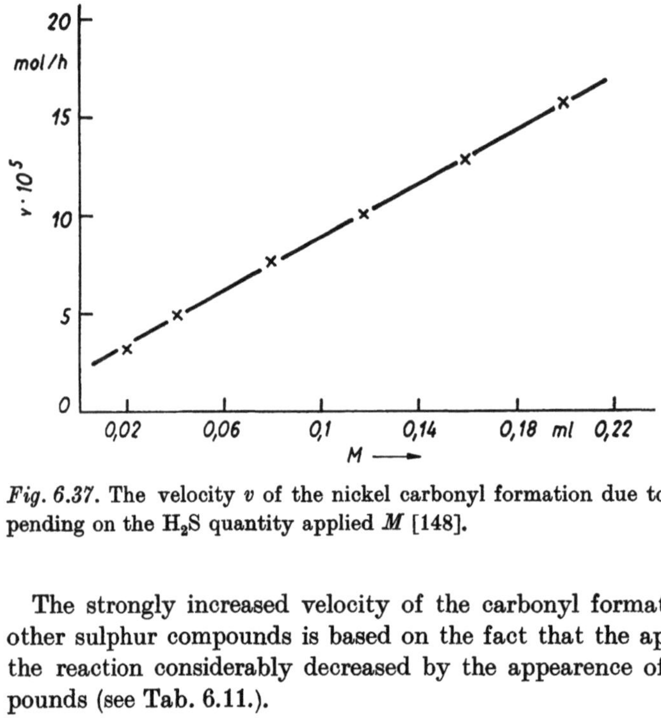

Fig. 6.37. The velocity v of the nickel carbonyl formation due to the mechanical treatment depending on the H_2S quantity applied M [148].

The strongly increased velocity of the carbonyl formation influenced by H_2S and other sulphur compounds is based on the fact that the apparent activation energy of the reaction considerably decreased by the appearence of intermediate sulphur compounds (see Tab. 6.11.).

Table 6.11. Reduction of the apparent activation energy E_A at the formation of nickel carbonyl influenced by hydrogen sulphide [150].

Gas applied	E_A, kJ/mole
CO	55
CO + 5% H_2S	27

The catalytic effect of the metal sulphide appearing as an intermediate product depends also to a large extent on the activity of the product. Thus the velocity of carbonylization of an untreated sulphurized sample is extremely small and during the treatment increases only by two orders of magnitude. The same behaviour has been observed during the formation of nickel carbonyl using nickel matte which is a nickel metal converted by sulphur. As Fig. 6.38. shows, here the high reaction velocity is also tied to the treatment. Accordingly after interrupting the treatment a fast reduction has been observed of the carbonylization velocity [148].

The formation of nickel carbonyl is sensitized by the addition of MoS_2, FeS, FeS_2, and so on, because the applied sulphides interact tribochemically with the nickel and form γ-NiS by rearrangement [148].

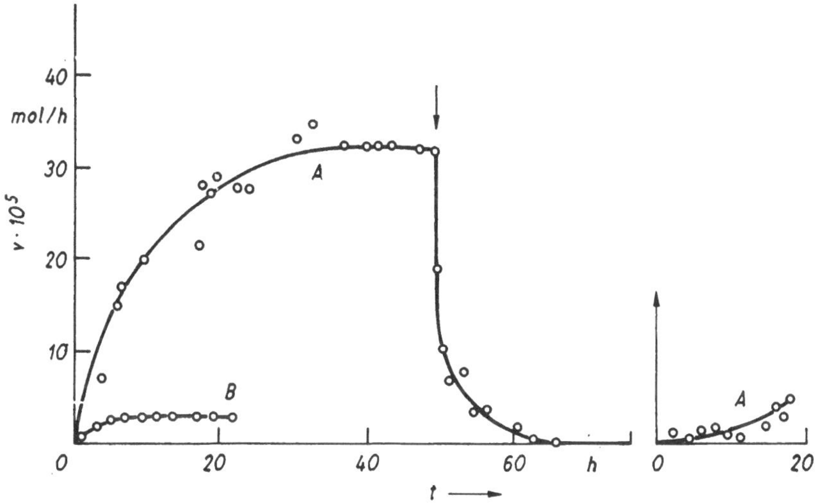

Fig. 6.38. The velocity v of the nickel carbonyl formation with and without treatment of the solid [148].

A — nickel matte, B — nickel metal, ↓ — interruption of the treatment for 1,700 hours, ↑ — recommencement of the treatment, t — time.

A further example for the catalyst only coming into existence in a tribochemical way is the reaction of titanium and titanium (IV) chloride in an inert gaseous atmosphere accompanied by the formation of various titanium chlorides which are used as catalysts for the low pressure polymerization of ethylene [151]. An effective catalyst for the polymerization of ethylene also comes into being with the milling of $TiCl_3$ in a propylene atmosphere [152]. The milling of a mixture of $TiCl_3$ and $AlCl_3$ in an inert atmosphere also leads to a catalyst suitable for the α-olefin polymerization [153].

Sometimes the tribochemical formation of the catalyst takes place via several reaction steps. For the α-olefin polymerization, for instance, a catalyst is used which is formed in first step by milling $TiCl_4$ and Al according to the equation:

$$3\,TiCl_4 + Al \rightarrow AlCl_3 + 3\,TiCl_3 \qquad (11)$$

and in second step by milling the formed product with triethylaluminium [154].

A catalyst is obtained by milling α-Al_2O_3 with quartz which after a longer treatment time is largely amorphized and consists of aluminosilicates [155].

An increase in the catalytic activity of contacts is also achieved by the doping of catalysts with certain impurities. This may be obtained, for instance, by coprecipitation of the promotor from the solution, or by bombardement with elementary particles. A so-called tribodoping of catalysts is expected to result in changed tribocatalytic properties in analogy to the tribodiffusion processes discussed in 4.7. and 5.7. However, systematic investigations have not been reported to date.

6.2.2. *The influence of tribomechanical treatment on the catalytically active surface area*

6.2.2.1. *Changes in surface size by mechanical stress*

The same laws as those generally discussed for solids in 5.5. are substantially applicable to the changes in the surface size of catalysts. The following specific conclusions may be drawn for catalysts:

— The value of the specific surface area, and consequently the catalytic activity may be increased up to the establishment of the HÜTTIG milling equilibrium during the mechanical treatment of catalysts having relatively small specific surface areas, e.g. supported catalysts on kieselgur base [100].

— If there is a relatively large specific surface area at the beginning of the mechanical stress, as is the case for the majority of the applied supported catalysts having values of some 100 m^2/g, then a reduction of the specific surface area and consequently of the catalytic activity will appear under the influence of the mechanical treatment. In many cases this reduction will not be compensated by the activity-increasing effect of the lattice defects (see 6.2.3.). For instance, the initial surface area of an α-Al_2O_3 catalyst in the range of about 200 m^2/g decreases to a value of 20···30 m^2/g after milling 10 hours [147]. However, the treatment proves to be sufficient to produce catalysts having a precisely predetermined surface area, corresponding with the intended purpose.

Special conditions often exist for catalysts which have been used for a longer time and in which the accessible surface area is reduced by deposits at the outer boundary layer of the catalyst grain. Such catalysts having partly lost their activity in this way may be reactivated again by milling.

A short-term milling causes an increase in the surface size in spite of the relatively large initial surface area of about 100 m^2/g. As is shown in 7.2. the regeneration of catalysts used up or lowered in their effectiveness is possible in this way [147].

6.2.2.2. *Tribomechanically caused changes in the catalytically active surface*

According to the remarks in 5.6., fresh surfaces are produced by deformation and fracture processes during the tribomechanical stress of solids, which are free of adsorbate and reaction layers during their formation. The production of catalytically acting fresh surfaces leads to an important increase in the activity of catalysts, in numerous cases totalling more than one order of magnitude. As an example the decomposition of water at pure germanium surfaces freshly produced by a fracture process is shown in Fig. 6.39., [107].

Whereas the reaction mentioned may be measurably observed on aged surfaces above 423 K only, it can be already found on pure surfaces at room temperature. Generally the ignition temperature of catalytic reactions is essentially lowered by mechanical treatment (Chapt. 5., [96, 110, 211], [107, 119]).

If the fresh surface produced by a solitary fracture process is covered by the interaction with substrate or molecules of the surrounding phase in the course of the catalytic reaction, the high catalytic activity drops to a value lower by 1···3 orders of magnitude (steady-state reaction phase in Fig. 6.39.) (Chapt. 4., [50], [107, 130]). With permanent treatment the high catalytic activity characteristic of fresh surfaces, remains stationary for a longer time (see Fig. 5.45.).

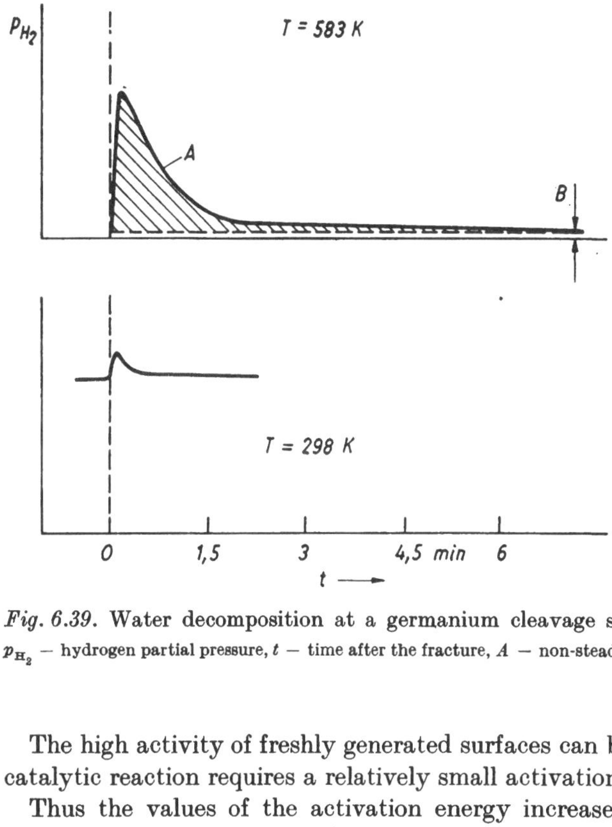

Fig. 6.39. Water decomposition at a germanium cleavage surface developed by fracture [107]. p_{H_2} — hydrogen partial pressure, t — time after the fracture, A — non-steady portion, B — steady-state portion.

The high activity of freshly generated surfaces can be attributed to the fact that the catalytic reaction requires a relatively small activation energy.

Thus the values of the activation energy increase for the N_2O decomposition on pure Ge-surfaces from 29 kJ/mole to 192 kJ/mole with growing coverage of the fresh surfaces, and a direct proportionality exists between the activation energy and the logarithm of the degree of coverage. The rise of the activation energy is connected with the growing blocking of the active centres at the fresh fracture area [106].

The blocking of the surface may be caused by the adsorption of the substrate, of the reaction product coming into existence during catalysis [107, 106], or of available impurities (Chapt. 4., [50]).

Thus, for instance, a hydrogenating catalyst treated in the presence of air is no longer effective so that no catalytic effect will be found without tribomechanical stress in a subsequent action of the substrate mixture. As Fig. 6.40. shows, the steady-state level measured before the catalyst was poisoned will be reached again after the revival of the treatment process. A similar curve course is obtained if a catalyst poisoned by thiophene is mechanically treated and thereby a wide regeneration of the catalyst facilitated (Chapt. 4., [50]).

According to 5.6.2. non-equilibrium states appear during the competition sorption of gaseous mixtures to mechanically produced fresh surfaces, leading to a sorption ratio of the individual gases different to the equilibrium state. The example of ethylene hydrogenation in 5.6.2. shows that the reaction behaviour at the catalyst surface is substantially changed by this (Chapt. 4., [50]).

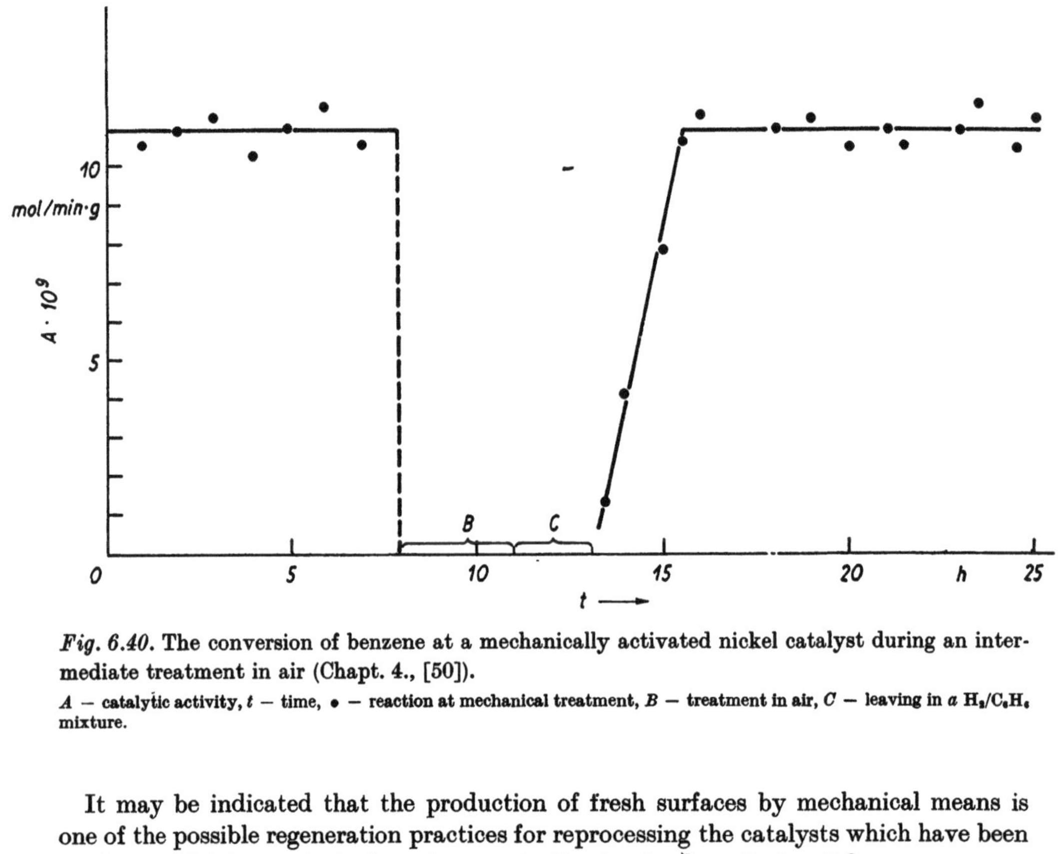

Fig. 6.40. The conversion of benzene at a mechanically activated nickel catalyst during an intermediate treatment in air (Chapt. 4., [50]).
A – catalytic activity, t – time, • – reaction at mechanical treatment, B – treatment in air, C – leaving in a H_2/C_6H_6 mixture.

It may be indicated that the production of fresh surfaces by mechanical means is one of the possible regeneration practices for reprocessing the catalysts which have been used up, which in particular have become ineffective by deposits (see 7.2.).

6.2.3. *The relation between disorder and catalytic activity of tribomechanically stressed catalysts*

The influence of the disorder produced by mechanical treatment on the catalytic activity has been particularly studied in metallic and oxidic catalysts (see Tab. 6.10.). Here a close relation has been found to exist between the changes in the specific catalytic activity and the lattice defects produced in the catalyst, as could be seen in several examples (e.g. Chapt. 4., [87], Chapt. 5., [211], [104, 123, 129, 141]).

This correlation is represented both for the activation process and the annealing process running with the number of applications, by the example of fat hydrogenation in Fig. 6.41.

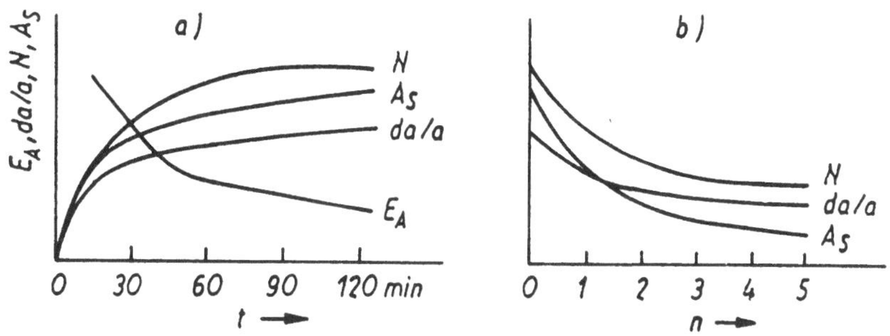

Fig. 6.41. Relations between the specific catalytic activity A_s in the fat hardening and mechanically produced structural disorders (Chapt. 4., [87]).

E_A — activation energy, da/a — lattice distortions, t — treatment time, n — number of uses, N — dislocation density.

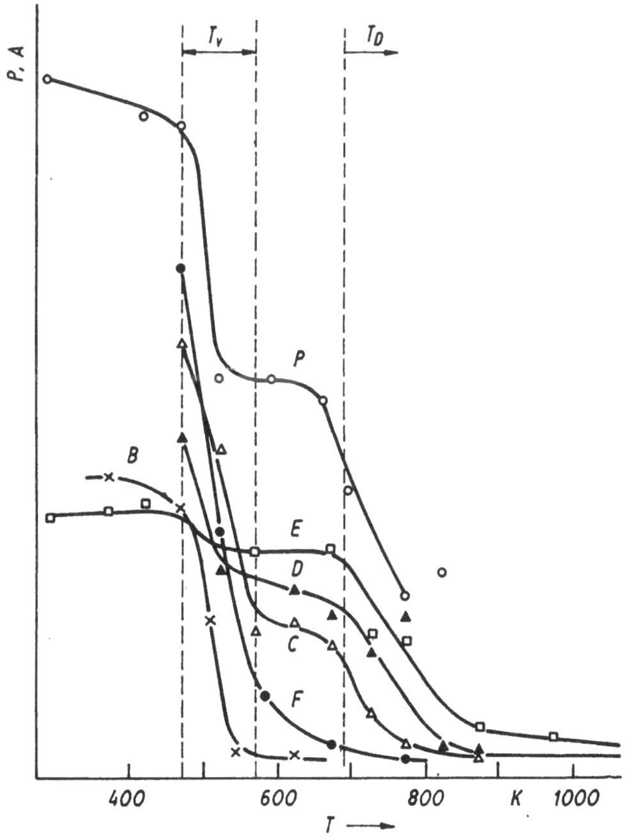

Fig. 6.42. Decrease of the thermoelectric force P and of the catalytic activity A of mechanically treated nickel depending on the heating temperature T [129].

P — thermoelectric force, B — ethylene hydrogenation, C, F — o-p-hydrogen transformation, D — dehydrogenation of ethyl alcohol, E — dehydrogenation of cinnamic acid, T_v — annealing of point defects, T_D — annealing of dislocations.

In both processes, the specific catalytic activity changes similarly in course to the changes of both the lattice distortions and the dislocation density. As Fig. 6.41. shows, a lowering of the activation energy occurs with increasing concentration of defects.

Table 6.12. Activation energy E_A of the SO_2 oxidation at a tribomechanically stressed V_2O_5 catalyst [135].

Activation time, t	Tests	E_A, kJ/mole
—	Without treatment	144
5	After interrupting the mechanical stress	139
10		135
30		132
5	During the mechanical stress	21

A particularly strong reduction of the activation energy and accordingly a strong increase in activity has been observed if the catalytic reaction runs during the treatment process (see Tab. 6.12.).

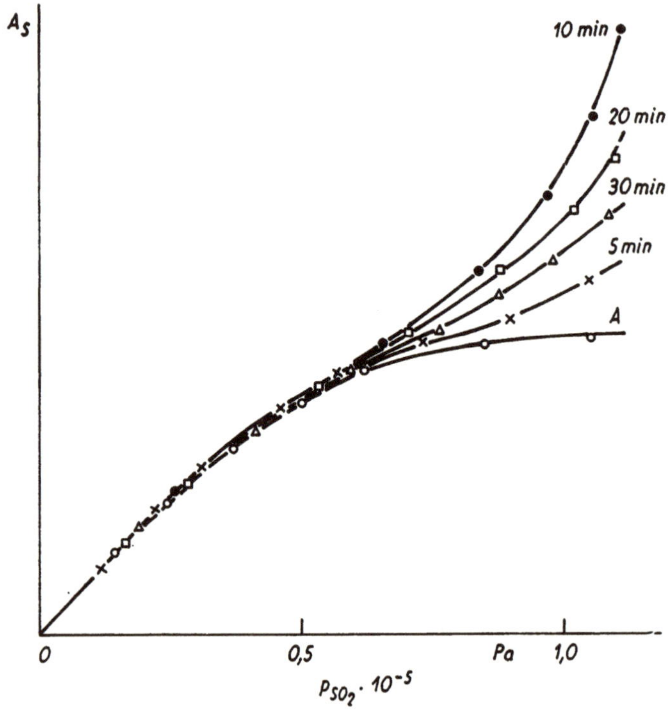

Fig. 6.43. Adsorption isotherms of SO_2 at V_2O_5 catalysts milled for differently long times at 298 K [135].
A_S — specific adsorption, p_{SO_2} — SO_2 partial vapour pressure, A — initial sample, t, min — activation time.

Investigations on the nickel catalyst showed that for several catalytic reactions, a decrease in the catalytic activity takes place in a temperature range during the annealing of the catalyst, where certain defects anneal. According to Fig. 6.42. the reduction of the thermoelectric force associated with these defects and of the catalytic activity runs similar in course to the annealing of point defects (T_v-range) and dislocations (T_D-range).

From this it is concluded that the structural defects mentioned are the main reason for the catalyst activity changed by treatment.

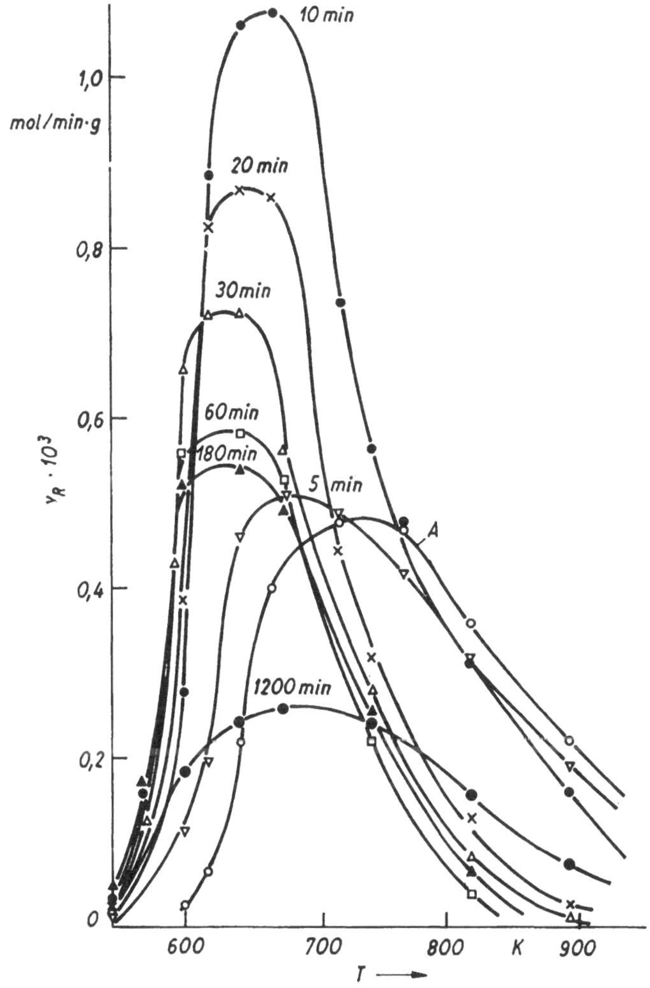

Fig. 6.44. Velocity V_R of the SO_2 oxidation depending on the temperature T at V_2O_5 catalysts milled for differently long times [135].
A – initial sample, *t*, min – activation time.

The question of how in detail the structural defects influence the elementary processes of the catalytic reaction at the catalyst is variously interpreted. Thus, for instance, the multiplet theory is used to explain the changes in catalyst properties appearing during the mechanical treatment [124, 134]. This requires the structural and energetic conformity between reactants and lattice of the catalysts in catalytic processes. As a result of mechanical treatment a change in the spacing of lattice atoms appears. Because of this possibility exists for new atoms spacings to arise having an optimum value in terms of the multiplet theory, and consequently leading to a lowering of the activation energy of the catalytic reaction. Other authors variously regard the increase in the energy content of the catalyst [121], the change of the FERMI level [134], and the changed adsorption of the substrate at the catalyst surface [135], as the reason for the increased specific catalytic activity of tribomechanically treated catalysts.

As an example the adsorption isotherms of SO_2 at cold worked V_2O_5 at 298 K are given in Fig. 6.43. The strongest sorption may be found in those milling samples (treatment time 10 min) which according to Fig. 6.44. also have the highest catalytic activity. There is definitely the possibility that the type and stability of the surface complexes appearing intermediarily during the catalysis and consequently also the velocity of the catalytic reactions are changed on the disturbed surface.

For instance, the hydrogenation of cyclohexane to hexane at platinum catalysts was comprehensively investigated. The hydrogenation velocity turned out to be higher by one order of magnitude at steps than at the smooth surface, and the reaction runs faster by one further order of magnitude at the so-called kinks within the steps, because they have proved to be very effective in the breaking of C—C bonds [156]. Such structure-sensitive reactions are much changed by the treatment process.

6.2.4. *Changes in catalytic activity by tribomechanically effected secondary structural changes*

The activity of supported catalysts rich in surface areas is to a large extent determined by the porosity of the carrier which above all influences the catalytic reaction of preceding and succeeding diffusion processes.

As has been shown by the example of kieselgur the secondary structure of a carrier may often be fundamentally changed by tribomechanical treatment (Chapt. 4., [50]). According to Fig. 5.32. the initial porosity of the catalyst is practically changed by milling in every case regardless of whether the stress process is conducted in vibration, jet, or other mills. For instance, an enlargement of the pore volume, and simultaneously a displacement of the mean pore radius, takes place during the treatment of V_2O_5 in a jet mill as Fig. 6.45. shows.

This displacement is directly related to the catalytic activity. It becomes apparent that the displacement of the mean pore radius to smaller values is accompanied by the lowering of the ignition temperature and by the increase of the catalytic activity (see Fig. 7.1.), [100].

The increase in the catalytic activity of mordenite by milling is also attributed to a secondary structure effect. The main channels of the structure are divided during the mechanical treatment and the resistance to the diffusion limiting the reaction velocity is reduced in this way [146].

It should be pointed out that a tribomechanically caused change in porosity also plays a leading part in the regeneration of used up catalysts (see 7.2.1.2.).

6.2.5. *Summary*

In summary, it may be stated that several reasons are responsible for the tribomechanically caused changes in the activity of a catalyst. The high concentration of defects in the solid produced by treatment, leads in most cases to a strong reduction of the activation energy and consequently to an increased catalytic activity. If at the same time inhibitions caused by covering films of every kind at the catalyst surface are removed by the treatment process (fresh surface effect), a particularly strong increase in the catalytic activity will appear through the superposition of the effects [157].

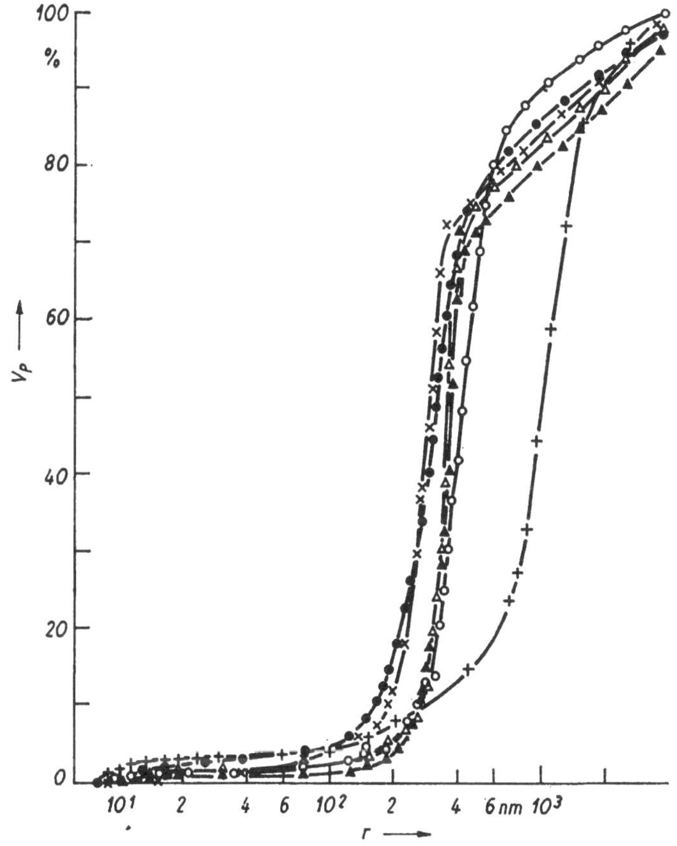

Fig. 6.45. Integral pore radius distribution curves of jetmilled V_2O_5 [100].

V_p — pore volume, r — pore radius, + — unmilled, ○ — milled one time, ● — milled for two times, △ — milled for three times, ▲ — milled for four times, x — milled for five times.

However, the utilization of this effect is only possible if the use of the mechanically activated catalyst takes place at relatively low temperatures, because the catalytic effect is particularly strongly increased at a reduced ignition temperature especially in the lower temperature range as compared with that of an untreated catalyst. The structural disorders anneal relatively quickly at higher temperatures, (see Fig. 6.41.b), and the activation effect based on structural imperfections of the solid is lost. The possibility to maintain the activity consists in a continuous treatment during the catalytic reaction [158], for instance by using a pulsing fluidized bed [100].

By contrast certain catalytic reactions, such as the formation of NH_3 from the elements [114, 158] only run during the treatment process. Evidently the reactions are connected to the highly initiated states in postplasma.

Tribomechanically caused changes in the secondary structure also lead to both an increase and a decrease of the catalytic activity. As opposed to the above-mentioned changes these also remain stable for a longer period in the temperature range where commercial catalysts are used. A mechanical activation based on a changed secondary structure is however only possible with catalysts having a relatively small initial surface area and associated with this, a small porosity. This effect is used for the regeneration of aged catalysts.

6.3. Tribochemistry of polymers

6.3.1. Introduction

The first systematic investigations in the field of the tribochemistry of macromolecular compounds were already conducted in the 1920s, when the mechanical treatment of polymers, e.g. the comminution of caoutchouc, wood, cellulose, and synthetic polymers such as polystyrene, was found to cause a diminution of the molecular weight and a degradation of the macromolecules [159—166]. As an example the degradation curves of various polymers obtained by using a vibration mill are represented in Fig. 6.46.

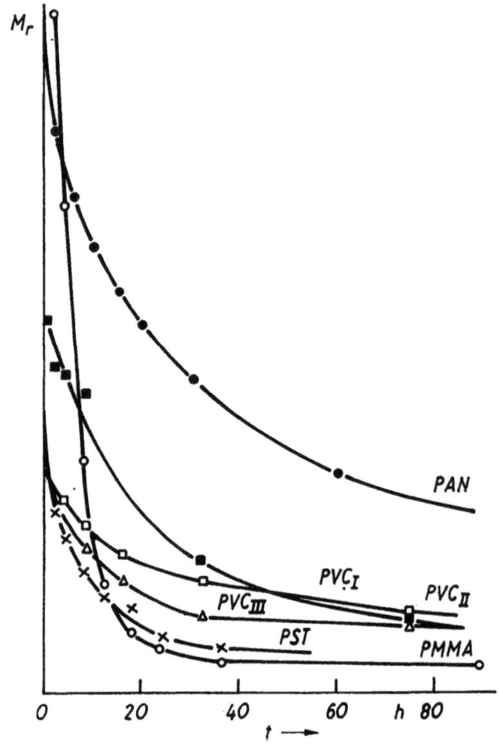

Fig. 6.46. Tribochemical degradation of various polymers [167].

M_r — average value of the relative molecular mass at time t of the mechanical treatment, PAN — polyacrylonitrile, PVC_I, PVC_{II}, PVC_{III} — different kinds of polyvinyl chloride, PST — polystyrene, PMMA — polymethylmethacrylate.

It becomes apparent that the degradation characteristics of the individual polymers are the same in spite of the gradual differences in the course of the curves, and that the dependence of the polymerization degree on the milling time may be described by exponential curves [167—169] as with the tribochemical synthesis of lower homologous hydrocarbons (see 6.1.1.):

$$M_r = (M_0 - M_\infty) \cdot e^{-kt} + M_\infty \tag{12}$$

(M_r — average value of the relative molecular weight to the time t, M_∞ — limiting value of the relative molecular weight, M_0 — relative molecular weight of the raw polymer, k — coefficient.)

The degradation of macromolecules and the production of other structural disturbances result in a change of the polymer properties, particularly of the solubility, the resistance to chemicals, the capability of swelling, the tensile, fatigue, and impact strengths as well as of elasticity and plasticity.

In practice the changes of properties connected with the tribochemical degradation are important in two different respects:

1. as an inevitable attendant phenomenon when mechanical energy acts on polymeric materials during their processing and use [170] and
2. as intended change of the properties of polymeric solids aiming at the production of materials having specific properties adapted to the purpose.

A number of papers have been published in the last two decades dealing with these two aspects. They have partly been reviewed already (Chapt. 1., [2]). With in the framework of this section, some characteristic aspects of the tribochemical changes of polymers will be discussed without dealing with the diversity of the phenomena.

6.3.2. *Kinds of mechanical influences on polymers*

The following possibilities of mechanical energy influences have become important for the technical treatment of polymeric bodies

- cutting and breaking
- milling in mills
- plasticization of polymers on rolls or in extruders
- pressing of polymeric solutions through narrow capillary tubes or nozzles
- treatment of polymeric solutions using high-speed stirrers
- cryolytic degradation of polymers
- degradation of polymers by ultrasonics [171]
- degradation of polymers by shock waves [172]

Caused by the high velocity of the changes in load induced by this during the degradation of polymers by shock waves completely different phenomena are often found to those found during the action of tribomechanical energy [173]. Therefore they will not be considered in the following survey.

6.3.3. *Changes in polymer structure and texture by the action of mechanical energy*

The action of mechanical energy on polymeric solids leads to radical changes in their structure and texture. These changes are characterized by an increase in excess free energy of the system (see 4.8.), by the breakage of intermolecular bonds which stabilize the overmolecular structure of polymers, by the reduction of density, the increase of the surface area, by the changes of the valence angles as well as of the intermolecular distances of polymeric chains, and by a decrease in crystallinity. These processes are summed up by the term tribochemical disaggregation [174].

6.3.3.1. *Comminution of polymers*

In the mechanical treatment of polymers, below the glass transition temperature there occurs first of all a comminution and thereby a reduction of the particle size of the material. The final value is determined both by the properties of the polymer and by the treatment conditions and is in the order of 10^{-3} cm. For example, polymeric powder having 15···40 μm is obtained if polyethylene is milled in a vibration mill at 80 K. However, under the same conditions at room temperature a squashing of the polyethylene particles takes place agglomerating very quickly to scales accompanied by the reduction of the specific surface area [175].

A series of polymers may be dispersed to surface areas of 100 m^2/g by vibration milling at lower temperatures. If, for example, PMMA is treated for a longer time an increase even takes place in the specific surface area up to 700 m^2/g [176].

As Tab. 6.13. shows the various polymers can be divided into two groups in dependence on the surface area size. Polymers whose surface areas increase to about 100 m^2/g belong to the first group and polymers reaching surface area values of only some m^2/g to the second one.

Table 6.13. Specific surface area S of polymers after a vibration milling at 80 K and after a fast heating up of the vibration milled samples to 300 K (Chapt. 5., [13]).

Polymer	S, m^2/g	
	80 K	300 K
Polymethylmethacrylate	130	4.5
Poly-α-methylstyrene	145	9
Polypropylene	90	< 5
Polyisobutylene	80	< 1
Polyester acrylate	4.8	4.1
Polystyrene	3.2	2.5
Polyethylene	3	< 1

However, the large surface areas obtained at low temperatures are not stable and decrease markedly during heating to room temperature. Surface areas having values of about 100 m^2/g even decrease by 1···2 orders of magnitude. This process represents a so-called low-temperature sintering and is thereby released, that the barrier to the rotations on the C—C bonds is overcome and the chains obtain a high mobility, in polymers having quaternary carbon atoms in the base chain. The temperature at the commencing mobility is at 130 K for first group polymers and apparently below 80 K for the second group polymers where the barrier to the inner rotation is smaller.

Accordingly, with the dispersal of polymers, in addition to the production of a new surface due to the reduction of the particle size, there occurs significant loosening of the secondary structure of the material. Under these conditions foreign molecules with a relatively high velocity are capable of diffusing into the inside of the substance. For instance a value of $D = 10^{-22}$ m²/s has been found for the tribodiffusion coefficient of certain compounds in vibration milled polyethylene [22] which corresponds approximately to the tribodiffusion coefficient of argon in nickel (see Tab. 5.20.). Dissolution processes are also strongly accelerated by this [177].

The comminution of polymers is above all useful for the production of fine polymeric powders and for the recovery of polymeric wastes [178].

6.3.3.2. *Tribochemical degradation of polymers*

A degradation of the polymer chains occurs beyond the comminution. The velocity of the tribochemical degradation of polymers is mainly determined by the intensity of the mechanical stress. Thus in vibration mills the intensity of treatment may be strengthened by increasing the amplitude of vibration (see Tab. 4.4.), ball size and density (see Fig. 4.12.), and by reducing the filling proportion. The different degradation rate of starch using various filling levels is shown as an example in Tab. 6.14.

Table 6.14. Effect of the filling ratio α and of the treatment time t on the tribochemical decomposition of starch in a vibration mill [168].

α, %	t, h	Degree of polymerization P_n, Number of the anhydroglucose units
18	240	123
1.25	70	125
1.25	120	75

A typical phenomenon for the milling of polymers is the appearance of a so-called limit of degradation, which cannot be reduced further however long the milling (see curves running parallel to the abscissa in Fig. 6.46.). The lateral mobility of the macromolecules is assumed to have become so large at the limit of degradation that the mechanical energy is exclusively utilized for the intermolecular displacement of polymer chains [179].

In conformity with the remarks in 4.5., a temperature independence is also measured for the polymer degradation in several systems, within certain limits (Chapt. 5., [13], [165, 180, 181]). It is apparent, however, that this condition applies essentially to the vitreous state of the polymer and that changes of the elastic behaviour of polymers caused by temperature can lead to a temperature dependence of the degradation of polymers.

As an example the dependence of the velocity coefficient of the PMMA (polymethylmethacrylate) degradation on temperature is shown in Fig. 6.47. The vitreous state dominates up to a temperature of 330 K. In this range the tribochemical reaction does not depend on temperature. Above this temperature the polymeric body changes into the highly elastic state. Since the flexibility and mutual mobility of the macromolecules grow with rising temperature an increasingly larger portion of the mechanical energy is utilized for the intermolecular displacement of the polymer chain and accordingly a negative temperature gradient is measured in the highly elastic region.

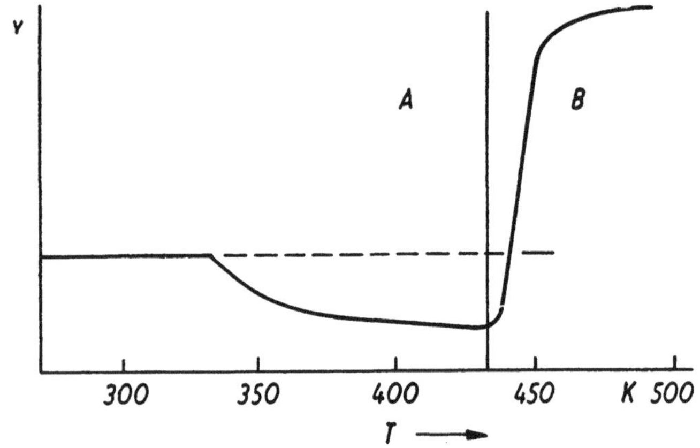

Fig. 6.47. Velocity *v* of the tribochemical degradation of polymethylmethacrylate depending on the temperature *T* [167].

A – region of the tribochemical degradation, *B* – region of the thermal degradation.

In some cases the specific design of a treatment device may cause special effects in tribochemical polymer conversion or only enable certain reaction species. For instance, it is not possible in the usual extruders to obtain a tribochemical cross-linking of polyethylene (PE) with maleic anhydride (MSA), because there are too few macro-

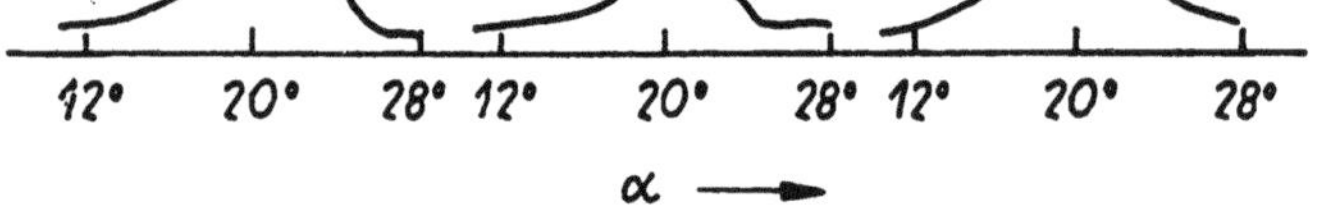

Fig. 6.48. X-ray diffraction diagrams for polyethylene [175].

a – initial sample, b – mechanically treated sample, c – melted sample, α – glancing angle.

radicals available for reaction due to an insufficient polymer degradation [182]. If however a so-called disc-type extruder [183] is used which leads to the production of macro-radicals through the appearance of high stress gradients, then an intensive cross-linking between PE and MSA may be obtained [184]. It is also a general characteristic of tribochemical reactions of polymers that the treatment devices must be precisely adapted to the particular polymer system and the purpose of application.

The dispersing of polymeric solids is simultaneously combined with a destruction of the crystalline structure and consequently leads to an amorphization of the polymers. This becomes apparent for instance through roentgenographic investigations of polyethylene (see Fig. 6.48.).

After the vibration milling, the reflexes (a) characteristic of the crystalline components of polyethylene are missing in the spectrum of the treated polymer (b) whereas a great similarity to the X-ray spectrum of the polyethylene melt exists (c). However, compared to the latter the spectrum of the mechanically treated sample still shows weakly developed diffraction maxima. From this the conclusion is drawn in conformity with the results obtained in mechanically treated quartz (see 6.1.2.), that in spite of extensive amorphization a certain order is maintained during the action of mechanical energy on polymers, e.g. in the form of deformed crystal fragments or simple overmolecular aggregations [175]. For the transition of crystalline polymers to amorphous bodies models have been developed for individual systems [174, 185]. The mechanical destruction of cellulose has in particular been studied as an example [186]. This polysaccharide consists of micro-fibrils combined in monomolecular, lamellar layers, and micro-fibrillar bunches. The aggregation is brought about by intramolecular and intermolecular hydrogen bridges. The destruction of the structure during the mechanical treatment takes place via the following steps:

Microfibrillar bunches → elementary micro-fibrils → crystallites → fragments of the lattice and of the macromolecular chains still containing H-bridge bonds (13)

In the first stage of destruction, intermediate surface areas appear which still preserve a fibrous character. In the course of treatment the strongly anisotropic crystallites change more and more into asymmetric particles. The final products of the degradation are highly disperse powder having completely lost the fibrous character and having an extensive amorphous structure. Under certain conditions, e.g. in a sliding process, the mechanical stress may lead to an increase in the crystallinity of the polymer. Thus an orientation of the macromolecules takes place accompanied by a growth of the crystalline portion to 50 per cent with the sliding stress of polytetrafluoroethylene having a crystalline portion of 40 per cent. The resulting abrasive particles are crystallized up to 100 per cent [187, 188].

6.3.4. Mechanism of tribochemical degradation

The causes of the tribochemical degradation of polymers have been comprehensively discussed in 5.2. and 5.9.3. In the following the most important aspects of the mechanism of the degradation reactions are presented in summary:

— Under the effect of mechanical energy, an elastic deformation of the bonds between the atoms of the polymer chain will appear and consequently a shift of their natural frequency (see 5.3.1.).

— Under the effect of a consistent stress, the strength of the macromolecules is reduced and the degradation consequently facilitated under the influence of the thermal fluctuation (see 5.2.1.).

— Because of the non-uniformity of the stress distribution inside the polymer the overloaded bonds are first of all broken after the application of stress (see 5.2.1.).

— With changing mechanical fields the elastic energy is constantly redistributed among the different macromolecule sections in the system. Excited vibration states in the non-equilibrium occur with the relaxation of the mechanical energy, leading to the breaking-up of main-valency bonds when a certain limiting level of elastic energy (E_{def}) is reached (see 5.2.2.).

— Activated radicals are formed by the tribomechanical degradation of both carbo-chained and hetero-chained polymers [189—192]. These activated radicals have excess free energy and form the raw material for a series of other secondary reactions (see 5.9.3.).

The reactions running on the activated radicals are to a large extent determined by the character of the polymers. Polyformaldehyde (PFA) and polyisobutylene (PIB) having a low polymerisation heat degrade due to the appearance of activated radicals accompanied by the formation of low-molecular products. The quantity of the degradation products essentially separated as monomers is many times larger than the concentration of the formed radicals (Chapt. 5., [21]).

A far smaller quantity of low-molecular compounds is separated ($<$ 0.1 per radical [192]) during the degradation of polymers with higher polymerisation heats and correspondingly larger activation energy of the degradation, e.g. polyethylene and polypropylene.

In this type of reaction, e.g. with all the vinyl polymers, the active terminal radicals pass into the more stable radicals with the free valency in the middle of the chain (chain radicals), accompanied by the separation of a hydrogen atom from the neighbouring polymer. This reaction already runs at 80 K in the vacuum. The chain radicals lead to the degradation of the chain by forming C=C double bonds according to the following equation (Chapt. 5., [240]):

$$-CH_2-\dot{C}H-\underset{\substack{| \\ R}}{CH}-CH_2-CH_2- \rightarrow -CH_2-CH=\underset{\substack{| \\ R}}{CH} + \dot{C}H_2-CH_2-, \qquad (14)$$

whereby the unsaturated bonds come into existence particularly in close proximity to branches.

If oxygen is present, then chain peroxide radicals develop which form aldehydes in the manner of an isomerisation degradation:

$$-CH_2-\underset{\substack{| \\ O-O\cdot}}{CH}-CH_2- \rightarrow -CH_2-C\overset{\large O}{\underset{\large H}{\lessdot}} + \cdot O-CH_2- \qquad (15)$$

Since 2 to 4 terminal radicals once again form during the cleavage reactions the possibility exists that these in turn react with neighbouring macromolecules to form chain-radicals and lead to the degradation.

A kind of chain reaction of the tribochemical degradation is present. Quantitative estimates show that the number of macromolecules having double bonds is in fact 10^3 times higher than the number of free radicals which are situated in the less stressed regions of the polymer after the breaking-off of the chain reaction (Chapt. 5., [240]).

Altogether, groups of methyls, vinyls, alcohols, acids, aldehydes, and esters could be proved by IR-spectroscopy in the tribochemical degradation of polyethylene (Chapt. 5., [11]). Accordingly a great number of the most varied chemical reactions can be expected to appear at the tribochemically produced polymer radicals depending above all on the type of polymers, on the prevailing tensions, on the temperature, and on the kind of surrounding medium, e.g. on the presence of water as a hydrolytically acting agent [193].

6.3.5. *Modification reactions at tribochemical degradation*

6.3.5.1. *Modification reactions in the polymer-polymer system*

Below, the main variants of the tribochemical synthesis of the two-component systems will be shown in the form of schematic drawings.

If two polymers are simultaneously degraded then side by side macroradicals of the two polymers develop according to pattern I [194]:

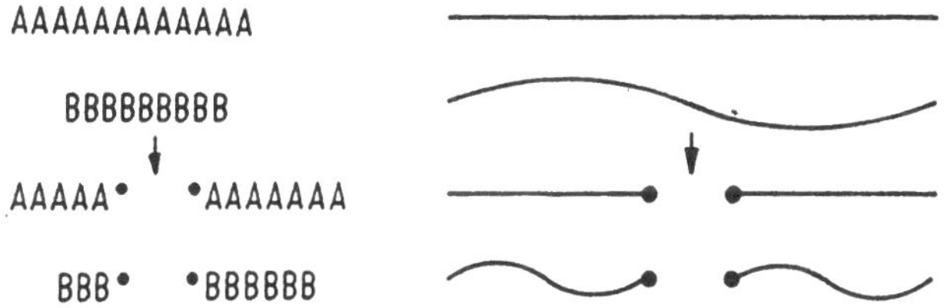

(I)

Apart from the recombination of homogeneous radicals the simplest transformation consists in the combination of two different macroradicals with the formation of block copolymers according to pattern II:

(II)

This process takes place when the respective macromolecules are approximately equally active and the polymer chains have no reactive centres which would be able to interact with the free macroradicals.

If only one of the polymers is split to macroradicals due to a varying stability of the two polymers in respect of mechanical stresses, the grafting of these macroradicals onto the active centres of the second component macromolecules is then the most probable secondary process, accompanied by the formation of grafted polymers according to pattern III:

(III)

For instance, the grafting points can be arbitrarily arranged methyl groups with mobile hydrogen which are capable of reacting with the free macroradicals of the component A. If the bonds in chain A are only weak and the chain has a great number of reactive points which are able to react with the free radical, then the grafting runs predominantly according to pattern III.

But it is also possible that a macroradical of the polymer A developed by cleavage is stabilized by the reaction of an atom eliminated from the chain B according to equ. (5.62) and subsequently the stabilization of the formed radical B with radical A takes place according to pattern IV:

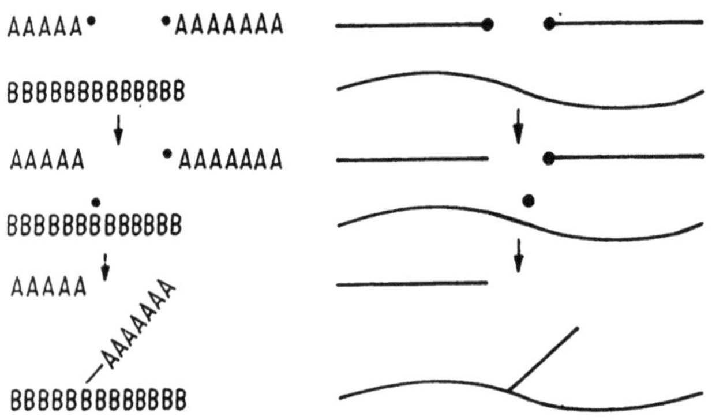

(IV)

Unlike pattern III, one obtains products by following pattern IV which in addition to the polymers B_n grafted with A_n, contain stable fragments of the polymer A in the mixture. The variant III is more probable if the polymer has double bonds, whereas variant IV has priority if these are methyl groups.

Also if only the chains of one of the two polymers are ruptured by the action of mechanical forces in the polymer mixture, the free radicals which occur can bring about a breaking of the other polymer according to pattern V, accompanied by the formation of block copolymers:

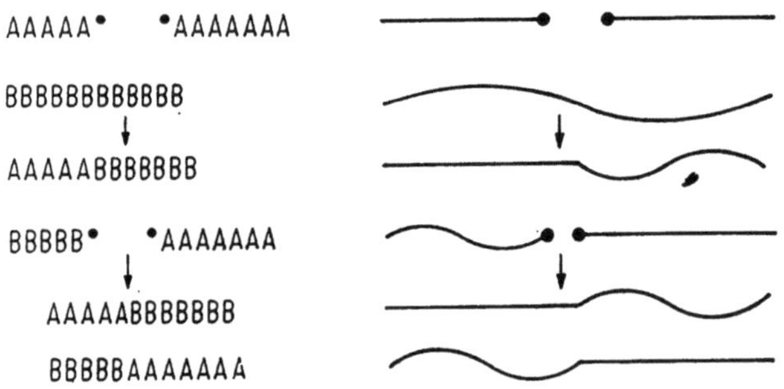

(V)

The difference between block copolymers developing according to patterns II and V consists in the fact that in II the blocks A and B cannot on principle be smaller than M_∞ for both polymers, whereas in V, the points of action produced by the radicals in the chain B are statistically distributed.

If chain processes running via radicals are initiated due to the mechanical treatment, a formation of three-dimensionally cross-linked products is possible. The cross-linking may, for instance, take place in a direct interaction between radicals and the active centres of the neighbouring chains according to pattern VI:

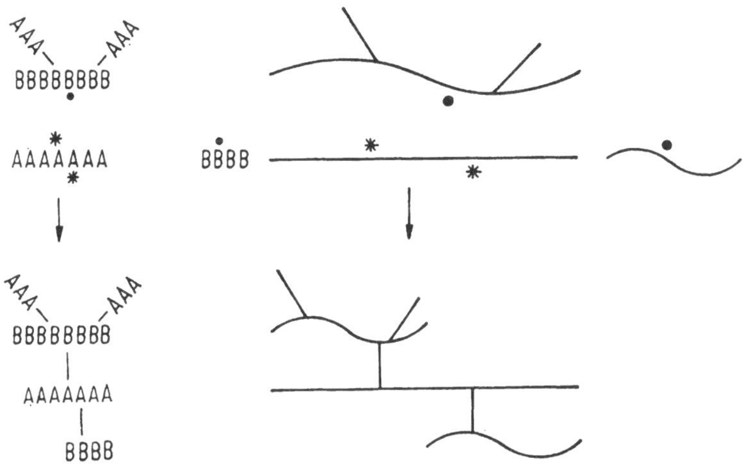

(VI)

If only one of the polymers tends to structure formation then only this polymer forms a net onto which chain fragments are grafted according to pattern VII:

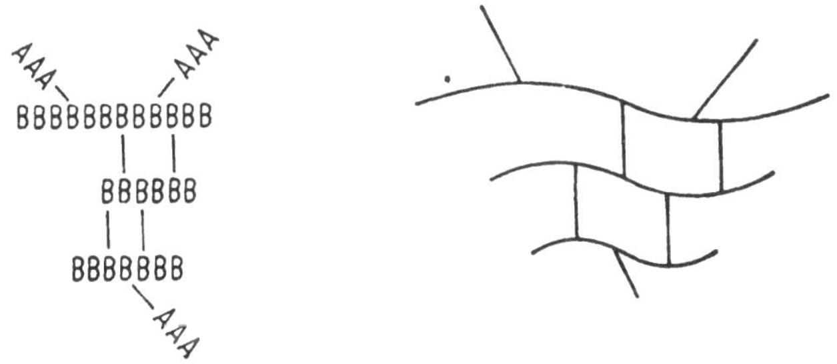

(VII)

On account of the schematic representation the tribochemical synthesis of two-component systems may be exhaustively interpreted. However, the products formed are again transformed tribochemically by the mechanical treatment. Thus, for instance, three-dimensional structures degrade to individual fragments under the influence of mechanical forces, whereby a complicated mixture of grafted and cross-linked products and of block fragments comes into being.

As an example reference may be made to the graft-block copolymerization of high-pressure-process polyethylene with ethylcellulose [195]. A mixture of grafted (a) and cross-linked (b) block copolymers whose formation velocities are determined by intensity and duration of the shearing and by the composition of the mixture, develops in a plastograph with the simultaneous mechanical stress of the two polymers, e.g. by shearing (see Equ. 16).

In this way, polymers are produced having changed physico-chemical properties, for instance in respect of moisture uptake and solubility. Such products are superior to the initial products in terms of some required technical properties, e.g. certain electric parameters (see Tab. 6.15.).

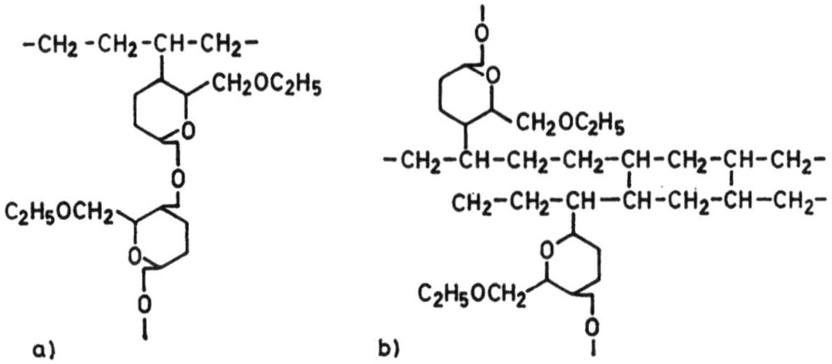

(16)

Table 6.15. Changes of various properties of polyethylene PE by graft block polymerization with 10 per cent ethylcellulose EZ [195].

Polymer	Solubility in methanol, per cent	Solubility in dichloroethane, per cent	Moisture uptake, per cent	Dielectric constant, E	Dielectric loss angle, $\tan \sigma$
PE	8.32	4.81	0.31	2.1	$1.7 \cdot 10^{-5}$
Graft block copolymer (90% PE, 10% EZ)	16.77	27.73	2.84	2.38	$3.2 \cdot 10^{-5}$

6.3.5.2. *Modification reactions in the polymer-monomer system*

The tribochemical degradation of polymers may also be carried out in the presence of one or several monomers. In this the formation of the most varied copolymers similarly occurs in the absence of acceptors. If in the simplest case the macro-radicals appearing during the mechanical degradation are capable of initiating the polymerization of the attendant monomer, block copolymers are formed according to pattern VIII [194]:

AAAAAAA• + nB
↓
AAAAAAA BBBBBB•
↓
AAAAAAA BBBBBB

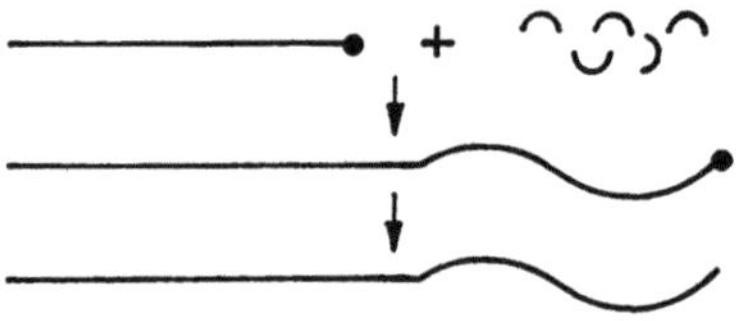

(VIII)

The growing block B is stabilized by a termination mechanism. There is also the possibility that the macroradical of polymer A is stabilized by the attachment of an atom of the monomer splitting off from B with the latter transferring into an excited state and releasing the polymerization of the attendant monomer according to pattern IX:

AAAAA• + B ⟶ AAAAA + B•

B• + nB ⟶ BBBBB•

$\xrightarrow{+B^\bullet}$ BBBBBBB

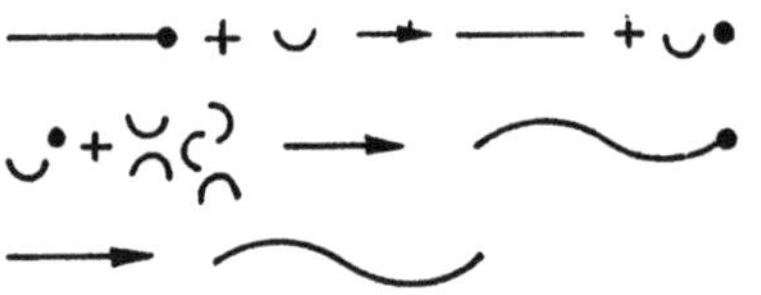

(IX)

If the reaction runs according to pattern IX, the formation of a mixture of products is expected to happen, consisting of linearly degraded chains of the polymer A and of all kinds of polymers which occur in the polymerization of the monomer B.

The mechanical degradation leads to the formation of copolymers with 3 components when two monomers with a not greatly different activity are available. According to pattern X these copolymers consist of blocks of the initial polymer and blocks of an irregular copolymer of the two monomers:

AAAAAAA• + nB + mC

AAAAAAABCBBCCCBC•

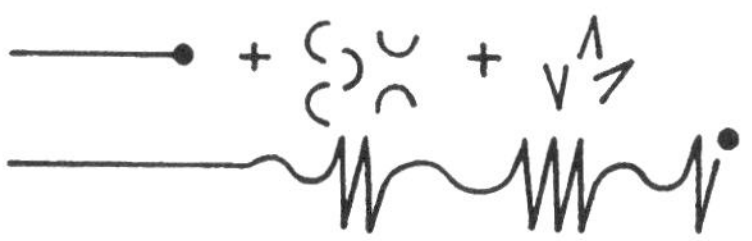

(X)

If the difference in the activity of the two monomers is very great, the monomer having the higher activity will react first with the macroradical. Only when its concentration lowers to a certain quantity or when it is completely used up will the polymerization of the less active monomer follow according to pattern XI:

AAAAA• + nB + mC

AAAAABBBBBBCCCCCC•

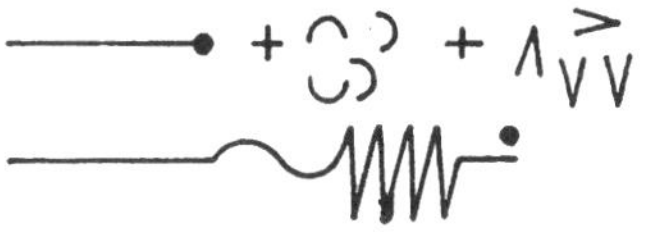

(XI)

The mechanical degradation of polymers may also take place in the presence of comeres, that is to say substances which are normally not subject to self-contained polymerization but which however, let themselves be copolymerized with other monomers. With this only one link of the chain grows which either recombines with another macroradical (pattern XII) or is grafted to a chain (pattern XIII):

AAAAA• + M → AAAAAM•

+ AAAA• → AAAAAMAAAA

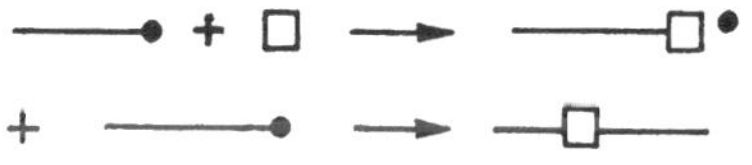

(XII)

AAAAM• + AAAAAAA

↓

AAAAM\
AAAAAAA

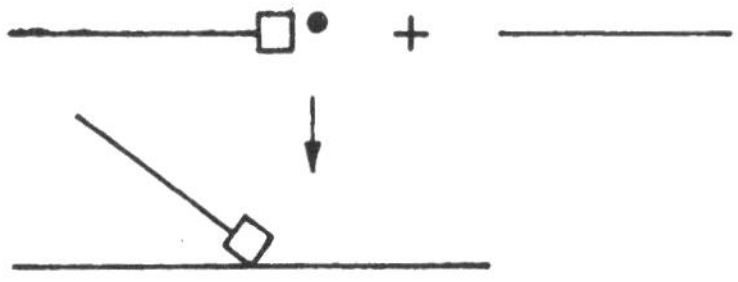

(XIII)

Modification reactions accompanying the tribochemical degradation of polymers in the presence of monomers have been investigated in numerous systems such as in polyethylene in the presence of maleic anhydride [175, 182, 184, 196], methacrylamide, acenaphthylene, acrylic acid, and acrylonitrile [175], in polypropylene with maleic anhydride [197], in polyisoprene with maleic anhydride [198], and with vinyl acetate [199], polymethylmethacrylate with acrylic acid [200], polyamide with acrylic acid [201, 202], polyacrylonitrile with vinyl chloride [203], etc.

As an example the copolymer yield appearing with the treatment of polyethylene in the presence of a few monomers is given in Tab. 6.16. A comparison of the copolymer yields in dependence on the choice of the monomer shows that their activity accords with their chemical properties and their stereochemical character.

The modification products of polyethylene represent mixed block and grafted polymers with cross-linked or linear structure (see Tab. 6.16.).

Table 6.16. Yield and structure of the copolymers developing due to the vibration milling of polyethylene and different monomers in air [175].

Monomer	Treatment time t, min	Copolymer yield, per cent	Copolymer structure
Acrylic acid	5	48.0	cross linked
Acrylonitrile	3	12.3	cross linked
Methacrylamide	6	13.3	cross linked
Acenaphthylene	5	—	linear
Maleic anhydride	5	0.3*	linear

* Quantity of the grafted maleic anhydride.

One of the most common techniques of tribochemical modification is the carboxylation of polymers with acrylic acid or maleic anhydride. Polymers with a graded hydrophilic character may be produced by this. This alteration of properties plays a special part in elastomers, because vulcanisation with metallic oxides may be attained due to the saline nature of the modified elastomer [204, 205].

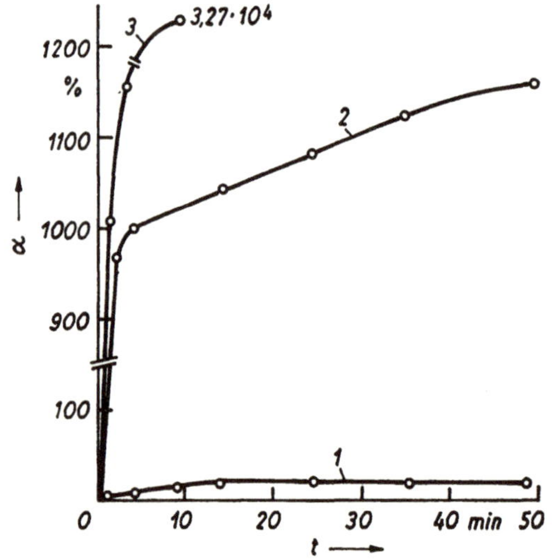

Fig. 6.49. Kinetics of the swelling of various polymers depending on the time *t* in water at 298 K [201].

α — degree of swelling, 1 — initial polyamide, 2 — tribocopolymer with 30 per cent polyacrylic acid, 3 — tribocopolymer with 50 per cent polyacrylic acid.

The adhesion effects, for instance, as against metals are improved by this and the solubility and swelling capacity are much changed, particularly in hydrophilic solvents. How considerably these properties can change in copolymerization with monomers is shown in Fig. 6.49. with the example of a polyamide grafted with acrylic acid.

A special method in the synthesis of tribochemically carboxylated polymers consists in the mechanical treatment of the frozen polymer solution in acrylic acid [33, 34]. The reaction mechanism is based on the initiation of the polymerization of the frozen monomer by free macroradicals formed by the tribochemical degradation of the polymer. Depending on the type of polymer, mixed grafted and block polymers with a linear or spatial structur come into existence during the modification. As Fig. 6.50. shows the solid phase reaction runs with a relatively high velocity. For instance, in the polyamide-acrylic acid system the tribochemical reaction towards the copolymer will already be concluded after a treatment time of 60 s.

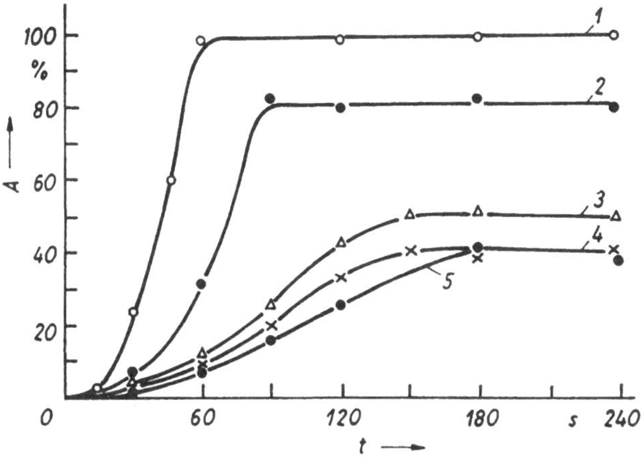

Fig. 6.50. Yield of tribocopolymers *A* for the vibration milling of various polymers in freezed acrylic acid solution depending on the treatment time *t* [206].
1 – polyamide, 2 – polymethylmethacrylate, 3 – cellulose, 4 – polyvinyl chloride, 5 – polyethylene terephthalate.

The mechanical activation of polymers is mainly carried out in a dry state, since the structural imperfections appear most markedly here.

In a liquid medium the fluids reduce the energy transfer onto the solid polymer particles. Hence not so much a macro-molecule chain degradation [171] is reached but on often-desired degradation of the aggregates and a good dispersity of suspensions.

The examples given here already show that the possibilities of the tribochemical degradation and of the tribochemical synthesis are very manifold and open up new ways for the production of modified polymers with selective properties adapted to purpose. The possibilities of synthesis may be still increased further by virtue of the fast that several kinds of energy such as radiation and mechanical energy act simultaneously on polymers and lead to strong non-additive changes in their properties [207].

6.4. Triboanalysis

As has been shown in 5.6.5. tribochemical reactions between two solids already run at room temperature. Dissociation reactions which like the decomposition of calcium carbonate run at 1,000 K, also take place tribochemically under standard conditions. Thus small amounts of CaO are already indicated with a single scratching action on a smooth calcite crystal surface.

If the formation of the reaction product can sensitively be proved analytically then the tribochemical reaction between solids may serve as an easily conducted method

Table 6.17. Triboanalytical reactions (Chapt. 5., [153, 160], [208, 209]).

Substance to be detected	Analytical reagency	Reaction product	Detection	Comments
Nickel	Dimethylglyoxim	Nickel dimethylglyoxim	Red colouring	Addition of ammonium carbonate increases sensitivity
Nickel cyanide	Dimethylglyoxim $AgNO_3$	Nickel dimethylglyoxim, AgCN	Red colouring	
Lead	Alkali iodide	Lead iodide	Yellow col.	Moisture favours the reaction, heat effect disturbes
Cerium	Sodium peroxide	Ceric peroxide	Yellow col.	Addition of sodium nitrate increases sensitivity
Iron (III)	Potassium thiocyanate	Ferric thiocyanate complex	Red. col.	Complex extractable with ether
	Potassium cyanoferrate (II)	Berlin blue	Blue col.	
Iron (II)	Potassium cyanoferrate (III)	Turnbull blue	Blue col.	
	Ferroin	Colour complex	Red. col.	
Uranium	Potassium cyanoferrate (II)	Uranium cyanoferrate (II)	Red-brown col.	In addition to iron and copper uranium detection possible with sodium thiosulphate
Chromium	Sodium peroxide Lead sulphate	Lead chromate	Yellow col.	
	Sodium peroxide Silver nitrate	Silverchromate	Red col.	
Cobalt	Alkali thiocyanate reductant	Cobalt thiocyanate	Blue col.	Sensitivity 10^{-6} g
Molybdenum	Alkali thiocyanate	Tungsten thiocyanate	Yellow col.	
Tungsten	Alkali thiocyanate	Silver thiocyanate	Yellow col.	
Silver	Sodium thiosulphate	Silver thiocyanate	Yellow col. with transition to black	

Table 6.17. (Continuation)

Substance to be detected	Analytical reagency	Reaction product	Detection	Comments
Thiosulphate beside sulphite, sulphate	Silver nitrate	Silver thiosulphate	Yellow col. with transition to black	
CaO	FEIGL reagency (silver sulphate, manganese (II) sulphate)	Manganese dioxide, silver	Black col.	

of detection for certain elements. Especially suitable for this are reactions in which a coloured reaction product comes into being (see Tab. 6.17.). In many cases such reactions permit an extremely sensitive detection and therefore are suitable for microanalysis.

The analytic detection is generally performed by the trituration of the solid substances in a mortar. Frequently only extremely small quantities of substance are required for this. But there is also the possibility to precipitate the analytic reagent to be used for the reaction by steeping the solutions onto hard, porous ceramic material. In this the ceramic may be used in the form of sheets for stroking or little rods. In the first case the material abraded by a thin stroke from a solid is sufficient to perform the reaction. If little rods are used under certain conditions even a point-shaped rubbing is sufficient for tribochemical microdetections.

If the colour of the substance to be proved is overlapped by the presence of dark-dyed materials, then detection is often possible by applying a drop of solvent and absorbing it with a porous material. The extract transferred in this way onto a second porous plate can in turn be brought in contact with a reagent, e.g. with an oxidant or reductant, and after drying up of the substance be subjected to a second test of stroking. Thus materials containing cerium show an intensive yellow colour when triturated with sodium peroxide (see Tab. 6.17.) which, however, is stable in the alkaline range only and disappears again during further trituration with sodium hydrogensulphate.

The micro-analytic detection of the contained impurity additions may not succeed in very highly leaked ceramic materials, e.g. in magnetesite bricks. In this case the material may be digested very easily by triturating a small sample with sodium or potassium hydrogen sulphate, ammonium fluoride, sodium hydroxide, ammonium carbonate or sodium peroxide.

The sensitivity of triboanalytical detections also depends on the initial state of the solid. Thus, for instance, an active nickel oxide produced by the decomposition of carbonate at 673 K, already reacts with dimethylglyxine after a short trituration and leads to a conversion of several per cent by weight. However, nickel oxide annealed at 1,073 K reacts very little even with longer trituration. But if this product is tribochemically digested with ammonium carbonate then it also reacts very quickly with dimethylglyoxime. Generally tribochemical reactions are very sensitive. The detectability limits lie between 10^{-6} and 10^{-9} g. The highest sensitivity of the detection of all the tribochemical micro reactions investigated until now has been given for the formation of thiosulphate from sulphite and sulphur (see Tab. 6.17.). If the crystallization of an oversaturated solution of sodium thiosulphate is used as detection by the application of crystallization centres then already $10^{-15} \cdots 10^{-18}$ g are sufficient for the initiation of the process.

In individual cases the resultant coloured product may be extracted by means of a solvent, e.g. nickel dimethylglyoxime with chloroform. In such systems the tribochemical reaction can be quantitatively followed by determining the intensity of colours.

In summary it may be stated that the triboanalytical method is applicable to very many reactions known in analytical chemistry and that it is extremely sensitive. It is also extraordinarily variable and offers the possibility for the development of completely new microanalytical detections (Chapt. 5., [153, 160], [208, 209]).

6.5. Tribochemical reactions for preparation purposes

With appropriate mechanical treatment a series of tribochemical reactions runs with such a high velocity that a quantitative reaction of the used solid takes place in a time comparable with thermal reactions. In such cases tribochemical reactions can be used for the synthesis of chemical products, too [210].

The specific particularities of the course of tribochemical reactions often lead to advantages in experimental outlay relative to traditional preparation methods, so that in literature numerous suggestions have been made for the utilization of tribochemistry for preparation purposes. The following aspects speak for the utilization of tribochemical reactions in material conversion processes:

— Numerous reactions may already be initiated at room temperature, which without any treatment run with measurable velocity above 500 K only. A thermal source of heat can be dispensed with. As an example the production of $SiCl_4$ may be mentioned [211]:

Coarse lumps of ferrosilicon of 30 g (Si content about 77 per cent) are fed into the milling vessel (800 ml contents) of a vibration mill capable of being sealed fast. The vessel is equipped with a feed pipe and an offtake pipe. Dried chlorine is introduced and the vibration mill is run after having displaced the air out off the reaction vessel. The reaction starts some minutes later accompanied by a large emission of heat. This milling action is continued for about 1.5 to 2 hours. The silicon (IV) chloride produced is drained off with excess chlorine by means of the offtake pipe and condensed using a trap cooled with the help of ice and a common salt mixture. The condensate, which still contains a small portion of higher silicon chlorides in addition to silicon (IV)chloride, is cleaned in the usual way by mercury treatment (removal of possibly dissolved free chlorine) and subsequent fractionating distillation. The yield amounts to about 125 g.

Other halides such as $TiCl_4$ or BCl_3 may also be prepared in similar fashion, with a yield of $90 \cdots 95$ per cent [211].

— Numerous chemical syntheses run with a justifiable velocity only with a correspondingly high pressures. Among these for example is the formation of certain carbonyls from metal and carbon oxide. Thus molybdenum hexacarbonyl may only be produced with thermal initiation at temperatures of $500 \cdots 600$ K and pressures of about 2×10^7 Pa. But this requires the use of a sophisticated discontinuous production process. As is shown by the example of molybdenum hexacarbonyl this compound can already be prepared tribochemically with correspondingly good yield at normal pressure [212]:

200 g molybdenum powder are given, together with porcelain balls, into a 5/*l* milling vessel capable of being closed fast and made of porcelain. The quantity of porcelain balls is so determined that the degree of filling totals about 80 per cent. The vessel is filled with CO and connected with a 10*l* supply vessel from which carbon oxide may flow if used. About 1 per cent H_2S is added to the carbon oxide as sensitizer. After a treatment time of 100 hours at room temperature the molybdenum hexacarbonyl is extracted by means of ether and recrystallized from the latter again.
The yield amounts to 35 g.

It is also possible during the treatment to desublimate the molybdenum hexacarbonyl at an increased temperature in flowing medium, allowing the procedure to be conducted continuously.

— Further advantages in using tribochemical reactions can be seen in the fresh surface effect appearing by the action of mechanical energy (see 5.6.). Reaction and adsorption layers coming into existence are removed by treatment and the passivation of surfaces occurring through side reactions and reaction poisons is offset again. For instance, nickel carbonyl is also formed without treatment on freshly reduced nickel surfaces in the presence of carbon oxide. However, this reaction will soon come to a standstill, since inter alia traces of oxygen act as a reaction poison, and the oxidation of nickel by CO running as a side reaction will contribute to the formation of a covering film. This requires the intermittent reduction of the nickel surface if the reaction is thermally conducted. However by contrast, with mechanical treatment a nearly complete reaction takes place if the reaction is conducted continuously [212].

165 g pyrophoric nickel, which can be produced for instance by the reduction of nickel oxalate, is treated in a vibration mill for 6 hours as given in the previous preparation. The nickel carbonyl is transferred together with flowing carbon oxide into a cooling trap and condensed at 553 K. The metal powder simultaneously carried away from the reaction vessel must first be removed from the gas stream by a dust precipitator.
The yield amounts to 400 g nickel carbonyl.

— As already shown in 5.6., the fresh surface effect is also the basis of the tribooxidation processes. Thus most of the usual metals can be converted quantitatively into the corresponding oxides by milling them at room temperature in an oxygen atmosphere (Chapt. 4., [57]).

— Most reactions between two and more solids run with a measurable reaction at higher temperatures only, since these reactions nearly always depend on diffusion and consequently the transport of lattice components needed for a quantitative reaction over long distances presupposes their sufficiently large mobility of diffusion (transport effect). As is seen in 5.6.5. under mechanical treatment, reactions between solids mainly run according to diffusionless kinetics already at room temperature. In some cases the reaction takes place quantitatively, as is shown by the example of barium wolframate production (Chapt. 5., [149]):

Molare amounts of anhydrous BaO free of peroxide, $Ba(OH)_2$ with exact stoichiometric composition, $BaCO_3$ or $BaSO_4$ are mixed 1 : 1 with WO_3 and milled in a vibration mill for about 5 hours. The following conditions should be observed: degree of filling 70 per cent, filling relation 1 to 2 per cent, ball diameter 10 to 13 mm, amplitude 3 mm.
Yield: 100 per cent.

Detection of the reaction: the substance dissolves in 0.1 n acid; additionally roentgenospectroscopy

Reference substance: a sample annealed at 573 K for 3 hours.

— There are numerous chemical systems which show a different course of reaction during the tribochemical stress than to that during the application of heat. To these belongs the decomposition of bromates as already shown in 4.1. and 5.3.

Another example is the reaction of tin with benzyl halides. Under exclusively thermal initiation conditions polybenzyls come into existence accompanied by the release of hydrogen halides. Compared to this, with mechanical stress metalloorganic compounds develop, as will be shown below by the reaction of dibenzyl dichlorostannane (Chapt. 4., [75]):

44,2 g tin and 32 g benzyl chloride are fed together with 10 kg porcelain balls into a milling vessel capable of being closed fast. After exchanging the air atmosphere for nitrogen the reaction mixture is stressed tribomechanically in a vibration mill for 3 hours.

The following reaction takes place:

$$Sn + 2\,C_6H_5CH_2Cl \rightarrow SnCl_2 + 2\,C_6H_5\dot{C}H_2 \quad (17)$$

$$Sn + 2\,C_6H_5\dot{C}H_2 \rightarrow (C_6H_5CH_2)_2Sn \quad (18)$$

$$(C_6H_5CH_2)Sn + 2\,C_6H_5CH_2Cl \rightarrow (C_6H_5CH_2)_2SnCl_2 + 2\,C_6H_5\dot{C}H_2 \quad (19)$$

$$2\,C_6H_5\dot{C}H_2 + SnCl_2 \rightarrow (C_6H_5CH_2)_2SnCl_2 \quad (20)$$

The dibenzyl dichlorostannane which is formed is recrystallized from glacial acetic acid: yield: 90 per cent or 40 g, respectively.

Further examples can be found in 7.2. where the utilization of tribochemistry is discussed in relation to the design of economically working material transformation procedures.

6.6. Tribochemistry of solid state electrode

6.6.1. Introduction

The metal — electrolyte system, the solid state electrode, is one of the most common reaction systems in practice. This is by virtue of the fact that metals belong to the most used engineering construction materials with an electrolytic medium nearly always being in contact with the metal surface area and may be in the form of an adsorbed aqueous layer. At the same time this system is always subjected to the influence of mechanical forces from the manufacturing process up to use. Therefore the tribochemistry of the solid state electrode (in terms of the definition of this book) is an important subsystem of the material behaviour of metals.

Whereas a great number of works have been published on the field of corrosion processes under the influence of mechanical energy, until now, relatively few have been published with regard to the thermochemistry of solid state electrodes from the tribochemical point of view. Therefore in this section special attention is given to a comprehensive representation of thermodynamics. In this particular emphasis is put on the mechanically induced local processes the knowledge of which contributes to the understanding of all the technically relevant processes of the tribochemistry of solid state electrodes.

Today electrode kinetics are theoretically extremely well founded. Therefore only some examples will be given in the present section which are to serve the complex consideration of the metal-electrolyte interface layer as a whole.

There is such a wealth of knowledge in the field of triboelectrochemical processes in engineering, that a key-word like enumeration of selected problems and approaches can given only.

6.6.2. *Thermodynamics*

6.6.2.1. *Integral processes*

6.6.2.1.1. *Processes at solid state electrodes*

The knowledge that in comparision to exclusively thermal reactions heterogeneous chemical reactions at solid surfaces with application of mechanical energy, or tribochemical reactions, are characterized by a fundamentally changed energy and reaction behaviour, may be extended to the metal (solid) — electrolyte system. The metal (solid) — electrolyte system, the solid state electrode, differs from other heterogeneous solid-reaction medium systems by the feature that in this system the matter transfer by chemical reactions is a transfer of electrically charged components through an electrically conducting medium. In the case of a (solid state) electrode, energetic changes are accompanied by an alteration of parameters in the (inner and outer) electric field, so that the increase in the chemical potential in the case of mechanically activated "common systems" is faced with an increase in the electrochemical potential in the mechanical activation of metal-electrolyte systems.

The increase in the electrochemical potential of the solid state electrode through the supply of mechanical energy takes place by a simultaneous change in the chemical potential μ and in the potentials inside the phase φ (GALVANI potential) both at the electrolytes and at the solids side of the interface layer. At the electrolytes side changes arise in the chemical potential by changes in the concentration conditions of the system components; changes in the phase potentials are caused by changed charge distributions within the individual partial phases of the electrolyte (electrolyte interior, phase boundary layers).

At the solid's side changes arise in the chemical potential through the supply of excess free enthalpy due to the effect of mechanical energy, whereas changes in the phase potentials arise by the deformation of the electron levels of metallic atoms due to the development of lattice defects.

Changes within the phase boundary layer, i.e. both the electrolytic double layer and the reaction layers (covering films, sorption layers, etc.), have a particular importance for changes in the electrochemical potential of the metal (solid) — electrolyte system by the supply of mechanical energy.

6.6.2.1.2. *Static or dynamic application of mechanical energy by macroscopic deformation*

Studies of changed solid state electrode reactions where the metal (solid) — electrolyte system is subjected to macroscopic stresses (compressive or tensile stresses) have mainly been carried out in connection with the search for the reasons for stress corrosion cracking and corrosion fatigue (see 6.6.4.3. and references cited there). The changed reactivity of metals influenced by macroscopic mechanical stresses may be interpreted as an interaction between formation and motion of defects and electrochemical processes. In addition changes in sorption behaviour as well as covering film effects (rupture of passivating or semi-passivating reaction layers, formation of local fresh surfaces etc.) play a dicisve part (see 6.6.4.3.). All the processes mentioned may manifest themselves in changes of the electrode potential (both in a positive and a negative direction).

Energetic changes of the metal (solid) — electrolyte system while excluding covering film effects (e.g. by the suitable choice of pH and electrode potential according to [213]) always manifest themselves in a negative deviation of the electrode potential. Many authors have found a steady state negative shift of electrode potentials following a tensile stress at various electrodes [214—216]. Short-term non-steady negative shifts of electrode potentials could be observed during the stress, which nearly or completely return to the initial value of the electrode potential after few seconds [227 to 234].

The negative shift of electrode potentials by macroscopic deformation of metals (compressive or tensile stress of the electrode as a whole) ranges in the order of $1 \cdots 10 \times 10^{-3}$ V. If the covering film destruction effect occurring at the metal (solid) — electrolyte interface is eliminated by the choice of the system as well as by the creation of corresponding boundary conditions, then the excess free enthalpy supplied to the system by the mechanical treatment is left over as the main reason for the changes in potential.

This is characterized by a disturbance of the double layers of the phase boundary as well as by dislocations, stacking faults, and other structural disorders at the solid's side of interface layer. The influence of the deformation of double layers and the reduction of the potential barrier at the electrode surface thus caused, are accentuated as one important reason for changes in the electrode potential [231, 235]. At the same time negative shifts of the equilibrium potential of copper electrodes were found to occur with increasing lattice distortion [236—238]. The influence of the disturbed solid structure on the electrode reactions takes place via an activation of the anodic dissolution of metal at active centres of the metal surface areas. In the case of a mechanical deformation, the latter are mainly formed at the intersection sites of solide surfaces, dislocations as well as at moving grain boundaries. Finally as a result the changed reactivity of the electrode under the static and dynamic effect of mechanical energy during macroscopic deformation is moulded by the occurrance of microscopic local elements "disorder lattice regions/less disordered lattice regions" at the metal surface area [232, 235, 239].

On account of this assertion it has to be noted that both in the measuring of non-steady or steady-state electrode potentials and in the investigation of cathodic or anodic integral reactions at mechanically stressed electrodes, local reactions within the electrode surface always run, which are not, or only partly, measurable by the measurements of integral electrochemical properties of the electrode.

The causal connection between the production of lattice disorders in metal as a result of the macroscopic supply of mechanical energy, the appearance of dislocations at the interface and the special electrochemical reactions thus caused, represents a simple example for the thermodynamic coupling of mechanical and chemical processes. The dislocation current and the electric current are vectorial processes capable of coupling (see 4.8.2.).

Thus the entropy production of triboelectrochemical processes may be written as (while neglecting adsorption effects), [235, 236, 240]:

$$T\sigma = J \frac{n\Delta\tau}{\bar{\alpha}} + I \cdot \tilde{A} \tag{21}$$

(T — absolute temperature, σ — entropy production, J — dislocation current, n — number of dislocations in a planar accumulation, $\Delta\tau$ — macroscopic stress difference (strain hardening), I — electric current, $\tilde{A}$ — electrochemical affinity, $\bar{\alpha}$ — dislocation concentration).

The product $\frac{n\Delta\tau}{\bar{\alpha}}$ represents the affinity of the "chemical" reaction for the development of dislocations (see 6.6.2.2.4.). From equ. (21) result the linear phenomenological equations for the currents:

$$J = L_{11} \frac{n\Delta\tau}{\bar{\alpha}} + L_{12}\tilde{A} \tag{22}$$

$$I = L_{21} \frac{n\Delta\tau}{\bar{\alpha}} + L_{22}\tilde{A} \tag{23}$$

whose phenomenological linearity coefficients L_{12} and L_{21} are equal according to ONSAGER

$$L_{12} = L_{21} \tag{24}$$

Some effects having interesting physical interpretations can be derived from the equations (22) and (23) while taking into account equ. (24):

The electric current (corrosion current) per unit of dislocation current in electrochemical equilibrium:

$$\left(\frac{I}{J}\right)_{\tilde{A}=0} = \frac{L_{21}}{L_{11}} \tag{25}$$

The dislocation current per unit of corrosion current in the absence of mechanical stresses

$$\left(\frac{J}{I}\right)_{\Delta\tau=0} = \frac{L_{12}}{L_{22}} \tag{26}$$

The electrochemical potential difference per unit of mechanical stress difference in the absence of a corrosion current

$$\left(\frac{\tilde{A}}{\frac{n\Delta\tau}{\bar{\alpha}}}\right)_{I=0} = -\frac{L_{21}}{L_{22}} \tag{27}$$

The mechanical stress difference per unit of the electrochemical affinity in the absence of a dislocation current

$$\left(\frac{\frac{n\Delta\tau}{\bar{\alpha}}}{\tilde{A}}\right)_{I=0} = -\frac{L_{12}}{L_{11}} \tag{28}$$

Equ. (25) characterizes the mechanical activation of corrosion processes by macroscopic deformation, equ. (26) a so-called "chemomechanical effect", a kind of "environment sensitive behaviour of materials" (comp. 6.6.4.3.2.). Equ. (27) characterizes the transfer potential of dislocations and is consequently closely connected with the phenomenon of negative shift of electrode potential by the deformation of the solid state electrode. It may easily be seen that the effects mentioned with their possible interaction using the ONSAGER principle of reciprocity (equ. (24)), represent complete analogues to the known electrokinetic phenomena [241]:

In this equ. (25) corresponds to the streaming current, equ. (26) to the electro-osmosis (in the given case of dislocations), equ. (27) to the streaming potential or mechano-electric effect, equ. (28) to the electro-osmotic pressure.

6.6.2.1.3. *Dynamic supply of mechanical energy by multiple impacts*

The metallic solid surface is subjected to the action of multiple impacts (that is a more or less frequent sequence of impacts) when stressed by erosion, abrasion, or friction. An analysis of the activation of electrochemical reactions by the multiple impact effect shows both fundamental differences and certain parallels to the activation of electrochemical reactions by the supply of mechanical energy in macroscopic deformation:

The spatial and temporal dissipation of the mechanical energy supplied to the metal (solid) — electrolyte systems runs via a sequence of energy "states" which may be systematized according to the "hierarchy" developed in 4.1.

The "state" of the edge or post-plasma which is reached via the complex of recombination of the plasma products after the decay of the triboplasma is without doubt the most important for the changed reactivity of the solid state electrode under the action of impact. This "state" is characterized by an essentially higher degree of excess energy than the one caused by macroscopic deformation. This can be proved by a simple comparison. Whereas the energy stored in the form of point lattice defects and dislocations is in the order of some 10^1 J/mole [242, 243], and the energy excess produced by the plastic straining of wires does not exceed the order of some 10^2 J/mole (e.g. [227]), the yield of post-plasma excess energy should extend to the sublimation enthalpy (Chapt. 1., [4]) via the fusion enthalpy of metals i.e. an order of magnitude of 10^3 J/mole.

With the effect of impacts on solid state electrodes, correspondingly similar potential shifts are expected to appear which have higher absolute values than in the case of macroscopic deformation. Thus a negative shift of the electrode potential of a Cu/Cu^{2+} electrode of about 40 mV as well as a similar effect on a Zn/Zn^{2+} electrode effected by impact has been found, in dependence on the treatment intensity [242, 243]. Measurements on Cu/Cu^{2+} electrodes with simultaneous grinding (silicon-carbide paper) show negative shifts of electrode potential up to 100 mV in dependence on the intensity of mechanical treatment as well as on the composition of the electrolyte [244].

If the potential of 100 mV is converted into energetic quantities then an excess free enthalpy yield of about 20 kJ/mole would result. Indeed, this value exceeds the energy excess produced by strain (macroscopic deformation) by one order of magnitude.

The "excess energy rates" calculated from the integral negative shift of electrode potentials in this section are subject to a restriction. The integrally measurable potential alterations represent only a fraction of the local potential peaks, since the latter are vastly reduced by local electrochemical reactions (local elements), [241, 244, 245], see 6.6.2.2.4.

Finally let us sum up once again the common and the different processes appearing due to the activation of electrochemical reactions by macroscopic deformation and by the action of multiple impacts, respectively:

— Common processes:

The changes in the electrode potential chiefly manifest themselves by a negative shift of electrode potentials. The fundamental reactions take place on the basis of micro-local elements: disturbed structure/electrolyte/less disturbed structure and disturbed covering (reaction, sorption) film/electrolyte/less disturbed covering film, respectively.

— Different processes:

The magnitude of the activation effects differs strongly. The most highly activated "states" due to macroscopic deformation exist in the form of disturbed covering films and simple lattice disorders on the metal surfaces, respectively, but with activation by multiple impacts, in the form of a structure deformed up to a near total destruction of the lattice. Therefore there is a difference in the integral alteration of the electrode potential by one order of magnitude.

6.6.2.2. *Local processes*

6.6.2.2.1. *Local elements produced by impact: "Disturbed interface layer/less disturbed interface layer"*

Reference has already been made to the fact that the increased reactivity of solid state electrodes by the addition of mechanical energy may be attributed to localized processes in the interface layer. In the case of macroscopic deformation the emergence sites of dislocations, and in the case of the impact effect the respective "contact or target areas", are the active sites. These favoured places of the phase boundary layer are places having the highest energy excess. This leads to the highest local solution tension. Here the question arises whether local reactions take place under the reduction of excess energy at these sites or whether the active site represents a "reversible" activated microelectrode. Some experiments performed on pre-activated metal surfaces [226, 246] indicate the last-mentioned possibility [235].

These results indicate that a precipitation mainly occurs on the most-strongly disturbed surface regions, accompanied by the re-arrangement of a disordered crystal lattice [247] (smallest kinetic input for electrocrystallization) or that an "electrochemical annealing" of lattice disorders near the surface takes place very slowly (over several days). The presence of a "reversible" activated electrode provides justification for the so-called "electrochemical method for the determination of residual strain energies" with concluding the existing mechanical stresses from the negative shift of electrode potentials [229, 230, 231, 236, 139, 244].

The first possibility, that is the spontaneous reduction of the excess energy of the system produced by mechanical supply of energy, brought-about by local reactions "disturbed interface layer/less disturbed interface layer", is similarly discussed [245, 248].

Here the integral electrode potential of a mechanically treated electrode (during the treatment) is regarded as a mixed potential between irreversible local reactions and reversible reactions [244]. Furthermore the spontaneous local reactions occurring during the cutting deformation of Cu/Cu^{2+} electrodes has been directly detected radiochemically [249, 250], (see 6.6.2.2.4.), (Fig. 6.52.–6.54.). In this in contrast to the abovementioned phenomenon [248], a spontaneous precipitation of metal from the electrolyte takes place on the little deformed, and consequently little activated, edges of the cut. From this the following greatly simplified scheme of a local element (Fig. 6.51.) produced by impact may be derived [249].

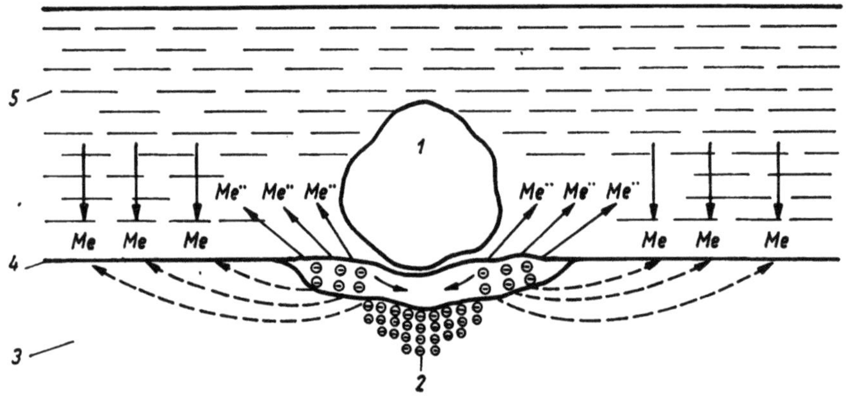

Fig. 6.51. Model of local cell $Me_{disturbed}/Me_{undisturbed}$ produced by impact [249].
1 – impact body, 2 – impact zone, 3 – solid metal, 4 – phase boundary Me (solid) electrolyte, 5 – electrolyte containing $Me^{\cdot\cdot}$ ions.

Here the highly active "target site" of an impact acts as local anode and the less disturbed environment as local cathode. The transport of electrons is carried out by the metal.

Evidently activation barriers have to be overcome for such local reactions (e.g. crystallization overpotential in the cathodic subprocess) which can only be brought about by the far higher energy during the deformation but hardly by the relatively small stored energy of the lattice disorders after the deformation. As has been detected by experiments the two kinds of processes actually run side by side. The irreversible electrode activated mechanically dominates during the treatment, but the "reversible" mechanically activated electrode, having increased interface layer dynamics without gross local reactions, dominates after the treatment. By this the behaviour of triboelectrochemical reactions corresponds from a thermodynamic point of view to the general behaviour of tribochemical reactions discussed in 4.8.

6.6.2.2.2. *Analysis of the mechanism of the formation of impact produced local elements*

Every electrode potential, and consequently also the partial potentials within a local element, represent an algebraic sum of Galvani or inner phase potentials which in turn is composed from the outer or Volta potentials and the surface potentials of the respective participating phases.

The piercing of the electrolytic layer by a tool with the subsequent impact braking at the solid, already causes changes in the surface potentials of the electrolytic phase in each case.

Further more, in the moment of impact, the strongly disordered lattice structure of the metallic phase at the target provided with an energy excess, will with certainty undergo an alteration of the electrochemical potential of the electrons (equivalent to a change of the FERMI level position). Between the disordered and undisordered lattice regions of the metal, similar to two different metallic phases, a GALVANI potential, the so-called deformation potential [242], develops through the instantaneous formation of an equilibrium-FERMI level.

The alteration of the GALVANI potential of the disordered lattice structure takes place by the changes in the surface and VOLTA potentials of the metal at the moment of deformation and dissipation of mechanical energy (comp. [236]). A change in the work function of the target site is closely associated with the alteration of these quantities.

Additionally the alteration of the potentials of the metallic phase in turn causes a change in the respective electrolytic phase potentials in the impact region.

The increased energy of the interface layer at the target site manifests itself by an increase of the metal ion detachment into the electrolyte as well as by a mutual transfer of electrostatic charges between metal and electrolyte. Through this a change in the zero-charge potential is caused at the target site.

The impact-produced local element [(metal/electrolyte)]*/metal/electrolyte[1] results from the sum of the mentioned changes of the electric phase potentials and from their interaction.

6.6.2.2.3. *Interaction between local and integral processes*

The mixed potential of a real solid surface is a mixed potential through all the local single-electrode potentials integrated through the entire electrode surface. Such local elements appear by energetic inhomogeneities of the real surface caused by variously indicated crystal areas, defects, inhomogeneous sorption and reaction layers, chemical impurities in the metal surface and others.

The same is to apply to the integral steady-state electrode potential of an impact-treated solid state electrode. This is a mixed potential through all the "natural" and impact-induced local elements "disturbed interface layer/less disturbed interface layer" (comp. [245]).

It is noticable that the maximum achievable changes in the integral steady-state electrode potential of solid state electrodes through impact effects lie in the order of some 10^4 J/mole, whereas known tribochemical reactions in other systems require an expenditure of energy in the order of some 10^5 J/mole [243, 251]. Among others the kinetics of the local processes is responsible for this discrepancy as indeed the mixed potential of an electrode mainly depends on kinetic factors. The most important factor for the observability of local alterations of the electrode potential under impact is evidently the relationship between the decay time of local potential peaks via local reactions, and the production of new impacts. In addition to the impact energy the impact frequency is also a substantial quantity influencing the alteration of the integral steady-state potential.

With the quantity of the treatment parameters being constant, the reactivity of the local element system plays an decisive role. If, for instance, the reactivity is substantially inhibited, then the destruction of the local potential peaks occurs relatively slowly. An increased detachment of cations from the metallic phase appears through the increased electrolytic solution tension of the metal at the target site (local anode)

[1] The symbol [ ]* means a system changed by impact.

without the resultant electron excess in the metal (electron accumulation) being able to be rapidly destroyed by local cathode processes. An electron excess in the solid state electrode developed in this way is the reason for an integrally measurable negative shift of the steady-state steady potential under impact action, that is the system tends to have a "reversible" mechanically activated electrode (see 6.6.2.2.1.).

It is easy to discern that with the constant increase of the target site reactivity (mechanical activation of the partial anode processes) the "efficiency" of the partial cathode processes has a decisive importance. The efficiency of the partial cathode depends on

— the number of cathodically reactive species in the electrolyt and on their activity,
— the deposition potential of cathodically reactive species, and thereby on the magnitude of the electrochemical affinity of the local element,
— the overpotential of cathodically reactive species at the less disturbed metal surface,
— the presence of sorption, reaction, and other covering films,
— the area ratio partial anodes/partial cathodes, that is on the ratio of treated to untreated or of mechanically activated to non-activated areas, in the appropriate period (a.o. on the impact density per unit of area).

In summary it may be said that the more reactive the local element system, the smaller is the integral perceptibility of an alteration of the electrode potential.

The laws of the interaction between local and integral processes at solid state electrodes derived for the effect of impact in the section in question, are generally valid for all kinds of real metal (solid)—electrolyte systems where local processes (preferably induced by mechanical action) play a decisive part. This applies particularly to stress corrosion cracking and corrosion fatigue (local processes between crack bottom and crack edge) (see 6.6.4.3.1. and 6.6.4.3.2.).

6.6.2.2.4. *Reactions of impact-produced local elements*

An increased metal deposition without external current, occurs under the impact effect on solid state electrodes in the presence of concentrated metallic salt solutions the so-called tribogalvanic deposition [249, 225]. This applies both to the case where the dissolved metal corresponds to the metal of the electrodes and to the case of other dissolved metals less noble in relation to the metal of electrodes. The tribogalvanic deposition takes place with electrode potentials which are more positive than the deposition potential of the corresponding metal in the corresponding electrolyte. Thus actually a metal deposition, induced or increased by the mechanical treatment of the solid state electrode exists, "against" the electromotive series.

In miscible coating substrate systems this deposition is accompanied by an abnormally strong mechanically activated diffusion of the deposited metal inside the matrix of the solid state electrode [249, 252], (see 6.6.3.1.) whereby the precipitation is not linked to the conditions of miscibillity [225].

An analysis of possible deposition mechanisms (formation of mixed crystals [253—257, 335—337], underpotential deposition [258—263]) indicates the reactions of impact-produced local elements in addition to the possibility of a precipitation by the mechani-

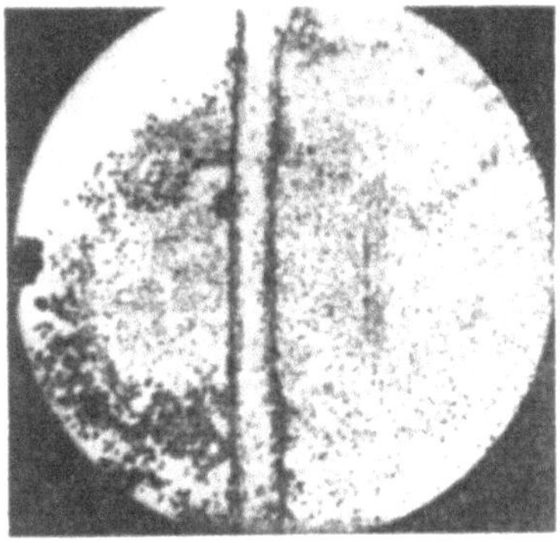

Fig. 6.52. Scratch tracks in the $Cu/Cu^{64}Cl_2$ system [249, 250] Magnification approx. 330×.

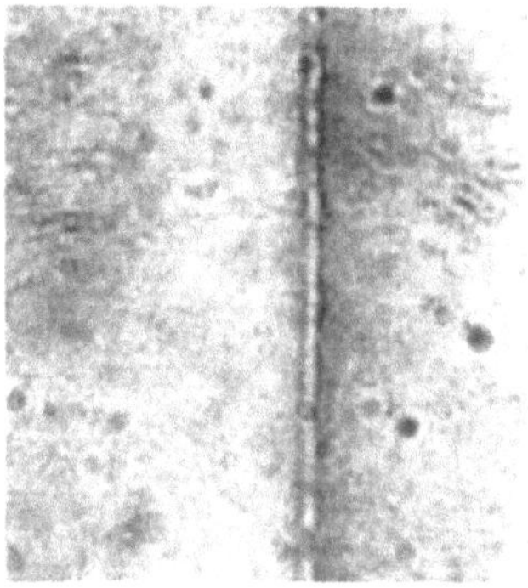

Fig. 6.53. Scratch tracks in the $Cu(Zn)/Zn^{65}Cl_2$ system [249, 250] Magnification approx. 260×.

cally activated formation of mixed crystals [249]. Cu single-crystal surfaces, low in lattice defects, scratched in labelled electrolytic solution (Cu^{64} or Zn^{65}), provide autoradiographically proof of local reactions according to Fig. 6.51. (6.6.2.2.1.) referred to as impact-induced local electrolysis. Hereby metal is dissolved on the bottom of the scratch trace (most strongly disturbed solid region), but it is precipitated at the edges of the scratch trace (weakly disturbed solid region). Fig. 6.52. shows the radioautograph of a local Cu precipitation, Fig. 6.53. that of a local Zn precipitation. Fig. 6.54. shows the photometric analysis (densitometry) of an autoradiograph in accordance

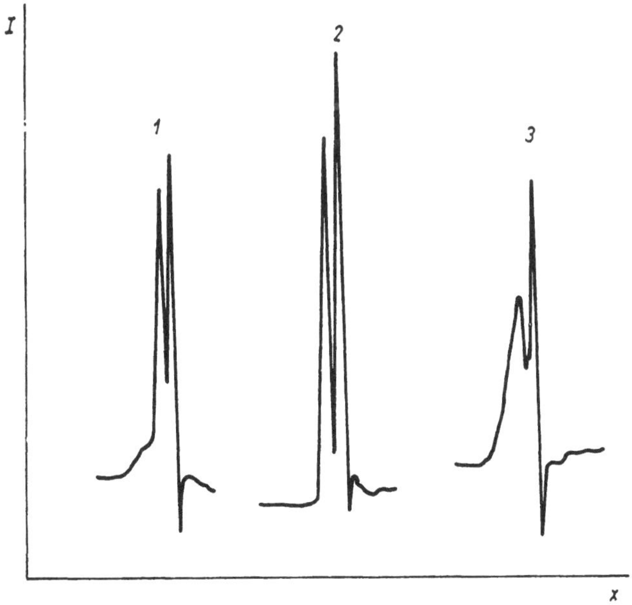

Fig. 6.54. Densitometer curves of radioautographs according to Fig. 6.52. [264].
I – blackening, x – space co-ordinate, slit width 10 μm, load of the cutting tool: 1 – 1.23 N, 2 – 2.2 N, 3 – 4.9 N.

with Fig. 6.52., in dependence on the load of the scratching tool [249, 250, 264]. Figs. 6.52.—6.54. completely confirm the local element scheme of Fig. 6.51.

In order to quantitatively determine the mechanically activated local processes at solid state electrodes, the formally calculated chemical potential of dislocations [236] may for instance be used:

$$\mu_d = \frac{\tau}{\bar{\alpha}^{\frac{1}{m}}} \tag{29}$$

μ_d — chemical potential of a dislocation, τ — mechanical load, $\bar{\alpha}$ — dislocation concentration ($\bar{\alpha} \approx 10^9$—$10^{11}$), m — coefficient ($m \approx 1{,}0 \pm 0.5$).

If the maximum dislocation concentration $N_{\max}$ (10^{12} dislocations $\times$ cm^{-3}) introduced and treated as an Avogadro number, then on average one partial crystal volume $N_{\max}^{-1}$ occurs for one dislocation and according to [265] one can speak of 1 mole dislocations with the number of moles $\frac{N}{N_{\max}}$.

With

$$N \ll N_{\max} \tag{30}$$

the following may be derived [236]:

$$\mu_d = \Delta G^+ + R'T \ln \frac{N}{N_{\max}} \tag{31}$$

(ΔG^+ increase of the partial thermodynamic potential of dislocations under subtraction of the configuration entropy, $R' = kN_{\max}$, K — Boltzmann constant; other symbols see equ. (21) and (29).)

The "chemical affinity" in the "chemical reaction" of the formation of dislocations and their movement A is then

$$A = \mu_d = \frac{\Delta\tau}{\bar{\alpha}^{\frac{1}{m}}} \tag{32}$$

(Symbols see equ. (21) and equ. (29).)

With $m = 1$, from the mentioned relationship results (comp. equ. (21) and (28)):

$$A = \frac{\Delta\tau}{\bar{\alpha}} = R'T \ln \frac{N}{N_0} \tag{33}$$

(Here N_0 is normalized so that $N = N_0$ for $\Delta\tau = 0$.)

On the other hand the electrochemical fundamental equations may be established for local electrolytic processes in the system $Me(I)_{solid}/Me(I)^{Z_i+}$ (nobler solid state electrode as substrate) and $Me(II)/Me(II)^{Z_i+}$ (as less nobler electrode of the electrolytic metal), [242]:

For the "natural" local reaction in the above system, that is for the cementation of ions of the nobler metals under the dissolution of ions of less nobler metals, holds:

$$\frac{1}{Z_i}\,\mathrm{Me(II)} + \frac{1}{Z_i}\,\mathrm{Me(I)}^{Z_i+} \leftrightharpoons \frac{1}{Z_i}\,\mathrm{Me(II)}^{Z_i+} + \frac{1}{Z_i}\,\mathrm{Me(I)} \tag{34}$$

with the electrochemical affinity $\tilde{A}_{\mathrm{LE}}$.

(Z_i — charge number.)

For the impact-produced local electrolysis with the precipitation of electrode metal ions Me(I) holds:

$$\left[\frac{1}{Z_i}\,\mathrm{Me(I)}\right]^* + \frac{1}{Z_i}\,\mathrm{Me(I)}^{Z_i+} \underset{\leftarrow}{\xrightarrow{\text{impact}}} \left[\frac{1}{Z_i}\,\mathrm{Me(I)}^{Z_i+}\right]^* + \frac{1}{Z_i}\,\mathrm{Me(I)} \tag{35}$$

(The square brackets and the symbol* express that here is involved a cluster of atoms disturbed by impact or a variety of ions with local concentration differences relative to the surrounding electrolyte).

The electrochemical affinity of the local reaction according to equ. (35) $\tilde{A}^*_{\mathrm{LE}}$ is defined by

$$\tilde{A}^*_{\mathrm{LE}} = \Delta\tilde{\mu}_i{}^* \tag{36}$$

under the necessary condition

$$\Delta\tilde{\mu}_i{}^* > \Delta\tilde{\mu}_i{}^0 \tag{37}$$

($\Delta\tilde{\mu}_i{}^*$ — electrochemical potential difference of the impact-disturbed solid state electrode ("target site"), $\Delta\tilde{\mu}^0$ — electrochemical potential difference of the (homogeneous) undisturbed solid state electrode.)

For the impact-induced local electrolysis of the ion deposition of a less nobler metal Me(II) on the nobler solid state electrode Me(I), the so called "tribogalvanic effect" is valid:

$$\left[\frac{1}{Z_i}\,\mathrm{Me(I)}\right]^* + \frac{1}{Z_i}\mathrm{Me(II)}^{Z_i+} \underset{\leftarrow}{\xrightarrow{\text{impact}}} \left[\frac{1}{Z_i}\,\mathrm{Me(I)}^{Z_i+}\right]^* + \frac{1}{Z_i}\,\mathrm{Me(II)} \tag{38}$$

(Symbols see equ. (35).)

with the electrochemical affinity $\tilde{A}^*_{\mathrm{LE\,het}}$ according to

$$\tilde{A}^*_{\mathrm{LE\,het}} = \tilde{A}^*_{\mathrm{LE}} - \tilde{A}_{\mathrm{LE}} \tag{39}$$

(See equ. (34)—equ. (36).)

under the necessary condition

$$\tilde{A}^*_{\mathrm{LE\,het}} > 0 \tag{40}$$

Equ. (39) shows that the work for the local metal precipitation forced by impact "against" the electromotive series (expressed by $\tilde{A}^{*}_{\mathrm{LE\,het}}$) consists of the difference between the work for the dissolution of the solid state electrode at the target site of an impact with local electrode metal deposition (expressed by $\tilde{A}^{*}_{\mathrm{LE}}$) and the work of the local recorrosion of the metal deposited "against" the electromotive series with the cementation of the nobler component (expressed by $\tilde{A}_{\mathrm{LE}}$). As soon as equ. (40) is no longer fulfilled through the temporal decay of the local energy excess of the target site, only the latter process will still take place. It could be proved by an analysis of equ. (34—40) that the minimum rate of necessary work for tribogalvanic effect does not depend on the potential over wide ranges of the electrode potential of the substrate electrode [242]. An estimation of the order of magnitude of impact-produced "potential peaks" occurring locally from experimental data made on the basis of this knowledge, will result in a minimum amount of 300 kJ/mole. This amount exceeds by one order of magnitude the energy excess calculated from direct measurements of the electrode potential and approximates to that of other tribochemical reactions (see 6.6.2.1.3.).

6.6.3. *Kinetics*

Mechanical treatment of the solid state electrode influences the kinetics of the interface layer processes. The influence on the transport processes and reactions starts on the electrolyte's side and extends up to the near surface layers of the solids in the μm range. Some characteristic aspects are presented for all the individual regions of the interface layer.

6.6.3.1. *Effect of the impact treatment of solid state electrodes on the electrolyte's side diffusion conditions in the phase boundary layer*

The delivery and removal of reactants and reaction products respectively, is frequently a rate-determining step (r.d.s.) of solid reactions. The impact body stress of solid state electrodes is expected to cause changes in the electrolyte's side diffusion conditions. In order to study this phenomena a pure electron transfer reaction limited in diffusion, the oxidation of $K_4[Fe(CN)_6]$ and the reduction of $K_3[Fe(CN)_6]$ on a rotating Pt disc electrode were selected (266). The electrode could undergo an impact treatment by submerging in a bed of impact bodies.

In Fig. 6.55. the limiting current densities of the oxidation of $K_4[Fe(CN)_6]$ are given, in dependence on the square root of the number of revolutions with and without treatment [267].

The results shown in Fig. 6.55. also correspond totally to the reduction of $K_3[Fe(CN)_6]$ under the same conditions [267].

It is seen that the $i_{\mathrm{lim}}/U^{1/2}$ dependence without treatment may be reproduced by a straight line in the whole number of revolution range from 9—64 rps. In the range from 22.5···64 rps the $i_{\mathrm{lim}}/U^{1/2}$ dependence is well reproduced by a straight line even with treatment. An increase is observed in the limiting diffusion currents for > 22.5 rps and a decrease for < 22.5 rps.

In Fig. 6.55. the linearity of the $i_{\mathrm{lim}}/U^{1/2}$ dependences of the redox reactions in the system released by electron transfer indicates according to [268, 269] the existence of pure diffusion control and of laminar flow conditions respectively.

In order to explain the increase in the limiting diffusion current of electron transfer reactions under impact stress with quasilaminar flow behaviour, it is assumed that the impact bodies carried

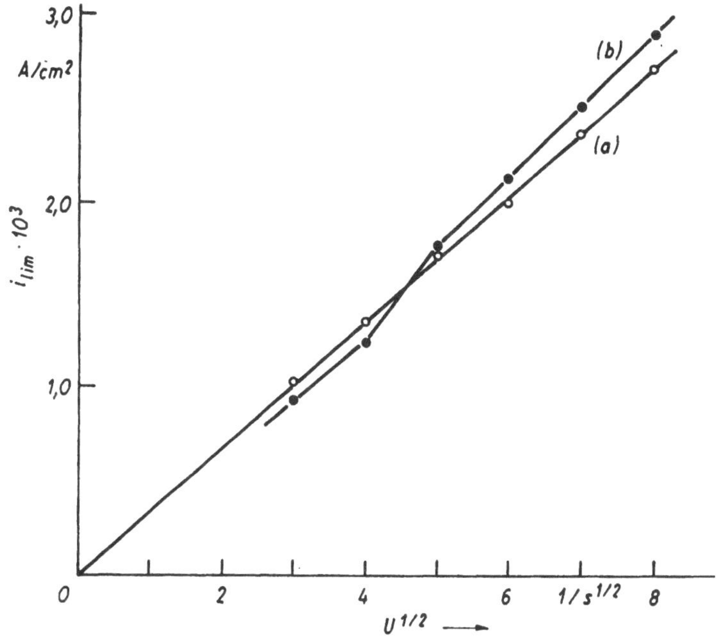

Fig. 6.55. Oxidation of $K_4[Fe(CN)_6]$ at Pt [267].

Dependence of the limiting current density i_{lim} on the square root of the speed of revolution U of the rotating disk-type electrode, (a) — without impact treatment, (b) — with impact treatment.

along by the electrolytic flow are surrounded by an adhering electrolytic film, in this way causing during the intersection of the diffusion boundary layer, an electrolytic transport exceeding the maximal stirring effect of the rotating disc electrode, in addition to the generation of microturbulences.

The lowering of the limiting diffusion currents with low speeds of revolution is caused by the partial covering of the electrode surface by impact bodies in analogy to [270 to 272].

An increase in the limiting diffusion currents under the action of impacts at solid state electrodes has also been observed under turbulent flow conditions (lateral surface of a rotating cylinder electrode) [273]. In this, however, the exponent of the i_{lim}/U dependence is changed by the impact bodies. Here too the activation of the transport processes in the electrolyte's side part of the interface layer is accounted for by a penetration of the diffusion boundary layer by means of the suspended impact bodies [274 to 277].

6.6.3.2. *The effect of the impact treatment of solid state electrodes on the electrocrystallization of metals*

Lattice imperfections are generated permanently and directly at the surface through the impact treatment of solids. Thus they are directly available to the interface layer for changes in reactivity. The continuous multiple generation of lattice imperfections is expected to lead to an alteration of the electrocrystallization behaviour at solid state electrodes stressed by impact. Therefore the electrocrystallization of zinc was investigated on a rotating Cu electrode treated by impact.

On this potentiodynamic polarization curves were made on the cathodic zinc deposition with and without mechanical treatment [278], (Fig. 6.56.).

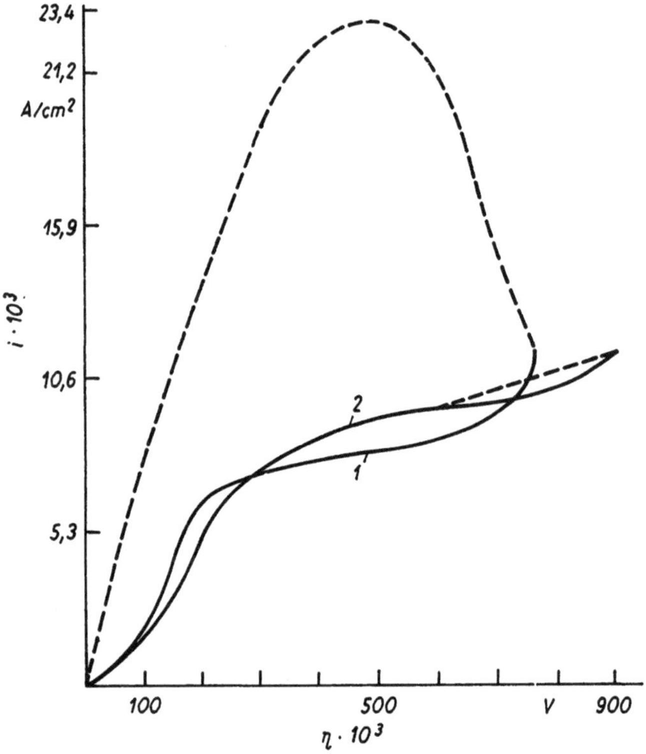

Fig. 6.56. Electrolytic zinc precipitation on copper [278].
Current density-potential curve at a rotating disk-type electrode, 1 — without impact treatment, 2 — with impact treatment, i — current density, η — overvoltage.

Curve 1 shows a marked hysteresis with a negative current/voltage characteristic, the so-called superpolarization, without treatment and with cyclic polarization [268]. With treatment (curve 2) an approximately anhysteretic course of the polarization curve results without a negative current/voltage characteristic. The phenomenon of the superpolarization being characteristic of the Zn—Cu system with cathodic Zn deposition fails to appear in the case of a solid state electrode treated by impact. The difference found experimentally of the cyclic polarization curve of cathodic zinc deposition onto copper without and with impact stress of the substrate may be explained as follows by extension of the interpretation given in [278]:

It is known that the electrocrystallization takes place according to the Kossel-Stranski mechanism after the following scheme [279—281], (Fig. 6.57.).

The metallic ion may be precipitated at three points on the electrode surface. That is as ad-atom "*b*" at a step "*c*" or dislocation "*d*". The transgression of a solvated ion "*a*" into the position of the ad-atom "*b*" under desolvation and partial charge transfer, is the process with the smallest activation energy. From the position "*b*" the ad-atom has to move by surface diffusion into the positions "*c*" or "*d*" in order to be incorporated into the lattice after the complete charge transfer. This surface diffusion is frequently the rate-determining step in electrocrystallization which becomes electrode-kinetically apparent in the crystallization overpotential. In this it has to be noted that the deposition of ad-atoms already happens with more positive potentials than corresponds to the reversible NERNST potential of the metal deposition in the given system (underpotential deposition [259, 260, 261–263]).

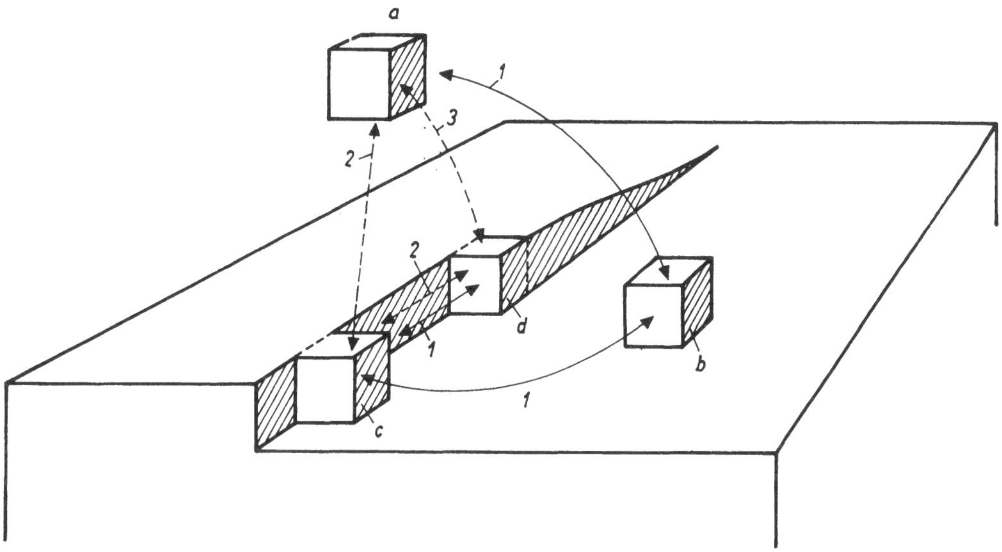

Fig. 6.57. KOSSEL-STRANSKI model [268].
1, 2, 3 – possible attachment and detachment paths of an atom, a–d – comments in the text.

If the potential is led to the first metal deposition (indicated by the rise of current) in the cathodic polarization of the substrate according to Fig. 6.56. then the crystallization overpotentials is overcome. If the potential is returned directly after it, then a strong (unrestricted) precipitation of metal occurs expressing itself in a rise of current with a negative current/voltage characteristic, that is, superpolarization. The newly formed crystallites facilitate the further crystallization after having once formed the first nuclei.

If this experiment is carried out with mechanical impact stress, a crystallization inhibition caused by necessary long surface diffusion paths of the ad-atoms into mainly lattice inclusion sites (see model Fig. 6.57.), will not occur because of the permanent generation of metal surface disturbances. Through this the unhysteretic course of the cyclic polarization curve with treatment may be accounted for. Simultaneously the relatively small deposition current with treatment, shows (in comparison to the much higher superpolarization current without treatment) that a growth inhibition of the nuclei is released by the multiple mechanical impact stress. Evidently a per se energetically favoured growth of larger crystallites is prevented by the continuously occuring destruction of primary crystallites. This is a process which with a smaller gross-exchange leads to non-oriented, fine grained metal depositions. Such differences in the macrostructure of the zinc precipitation on rotating Cu electrodes at 3,000 rpm are clearly illustrated in Fig. 6.58. and 6.59. (comp. 6.6.4.1.).

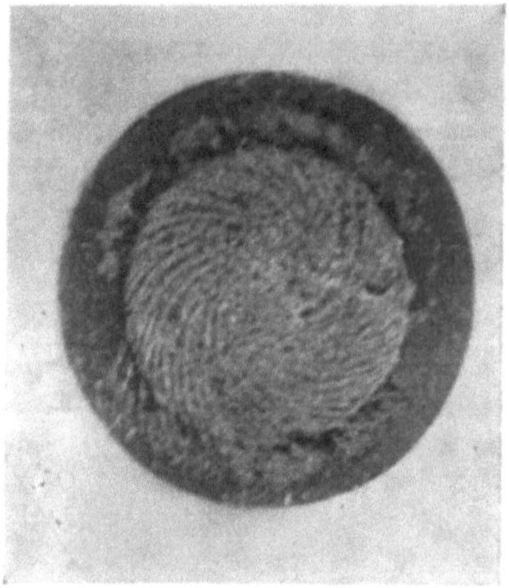

Fig. 6.58. Electrolytic zinc precipitation on copper at a rotating disk-type electrode without impact treatment [278] Magnification approx. 5 ×.

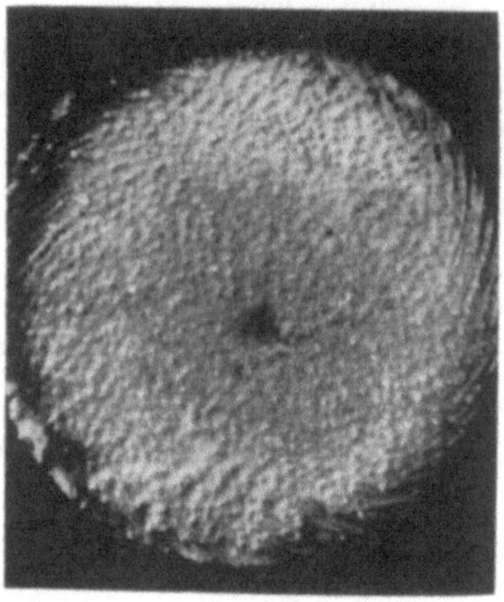

Fig. 6.59. Electrolytic zinc precipitation on copper at a rotating disk-type electrode with impact treatment [278] Magnification approx. 9×.

6.6.3.3. *The effect of macroscopic deformation on the electrode kinetics of the dissolution of metals*

Investigations of the change of the dissolution rate of metals during macroscopic deformation (strain, torsion) showed differently strong effects [215, 216, 220, 222, 240, 249, 282, 285–312]. Relatively small effects appear with strain in the elastic region

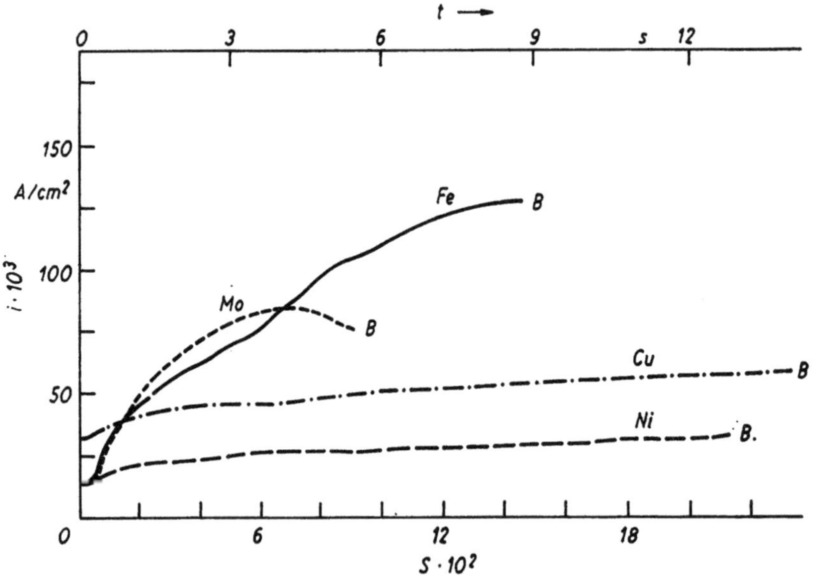

Fig. 6.60. Dissolution of various metals in dependence on uniform strain [298].
i – current density, t – time, S – strain, B – modulus of rupture.

[215, 216, 220, 222, 249, 282, 283]. Thus, for instance, it has been found in iron [298] that only the cathodic component current of the local hydrogen depostion is somewhat accelerated due to the effect of mechanical energy in the elastic region (reversible strain). From this results a positive shift of electrode potential of 0.1 mV with a loading of 9.8×10^3 N cm^{-2}. On the other hand an activation of the dissolution current has been found for plastic (irreversible strain) corresponding approximately to the factor 30 in relation to the unloaded material [284—312]. Fig. 6.60. and 6.61. serve as an example [298].

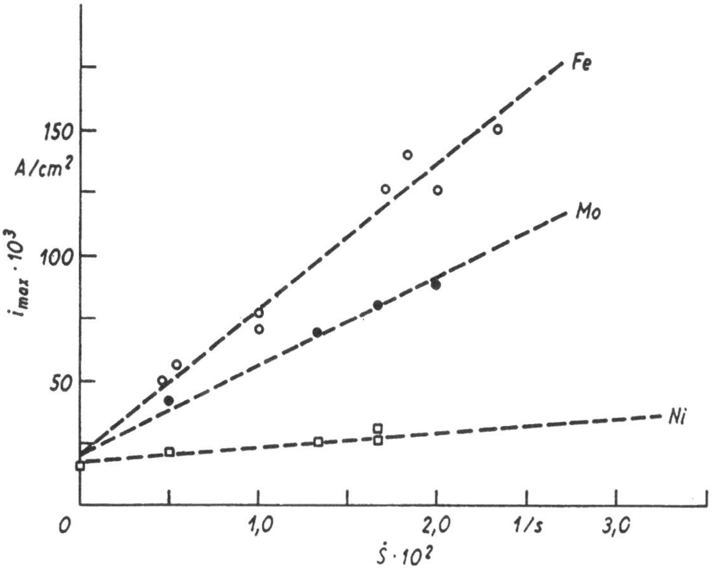

Fig. 6.61. Dissolution of various metals in dependence on the strain velocity [298].
i_{max} — maximal current density, $\dot{S}$ — strain velocity.

Fig. 6.60. shows the dependence of the dissolution current on the degree of deformation for the metals Fe, Mo, Ni, and Cu with constant deformation rate. In this the current density increases with the same degree of deformation much more for Fe and Mo than for Ni and Cu. Fig. 6.61. shows the maximum dissolution current density of mentioned metals in dependence on the strain rate. This dependence is linear for all the metals referred to.

A comparison of the dissolution behaviour of Armco iron in the passive potential range with that of stainless steels shows an activation of the dissolution current of the same kind as for plastic deformation [312]. A comparison of the mechanically activated dissolution behaviour of stainless steel shows in all the potential ranges of the dissolution (active, passive, transpassive) [313] a distinct influence on all the ranges. Thus the current density (potential curve in the range of the active dissolution and in the transpassive range, respectively) is shifted to the more negative electrode potentials, and the passivation current and the passive current increase with growing degree of deformation.

The explanation of the activation effects represented may be given in different ways. Either high index crystal planes generated by strain [305, 314—317], or the local increase of the chemical potential of the metals by mechanical stress (comp. 6.6.2.2.4.) are considered to be the primary reason for the effects mentioned [236]. Neither of the variant represent any alternative from the lattice-thermodynamic point of view. However, it is frequently difficult to distinguish activation effects determined by the structure of the solid from fresh surface effects, in the case of those solid state electrode processes in which secondary partial reactions represent the rate-determining step, such as adsorption, chemisorption, the formation of reaction layer, etc.

However, the electrode kinetics of such systems may not only be influenced by unsteady or steady fresh surface formation but also by the mechanically activated formation of reaction layers [318].

An example for this is the iron electrode, where as is known not only the passivation reaction is controlled by chemisorption processes but already the reaction kinetics of the active dissolution.

A formal equation derived on the basis of lattice thermodynamics and solid mechanics for the local dissolution current mechanically activated by macroscopic deformation is [236]:

$$I = i_a \left(\exp \frac{n \Delta\tau}{\bar{\alpha} R' T} - 1 \right) - i_k \tag{41}$$

(I — total current, i_a — anodic partial current density, i_k — cathodic partial current density, n — number of dislocations in a plane dislocation accumulation, $\bar{\alpha}$ — dislocation concentration, $\Delta\tau$ — mechanical stress difference (hardening), $R' = kN_{\max}$; k — BOLTZMANN constant, $N_{\max}$ — maximum dislocation concentration.)

With a sufficient deviation of the equilibrium potential, i_k may be neglected and we obtain for the increase of the anode current density Δi_a induced by deformation:

$$\Delta i_a = i_a \left(\exp \frac{n \Delta\tau}{\bar{\alpha} R' T} - 1 \right) \tag{42}$$

in the case

$$n \Delta\tau \ll \bar{\alpha} R' T \tag{43}$$

in a first approximation we obtain a linear dependence of the deformation-determined increase in dissolution current on the hardening:

$$\Delta i_a = i_a \frac{n \Delta\tau}{\bar{\alpha} R' T} \tag{44}$$

With a dynamic regime of the plastic flow, the magnitude of the mechanical activation depends on the deformation rate for each value of activation:

$$\Delta i_a = i_a \left(\frac{\dot{\varepsilon}}{\dot{\varepsilon}_0} - 1 \right) \tag{45}$$

($\dot{\varepsilon}$ — deformation rate, $\dot{\varepsilon}_0$ — rate of the relative deformation corresponding to the elastic limit.)

This means that the activation depends linearly on the deformation velocity (comp. Fig. 6.61.).

The local negative shift of electrode potentials is calculated according to [236]:

$$\Delta\varphi = -\frac{n\Delta\tau R}{\bar{\alpha}R'zF}; \tag{46}$$

(R — gas constant, z — charge number, F — FARADAY constant.)

However, in order to derive equ. (46) the "reversibe" mechanically activated electrode is proceeded from (comp. corresponding qualifications in 6.2.).

6.6.3.3.1 Fresh surface effects on solid state electrodes

Fresh surface effects may decisively contribute to the activation of tribochemical processes (see 5.6.). Numerous fresh surface effects are also known for solid state electrodes which occur in the grinding or scraping of the solid with very high abrasion intensities [319—352]. It is a basic experience of studies of fresh surface effects that fast reaction steps such as e.g. the transfer of electrons, are not influenced by the renewal of surfaces, whereas kinetic inhibitions being based on slow reaction steps are to a large extent eliminated [454—456]. Thus, for instance, the H_2 overpotential could be reduced on metals having a high H_2 adsorption capacity, such as Fe, Ni, Pd. Furthermore a decrease was observed in the overpotential of the oxygen deposition at Pd cathodes. A fresh surface activation due to the anodic dissolution of Fe and Ni in the active range shows a considerable decrease in the polarizability of these electrodes. The passive currents of easily passivatable electrodes such as Ti, Ni, Cr increase through grinding treatment with high velocities of the renewal of surfaces [340]. Fig. 6.62. illustrates the difference between the passive current of an Ni electrode treated abrasively (abrasion 1,200 atomic layers x s^{-1}) and without treatment. Here the difference amounts to about 3 orders of magnitude [335].

6.6.3.4. Tribodiffusion in the solid part of the interface layer

An important kinetic partial aspect of solid reactions comprises the transport processes of the solid phase preceding and following in time. Because of the mobility of the lattice components a thermal influence on these transport processes only occurs in temperature ranges of the order of two thirds of the absolute melting temperature (TAMMANN temperature) in metallic solids, defects (dislocations, vacancies) are already set in motion or mobile defects initiated, at room temperature, with deformation processes. These defects may considerably influence the transport processes in the solid phase (see 5.7.). The investigation of mechanically activated transport processes in metallic solids at room temperature has a great importance for a knowledge of the course of heterogeneous reactions in the metal-electrolyte system in deposition and dissolution (corrosion) processes as well as in the complex of friction, lubrication, and wear.

6.6.3.4.1. Metal/metal tribodiffusion at room temperature under the influence of impact

A specific type of tribodiffusion appears in the metal-electrolyte system which is distinguished by a particular intensity of the processes. A particular property of metal coatings deposited tribogalvanically (see 6.6.2.2.4.) is their interaction with the matrix under continuous impact treatment of the substrate. By this the coating metal is mainly

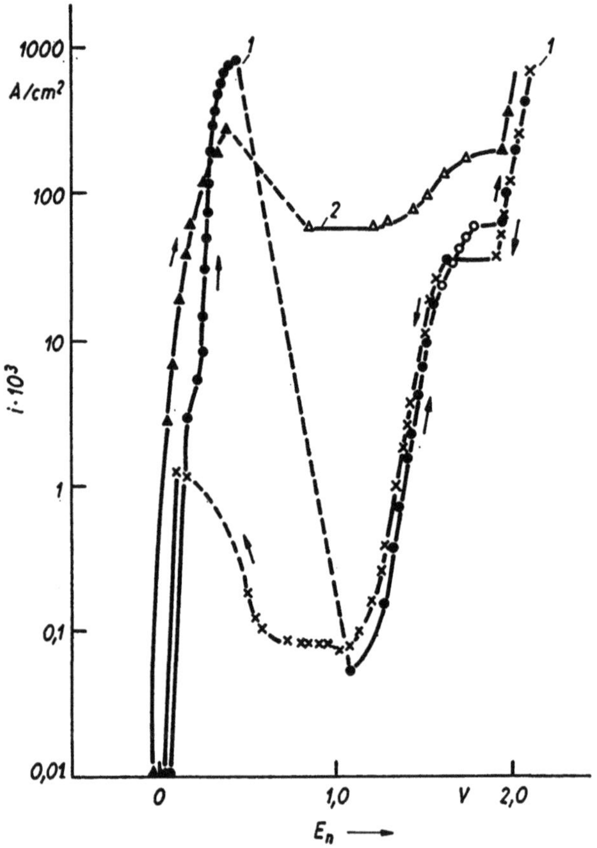

Fig. 6.62. Fresh surface effects at nickel electrodes [335].

1 – unground, 2 – ground, *i* – current density, E_h – electrode potential.

worked into the base metal accompanied by the formation of alloys at room temperature if the deposition of metal takes place slowly under intensive impact treatment [249, 353]. As investigations have shown, the formation of a surface containing Zn in the Cu—Zn system treated by impact depends on the sample potential, the electrolytic composition and concentration as well as on the treatment parameters [354].

Cu samples show a yellow colour of the treated areas after having been treated by impact in the Zn electrolyte. The changes in colour of the base metal are attributable to changes in the (light) absorption behaviour, which is dependent on the electron structure of the alloy. However for its part the electronic composition only changes with phase change or formation of mixed crystals [355]. The attendant colour change can be attributed to the formation of a Cu—Zn mixed crystal as has been confirmed by electron microprobe measurements [252].

A ternary phase having ferromagnetic properties is formed by the tribodiffusion of Mn and Al in Cu at room temperature [353]. The property of ferromagnetism also points to a certain state of order of the foreign atoms in the matrix lattice implanted by impact treatment. Until now the mutual penetration described between deposited metal and substrate has only been found in miscible metal systms.

In order to characterize the tribodiffusion in the Cu—Zn system, a deposition of metals was performed with and without mechanical treatment of the cold drawn and therefore rich in defects Cu-substrate in Zn^{2+} electrolyte labelled by Zn^{65}. A radiometallographic tracer technique (356) permitted the taking of Zn-distribution profiles in Cu in dependence on the distance from sample surface (Fig. 6.63.).

Fig. 6.63. clearly shows that it is not the number of defects per se which is responsible for the abnormal diffusion behaviour at room temperature but that the process depends on the motion of the defects (comp. Fig. 4.44). The effect of impact leads to dislocation motions and multiplications, to the formation of excess vacancies, and also, upon longer deformation, to grain destruction. It has been established that a destroyed and a deformed zone can be distinguished in the structure of Cu treated by impact. Such a structure is known for metal treated by cutting [357] whereby the destroyed zone is also still crystalline at particle sizes of 20···60 nm [358].

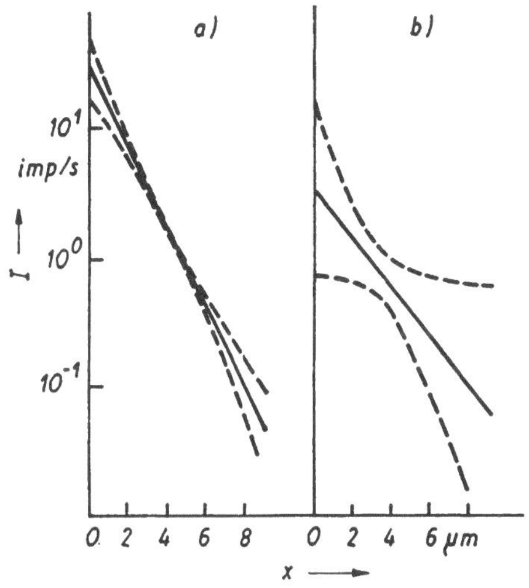

Fig. 6.63. Diffusion profiles of Zn^{65} in Cu [252]. (a) – with impact treatment, (b) – without impact treatment, I – residual activity, x – penetration depth, ---- – error ranges.

All the mentioned structural disorders are in part only formed by the mechanical treatment process. Hence they are in motion as are the dislocations, grain boundaries, etc., which are already present in the material. According to the information of various authors [359–361], moving defects considerably increase the diffusion rate. A one hundredfold increase in the pre-exponential factor of the diffusion equation has been found for the diffusion at moving grain boundaries [362].

6.6.4. *Tribochemistry of solid state electrodes in engineering*

Metals as important construction materials are frequently exposed to a medial stress by electrolytes in practice. In addition there is a series of technologies in which the metal-electrolyte system undergoes manifold effects of different forms of energy. In the following some important, technically relevant, mechanically activated solid state electrode processes are briefly dealt with. In this the qualifaction is made that mainly changes in system occurring during the supply of mechanical energy will be dealt with.

6.6.4.1. *Mechanically activated galvanic precipitation*

Mechanical procedures of pretreatment and aftertreatment of surfaces form the traditional know-how of galvanic metal deposition and are part of numerous advanced electroplating flow-line processes. In spite of the state of basic knowledge (grain destruction at mechanically treated metal surface, facilitating of diffusion processes, improvement of the quality of precipitations by preventing dendrite formation, etc.; comp. also 6.6.3.2.) until now relatively little use has been made of a large-scale industrial utilization of mechanically activated processes in situ [363—365], comp. [366].

Fig. 6.64. serves as an example for the changed precipitation parameters and layer properties. Presented here is the mechanical activation of a nickel deposition from WATTS electrolyte by friction stress of the base with variable pressure of the working tool [365].

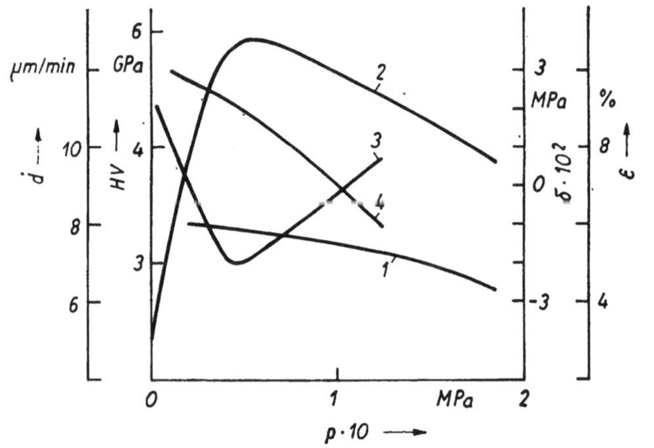

Fig. 6.64. **Mechanically activated galvanic precipitation of nickel [365].**
1 — precipitation velocity d, 2 — micro-hardness HV, 3 — inner stress δ, 4 — elasticity ε, p — pressure of the tool.

It has been found that the mechanical activation yields highgrade precipitations in the range of high current densities (50 ··· 150 A × dm^{-2}). Simultaneously deposition rates could be obtained in the order of 12.5···15 µm/min., instead of the usual ones of 1···2 µm/min. High pressure of the activation tool simultaneously increases the wear of the layers produced whereby the respective optimal activation intensity depends on the current density used.

The surface activation of the cathode by friction markedly shortens the duration of the crystal growth and increases the nucleation rate [365], (comp. 6.6.3.2.). Here a finegrained structure results having grain sizes below 1 µm, that is to say a reduction by the factor 10···50 relative to non-activated deposition (comp. Fig. 6.50.).

A relatively high degree of hardness with minimum values of elasticity and inner tensions exists where the activation tool is loaded by high pressure. This shows the greatest possible degree of orientation of crystallites. The further reduction of the hardness which takes place with increasing inner tensions, points to the destruction of the texture due to higher mechanical stresses. Metallographic investigations no longer show any grain structure here; the latter is in the range of the electron-microscopical resolution (comp. 6.6.3.4.1.).

The mechanically activated galvanic precipitations may certainly not be included at will into the technological sequences of today's mainly highly mechanized equipments for mass coating. Nevertheless, as a result of the effects mentioned it will be incorporated into engineering where the increased technological outlay of the mechanical surface treatment during the galvanic precipitation process is justified by the practical value of the parts and the specific material alteration obtained, as is the case e.g. in magnetic memories, components of friction pairings, etc.

6.6.4.2. *Cutting processes on metals in the presence of electrolytic media*

Increases in tool life and in the values of cutting parameters are possible through cutting with the effect of appropriate active media. Such active media include aqueous and non-aqueous electrolytes in addition to atmospheric oxygen [367, 457, 458] and surface active substances, emulsions, cutting oils, etc. As an example the measurement of the cutting force is given in dependence on the drilling distance (successive length of the drilling paths) of twist drills (X 82 W Mo 6.5) in the cutting treatment of steel (st 38) in various active media [368], (Fig. 6.65.):

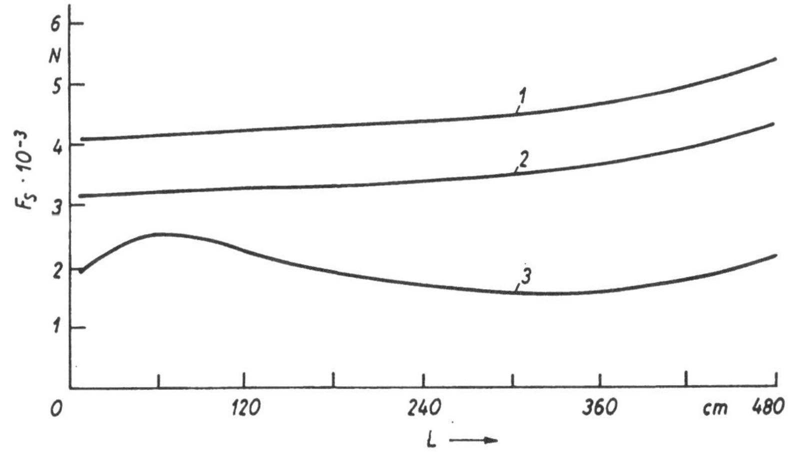

Fig. 6.65. Cutting treatment of St 38 in $MnCl_2$ solution [368].
Cutting force of the drill F_s depending on the drill path L in different media, 1 — air, 2 — water, 3 — $MnCl_2$.

An analysis of the cutting behaviour in $MnCl_2$ solution shows two phenomena: a momentary decrease in the cutting force and an improvement of the cutting properties by about 30 per cent, and a changed long-term behaviour.

The first indicates the existence of another fundamental wear mechanism. The interaction between ions having the electrochemical double layer and the specific adsorption to the solid state electrode, respectively, may play a part here. Moreover there may be an influence of the wear behaviour in dependence on the zeta potential of the double layer [369] and other chemomechanical effects, such as the alteration of the plastic behaviour of solids through active tribocorrosion [236, 370–372].

The long-term course of the cutting force with a growing drilling path is abnormal (comp. curves 1 and 2, Fig. 6.65.). Following an initial maximum of F_s with $L \approx 600$ mm the cutting force decreases with increasing drilling, that is to say, the practical value of the tool improves with constant use. The cutting force amounts to only 1,500 N with $L \approx 3{,}500$ mm, that is 33 per cent relative to the corresponding value of air and 42 per cent relative to water as active media. This course can only be explained by the changes in the material properties of the tool. Whereas the fundamental mechanisms illustrated first must act spontaneously, the long-term processes indicate a tribogalvanic deposition of manganese with the latter simultaneously incorporated by tribodiffusion into the material steel.

6.6.4.3. *Electrochemical tribocorrosion*

By tribocorrosion [373] is understood the damage of destruction of materials by chemical or electrochemical reactions with their environment under the effect of mechanical energy (see 7.4.4.2.). The mechanical energy may act in a macroscopic way (static, dynamic) by strain, bending, or torsion, cutting and non-cutting shaping, inner tensions, or in the form of local mechanical pulses by friction, impact actions by solids, liquids, gases, vacuum bubbles (cavitation) and by fluids (gases, liquids) with high flow velocity under non-laminar conditions. The following remarks deal exclusively with tribocorrosion in the metal-electrolyte system, that is to say with electrochemical tribocorrosion.

Although defects caused by the previous action of mechanical forces upon metallic solids (e.g. cold working) markedly influence corrosion behaviour and often decisively determine material behaviour in relation to certain media [341–343, 374–382] or the chemical potential of solid state electrodes, it is estimated that by this the integral corrosion rate [344] only increase by the factor 2 to 3.

The complex phenomena and effects of corrosion in technology with the simultaneous addition of mechanical energy may nevertheless extend far beyond the order of magnitude mentioned. All the fundamental processes of a thermodynamic and kinetic nature presented in the previous chapters may combine to bring about much higher activations, particularly with a localized attack (compare e.g. the activation of the dissolution current by the factor 1,000 in the abrasive treatment of passive solid state electrodes/6.6.3.3.1.). On account the great variety of electrochemical tribocorrosion processes in technology such as corrosion in friction [319–322, 323–340], erosion corrosion or corrosive wear by solid particles [345, 384–387], vibration [388, 389], and by cavitaton [390–401] it is only possible to treat in somewhat more detail the kinds of corrosion which are summarized by the term "environment-sensitive fracture" [402]. These processes are generally designated as "stress corrosion cracking", if the mechanical stresses are unidirectional, and as "corrosion fatigue", if the stresses fluctuate, that is to say, have a cyclic course (alternation of compressive and tensile stresses). In spite of all the differences in the appearance of both kinds of corrosion they have a number of common fundamental mechanisms and causes and merge partly into one another.

6.6.4.3.1. *Stress-corrosion cracking*

Stress-corrosion cracking (SCC) is a crack formation in metals with the simultaneous action of certain attack media and tensile stresses. Characteristic is a deformationless separation with an intercrystalline or transcrystalline course and often without the formation of visible corrosion products. Tensile stresses occasionally exist as internal stresses in the workpiece. The necessity of the simultaneous action of mechanical forces with certain reaction media upon certain metals is characteristic. According to the definitions of this book, SCC represents a tribochemical phenomenon [236, 240, 403].

The frequent disastrous damage in technology because of SCC has attracted extensive interest of research and general practice in this field [404—408]. In addition to individual factors which are the possible causes of SCC, such as hydrogen embrittlement [408—414], loss of cohesive strength of the metal by specific adsorption [415, 416, 420], and adsorptive lowering of strength [417] the so-called "film rupture mechanism" combined with active anodic dissolution of the crack bottom is regarded as a frequent fundamental mechanism of the SCC [344, 404—408, 417—426].

SCC is assumed to run approximately in the following way: After a mechanically induced rupture of the oxide film at the crack bottom a fresh surface is formed which corrodes locally very strongly for a short period of time (about $10^{-3} \cdots 10^{-1}$ s) on account of its increased electrochemical potential, before a repassivation occurs. The corrosion has now enlarged the fissure by a small rate ($10^{-7} \cdots 10^{-5}$ cm); as a result the material is more stressed before the fissure tip and strains more quickly. After some time the strain will be sufficiently great to go beyond the fracture strain limit of the reaction product (oxide film), and the cycle will start again.

In addition to material constants and the repassivation ability of the given metal-electrolyte system (particularly of the repassivation rate) the distribution of the electrolyte composition [411] and the electrode potential along the fissure also plays a leading part [236, 427]. Due to the pH of the crack bottom being shifted by some values to the acid range in comparison with the pH of the basic electrolyte, and due to the irregular potential distribution in the crack, respectively, both the formation of hydrogen and the anodic dissolution will occur locally even if the thermodynamic conditions of the integral electrode potential do not permit it [428—432]. Additionally there is the changed reactivity of mechanically induced local elements, in this case between crack bottom and crack edges (comp. 6.6.2.2.4.). Frequently all the fundamental processes mentioned above appear alternately or simultaneously [433].

Numerous preconditions for SCC may be derived from fracture mechanics [410, 434]. SCC processes could be detected in many alloys which clearly cannot be based on the film rupture mechanism (mixed crystal effects), [435, 438]. In spite of the variety of investigations of the phenomenon of SCC, a uniform theory cannot be foreseen yet [344, 402, 406, 433]. On the other hand a number of preventive measures may be taken on the basis of the knowledge of the system already available.

A large proportion of the cases of damage by SCC for instance, are not so much based on the service load but far more on the presence of internal stresses induced by the manufacturing process. This can be remedied by thermal aftertreatment [407, 439, 440] or by the additional application of compressive stresses by shot peening into the surface [440].

A further example is the SCC of natural gas pipelines which is induced by crevice corrosion between rolling skin and steel surface [441, 442] (cinder is frequently considered as a good temporary protection against corrosion). This damage could be completely excluded by the previous blasting of the cinder.

It has become known during recent years that the SCC of alloyed steels can be sharply reduced by the reduction of sulphur and phosphorus impurities. The presence of such occlusions favour a number of SCC mechanisms.

Frequently in addition to a selective thermal treatment small changes in composition cause a drastic reduction of the crack propagation rate in environmental sensitive cracking. This is a result of systematic basic research [443].

6.6.4.3.2. *Corrosion fatigue* (*CF*)

The most frequent kind of mechano-corrosive material destruction in engineering is CF. This is a process resulting from a combination of dynamic mechanical alternating stress and electrochemical corrosion. In contrast to SCC (specific material/medium combination, preferably static stresses), CF is not system specific, that is it can appear on any material in any corrosive media [444—447].

In this a process of gradual material destruction takes place leading to changes in the material properties, the formation and propagation of cracks, and finally to destruction. CF may be described as a synergetic interaction between corrosion and material fatigue. A simple additive overlapping of material fatigue and SCC is not permissible for the description of CF since due to the essentially non-specific dependence on the media CF may also appear if SCC does not come into consideration for the system. An important feature is the absence of the actual stressability limit otherwise available for materials (in air).

As fundamental mechanisms for CF the same are discussed as for SCC:

— Stress concentration at the bottom of corrosion pits,
— electrochemical attack in the plastically deformed crack bottom with less deformed regions as cathode (comp. 6.6.2.2.)
— electrochemical attack with film rupture,
— lowering of the surface energy by sorption processes (Rehbinder effect).

Similar fundamental causes also lead to a frequent overlapping of SCC and CF [448, 449]. Substantial impulses for clarifying CF stem from fracture mechanics [450].

Chemomechanical effects play a leading part in addition to adsorptive effects. The first ones are based on an alteration of the plastic material properties in electrochemically active media. On the one hand, an interaction between lattice defects and the electrochemical double layer within the metal-electrolyte phase boundary layer is discussed here and on the other hand the effect of the electrochemical dissolution of metals under corrosion conditions on the facility of their creeping [236, 370—372]. The quantitative interpretation of a chemomechanical effect on the basis of the thermodynamics of irreversible processes was already given in 1968 [240], (Comp. 6.6.2.1.2., particularly equ. (25)). This effect was rediscovered as the influence of the chemical potential on the deformation of metals in a simplified form under the name "chemostress coefficient" in 1978 [451].

In analogy to SCC a generalizing theory of CF has still to be established in spite of a wealth of knowledge. Advances in this field require an intensive co-operation of experts in the field of microplasticity and dislocation theory, of electrochemical corrosion theory, and of fracture mechanics [450].

6.6.5. *Final remarks*

The remarks in 6.6. show that the tribochemistry of the solid state electrode represents a complete analogue to the tribochemistry of other solid-reaction medium systems despite system-determined particularities. All the known processes may be classified into the hierarchic system of the energy "states" [452, 453, 461] although this classification is not always clear and readily apparent.

Triboplasmatic processes should exist in the case of electrochemical reactions induced by friction and impact. In addition such processes are also possible by macroscopic energy supply, if a high local concentration of mechanical energy dissipates through material destruction (fracture processes) (SCC, CF). A directly discribale appearance of triboplasmatic processes in the metal-electrolyte system has however not been known yet. Due to the high reactivity of the components of the very complex system, which is able, for instance, to directly process all the known triboemission processes by reactions within the interface layer, triboplasmatic processes are "wasted" by following processes.

Postplasmatic processes having a strongly irreversible character can be detected with certainty in the metal-electrolyte system.

The tribogalvanic effect as an example of reactions with negative affinity, is one of these, as well as the abnormal tribodiffusion in the interface layer metal/electrolyte.

Stress corrosion cracking (film rupture mechanism) with the alternating consequence of tribochemical and mechanical destruction processes may also very much be regarded as a triboinduced dissipative structure.

Changes in the reactivity of solid state electrodes due to stored lattice disorders occupy an extremely large place in the number of the processes investigated, since they decisively contribute to the material behaviour of metals in the reactive media, that is to say in practical application.

A particularity of numerous solid state electrodes is their great tendency to form reaction layers. In such systems triboenergetic processes can often no longer be seen clearly because of the changed kinetics in the formation of fresh surfaces. A further particularity of solid state electrodes is the system-determined spontaneous interaction between local and integral processes which influences decisively the appearance as well as the identifiability of tribochemical processes in the metal-electrolyte system.

Due to the very great technological importance of the metal-electrolyte system, the incorporation of tribochemical approaches and findings in the derivation and evaluation of fundamental mechanisms of intended (cutting, electroplating) and unintended mechanical activation (corrosion, corrosive wear, etc.) of electrochemical processes, should contribute to the better conservation and utilization of our resources.

References

[1] THIBAULT, N. W.: Amer. Min. **29** (1944), 250, 328

[2] KNIPPENBERG, W. F.: "Growth Phenomena in Silicon Carbide", Philips Research Report 18, p. 185, Eindhoven 1963

[3] PAULING, L.: „Die Natur der chemischen Bindung“, p. 89, 94, Weinheim/Bergstr. 1964

[4] GMELIN, Syst. 15: Si, Teil B, 8. Aufl. 1959, p. 793

[5] AGTE, C., MOERS, K.: Z. anorg. allg. Chem. **198** (1931), 233

[6] HEINICKE, G., HENNIG, H.-P., STEINIKE, U.: Kristall u. Technik 8 (1973), 379

[7] EBERT, I., HENNIG, H.-P.: chapt. 4., [10], p. 79

[8] WALTERS, G. K., ESTLE, T. L.: J. appl. Phys. **32** (1961), 1854

[9] BRODSKY, M. H., TITLE, R. S.: Phys. Rev. Letters **23** (1969), 581

[10] KUNATH, D., HENNIG, H.-P.: chapt. 4., [10], p. 124

[11] STEINIKE, U., HENNIG, H.-P.: chapt. 4., [10], p. A 31

[12] ICHIOKA, K., ITO, T., SHIRAKAWA, T., TOI, K., TOKUDA, T.: Bull. Chem. Soc. Jap. **49** (1976), 1199

[13] SHIRAKAWA, T., SHIMIZU, K., TOI, K., ITO, T.: Chem. Lett. (1975), 1097

[14] HENNIG, H.-P.: unpublished

[15] LIDSTRÖM, L.: Acta Polytechnica Scandinavica **Ch 75** (1968)

[16] CHODAKOV, G. S.: „Fizika izmel'čenija“, Moskva 1972

[17] LINDNER, K.-H.: Dissertation, Berlin (West) 1966

[18] BUTJAGIN, P. Ju., BYSTRIKOV, A. V.: chapt. 1., [9], 1, p. 63

[19] SCHLEGEL, E.: Dissertation B, Freiberg 1976

[20] SAKABE, H., KAWAI, K., KOSHI, K., SODA, R., HAMADA, A., SHIMAZU, M., HAYASHI, H.: Bull. Nat. Inst. Industr. Health. **4** (1960), 1

[21] CHARBONNIER, J., COLLET, A., LE BOUFFANT, L., DANIEL-MOUSSARD, H., DURIF, S.: Beiträge Silikose-Forschung Nr. 66 (1960), 1

[22] ROBOCK, K.: Staub-Reinhalt. Luft **28** (1968), 148; Beiträge Silikose-Forschung Nr. 92 (1967), 1

[23] ZACHARIASEN, W. J.: J. Amer. Chem. Soc. **54** (1932), 3841

[24] BLANKENBURG, H.-J.: „Quarzrohstoffe“, Leipzig 1978

[25] MOREY, G. W., FOURNIER, R. O., ROWE, J. J.: Geochimica Cosmochimica Acta **26** (1962), 1029

[26] STÖBER, W., ARNOLD, M.: Kolloid-Z. **174** (1961), 20

[27] VAN LIER, J. A.: Dissertation, Utrecht 1959; chapt. 4., [250], p. 318

[28] FRONDEL, C.: "Silica minerals", in: Dana's System of Mineralogy, III, New York/London 1962; ref. in chapt. 4., [250], p. 318

[29] ILER, R. K.: "The Colloid Chemistry of Silica and Silicates", New York 1955

[30] HEYDEMANN, A.: Beitr. Miner. Petrogr. **10** (1964), 242

[31] ALEXANDER, G. B., HESTON, W. M., ILER, R. K.: J. Phys. Chem. **58** (1954), 453

[32] BAUMANN, H.: Beiträge Silikose-Forschung H. 37 (1955), 47

[33] Ref. [28]; ref. in [17], p. 9

[34] STEINIKE, U., et al.: Crystal Res. Technol. **17** (1982), 1585

[35] HEEGN, H., BERNHARDT, C., GOTTSCHALK, J., HUSEMANN, K.: Chem. Techn. **26** (1974), 696

[36] STEIER, K., FUERSTENAU, D. W.: Chemie-Ing.-Techn. **48** (1976), 569; Chemie-Ing.-Techn. MS 376/76

[37] HENNIG, H.-P., EBERT, I., STEINIKE, U., GEISSLER, H., KRETZSCHMAR, U.: Kristall u. Technik **15** (1980), 353

[38] STEINIKE, U., EBERT, I., GEISSLER, H., HENNIG, H.-P., KRETZSCHMAR, U.: Kristall u. Technik **13** (1978), 597

[39] TALBOT, J. H., BRADLEY, A. A., KEMPIS, E. B.: Brit. J. appl. Phys. **14** (1963), 399
[40] BERGMANN, I., CARTWRIGHT, J., CASSWELL, C.: Brit. J. appl. Phys. **14** (1963), 399
[41] STEINIKE, U., MÜLLER, B., EBERT, I., HENNIG, H.-P., RICHTER-MENDAU, J.: Ber. Akad. Wiss. der DDR, Berlin 3N/1981, p. 150
[42] WEYL, W. A., MARBOE, E. C.: Sprechsaal **107** (1974), 1030
[43] BURTON, Y. J.: Trans. Inst. Chem. Engrs. **44** (1966), 37
[44] STEINIKE, U., EBERT, I., JEDAMZIK, J., KRETZSCHMAR, U., HENNIG, H.-P.: Kristall u. Technik **13** (1978), 1229
[45] KING, E. J., NAGELSCHMIDT, G.: "Proc. Pneumoconiosis Conf.", Johannisburg 1959; ref. [17], p. 9
[46] VAN LIER, J. A., DE BRUYN, P. L., OVERBECK, J. T. G.: J. Phys. Chem. **64** (1960), 1675
[47] GORDON, R. L., HARRIS, G. W.: Nature **175** (1955), 1135
[48] LE BOUFFANT, L.: Bull. Soc. franc. ceramique **71** (1966), 65
[49] DEMPSTER, P. B., RITCHIE, P. D.: Nature **169** (1952), 538; J. appl. Chem. **3** (1953), 182
[50] HOFMANN, U., ROTHE, A.: Z. anorg. allg. Chem. **357** (1968), 196
[51] CLELLAND, D. W., RITCHIE, P. D.: J. appl. Chem. **2** (1952), 42
[52] STEINIKE, U., MÜLLER, E., RICHTER-MENDAU, J., HENNIG, H.-P.: Kristall u. Technik **14** (1979), K 37
[53] EDELMAN, L. I., CHODAKOV, G. S.: Dokl. AN SSSR **202** (1972), 391
[54] RADCIG, V. A., BYSTRIKOV, A. V.: Kinetika i Kataliz **19** (1978), 713
[55] ADAMTSCHIK, K. A.: Silikattechnik **3** (1952), 557
[56] RUŽEK, J.: Silikáty, Praha **14** (1970), 263
[57] RADCIG, V. A.: Kinetika i Kataliz **20** (1979), 448, 456
[58] SILIN, A. R., SKUJA, L. N., ŠENDRIK, A. V.: fizika i chimija stekla **4** (1978), 405
[59] RUŽEK, J.: Silikáty, Praha **18** (1974), 295
[60] PAĖ, A. Ja., PAĖ, P. I., REALO, Ė. Ch., UJBO, L. Ja.: Dokl. AN SSSR **200** (1971), 1066
[61] HENDERSON, J. H., SYERS, J. K., JACKSON, M. L.: Isr. J. Chem. 8 (1970), 357
[62] BECKWITH, R. S., REEVE, R.: Geochimica et Cosmochimica Acta **33** (1969), 745
[63] BUTENUTH, G., WOLFF, S., WEIHER, J., REKER, G.: Ber. dtsch. keram. Ges. **51** (1974), 33
[64] BERGMANN, I., PATERSON, M. S.: J. appl. Chem. **11** (1961), 369
[65] EBERT, I., HENNIG, H.-P.: Z. phys. Chem. **255** (1974), 812
[66] RADCIG, V. A., CHALIF, V. A.: Kinetika i kataliz **20** (1979), 705
[67] SCHÖNHERR, S., SCHRADER, R., STEINERT, K.-H.: Z. anorg. allg. Chem. **432** (1977), 215
[68] BARKER, C., TORKELSON, B. E.: Geochimica et Cosmochimica Acta **39** (1975), 212
[69] TILLER, H.-J., APFEL, C., RUDAKOFF, G.: Wiss. Zschr. Friedrich-Schiller-Univ. Jena, Math.-Nat. R. **27** (1978), 621
[70] STRELECKIJ, A. N., BUTJAGIN, P. Ju.: Dokl. AN SSSR **225** (1975), 1118
[71] HENNIG, H.-P., EBERT, I., GEISSLER, H., STEINIKE, U., KRETZSCHMAR, U.: ref. [41], p. 114
[72] VITKO, Jr., J.: J. appl. Phys. **49** (1978), 5530
[73] SHENDRIK, A. V., YUDIN, D. M.: phys. stat. sol. (b) **85** (1978), 343
[74] CHERENDA, N. G., SHENDRIK, A. V., YUDIN, D. M.: phys. stat. sol. (b) **69** (1975), 687
[75] GEISSLER, H., et al.: Crystal Res. Technol. **17** (1982), 1259
[76] MARX, G., FINK, P.: Wiss. Ztschr. Friedrich-Schiller-Univ. Jena, Math.-Nat. R. **26** (1977), 597
[77] MORROW, B. A., CODY, I. A.: J. Phys. Chem. **80** (1976), 1995
[78] BRD-Patent 2 204 707
[79] YASUE, T., ARAI, Y.: Sekkooto sekkai **139** (1975), 7
[80] TROMBE, J. C., MONTEL, G.: "8. Intern. Symp. on the Reactivity of Solids", Gothenburg 1976
[81] NAREY-SCABO, S.: Z. Krist. **75** (1930), 387
8[2] MEHMEL, M.: Z. Krist. **75** (1930), 323

[83] ELLIOT, J. C., YOUNG, R. A.: Nature **204** (1967), 904
[84] KLEBER, W.: „Einführung in die Kristallographie", Berlin 1961
[85] GORDEEVA, G. I., KOLOSOV, A. S., ČAIKINA, M. V., BOLDYREV, V. V.: chapt. 1., [11], p. 145
[86] MONTEL, G.: Bull. Soc. Chim. France (1968), 1693
[87] HEINICKE, G., PAUDERT, R., HARENZ, G.: chapt. 1., [9], p. 89
[88] LASIEWICZ, K.: Przemysl Chemiczny **47** (1968), 15
[89] Chapt. 3., [44], p. 119
[90] ARAI, Y., YASUE, T., HONDA, T.: Nippon Kagaku Zaishi (1976), 591
[91] PÖTHIG, R., DÜNKEL, L., PAUDERT, R., HARENZ, H., GRIMMER, A. K.: Kristall u. Technik **13** (1978), 879
[92] BOLDYREV, V. V., KOLOSOV, A. S., ČAIKINA, M. V., AVVAKUMOV, E. G.: Izv. sib. otd. AN SSSR ser. chim. nauk **10** (1978), 52
[93] ČAIKINA, M. V., GORDEEVA, G. I., KOLOSOV, A. S., BOLDYREV V. V.: chapt. 1., [11], p. 145
[94] PAUDERT, R.: unpublished
[95] HEINICKE, G.: chapt. 1., [11], p. 8
[96] BOLDYREV, V. V., KOLOSOV, A. S., ČAIKINA, M. V., AVVAKUMOV, E. G.: Dokl. AN SSSR **223** (1977), 892
[97] PÖTHIG, R.: Dissertation, Berlin 1979
[98] PAUDERT, R., HARENZ, H., HEINICKE, G., BORNMANN, H., MÜLLER, S., LINKE, E.: chapt. 1., [11], p. 142
[99] Chapt. 4., (116], p. 71
[100] HEINICKE, G., PAUDERT, R.: Technik **26** (1971), 643, 709
[101] ECKELL, J.: Z. Elektrochem. **39** (1933), 433
[102] RIENÄCKER, G.: Z. Elektrochem. **46** (1940), 369
[103] SCHEIBE, H.: Dissertation, Dresden 1957
[104] UHARA, I., YANAGIMOTO, S., TANI, K., ADACHI, G., TERATANI, S.: J. Phys. Chem. **66** (1962), 2691
[105] SHAW, M. C.: J. appl. Mechan. **15** (1948), 37; Lit. 5.147
[106] ERTL, G.: Z. phys. Chem. N.F. **50** (1966), 46
[107] ERTL, G., GIOVANELLI, T.: Z. Elektrochem. angew. physik. Chem. **72** (1968), 74
[108] ONO, T., OGINO, Y.: Catalyst Tokio 8 (1966), 284
[109] NAKAJIMA, S., OGINO, Y.: Catalyst Tokio 8 (1966), 287
[110] HEINICKE, G.: chapt. 1., [1], p. 132
[111] THIESSEN, P. A., HEINICKE, G., MEYER, K.: „Mitteilungsblatt d. Chem. Gesellsch. in der DDR, Sonderheft Physikalische Chemie von Grenzflächenvorgängen", **16** (1959), 56
[112] Chapt. 3., [1], p. 134
[113] HEINICKE, G., MEYER, K., SENZKY, U.: Z. anorg. allg. Chem. **312** (1961), 180
[114] THIESSEN, P. A., HEINICKE, G., BOCK, N.: Z. Chem. **14** (1974), 76
[115] THIESSEN, P. A., HEINICKE, G., MEYER, K.: DDR-WP 27596
[116] SCHRADER, R., NOBST, P., TETZNER, G., PETZOLD, D.: Z. anorg. allg. Chem. **365** (1969), 255
[117] LIHL, F., ZEMSCH, P.: Berichte d. Bunsenges. phys. Chem. **56** (1952), 979
[118] KEATING, K. B., ROZNER, A. G., YOUNGBLOOD, J. L.: J. catalysis **4** (1965), 608
[119] SCHRADER, R., GRUND, H., TETZNER, G.: Z. Chem. **3** (1963), 356
[120] GRUND, H.: Neue Hütte **10** (1965), 442
[121] SCHRADER, R., STÄDTER, W.: Acta Chimica Academiae Scienticum Hungaricae Tomus **55** (1968), 39
[122] STÄDTER, W.: Dissertation, Freiberg 1970
[123] SCHRADER, R., STÄDTER, W., OETTEL, H.: Chem. Techn. **23** (1971), 363
[124] KIPERMAN, S. L., BALANDIN, A. A., DAVYDOVA, I. R.: Dokl. AN SSSR ser. chim. 113 (1957), 335, 1482
[125] CARRION, J., CRIADO, J. M., HERRERA, E. J.: ref. [80], p. 211

[126] SENZKY, U.: Ingenieurarbeit, Berlin 1960
[127] SCHRADER, R., TETZNER, G., GRUND, H.: Z. anorg. allg. Chem. **342** (1966), 212
[128] SCHRADER, R., TETZNER, G., GRUND, H.: Z. anorg. allg. Chem. **343** (1966), 308
[129] KISHIMOTO, S.: J. Phys. Chem. **66** (1962), 2694
[130] BADDOUR, R. F., SELVIDGE, Ch. W.: J. Phys. Chem. **71** (1967), 2536
[131] SCHRADER, R., HENNEK, H.: J. prakt. Chem. **14** (1961), 234
[132] SCHRADER, R., KROLL, Z., SEIFERT, S.: Z. anorg. allg. Chem. **336** (1965), 11
[133] SCHRADER, R., JAKOB, G.: Chem. Techn. **19** (1966), 414
[134] SCHRADER, R., VOGELSBERGER, W.: Z. anorg. allg. Chem. **368** (1969), 187
[135] PAUDERT, R.: Chem. Techn. **17** (1965), 449
[136] SWINARSKI, A., LAZIEWICZ, K., JERZY, K.: Chem. Stosowana Ser. A **10** (1966), 271
[137] TAKAHASHI, H., TSUTSUMI, K.: Kogyo Kagaku Zasshi **71** (1968), 1345
[138] TAKAHASHI, H., TSUTSUMI, K.: Dechema-Monogr. **57** (1967), 475
[139] SADAHIRO, Y.: J. Soc. Materials Sci. Jap. **21** (1972), 520
[140] SCHRADER, R., THIEM, E.: Z. phys. Chem. **234** (1968), 41
[141] SCHRADER, R., DEREN, J., FRITSCHE, B., ZIOKOWSKI, J.: Z. anorg. allg. Chem. **379** (1970), 25
[142] SCHRADER, R., FRITSCHE, B.: Z. anorg. allg. Chem. **379** (1970), 17
[143] UDA, A., ONO, T., OGINO, Y.: Kogyo Kagaku Zasshi **71** (1968), 1352, 1398
[144] SHIRASAKI, T., OCHIAI, Y., MURANAKA, K.: Kogyo Kagaku Zasshi **71** (1968), 1368
[145] AKAMATSU, H., TAKAHASHI, H.: Advances in Clay Science **5** (1965), 197
[146] NAKANO, N., NISHIMURA, Y., TAKAHASHI, H.: Bull. Jap. Petroleum Inst. **13** (1971), 205
[147] DSHJING-JUNG, H.: Diplomarbeit, Merseburg 1960
[148] HEINICKE, G.: Chem. Techn. **15** (1963), 197
[149] HEINICKE, G., HARENZ, H., SIGRIST, K.: chapt. 4., [43], p. 180
[150] HEINICKE, G., BOCK, N., HARENZ, H.: Z. anorg. allg. Chem. **372** (1970), 162
[151] USP 2965630
[152] USP 2927915
[153] USP 2968652
[154] USP 2878373
[155] SHIRASAKI, T., OCHIAI, Y., MURANAKA, K.: Kogyo Kagaku Zasshi **71** (1968), 1363
[156] BLAKELEY, D. W., SOMORJAI, G. A.: J. catalysis **42** (1976), 181
[157] THIESSEN, P. A., HEINICKE, G., MEYER, K.: „III. Intern. Kongr. f. grenzflächenaktive Stoffe", II., p. 514, Köln 1961
[158] THIESSEN, P. A., HEINICKE, G., MEYER, K.: „Mitteilungsblatt d. Chem. Gesellsch. in der DDR, Sonderheft, Physikalische Chemie von Grenzflächenvorgängen", (1959), p. 57
[159] WAENTIG, P.: Textile Forsch. **3** (1921), 154
[160] STAUDINGER, H.: Ber. dtsch. chem. Ges. **63** (1930), 921
[161] STAUDINGER, H., HEURER, W.: Ber. dtsch. chem. Ges. **67** (1934), 159
[162] STAUDINGER, H., DREHER, E.: Ber. dtsch. chem. Ges. **69** (1936), 1096
[163] HESS, K., GRAMBERG, W.: Kolloid-Z. **97** (1941), 87
[164] HESS, K., GUNDERMANN, J.: Z. phys. Chem. **49** (1942), 64
[165] HESS, K., STEURER, E., FROMM, H.: Kolloid-Z. **98** (1942), 290
[166] GROHN, H.: Habilitationsschrift, Potsdam 1955
[167] GROHN, H., BISCHOF, K., LÖSCHE, M., MÖCKEL, K.: Plaste u. Kautschuk 8 (1961), 593
[168] AUGUSTAT, S.: Die Stärke **14** (1962), 39
[169] MATEVOZJAN, R. O., POSTOVSKIJ, I. Ja., ČIRKOV, A. R.: Ž. ob. chim. **29** (1959), 859
[170] HOLMSTRÖM. A.: Polymer. Enging. Sci. **17** (1977), 728
[171] SIMIONESCU, C., OPREA, C. V., NEGULIANŬ, C., TOPA, M.: Plaste u. Kautschuk **24** (1977), 604, 689
[172] KLIMENKO, V. Ju., DREMIN, A. N., BABARĖ, L. V.: „5. Vsesojuznyj simpozium po Mechanoemissii i Mechanochimii tverdych tel", Tezisy, p. 76, Tallin 1975
[173] BABARĖ, L. V., DREMIN, A. N., KLIMENKO, V. Ju.: ref.[172] p. 74

[174] Simionescu, C., Oprea, C. V.: Cell. chem. Technol. 4 (1969), 361
[175] Protasov, V. G., Baramboim, N. K.: Plaste u. Kautschuk 19 (1972), 89
[176] Ref. 5., [13], p. 82
[177] Grohn, H., Bischof, K., Heusinger, H.: Plaste u. Kautschuk 9 (1962), 182
[178] Schönert, K.: „4. Europäisches Symposium Zerkleinern", Nürnberg 1975
[179] Steurer, E., Hess, K.: Z. phys. Chem. 193 (1944), 248
[180] Ref. [165], p. 148
[181] Barambojm, N. K.: Dokl. AN SSSR 114 (1957), 568
[182] Protasov, V. G., Baramboim, N. K.: Plaste u. Kautschuk 23 (1976), 187
[183] Morosov, V. I., Starkman, V. P., Rylov, Ju. Ju.: Plastmassy (1962), 56
[184] Protasov, W. G., Baramboim, N. K.: Plaste u. Kautschuk 23 (1976), 267
[185] Patrikeev, G.: Dokl. AN SSSR 146 (1962), 402
[186] Simionescu, C., Oprea, C. V.: Plaste u. Kautschuk 18 (1971), 484
[187] Jost, H.: Schmierungstechnik 5 (1974), 6
[188] Jost, H., Richter-Mendau, J.: Plaste u. Kautschuk 18 (1971), 436, 512
[189] Pike, M., Watson, W. F.: J. Polymer Sci. 9 (1952), 229
[190] Kargin, V. A., Plate, N. A.: Vysokomol. soed. 1 (1959), 330
[191] Berlin, A. A.: Dokl. AN SSSR 110 (1956), 401
[192] Radcig, V. A., Pudov, V. S., Butjagin, P. Ju.: Vysokomol. soed. 98 (1967), 414
[193] Barambojm, N. K., Aleksandrov, V. I.: Vysokomol. soed. 14 (1972), 1578
[194] Baramboim, N. K., Protasov, V. G.: Technik 30 (1975), 76
[195] Oprea, C. V., Simionescu, C.: Plaste u. Kautschuk 19 (1972), 897
[196] Protasov, V. T., Barambojm, N. K.: Vysokomol. soed. 14 (1972), 1116
[197] Romanov, A.: Plaste u. Kautschuk 12 (1965), 521
[198] Machmudbekova, N. L., Barambojm, N. K.: Vysokomol. soed. 16 (1974), 1296
[199] Barambojm, N. K., Machmudbekova, N. L.: Vysokomol. soed. 13 B (1971), 428
[200] Simakov, Ju. S.: ŽPK 47 (1974), 2137
[201] Barambojm, N. K., Simakov, Ju. S.: Vysokomol. soed. 8 (1966), 235
[202] Simakov, Ju. S., Barambojm, N. K.: Mitteilungsblatt d. Chem. Gesellsch. in der DDR 21 (1974), 268
[203] Grohn, H., Bischof, K., Heusinger, H.: Plaste u. Kautschuk 9 (1962), 222
[204] Barambojm, N. K., Machmudbekova, N. L.: Vysokomol. soed. 14 (1972), 418
[205] Barambojm, N. K., Machmudbekova, N. L.: Vysokomol. soed. 13 B (1971), 428
[206] Simakov, Ju. S., Barambojm, N. K.: Naučnye trudy MTILP 37 (1972), 105
[207] Barambojm, N. K., Širjaeva, L. L.: Vysokomol. soed. 13 B (1971), 357
[208] Peters, K., Pajakoff, S.: Microchim. Acta (1964), 429
[209] Chapt. 4., [2], p. 89
[210] Heinicke, G.: „Tribochemie": in „Methoden des anorganisch-chemischen Experimentierens", p. 259, Berlin 1976
[211] Köster, A.: DBP 1067784, DBP 1045985
[212] Bock, N.: Diplomarbeit, Berlin 1963
[213] Pourbaix, M.: „Atlas d'Equilibres Electrochimiques á 25 °C." Paris 1973
[214] Walker, W. H.: Trans. Am. Electrochem. Soc. 11 (1907), 153
[215] Endo, H., Kanazawa, S.: Sci. Rpts. Tohoku Imp. Univ. 20 (1951), 124
[216] Giulotto: Nuovo cimento 13 (1936), 220
[217] Simnad, H. T., Evans, U.: Iron-Steel-Inst. London 156 (1947), 531
[218] Simnad, H. T., Evans, U.: Trans. Faraday Soc. 97 (1950), 31
[219] Simnad, H. T.: J. Electrochem. Soc. 97 (1950), 31
[220] Fryxell, R. F., Nachtrieb, R. H.: J. Electrochem. Soc. 99 (1952), 495
[221] Nobe, K., Miniato, O. K., Seger, W. F.: Trans. Met. Soc. AIME 212 (1958), 884
[222] Ling Yang, Horne, U. T., Pound, G. M.: Physical Metallurgy of Stress Corrosion Fracture, London 1959

[223] NIKITIIV, L. V.: Ž. ob. chim. **9** (1939), 794; **11** (1941), 146
[224] ROSEANU: Wear **10** (1967), 127
[225] THIESSEN, K. P., SIEGLING, W.: Z. Chem. **18** (1978), 79
[226] HEIDEMEYER, I.: „II. Europäischer Tribologie-Kongreß Eurotrib 77", Düsseldorf 1977
[227] GAUTEM, B., IKA, B.: Proc. Indian Acad. Sci., **18** (1943), 350
[228] GERISCHER, H.: Werkstoffe u. Korrosion **8** (1957), 314
[229] KOCH, R.: Metalloberfläche (1954), 37, 74
[230] WINDFELDT, A.: Electrochem. Acta **9** (1964), 1139, 1295
[231] NOBE, K., BAUM, E., SEYER, W. F.: J. Electrochem. Soc. **108** (1961), 97
[232] NOBE, K., SEYER, W. F.: J. appl. Phys. **29** (1958), 1632
[233] HARUYAMA, S., ASUWA, S., NAGASAKI, K.: Denki Kagaku Repts. **39** (1971), 564
[234] VOLČKOVA, L. M., i sotr.: ŽFCH **39** (1971), 546
[235] KARPENKO, G. V., i sotr.: Ėlektrochimija **7** (1971), 220
[236] GUTMAN, Ė. M.: Mechanochimija metallov i zaščita ot korrozii", Moskva 1975
[237] LORENZ, W. J., EICHKORN, G.: Berichte d. Bunsenges. phys. Chem. **70** (1966), 856
[238] PAUL, M., WIELAND, M.: Electrochim. Acta **14** (1969), 10
[239] LEWIS, D., NORTHWOOD, D. O., PEARCE, C. E.: Corr. Sci. **9** (1969), 779
[240] GUTMAN, Ė. M.: fiz.-chim. Mech. Mat. **2** (1966), 188; **1** (1965), 339; **3** (1967), 261; **4** (1968), 87
[241] DE GROOT, S. R.: „Thermodynamik irreversibler Prozesse", Mannheim 1960
[242] THIESSEN, K. P.: Z. Chem. **16** (1976), 9
[243] ZIMAN, J. M.: "Principles of the Theory of Solids", Cambridge 1965
[244] LEWIS, D., NORTHWOOD, D. O.: J. appl. Phys. **39** (1968), 4466
[245] CLARKE, M.: Corr. Sci.: **10** (1970), 671
[246] LEWIS, D., PEARCE, C. E.: Electrochim. Acta **16** (1971), 747
[247] VERMILYEA, D. A.: J. Chem. Phys. **27** (1957), 814
[248] GERSTNER, H., RICKERT, H.: Z. f. Metallkunde **46** (1955), 681
[249] THIESSEN, K. P.: Z. Chem. **14** (1974), 135
[250] PINKOWSKI, A., THIESSEN, K. P.: Nukleonika **21** (1976), 161
[251] THIESSEN, K. P.: Mber. Dt. Akad. Wiss. **10** (1968), 200
[252] HÜBNER, W., THIESSEN, K. P.: Wear **55** (1979), 305
[253] KABANOV, B. I.: Usp. Chim. **34** (1965), 1813
[254] KABANOV, B. I., KISELEVA, I. G., ASTACHOV, I. I.: Ėlektrochimija 8 (1972), 955
[255] DETTNER, H. W., ELZE, I.: „Handbuch der Galvanotechnik", München 1966
[256] DIRKSE, T. P., VRIELAND, E. G.: J. Electrochem. Soc. **106** (1959), 997
[257] DE GEISO, R. C., ROGERS, L. B.: J. Electrochem. Soc. **106** (1959), 433
[258] SCHULTZE, J. W.: Berichte, d. Bunsenges. phys. Chem. **78** (1974), 693
[259] KOLB, D. M.: Advances in Electrochemistry and Electrochemical Engineering **11** (1978), 125
[260] KOLB, D. M., PRZASNYSKI, M., GERISCHER, K.: J. electroanal. Chem. **54** (1974), 25
[261] SCHMIDT, E., GYGAX, H. R.: J. electroanal. Chem. **13** (1967), 378
[262] BRAININA, K. Z. et al.: J. electroannal. Chem. **35** (1972), 165
[263] BEWICK, A., JOVIČEVIČ, J., THOMAS, B.: Faraday Symp. of the Chemical Society 1977 No. 12, 24
[264] WALIŠ, L.: private communication
[265] SHEWMON, P. G.: "Diffusion in Solids", New York 1963
[266] MÜLLER, L., SOHR, R.: Z. Chem. **13** (1973), 390
[267] LEITEL, E., PINKOWSKI, A., THIESSEN, K. P.: Z. phys. Chem. **260** (1979), 448
[268] VETTER, K. J.: „Elektrochemische Kinetik", Berlin 1961
[269] LEVIČ, V. G.: „Fiziko-chimičeskaja gidrodinamika", Moskva 1959
[270] LANDSBERG, R., THIELE, R.: Electrochim. Acta **11** (1966), 1243
[271] SCHELLER, F., LANDSBERG, R., WOLF, H.: Electrochim. Acta **15** (1970), 525
[272] SCHELLER, F., MÜLLER, S., LANDSBERG, R., SPITZER, H.-J.: J. electroanal. Chem. **19** (1968), 187

[273] PINI, G. C., DE ANNA, P. L.: Electrochim. Acta **22** (1977), 1423
[274] HEITZ, E.: „IV. Kongreß der Europäischen Gemeinschaft für Kohle und Stahl", Frankfurt/Main 1968
[275] POSTLETWAITE, J., HOLDNER, O. N.: Can. J. chem. Engng. **54** (1976), 255
[276] SMITH, J. W., KING, D. H.: Can. J. chem. Engng. **53** (1975), 41
[277] HEITZ, E., KREYSA, G., DEIWIKS, W.: 27th Meeting of International Society of Electrochemistry, Zürich 1976
[278] PINKOWSKI, A., THIESSEN, K. P.: Ref. 1., [9], III, p. 3
[279] KOSSEL, W.: Nachr. Ges. Wiss. Göttingen, math.-physik. Kl. (1927), 135
[280] STRANSKI, J. N.: Z. phys. Chem. **136** (1928), 259
[281] KNACKE, O., STRANSKI, J. N.: Erg. exakt. Naturwiss. **26** (1952), 383
[282] TAN, S., NOBE, K.: Can. J. Chem. **41** (1953), 495
[283] EVANS, U. R., SIMNAD, M. E.: Proc. Reg. Soc. London **A 188** (1947), 372
[284] NOBE, K., TAN, S.: Corrosion **18** (1962), 391 t
[285] HOAR, T. P., HINES, J. G.: J. Iron-Steel-Inst. **182** (1956), 124
[286] HOAR, T. P., WEST, J. M.: Nature **181** (1958), 835
[287] HOAR, T. P., WEST, J. M.: Proc. Roy. Soc. London **A 286** (1962), 304
[288] HOAR, T. P., SCULLY, J. C.: J. Electrochem. Soc. **111** (1964), 348
[289] HOAR, T. P.: Corrosion **19** (1963), 311
[290] VAN ROYEN, D.: Corrosion **16** (1960), 421
[291] VAN ROYEN, D., BARNART, S.: J. Electrochem. Soc. **108** (1961), 222
[292] LOGAN, H. L.: J. Res. Natl. Bur. Std. **61** (1958), 503
[293] LOGAN, H. L.: J. Res. Natl. Bur. Std. **65 C** (1961), 165
[294] FUNK, A. G., GIDDINGS, J. C., CHRISTENSEN, C. J., EYRING, H.: J. phys. Chem. **61** (1957), 1179
[295] GIDDINGS, J. C., FUNK, A. G., CHRISTENSEN, C. J., EYRING, H.: J. Electrochem. Soc. **106** (1959), 9
[296] RAICHEFF, R. G., DAMJANOVIČ, A., BOCKRIS, J. O'M.: J. Chem. Phys. **47** (1967), 2198
[297] RAICHEFF, R. G.: Rev. Sci. Instr. **38** (1967), 919
[298] DESPIČ, A. R., RAICHEFF, R. G., BOCKRIS, J. O'M.: J. Chem. Phys. **49** (1968), 926
[299] GRUN, N. D., SALTZMANN, G. H.: Corrosion **20** (1964), 27
[300] GUTMANN, Ė. M., STORONSKIJ, V. I., KARPENKO, G. V.: fiz.-chim. Mech. Mat. **4** (1968), 324
[301] RJABČEKNOV, A. V., NIKIFOROVA, V. M.: „Korrozija i zaščita Metallov v mašinostroenii", Moskva 1959
[302] GERASIMOV, V. V., ŠUVALOV, V. A., EMEL'JANCEVA, E. I.: Zaščita metallov **7** (1971), 178
[303] GERASIMOV, V. V., TIMOŠIK, V. A., FOKIN, M. N.: Zaščita metallov **7** (1971), 591
[304] GRAFEN, H.: Corrosion Sci. **7** (9167), 177
[305] MIGAI, L. L., VEDONEEVA, M. N., ĖLSUK, N. P.: Chimičeskoe i neftjanoe mašinostroenie (1971), 22
[306] FRANCE, W.: Anti-corrosion methods and materials **16** (1969), 15
[307] GARZ, I., HAFKE, V.: Corrosion Sci. **11** (1971), 329
[308] FRANKENTHAL, R. P.: Corrosion Sci. 8 (1968), 350
[309] FREIMAN, L. I., KOLOTYRKIN, Ja. M., GIVENTAL, A. Ja.: Zaščita metallov **1** (1965), 286
[310] BEN BACHIR, A., ARIES, L., TRICKE, H.: Mem.-Sci. Rev. Met. **68** (1971), 687
[311] BEN BACHIR, A., ARIES, L., TRICKE, H.: Mem. Sci. Rev. Met. **69** (1972), 35
[312] RAICHEFF, R. G., DESPIČ, A. R.: Dokl. Bulg. AN **23** (1970), 535, 1091
[313] GUTMAN, Ė. M., ŠESTOPALOV, V. E.: Ref. [236], p. 80
[314] BARRETT, C.: "Structure of Metals", New York 1952
[315] TSIEN, L. S., CHOW, Y. S.: Proc. Roy. Soc. London A **163** (1937), 19
[316] DAMJANOVICH, A., SETTY, T. H., BOCKRIS, J. O'M.: J. Electrochem. Soc. **113** (1966), 429
[317] NAGASHIMA, T., TAMAMOTO, T.: Acta Met. **4** (1956), 94
[318] LEITEL, E., CZAJA, P., THIESSEN, K. P.: Z. phys. Chem. **263** (1982), 1178

[319] LANGE, E., BERGER, R.: Z. Elektrochem. **36** (1930), 980
[320] OHSE, R. W.: Z. Elektrochem. **63** (1959), 1063
[321] SCHWABE, K., DIETZ, G.: Z. Elektrochem. **62** (1958), 751
[322] DIETZ, H., WEICHINGER, H. G.: „Thermodynamische Elektrochemie", Heidelberg 1962
[323] NAGEL, K., FRANZ, D.: Naturwiss. **48** (1961), 156
[324] NAGEL, K.: Sitzungsber. Physik med. Soc. Erlangen **72** (1940/41), 127
[325] SCHIEWE, G.: Diplomarbeit, Erlangen 1961
[326] POPP, W.: Dissertation, Erlangen 1962
[327] WEIDINGER, H., LANGE, E.: Z. Elektrochem. **64** (1960), 468, 1165
[328] KLARK, G. B., AKIMOV, G. F.: Dokl. AN SSSR **30** (1941), 9
[329] VÖGTLE, G.: Schmiertechnik **2** (1971), 55
[330] CLARK, G. B., AKIMOV, G. V.: "Proceedings of 2 nel Conference on Corrosion of Metals", Vol. 2, p. 1,43, Moscow 1943
[331] TOMAŠOV, N. D., ALTOVSKIJ, R. M.: „Korrizija metallov i splavov", Moskva 1963
[332] TOMAŠOV, N. D., ČERNOVA, G. P., ALTKOVSKIJ, R. M., BLINČEVSKIJ, G. K.: Zavodskaja Laboratorija **24** (1956), 229
[333] KABANOV, B. N., BARELKO, E. V.: Dokl. AN SSSR **90** (1953), 1059
[334] SCHWABE, K.: J. Phys. Chem. **214** (1960), 6; **205** (1955), 304
[335] KUNZE, K., SCHWABE, K.: Corros. Sci. **4** (1964), 109
[336] SCHWABE, K., Werkstoffe u. Korrosion **15** (1964), 70
[337] WOLF, G.: Z. phys. Chem. **223** (1963), 4
[338] DIETZ, H., GOHR, H.: Z. phys. Chem. **223** (1963), 17
[339] TOMAŠOV, N. D., STRUKOV, N. M., VERŠININA, L. P.: Dokl. AN SSSR **171** (1966), 5
[340] TOMAŠOV, N. D., VERŠININA, L. P.: Ėlektrochim., akta **15** (1970), 501
[341] HARWOOD, I.: Corrosion **6** (1950), 249, 290
[342] TAMMAN, G., NEUBER, F.: Z. anorg. allg. Chem. **207** (1932), 87
[343] FRIEND, J. A. N.: J. Iron-Steel Inst. **117** (1928), 639
[344] VERMILYEA, D. A.: Physics today **29** (1976), 9 Materials Science and Engineering **25** (1976), 29
[345] HEIDEMEYER, J.: Schmiertechnik u. Tribologie **22** (1975), 84
[346] WATERHOUSE, R. B.: Tribology **3** (1970), 158
[347] WATERHOUSE, R. B.: "Fretting Corrosion", Oxford/New York, 1976
[348] TAYLOR, D. E., WATERHOUSE, R. B.: Corr. Sci. **14** (1979), 11
[349] WIEGAND, H., BROSZEIT, E., HIRTH, F. W., SPECKHARDT, H.: Werkstoffe u. Korrosion **23** (1972), 87
[350] BROSZEIT, E., HESS, F. J., WAGNER, E.: Wear **30** (1974), 311
[351] BODEY, A.: Deutsche Kraftfahrtforschung **84** (1954)
[352] FURSUND, K.: Wear **1** (1957), 104
[353] THIESSEN, K. P.: Mber. Dt. Akad. Wiss. **13** (1971), 46
[354] HÜBNER, W., THIESSEN, K. P.: ref. [226]
[355] DEHLINGER, U.: „Theoretische Metallkunde", Berlin 1955
[356] HÜBNER, W., WALIŠ, L.: Isotopenpraxis **11** (1975), 317
[357] TURLEY, D. M.: J. Inst. Met. **99** (1971), 271
[358] VAN DIJCK, I. A. B.: Wear **42** (1977), 109
[359] BALUFFI, R. E., RUOFF, A. L.: phys. stat. sol. (B) **42** (1970), 11
[360] HÄSSNER, A.: Freiberger Forschungsh. B. 168 (1972)
[361] KULEMIN, A. B., MICKEVIČ, A. M.: Dokl. AN SSSR **189** (1969), 518
[362] SMIDODA, K., GOTTSCHALK, W., GLEITER, H.: Acta met. **26** (1978), 1833
[363] DAVE, R.: Machinery **78** (1972), 37
[364] EISNER, S., WISELOM, N. E. jr.: Platiny **58** (1971), 913, 1094, 1183
[365] TERENE, N. S., RAGAUSKAITE, R. A., TAICAE, L. I.: Intensifikacija technologičeskich processov pri osaždenii metallov i splavov", Moskva 1977
[366] MEYER, W.: „Jahrbuch der Oberflächentechnik" **35** (1979), 39

[367] Ohgo, K., Nakayima, K., Awano, T.: Wear **40** (1976), 85
[368] Lochmann, K., Czekalla, B., Köhler, H., Hübner, W., Thiessen, K. P.: Fertigungstechnik und Betrieb **28** (1978), 97
[369] Westwood, A. R. C.: J. Mater. Sci. **9** (1974), 1871
[370] Uhlig, H. H.: J. Electrochem. Soc. **123** (1976), 1699
[371] Duquette, D., Uhlig, H. H.: Trans. Am. Soc. Metals **61** (1968), 449; **62** (1969), 839
[372] Revie, R., Uhlig, W.: Acta Met. **22** (1974), 619
[373] Heinicke, G., Harenz, H.: Technik **23** (1968), 236
[374] Mazza, B., Pedefferri, P., Sinigaglia, D., Della Sala, U., Lazzari, L.: Werkstoffe u. Korrosion **25** (1974), 239
[375] Heusler, K. E.: Z. Elektrochem. **62** (1958), 582
[376] Bockris, J. O'M, Dražič, D., Despič, A. R.: Electrochim. Acta **4** (1961), 325
[377] Kelly, E. J.: Electrochem. Soc. **112** (1965), 124
[378] Naeser, G., Fiedler, A.: Arch. Eisenhüttenwesen **38** (1967), 345
[379] Foroulis, Z. A.: J. Electrochem. Soc. **113** (1966), 532
[380] Foroulis, Z. A., Uhlig, H.: J. Electrochem. Soc. **111** (1964), 522
[381] Straumanis, M., Wang, Y.: Corrosion **12** (1956), 177 t
[382] Zareckij, E.: Ž. prikl. chim. **24** (1951), 24
[383] Wiegand, H., Broszeit, E., Hirth, F. W., Speckhardt, H.: Z. Werkstofftechnik **3** (1972), 2
[384] Bitter, J. G. A.: Wear **5** (1963), 169
[385] Jackson, L. D. A.: C. I. M. Transactions, **70** (1967), 219
[386] Postlethwaite, J., Tinker, E. B., Hawrylak, M. W.: Corrosion-NACE **30** (1974), 285
[387] Postlethwaite, J., Hawrylak, M. W.: Corrosion-NACE **31** (1975), 237
[388] Salamaženko, A. G., Karpenko, G. V.: fiz.-chim. Mech. Mat. **12** (1976), 98
[389] Kobzaruk, A. V., Karpenko G. V.: fiz.-chim. Mech. Mat. **14** (1978), 75
[390] Hollnagel, M., Grau, W., Siegler, K.: Z. phys. Chem. **225** (1964), 209
[391] Hollnagel, M., Siegler, K.: Z. phys. Chem. **242** (1969), 51
[392] Wiegand, H., Piltz, H.-H.: Werkstoffe u. Korrosion **15** (1964), 212
[393] Wiegand, H., Hirth, F. W., Speckhardt, H.: Werkstoffe u. Korrosion **11** (1971), 934
[394] E. D., Nečesa, V.: in „Zaščita ot korrozii gidrotechničeskich sooruženij v rečnych vodach", Moskva 1968
[395] Grivnin, Ju. A., Ėgel, Ju. U.: Ėnergomašinostroenie (1970) No 1
[396] Stenk, E. P., Trifel, M. S., Chanlraova, A. G., Mechmandarov, S. A.: Gidrotechničeskoe stroitel'stvo, (1967) No 2
[397] Wiegand, H., Piltz, H.-H., Schalmeister, R.: Metalloberfläche **16** (1962), 117
[398] Wheeler, W.: "Cavitation in Hydrodynamics", London 1956
[399] Glikman, L. A.: „Korrozionno-mechaničeskaja pročnost metallov," Moskva 1955
[400] Karasjuk, Ju. A., Kočerov, V. I., Benino, V. V., Galaktionova, N. L.: fiziko-chimičeskaja mechanika materialov **12** (1976), 87
[401] Kočerov, V. I. i sotr.: Zaščita metallov **12** (1976), 582
[402] Parkins, R. N.: Brit. Corr. J. **14** (1979), 5
[403] Heinicke, G., Harenz, H.: Technik **24** (1969), 252
[404] Rhodin, E.: "Physical Metallurgy of Stress Corrosion Cracking", New York 1959
[405] Staehle, R. W.: "Fundamental Aspects of Stress Corrosion Cracking", Houston 1969
[406] Kaesche, H.: „Spannungsrißkorrosion", Weinheim 1969
[407] Scully, J. C.: "The theory of Stress Corrosion Cracking", Brussels 1971
[408] „Korrozija pod naprjaženiem i vodorodnoe ochrupčivanie", sbornik dokladov, Dresden 1975
[409] Graf, L., Becker, H.: Z. Metallkunde **62** (1971), 685
[410] Kaesche, H.: Z. Metallkunde **67** (1976), 439
[411] Brown, B. F., Fujii, C. T., Dahlberg, E. P.: J. Electrochem. Soc. **116** (1969), 218
[412] Barth, C. F., Steigerwald, E. A., Troiano, A. R.: Corrosion **25** (1969), 353

[413] BARTH, C. F., TROIANO, A. R.: Corrosion **28** (1972), 259
[414] LUNARSKA, L.: zaščita metallov, **13** (1977), 141
[415] ORIANI, R. A.: Berichte d. Bunsenges. phys. Chem. **76** (1972), 848
[416] ORIANI, R. A., JOSEPHIC, P. H.: Acta Met. **22** (1974), 1065
[417] PARKINS, R. N.: Brit. Corros. J. **7** (1972), 15
[418] ROSENFELD, I. L., i sotr.: Itogi nauki, korrozija i zaščita ot korrozii, **1** (1971), 103
[419] VERMILYEA, D. A., TEDMON, C. S.: J. Electrochem. Soc. **117** (1970), 437
[420] HOAR, T. P., FORD, F. P.: J. Electrochem. Soc. **120** (1973), 1013
[421] HOAR, T. P., JONES, R. W.: Corr. Sci. **13** (1973), 725
[422] AMBROSE, J. R., KRÜGER, J.: J. Electrochem. Soc. **121** (1974), 549
[423] SCULLY, J. C.: Corr. Sci. **15** (1975), 207
[424] ARMSTRONG, R. D., COATES, A. C.: Corr. Sci. **16** (1976), 423
[425] SCHWENK, W.: Werkstoffe u. Korrosion **19** (1968), 741
[426] PRAŽÁK, M., HOLINKA, M.: ref. [408], p. 241
[427] DOIG, P., FLEWITT, P. E. J.: Proc. Roy. Soc. London A **357** (1977), 439
[428] SMITH, S. A., PETERSON, H. H., BROWN, B. F.: Corrosion **9** (1973), 151
[429] SEDRICKS, A. J., GREEN, J. A. S'., NOVAK, D. L.: Corrosion **27** (1971), 198
[430] BROWN, B. F.: ref. [411], p. 220
[431] POURBAIX, M.: J. Less-Common-Metals **28** (1972), 51
[432] ROSENFELD, I. L., MARSHAKOV, I. K.: Corrosion **20** (1964), 115 t
[433] MARICHEV, Y. A., ROSENFELD, J. L.: Corrosion **21** (1976), 423
[434] VERMILYEA, D. A., DIEGLE, R. B.: Corrosion **32** (1976), 26
[435] BAKISH, R., ROBERTSON, W. D.: Acta Met. **3** (1955), 513
[436] BAKISH, R., ROBERTSON, W. D.: Trans. Met. Soc. AIME **206** (1956), 1277
[437] FRANK, W., GRAF, L.: Z. Metallkunde **66** (1975), 555
[438] GRAF, L.: Z. Metallkunde **66** (1975), 749
[439] PARKINS, R. N.: Welding Res. **3** (1949), 95
[440] PARKINS, R. N.: Brit. Weld. Jul. **8** (1961), 24
[441] WENK, R. L.: 5th Symposium on Pipe Line Research, Amer. Gas. Anoe., Cat. No 630174, Houston (1974)
[442] STOCKBRIDGE, C. D., SEWELL, P. B., COHEN, M.: J. Electrochem. Soc. **108** (1961), 928
[443] SPEIDEL, M. O.: Met. Trans. **A 6** (1975), 631
[444] POCHMURSKIJ, V. I.: fiz. chim. Mech. Mat. **11** (1974), 3
[445] "Proc. Conference on Corrosion Fatigue", Houston 1972
[446] "Corrosion Fatigue: Chemistry, Mechanics and Microstructure", Houston 1972
[447] "Mechanisms of Environment Sensitive Cracking of Materials", Guildford 1977
[448] COWLEY, W. E., ROBINSON, F. P. A., KURICH, J. E.: Brit. Corr. J. **3** (1968), 223
[449] DUQUETTE, D. J., GEIL, M.: Met. Trans. **3** (1972), 1899
[450] PELLOUX, R. M., STOLTZ, R. E., MOSKOVITZ, J. A.: Mat. Sci. Engng. **25** (1976), 193
[451] CIFTAN, M., SAIBEL, E.: Solid state comm. **27** (1978), 435
[452] THIESSEN, K. P.: Z. phys. Chem. **260** (1979), 403
[453] THIESSEN, K. P., SIEBER, K.: Z. phys. Chem. **260** (1979), 410
[454] LEES, D. J., HOAR, T. P.: Corr. sci. **20**, (1980), 761
[455] BURSTEIN, G. T., DAVIES, D. H.: Corr. sci. **20**, (1980), 1143
[456] BURSTEIN, G. T., DAVIES, D. H.: J. Electrochem. Soc. **128**, (1981), 33
[457] PURSCHE, G., Wiss. Z. TH Karl-Marx-Stadt **15** (1973), 303
[458] PURSCHE, G., JAHN, G.: Schmierungstechnik **5** (1979), 10, 293

7. Application of tribochemistry in technology

7.1. Introduction

Mechanical energy is one of the energies occurring frequently in technology. Thus substances are milled on a large scale in chemical, metallurgical, preparation, and food industries. As Tab. 7.1. shows on a world scale the quantity of products to be comminuted already exceeds the milliard t limit. About 4 per cent of the total world power consumption is utilized for the comminution of technical products [1]. As has been illustrated in 3.1. this process is mainly conducted by the action of mechanical energy on the solid to be comminuted.

Table 7.1. Average specific energy demand, quantities used, and energy consumption as part of the worldwide power consumption for some selected materials (1). (Average 1970–1974.)

Substances	Energy demand kWh/t	Quantity of products 10^6 t/a	Power consumption ‰
Coal	7	2,940	2.1
Cement	67.5	640	7.7
Iron ores	40	480	3.4
Non-ferrous ore	30	305	1.6
Phosphate ore	40	110	0.8
Potash ore	0.5	100	0.01
Rock salt	7	122	0.15
Inorganic pigments	880	3	0.5
Wood pulp, cellulose	1,500	23.3	6.2
Cereals	26.5	348	1.6
Sugar, cacao, powdered milk	45	5	0.04

Another major portion of mechanical energy is utilized in prime movers and machines. 30···50 per cent of the mechanical energy produced in the world is lost by friction both in these machines and in the transfer of mechanical energy [2].

With increasing rationalization of manufacturing processes machines and plants are subjected to increasingly greater static and whereby the dissipation processes appearing due to the conversion of mechanical energie are also intensified. It can be understood from this development that tribochemical processes are gaining an ever greater importance in technology. There are two main spheres characterizing the effect and utilization of tribochemical processes in technology:

— The first sphere relates to the practical application of tribochemical processes in various branches of industry, such as the chemical, metallurgical, and building

Table 7.2. The technical importance of tribochemical reactions in the individual branches of industry.

Tribochemical process	Branch of industry	Example	Ref.
Synthesis, decomposition reaction, etc.	Chemistry	Formation of carbonyls from CO and metals	Chapt. 7.2.2.1.
Reduction reactions	Metallurgy	Reduction of iron oxides	Chapt. 7.2.2.1.
Tribocatalytical reaction	Chemistry	Regeneration of fatigued catalysts	Chapt. 6.2.; 7.2.1.2.
Degradation of polymers, modification reactions	Organo-chemical industry	Decomposition of starch	Chapt. 6.3.
Dissolution reactions	Preparation industry	Extraction of tin from activated tin ores	Chapt. 7.2.2.2.2.; 7.2.2.2.3.
Dissolution reactions	Agriculture	Dissolution of apatite fertilizers in the soil	Chapt. 7.2.2.2.4.
Grafting	Chemistry	Composites between inorganic solids and organic substances	Chapt. 7.2.2.4.
Hydration reactions	Building material industry	Formation of CSH phases at the hardening of silicate concrete	Chapt. 7.2.3.1.; 7.2.3.2.
Sintering reactions	Fire-resistant and ceramic industry	Production of MgO sintered bricks	Chapt. 7.2.3.4.
Oxidation reactions	Oxygen industry	Tribochemical ignition of fires in oxygen machines and plants	Chapt. 7.3.
Decomposition reactions	Explosion engineering	Mechanically induced ignition of explosions, percussion priming at projectiles	Chapt. 5.3.
Tribodiffusion	Material engineering	Hydrogen embrittlement in iron	Chapt. 5.7.
Dissolution of metals	Machine building and systems engineering	Erosion corrosion at iron	Chapt. 6.6.
Interface reactions	Treatment of materials	Drilling, cutting	Chapt. 6.6.
Interface reactions	Machine building industry	Tribochemical conversion at friction, lubrication, and wear Conversion of MoS_2 with the Fe-bearing	Chapt. 7.4.

industries. Using the specific particularities of tribochemical reactions, new or more economical process steps or processes may be developed (see 7.2.).

— A second, no less important sphere, for technology deals with the inhibition or prevention of chemical processes which lead to the wear of machines and materials in the dissipation of mechanical energy. Chief among these are tribocorrosion (see 6.6. and 7.4.) and tribochemical ignition (see 7.3.).

A survey of important tribochemical processes and their application in the various branches of industry is given in Tab. 7.2. Selected examples are discussed in detail in 7.2.—7.4.

7.2. Tribochemical processes in the material-changing industry

7.2.1. Process engineering aspects in tribochemical processes

7.2.1.1. Triboreactors and activation sets for tribochemical processes

At present, mills are still used as triboreactors and activation sets, respectively, for the tribochemical processes in the material-changing industry as they are used for pulverization in comminution engineering. The most important types of mills used until now are described in 3.1.2.2. Among these vibration and planetary mills have proved to be particularly suitable for the mechanical activation of the solid. However, as Tab. 7.3. shows the throughput rate of these machines is relatively small so that technical mass products like cement are only activated for special purposes in the mills mentioned. On the other hand the vibration mill has proved to be successful for the activation of low-tonnage products and has widespread application. Today ball mills and disintegrators have already been developed for very large throughputs. However, the degree of activation realizable in these machines is behind the values possible in vibration and planetary mills. This means they cannot be used for a series of tribochemical processes. Therefore a widespread introduction of tribochemical procedures, particularly in the processing of mass products, is closely connected with the still nascent development of effective large-scale activation sets.

Table 7.3. Throughput of different technical milling sets.

Activation set	Throughput, t/h	Ref.
Ball mill	> 150	[3]
Disintegrator	50	[4]
Jet mill	≈ 5	[5]
Vibration mill	< 5	[3, 6]
Planetary mill	≈ 0.1	[7]

As has been shown in 3.1.2.3. and 6.1.2., the stress mechanism dominating in the individual activation sets is different and effects the type and distribution of the lattice defects in solids. The selection of an activation set suitable for a certain technological process also depends on the practical properties the product to be activated is to have. Therefore an adaptation to the concrete objective of the process is required in each case.

Moreover it has to be taken into consideration that the individual activation sets possess different efficiencies. Tab. 4.8. shows that this efficiency may vary within wide limits during the manufacturing of a product with the same activity. Also differences in the design of the set often very strongly effect the activation properties of the mechanically stressed product [8—10] for a certain type of sets. Therefore the adaption of a tribochemical reaction investigated under laboratory conditions to an industrial scale further requires a technological optimization in terms of a maximum degree of activation and a minimum energy input.

7.2.1.2. *Particularities of tribochemical processes in the material-changing industry*

The mechanical stress of solids for an industrial-scale activation is one of the most energy-consuming steps of a procedure. If nevertheless a tribochemical process is to prove to be superior in comparison to a classical reference process then the activation expenditure negatively affecting the energy consumption of a procedure must be exceeded by the simultaneously occurring technological advantages of the tribochemical procedure. The latter are highly varied and will be discussed in the following summary:

— The reactivity of solids is often substantially increased by mechanical treatment. The reaction velocity can sometimes increase by several orders of magnitude due to the effect of mechanical energy and in this way lead to an improved space-time yield. For instance, the formation of a certain quantity of nickel carbonyl may be shortened from 90 to 2 hours by a tribochemical reaction as compared to the MOND-nickel process [11].

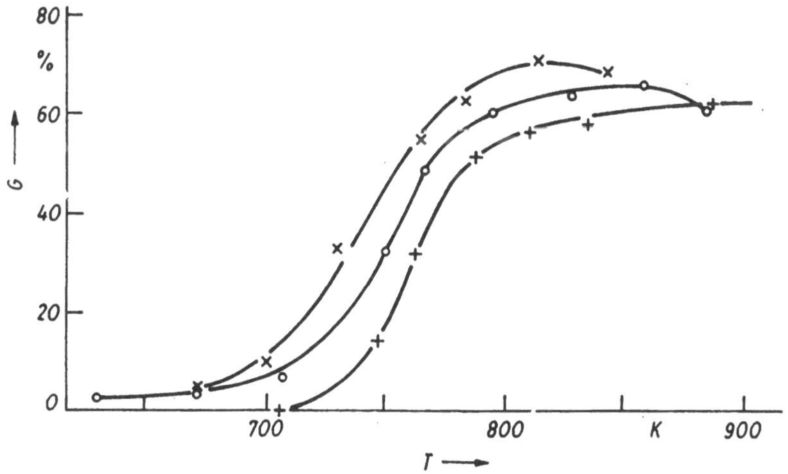

Fig. 7.1. SO_2 oxidation at V_2O_5 catalyst milled in a jet mill in dependence on temperature T [16]. G — degree of conversion, + — unmilled, o — milled for two times, x — milled for five times.

— Numerous tribochemical reactions such as the hydrogenation of oxides [12], the oxidation of metals ([13], Chapt. 4., [57]), the halogenation of elements [14, 15], etc. run, with the same velocity, at greatly reduced temperatures in comparison to the thermal reference procedure, so that a saving in heat energy occurs. If the reaction temperature is reduced, in addition to accelerated kinetics, a higher final yield may also be obtained on account of the equilibrium position displacement. As an example in Fig. 7.1. the reaction of SO_2 to SO_3 is shown by a V_2O_5 catalyst activated in a jet mill.

— Certain by-products which cannot be used up in classical procedures can be made into high-grade industrial products again after a mechanical activation. As an example may be mentioned, the mechanical recovery of fatigued hydrogenation catalysts, which are no longer usable due to a strongly reduced surface and a plugging of the pore necks by deposition products after a fivefold chemical regeneration and are dumped. As Fig. 7.2. shows, after an activation of 20 min. duration, such products will obtain the surface of a twice chemically regenerated catalyst and will thereby recover more than half of the lifetime of a new contact [17, 18].

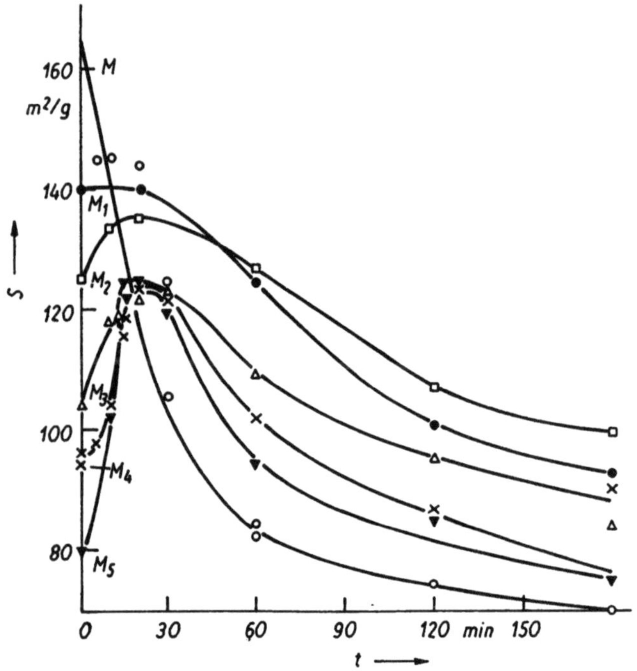

Fig. 7.2. Specific surface area S of a low-temperature hydrogenating catalyst in dependence on the treatment time t [17].

M — fresh catalyst, M_x — catalyst chemically regenerated by reduction for x times.

— In certain cases those raw materials can also be used in the mechanical activation which cannot or only with great difficalty, be made into the desired final products through other chemical procedures. Active Fe_2O_3 products are for instance used as a gas purifying medium in coking plants and gas works, which a.o. are obtained as filtration residue in the aluminium industry. With appropriate mechanical activation other cheaper iron oxides e.g. roasted pyrites, may also be used for this purpose. By treating the little active raw products, the H_2S absorptivity may be increased so much, (see Fig. 5.56.b), that they attain at least the quality of the active products [19].

As will be shown in 7.2.2.2.4., apatites may be made into fertilizers by mechanical activation, which in terms of efficiency correspond to superphosphate. Activation proves to be economical especially for low-grade apatites, since a particularly high portion of sulphuric acid is required for their chemical decomposition.

— Through treatment, the product obtains certain properties which only then facilitate or ameliorate further processing to the final product. Thus, for instance, roasted pyrites can only be pelletized with great difficulty. Mechanical activation not only effects a major increase in the activity of the product but also facilitate an easy pelletizibility of the powdery product and a high compactness of the pellets on account of active surface produced [19].

If mechanically activated products are used in the chemical decomposition of apatites then the acid reacts so fast with the rock phosphate that the established maturing time of several days and consequently the storage of the products in maturing cellars may be dispensed with. In this way a continuous production sequence comes into existence (see 7.2.2.2.4.).

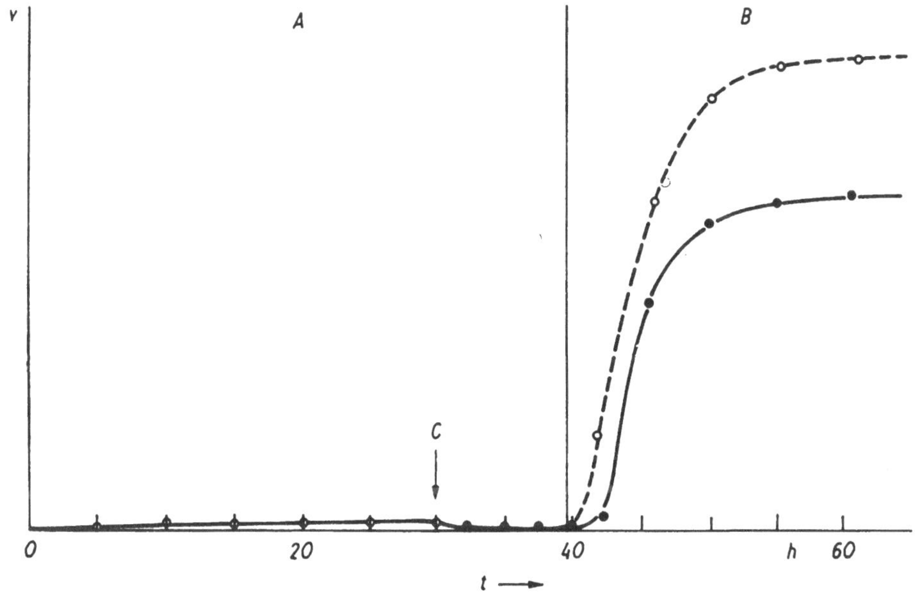

Fig. 7.3. The effect of oxygen on the course of the nickel carbonyl formation without (A) and with (B) treatment of nickel powder [20].

v – reaction velocity, t – time of reaction, ○——○, ○ – – ○ – CO, ●——● – CO + 3% O_2, C – addition of 3% O_2.

— Certain syntheses require very pure gases whose industrial manufacturing is often very expensive. For instance, highly purified carbon oxide is needed for the synthesis of nickel carbonyl, since traces of oxygen already stop the reaction. However, Fig. 7.3. shows that the velocity of the tribochemical reaction is only slightly decreased by an addition of 3 per cent oxygen so that the extreme purification of the synthesis gases may be dispensed with.

— A series of chemical reactions only run with an economically acceptable velocity at higher temperatures. Often the increase in temperature is combined with higher pressures. As an example the synthesese of metal carbonyls may be mentioned which run according to the high-pressure process. These synthesese are performed at $2 \cdot 10^7$ Pa and 450···500 K in most cases, and discontinuously because of the refilling of the high-pressure reactor with solid products. There are two advantages of the tribochemical carbonyl synthesis compared to the classical discontinuous high-pressure reaction: carrying out of the reaction at normal pressure and continuous passage of the reaction material through the triboreactor as precondition for an automatically controllable flow-line process (see 7.2.2.1. and 6.5.). As is seen in 7.2.3.1., using mechanically activated starting materials silicate concrete may also be manufactured at normal pressure and without discontinuous autoclave curing.

— In a series of procedures, with the introduction of mechanical activation, two or more technical processes may be combined into one process. For instance, in the manufacturing of filled polymers, the comminution and activation of the filler, the mixing with polymers, and the homogeneous distribution of an adhesive agent are required. The technology may be designed in such a way that the processes mentioned run simultaneously in the activation step [21]. In the manufacture of gas silicalcite, the activation of lime and quartz sand, the addition of water and aluminium powder as a blowwing agent as well as the processing of the four components to a homogeneous mixture, are also performed in the disintegrator, and this way several technological processes are combined in one (Chapt. 3., [42]).

— Through activation, a mixture consisting of similar elements or components, can change its properties in different ways, which then facilitates or ameliorates the subsequent separation of the constituents. In this way for instance, titanium may be extracted from activated titaniferous magnetite ores with subsequent thermal treatment and dissolution with hydrochloric acid, whereas iron remains behind in a non-soluble form [22], and pyrite may be separated from haematite in similar fashion [23].

— In numerous cases the quality parameters of the final products are changed due to the mechanical activation of the starting or intermediate products, whereby materials having a higher quality or new properties can come into existence. As an example reference may be made to the tribochemical polyamide modification by polyacrylic acid which leads to materials with an extremely high swelling capacity ($\alpha = 3.7 \cdot 10^4$) as given in 6.3.5.2. As shown in 7.2.3.1. and 7.2.3.2. the strength of cement and silicate concrete depends decisively on the respective degree of activation of cement and the silicate concrete raw material. Thus the strength of silicate concrete may be increased by the corresponding activation of the starting materials to values above $3 \cdot 10^8$ Pa [24].

The above particularities of a tribochemical procedure are material-specific to a high degree and are of varying importance to the technological design of material-changing processes. Therefore in every case the advantages and disadvantages of a tribochemical reaction course have to be compared and economically evaluated. In a series of cases tribochemical procedures have become generally accepted as a favourable variant in technology.

Some technologically important material-changing processes based on the tribochemical principle will be discussed in the following section.

7.2.2. *The importance of tribochemistry in the chemical and metallurgical industries*

7.2.2.1. *Tribochemical processes in the gas-solid system*

The velocity of reactions between gases and solids is limited in the first place by 2 factors: 1. by the inactivity of the solid and 2. by the formation of covering films at the solid surface inhibiting the reaction.

Table 7.4. Tribochemical gas-solid reactions in engineering.

Tribochemical reaction in chemistry	Ref.	Tribochemical reaction in metallurgy	Ref.
Nickel carbonyl synthesis from Ni and CO	[25, 26]	Nitride hardening	[34, 35]
Mechanical activation of iron oxides for the manufacture of gas purifying media	[19, 27]	Scaling	[34]
Regeneration of aged catalysts	[18]	Carburization	[34]
Oxidation of metals into metal oxides	[13, 4.57]	Decarburization	[34, 35]
Halogenation of elements	[14, 15]	Sulphurizing	[34]
Synthesis of nitrolime	Chapt. 5., [219]	Reduction of iron silicate in siliconized steels with H_2	[34]
Decomposition of tungsten ores by chlorine	[28]	Reduction of iron oxides with H_2	[34, 12]
Mechanical activation of technical Si for the direct synthesis of methylchlorosilanes	[29, 30]	Reduction of ore pellets with coke	[36]
Thermal decomposition of mechanically activated carbonates	[31, 32], Chapt. 4., [154]	Reduction of manganese ore	[37]
Reduction of mechanically activated sulphides into metals	[33]	Quantitative yield of metals from ores at reduction	[37]
		Manufacture of tungsten powders	[38]

As has been shown in 5.1.—5.7. and 5.9. for several reasons the reactivity of solids may be considerably increased by mechanical treatment. According to 5.6., for instance, with correspondingly intensive treatment the covering films are removed from the surface of the solids, so that because of this the heterogeneous reaction is no longer determined by a diffusion mechanism but by a fresh surface mechanism. Both effects common cause a very high acceleration of the reaction velocity. Through this, numerous tribochemical reactions run with such a high velocity that new technological procedures can be based on it. A selection of technologically significant gas-solid reactions is listed in Tab. 7.4.

In most cases the causes of the reaction increase have been little investigated in the individual tribochemical processes until now. By contrast, clear conditions exist in the synthesis of nickel carbonyl which will therefore be treated somewhat more fully as a model reaction in the following (4.6.). The scheme of the nickel carbonyl synthesis in technology is given in Fig. 7.4., [39].

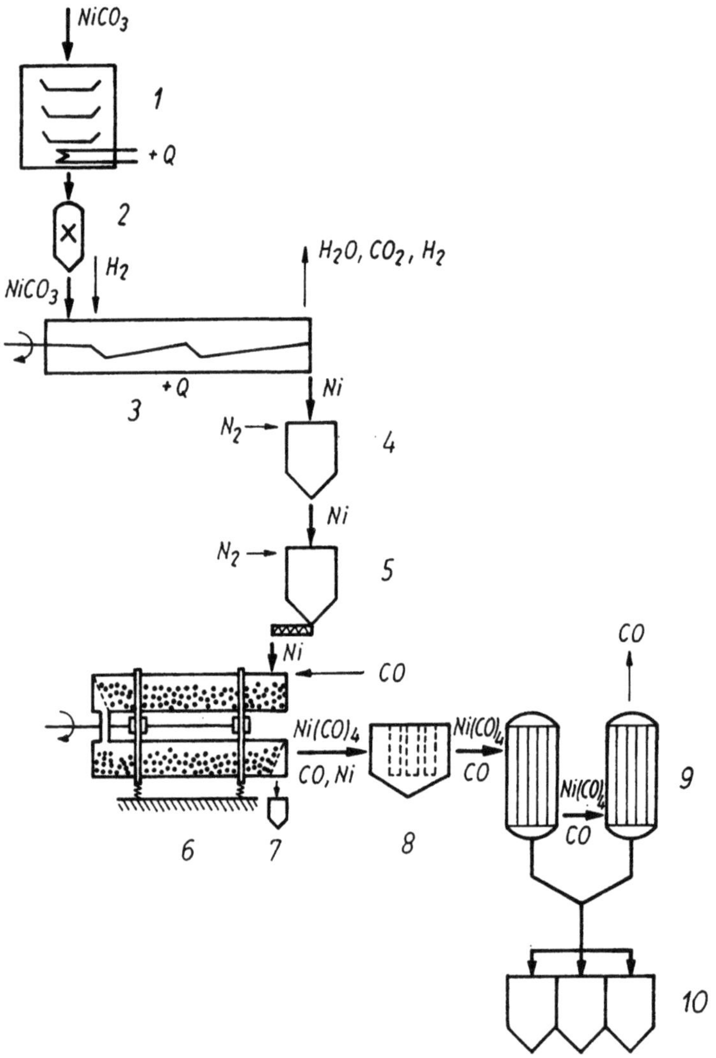

Fig. 7.4. Scheme of the technical plant for the tribochemical nickel carbonyl synthesis [25].
1 — drying, 2 — comminution, 3 — reducing, 4 — bunkering, 5 — dosing, 6 — tribochemical reactions, 7 — residue, 8 — dust removal, 9 — cooling, 10 — storing.

The heart of the plant is a vibration mill designed as a triboreactor. Here nickel is transformed into nickel carbonyl by continuous treatment of the metal powder in a CO atmosphere. Since the triboreactor works continuously the preconditions exist for a continuous procedure. This enables a stationary run over many months. On the other hand the synthesis may be interrupted at will by stopping the triboreactor, since the reaction velocity decreases by about one order of magnitude already after a short time and increases again with renewed mechanical treatment to the initial value.

The course of reaction using different nickel powders is represented in Fig. 7.5. Starting with annealed nickel powder the reaction velocity is extremely small and even with mechanical treatment only reaches medium values with an increase in reactivity by 2 orders (curve ▽, ▼, Fig. 7.5.).

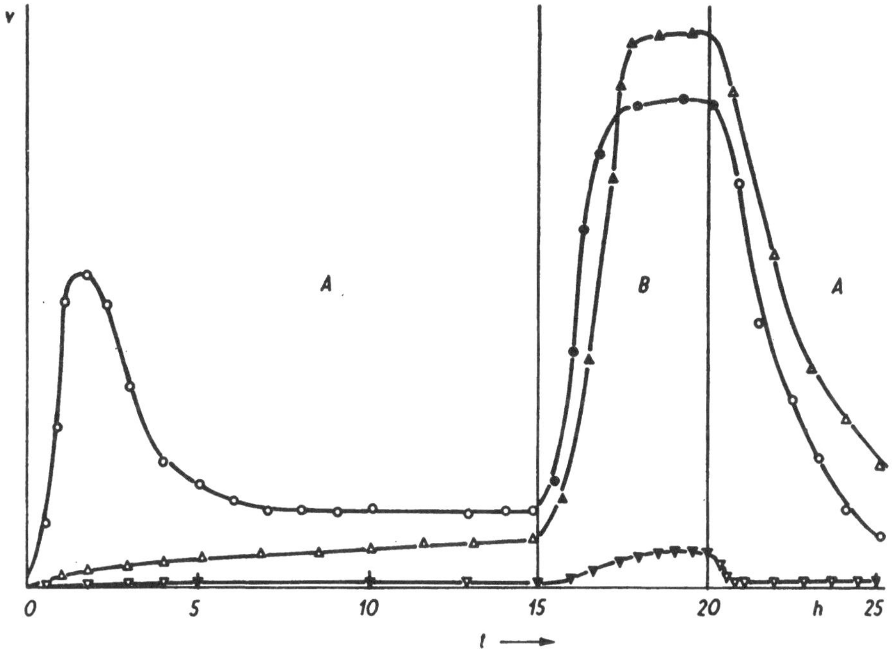

Fig. 7.5. Course of the nickel carbonyl formation without (A) and with (B) treatment using differently active nickel powders [40].

v – reaction velocity, t – time of reaction, ○ ● – pyrophoric powder, △ ▲ – pyrophoric powder covered by an oxide layer, ▽ ▼ – annealed powder.

Starting with freshly reduced pyrophoric powder according to the scheme in Fig. 7.4., then a relatively high transformation into nickel carbonyl already takes place without treatment (curve ○, ●; Fig. 7.5.). The classical MOND-nickel process operates in this region.

Differences in the course of reaction are obtained if the pyrophoric properties of the freshly reduced powder are removed through a short-term intermediate treatment with CO_2 (curve △, ▲; Fig. 7.5.). The reason for this consists in the formation of a covering film at the nickel surface which can only be slowly removed again by the nickel carbonyl reaction in the region *A* (Chapt. 4., [44]). But the relatively high reactivity of the pyrophoric product also only lasts for a short time, since parts of the surface are similarly covered by a film through impurity traces in the gas and through side reactions. For this reason an extensive reaction of the nickel to carbonyl is only obtained by a classical procedure if the covering film which develops is intermittently removed by reduction with hydrogen at increased temperatures. These differences are of no importance to mechanical treatment, since the covering films developing at the surface are repeatedly removed in the triboreactor.

The differences in reactivity of annealed and pyrophoric powders may be attributed to the different surface sizes of both kinds of nickel. As results from Fig. 7.6. this difference in surface size totals approximately 2 orders of magnitude. According to 5.5. the relatively large surfaces of phyrophoric powders are however not stable under treatment. Nevertheless, under certain circumstances, e.g. by a less than 100 per cent reduction of the nickel compound, it is possible to maintain stationary the relatively large surfaces, even with longer treatment. Starting with nickel powders with small surfaces the latter effect cannot be attained by milling (see Fig. 7.6., curve I). For this reason the application of pyrophoric nickel powders has proved to be useful for the tribochemical nickel carbonyl synthesis.

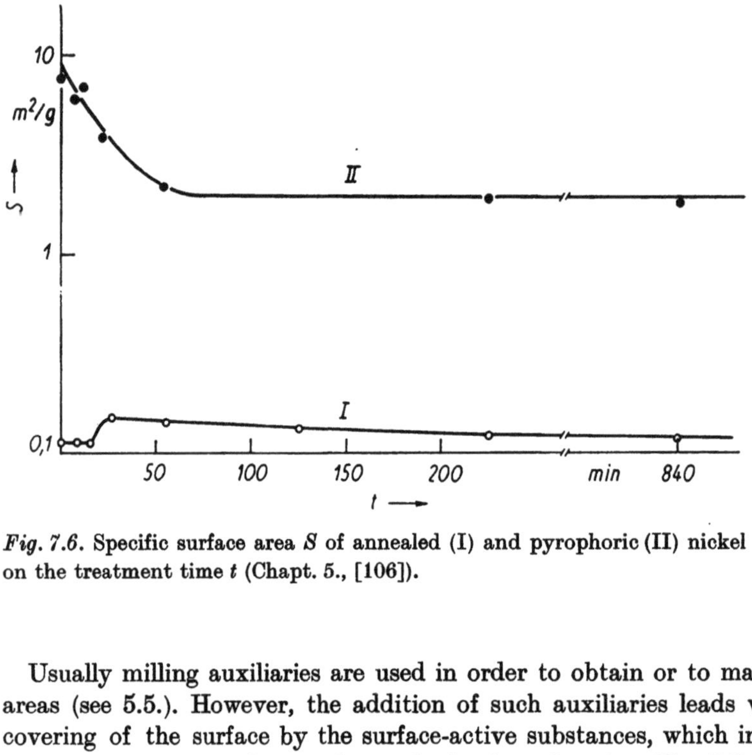

Fig. 7.6. Specific surface area *S* of annealed (I) and pyrophoric (II) nickel powder in dependence on the treatment time *t* (Chapt. 5., [106]).

Usually milling auxiliaries are used in order to obtain or to maintain large surface areas (see 5.5.). However, the addition of such auxiliaries leads very quickly to the covering of the surface by the surface-active substances, which inhibits the carbonyl reaction running simultaneously at the fresh surface (see Fig. 7.7.). In this case other ways have to be found in order to maintain a large surface (see above).

Although the reduction of the surface occurring due to the milling of pyrophoric powders means a loss of reactivity it is nevertheless overcompensated by the "structure" (see 5.9.) and "fresh surface" effects (see 5.6.) so that altogether the yield is higher by 1···2 orders of magnitude in comparison to the thermal reference procedure (Chapt. 5., [106]).

Finally it may be mentioned that other carbonyls such as iron (Chapt. 6., [212]), molybdenum (Chapt. 6., [212]), and tungsten carbonyls (Chapt. 4., [66]) can also be synthesized tribochemically. Moreover the process principles developed for the nickel carbonyl synthesis can be applied to these reaction system, too.

Altogether tribochemical synthesese in the gas-solid system have proved to be very favourable in economic terms if the specific process is performed at relatively low temperatures. Due to the extremely small dependence of tribochemical reactions on temperature, the differences in the velocity of thermal and tribochemical reactions become greater and greater with falling temperature (see 4.5.).

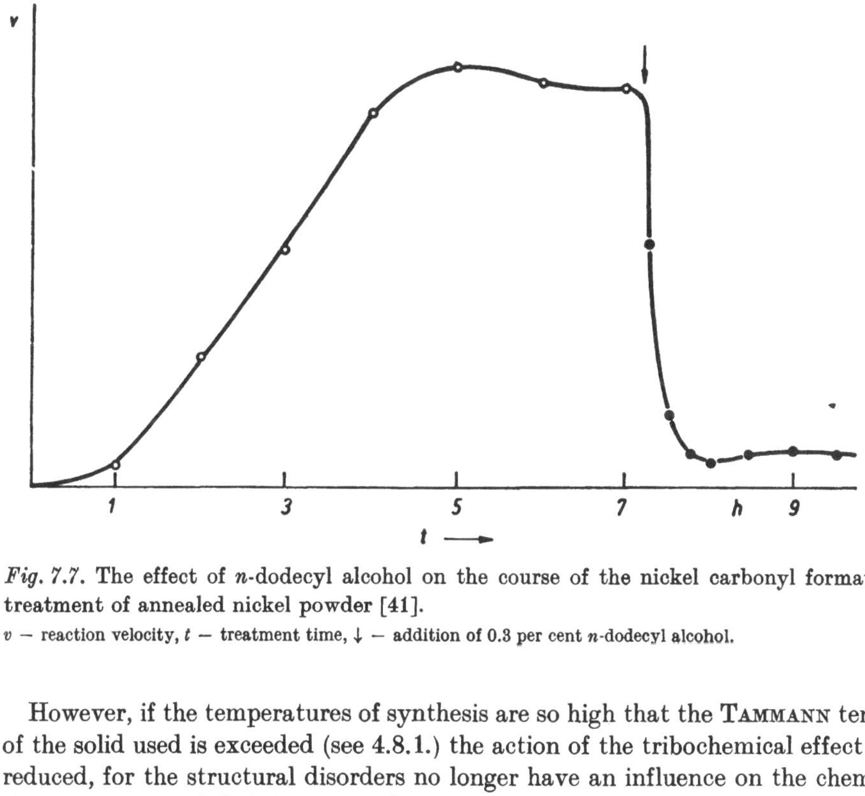

Fig. 7.7. The effect of *n*-dodecyl alcohol on the course of the nickel carbonyl formation at the treatment of annealed nickel powder [41].
v – reaction velocity, *t* – treatment time, ↓ – addition of 0.3 per cent *n*-dodecyl alcohol.

However, if the temperatures of synthesis are so high that the TAMMANN temperature of the solid used is exceeded (see 4.8.1.) the action of the tribochemical effect is greatly reduced, for the structural disorders no longer have an influence on the chemical reaction on account of their fast annealing.

7.2.2.2. Tribochemical processes in the liquid-solid system

7.2.2.2.1. Kinetics and thermodynamics of tribochemical dissolution processes

The remarks on the kinetics and thermodynamics of tribochemical processes given in 4.5. and 4.8. are also applicable to tribochemical dissolution processes. Studies, especially of metallic systems, have shown that with the combination of mechanical treatment and the dissolution process an increased dissolution and a shift of the potential take place in the liquid-solid system (see Fig. 5.57.). These processes play a leading part in the tribocorrosion of metallic systems in engineering and have been dealt with in detail in 6.6.

However kinetic and thermodynamic considerations with respect to the coupling of the action of mechanical energy with tribochemical processes are still at the beginning for non-metallic systems. Only few examples (e.g. [42, 43]) have been described in which under industrial conditions the digestion of minerals runs via a dissolution process directly during the mechanical treatment. From this it could be shown that the chemical digestion takes place essentially faster in comparison to the dissolution of preactivated products.

However, the dissolution of solids mechanically pretreated in aqueous media has been investigated both in scientific and in technological respects in numerous systems and has become important particularly to the processing of raw materials. In considering this group of reactions from kinetic and thermodynamic points of view emphasis has to be laid on 3 aspects:

— Through mechanical activation an increase of the surface, and consequently an accelerated dissolution of the solid, appears, due to comminution and deformation processes. This fact appears for instance from Tab. 6.8., according to which the dissolution velocity of mechanically non-treated apatites increases proportionally to the outer surface (see 5.5.).

— The excess energy of the system is increased and the dissolution of the solid made easier by the production of structural disorders. Most etching processes for the visualization of structural disorders are based on this effect. The structural disorders have proved in numerous cases to be the decisive factor for the increase of the rate of dissolution, so that even with a reduction of the outer surface by milling, the solubility of the product may considerably increase with growing degree of disturbance. As appears from Tab. 6.8. these conditions also apply to apatitic solids. The reason for the acceleration consists in the reduction of the activation energy E required for the dissolution process. The value of E may be lowered almost to zero by a correspondingly intensive treatment (Tab. 7.5.).

Table 7.5. Reduction of the activation energy E of the dissolution of apatites by mechanical treatment.

Substance	Activation set	E, kJ/mole	Ref.
Kola-apatite	(untreated)	29.8	Chapt. 4., [88]
Kola-apatite	Vibration mill	3.4	Chapt. 4., [88]

— With an increase of the degree of disturbance of a solid its equilibrium solubility is changed. According to Tab. 7.6. by this the value of the solubility constant of mechanically treated solids may be increased by some orders of magnitude in comparison to the untreated product. This effect often plays an important role in the industrial application of substances not easily soluble. Thus under certain conditions the solubility of apatites is a measure of their plant-physiological effectiveness and of their use as fertilizers (see Fig. 7.13.).

In practice an overlapping of the 3 effects discussed appears in most cases.

The type, intensity, and duration of treatment determine above all the degree to which each of these effects contributes to the increase in reactivity. In most cases at

Table 7.6. The displacement of the solubility constant K_L of Kola-apatite in water by mechanical treatment [45].

Pretreatment of Kola-apatite	K_L
Untreated	10^{-140}
Activated in a vibration mill	10^{-100}

the beginning of the treatment the process of the surface enlargement dominates. With increasing treatment time the effect of the structural disorders increases and in many cases even becomes decisive for the course of the reaction.

7.2.2.2.2. *Dissolution processes in the preparation of raw materials*

Although as a whole the thermal procedures prevail in the metallurgical industry, hydrometallurgical procedures have also been developed for the preparation of oxidic and sulphidic lowgrade raw materials needed for the manufacturing of the most

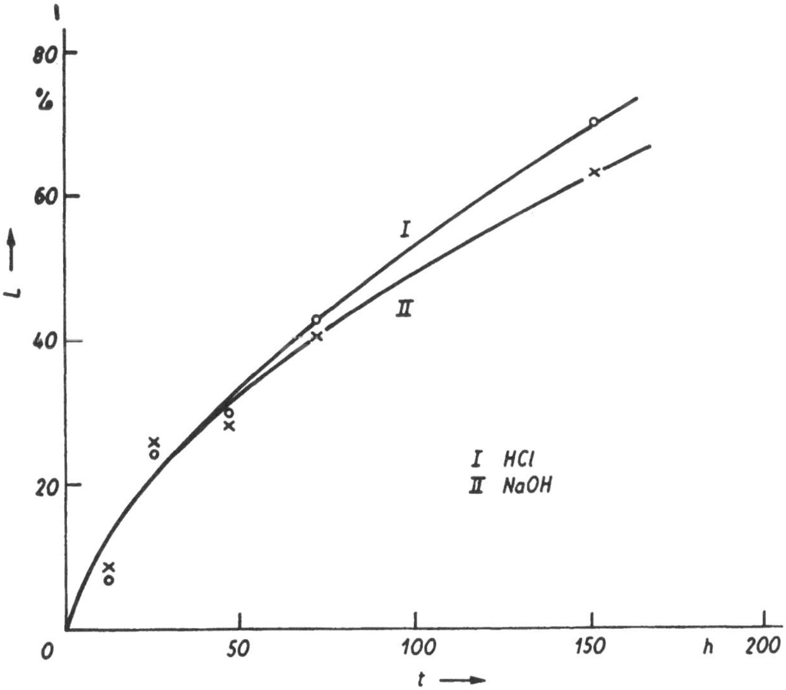

Fig. 7.8. The solubility L of SnO_2 in a prepared tin concentrate in dependence on the treatment time t [46].

varied metals, such as Cu, W, Al, Ti, Cr, Ni, etc. These procedures consist of four operation stages, namely the preparation of the raw materials, the real dissolution process, the enrichment, and the subsequent treatment of the metallic component.

Table 7.7. Tribochemical dissolution reactions in preparation industry.

No.	Tribochemical reaction	Ref.
1	Decomposition of tungsten ores with lyes	[28, 47]
2	Decomposition of wolframite in mineral acid	[48, 42, 43]
3	Decomposition of clay in mineral acid	[49, 50–53]
4	Dissolution of bauxites in sodium hydroxide solution	[22, 54, 55], Chapt. 5., [206], [56]
5	Decomposition of bauxites in acids and lays	[57]
6	Decomposition of kaoline in HCl	[58, 59]
7	Dissolution of tin stone and tin concentrates in acids and lays	[46, 60–65]
8	Dissolution of oxidic copper compounds	[66]
9	Pressure leaching of sulphidic copper ores with sulphuric acid	[67–69]
10	Pressure leaching of copper-nickel concentrates in the presence of alkali	[70]
11	Dissolutión of roasted pyrites in acids	[71]
12	Dissolution of quartz in lyes	Chapt. 6.1.2.
13	Dissolution of apatites in acids and complexing solvents	Chapt. 6.1.3.
14	Decomposition of fluor-spar in sulphuric acid	Chapt. 5., [100]
15	Activation of lime to lime fertilizers	[72]
16	Decomposition of chrome iron ore in hydrochloric acid and potassium hydroxide solution	Chapt. 5., [94]
17	Decomposition of titanum magnetite in mineral acid	[73]
18	Leaching of rutile	[74]

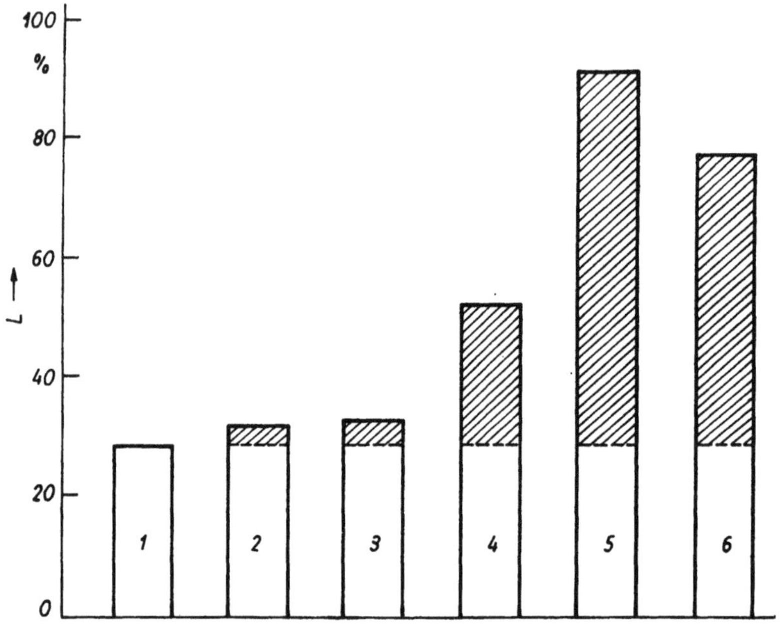

Fig. 7.9. The solubility L of aluminium contained in clay after the thermal and mechanical treatments in various activation sets [53]

1 – without mechanical treatment, 2 – disintegrator, 3 – attritor (wet), 4 – ball mill, 5 – vibration mill, 6 – thermal treatment.

As Fig. 7.8. shows, the "output" of the metal in the dissolution process depends to a major extent on the treatment time and on the degree of the raw material activation, respectively. Cassiterite is one of the raw materials which may only be digested with extreme difficulty by hydrometallurgy. The mechanical activation causes such an increase in reactivity that the dissolution of the tin ore by acid or basic leaching agents under usual leaching conditions becomes possible. A complete solubility may be attained with an appropriately long milling time.

A selection of tribochemical dissolution reactions particularly important to hydrometallurgy is presented in Tab. 7.7.

The works on the increase in the solubility of aluminium contained in aluminiferous raw materials by mechanical treatment in the most varied activation sets, have a special importance because the long-term an exhaustion of bauxite deposits can be anticipated,

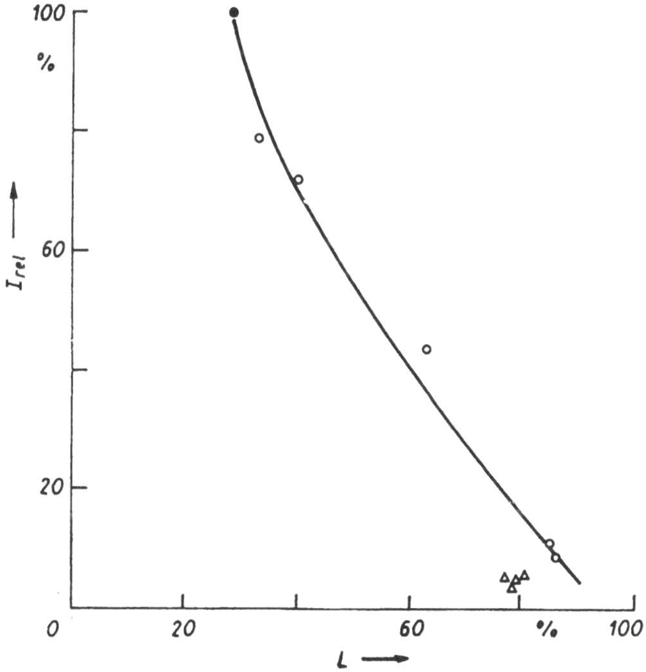

Fig. 7.10. Correlation between relative X-ray intensity I_{rel} and aluminium solubility L of differently activated clay samples [53].

• — raw material, ○ — mechanically activated, △ — thermally treated.

and other aluminiferous raw materials such as clays are gaining increasing importance. As Fig. 7.9. shows an increase in solubility is reached by the most different activation sets for the dissolution of activated kaolinite in hydrochloric acid. For instance, values over 90 per cent solubility are obtained using a vibration mill.

These values are even higher than for a thermal treatment of clay in which a transformation of kaolinite into soluble metakaolinite takes place. The reasons for the major increase in solubility are also based on structural changes which extend to the amorphization of the kaolinite. Fig. 7.10. shows that the solubility of the product increases nearly linearly at higher degrees of activation with the decrease of the relative X-ray intensity. The increase in solubility through mechanical activation is to a high extent determined by the degree of activation (see Fig. 7.9.) and by the stress conditions. Thus by using a vibration mill it is already possible today to optimize the stress parameters so that very low activation times are sufficient to obtain high solubilities (see Tab. 7.8.).

Table 7.8. The solubility of activated kaoline and clay, respectively, using vibration mills with different stress parameters.

Material	Treatment time h	Acid	Dissolution temperature K	Solubility of Al_2O_3, %	Ref.
Kaoline (Zettlitz)	≈ 500	HCl	(boiling)	80	[75]
Kaoline (Caminau)	25	HCl	298	90	[76]
	10		348		
Kaoline (Caminau)	50	H_2SO_4	—	84	[77]
Raw clay (58 per cent kaoline) (Guttau)	2	HCl	(boiling)	> 90	[53]

7.2.2.2.3. *Tribochemical cementation reactions*

Cementation reactions also play an important part for the enrichment and precipitation of the metals in the hydrometallurgical extraction of metals in the third and fourth treatment stage. Thus, for instance, in the obtaining of zinc by electrolysis, a careful purification of the zinc solution is required in order to extract the cadmium and copper contained [78].

Cupriferous solutions appear in different leaching processes in metallurgy from which copper has to be enriched by specific processes [79]. And already for a long time cementation reactions have proved to be suitable for this. Until now however, these reactions have had the substantial disadvantage of a too low reaction velocity, which has negatively affected the economy of the procedure. In addition they required a discontinuous mode of operation. With tribochemical conduct of the reaction basic improvements can be achieved for the course of the cementation process in terms of reaction velocity, specific consumption of metal, residual metal content in the solution, and the design of a continuous precedure. As an example, Fig. 7.11. shows the cadmium content of zinc samples milled for varying periods in dependence on cementation time. According to this the cementation process can be practically completely finished with mechanically activated powders. A high purity of zinc may be attained in this way.

In the case under consideration a continuous procedure may also be set up in analogy to nickel carbonyl formation (see Fig. 7.4.) using a continuous-type vibration mill (see Fig. 7.12.). The continuous-type vibration mill can also be applied to copper cementation where small-lumpy scrap or pig-iron granules are used as a treatment tool instead of balls. An extensive removal of copper takes place here, too, in spite of the application of low-grade scrap or pig-iron. A reduced consumption of scrap and a higher purity of the cemented products can be found as well. Thus the investment and operational costs are reduced with tribochemical conducting of the reaction in spite of the additional outlay for the construction of an activation unit [79].

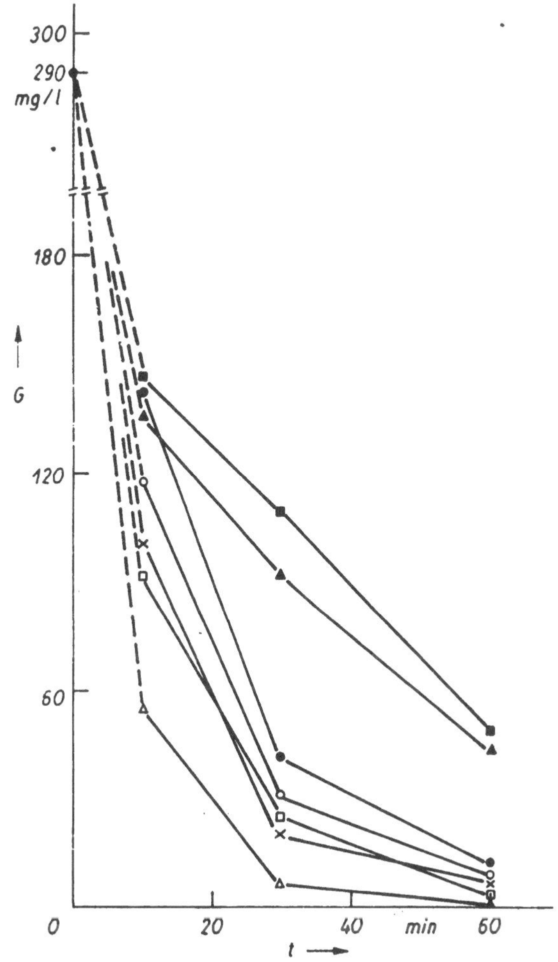

Fig. 7.11. Course of the cadmium cementation with the fivefold stoichiometric quantity of zinc powder treated tribomechanically for differently long periods of time (Chapt. 2., [29]).

G – cadmium content, t – cementation time, Treatment time: ● – untreated, ■ – 360 min, ▲ – 180 min, ○ – 5 min, x – 30 min, □ – 90 min, △ – 60 min.

7.2.2.2.4. *Tribochemical solubilization of rock phosphates*

Natural apatites have a very low solubility in neutral and alkaline environments and consequently a small availability to plants if used as phosphorous fertilizers. An increase in solubility results through comminution of apatites (see Tab. 6.8.). Hence rock phosphates are directly applied as fertilizers in a pulverized form in many countries (Rünaphos, Hyperphos, etc.), [80]. 2 Mill. t of pulverized phosphorite powder are for instance used as phosphorous fertilizer in the USSR [81]. However, these products are only available to plants to a limited extent on acid soils and have an insufficient initiating effect. A plant-physiological effect comparable to that of superphosphate is not obtained with these products. Therefore a large portion of rock phosphates is chemically converted into water-soluble monocalcium phosphate, particularly by decomposition with sulphuric and phosphoric acids.

An improvement of the fertilization effect of pulverized rock phosphates may also be obtained by a partial decomposition of these products by mineral acid. These products show a fertilization effect corresponding to that of a superphosphate–rock

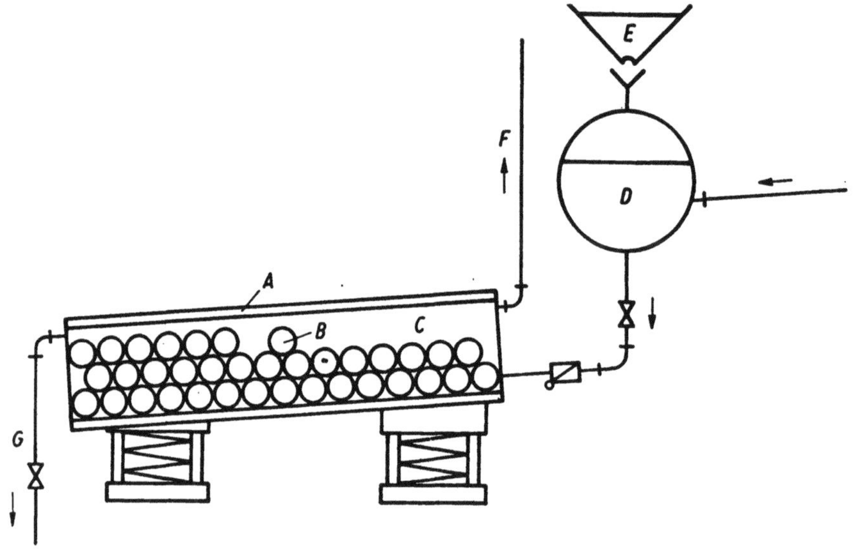

Fig. 7.12. Scheme of a triboreactor used for the cementation of zinc (Chapt. 2., [29]).
A – triboreactor, *B* – milling ball, *C* – reaction chamber, *D* – Zn/Cd solution, *E* – inflow, *F* – gas issue, *G* – Cd-free solution.

phosphate mixture [82]. The investigations described in 6.1.3. have shown that a series of structural disorders may be produced by intensive mechanical treatment going beyond comminution. These structural disorders lead to a considerable increase in the solubility. Apatites having lattice imperfections produced by tribomechanical activation and described as Tribophos [80] may have a complete solubility in alkaline ammonium citrate solution [80] and reach the fertilization effect of superphosphate [80, 83, 84]. In order to illustrate this, in Fig. 7.13. the equilibrium solubility before the activation of Kola-apatite is contrasted with that after the activation and compared with the solubility values of known fertilizers. The curve of the activated product lies far over that of the untreated Kola-apatite, and is in the range of the in water slightly soluble reference fertilizers. The activated product also exceeds the fertilizer Hyperphos [85] known as pulverized rock phosphate, particularly in the alkaline range.

Higher degrees of activation are not required for acid soils where the fertilization effect of the chemically decomposed phosphorous fertilizers is already in part obtained by pulverized rock phosphates.

With increasing pH of the soil a growing degree of activation leads to a marked increase in the fertilization effect, particularly in the transition to neutral and alkaline soils where pulverized rock phosphates are almost inactive [84]. If Tribophos is partially reacted by mineral acid, more favourable conditions result for the fertilization effect of Tribophos than by using pure activated products. A 50 per cent chemical solubilization of apatite has proved to be advantageous.

This fertilizer referred to as "Tribophos T" [87] has an initiating effect through the presence of water-soluble phosphates corresponding to that of superphosphate. Because of the acid environment in the fertilizer particle essentially improved conditions also exist for the solubility and dissolution velocity of the activated component.

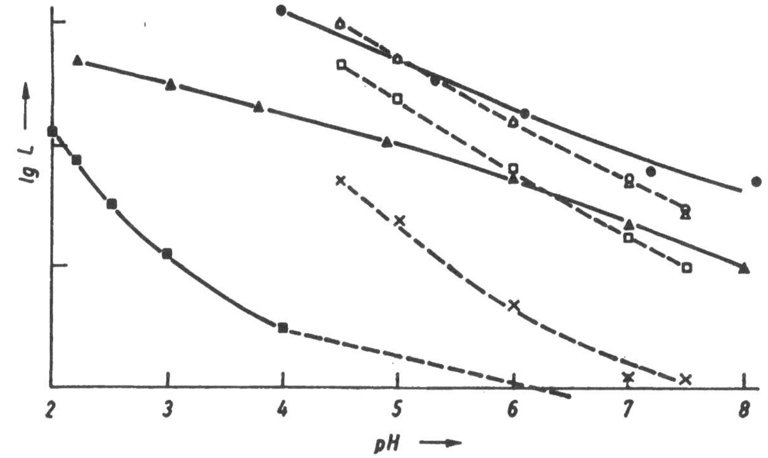

Fig. 7.13. Comparison of the solubility *L* of some *P*-fertilizers with activated Kola-apatite and a non-activated fraction of Kola-apatite in dependence on pH [85, 86].

o — dicalciumphosphate, △ — Rhenania phosphate, □ — Thomas phosphate, x — Hyperphos, • — Dicalciumphos, ■ — Kola-apatite, ▲ — Kola-apatite activated for 180 min.

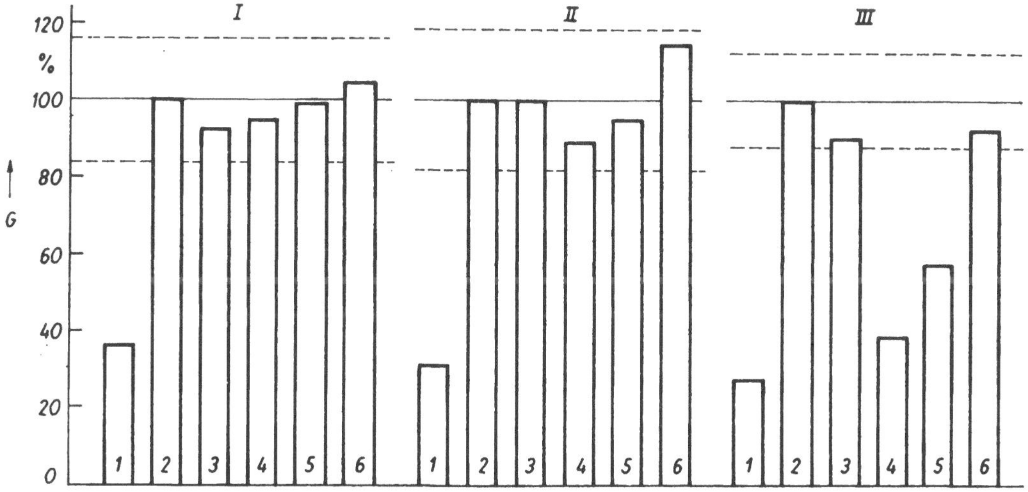

Fig. 7.14. Yield efficiency *G* of Tribophos and Tribophos T related to Superphosphate in comparison to other phosphate fertilizers on various soils in the year of application [85].

Soil I — pH = 5.5, Soil II — pH = 6.6, Soil III — pH = 7.1, 1 — without phosphorous, 2 — Superphosphate, 3 — alkali sintered phosphate, 4 — Rünaphos, 5 — Tribophos, 6 — Tribophos T. — — maximally admissible variation of the yields, *G* Superphosphate = 100 per cent.

Dicalcium phosphate and phosphoric acid develop through the partial hydrolysis of monocalcium phosphate whereby the phosphoric acid chemically decomposes a part of the activated rock phosphate. As Fig. 7.14. shows, such products have an effectiveness relative to pure Tribophos on neutral and alkaline soils comparable to that of superphosphate [87]. Plant-physiological investigations of several years' duration have also shown a fertilization effect corresponding to that of superphosphate.

Thus Tribophos T represents a phosphorous fertilizer having good initiating and long-term effects. It is important for the technological manufacturing of this product that already smaller degrees of activation are sufficient to reach the full fertilization effect than with Tribophos [87].

The portion of monocalcium phosphate contained in Tribophos T may be reduced to 35 per cent on weakly acid soils without causing the yield effect of the fertilizer to be lowered [80].

The increased reactivity of tribomechanically activated apatites also leads to an accelerated reaction of the components as compared to the application of non-activated raw materials, when fully decomposed by sulphuric or phosphoric acids [88—90]. In order to have a residual acid content of 5 per cent in superphosphate a curing-time of some weeks is required. As is seen in Fig. 7.15. this time may be shortened to some

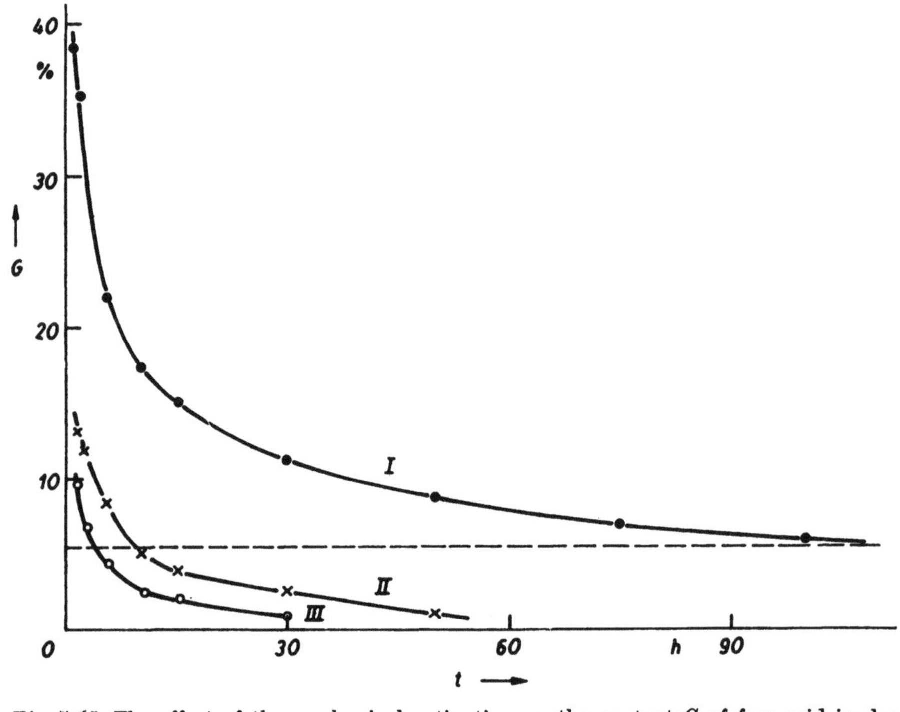

Fig. 7.15. The effect of the mechanical activation on the content *G* of free acid in dependence on the curing time *t* at a 50 per cent decomposition of Marocco-phosphate by sulphuric acid [80].
I — Marocco-phosphate mechanically non-activated (< 400 μm), II — Marocco-phosphate activated mechanically for 10 min, III — Marocco-phosphate activated mechanically for 30 min, — — admissible maximum content of acid in the commercial product.

hours if the apatite was mechanically activated before hand. A shortening of the curing-time also occurs when the mechanically activated phosphate is mixed with a chemically decomposed product having an acid content of above 5 per cent. As Fig. 7.16. shows, in this the free acid content is extremely rapidly decreased below the limit of 5 per cent laid down in the standard [80]. Therefore the curing process usually required for superphosphate may be dropped in the manufacturing of Tribophos T.

To sum up it can be said that the tribomechanical activation of rock phosphates represents a new technological principle for the manufacturing of fertilizers [91—95]. This principle may be used successfully for low-grade fertilizers in where a chemical decomposition is no longer economical or not possible for other reasons, e.g. through the presence of disturbing impurities. The combination of a chemical decomposition with a mechanical decomposition has proved to be useful, since fertilizers thus develop having initiating and long-term effects.

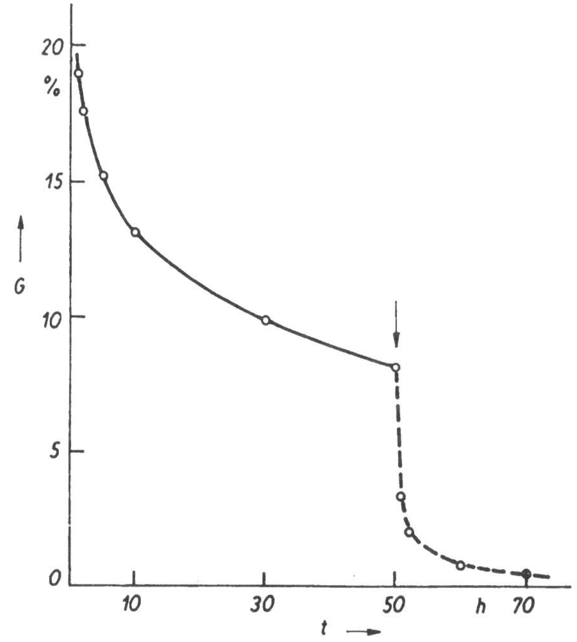

Fig. 7.16. Effect of admixing (↓) mechanically activated rock phosphate on the content *G* of free acid in Tribophos T [80].
t – curing time.

7.2.2.3. *Tribochemical processes in the solid-solid system*

Usually reactions between two or more solids run with a velocity sufficiently high for technical processes only at higher temperatures. In the material-changing industry important solid phase reactions are for instance the cement clinker formation (see 7.2.3.3.) and the formation of alkali sintered phosphates from rock phosphates, quartz sand, and soda [96] in which temperatures above 1,000 K are needed. These high temperatures are required because chemical reactions between solid phases are diffusion-controlled and the lattice components of the solid phase have a sufficiently great diffusion mobility at higher temperatures only [97].

As has been discussed in 5.6.5. under mechanical treatment the diffusion mechanism prevailing in thermal stress turns into the fresh surface mechanism so that the reaction between the solids is mainly determined by the real chemical reaction at the interface. For this reason a large number of solid reactions already run at room temperature when treated (see Tab. 5.19.).

For instance, the chemical decomposition of tungsten ores by soda is of technological interest [98]. Whereas the thermally activated reaction only runs above 600 K, under treatment the chemical decomposition already takes place at room temperature. If the solids are caused to react only after a mechanical pretreatment then the diffusion mechanism once again determines the reaction. As will be dealt with comprehensively in 7.2.3.4. diffusion processes run acceleratedly in disturbed solids. Therefore either an acceleration of the reaction or a reduction of the reaction temperature at the same reaction velocity may appear through mechanical activation of the solid phases [96, 98, 99].

The above-mentioned chemical decomposition of mechanically activated tungsten ores by soda takes place for instance at reduced temperatures and shorter retention times, in comparison to a thermal treatment of raw materials non-activated mechanically [98].

Sometimes a change in the reaction mechanism occurs through the mechanical activation of the parent substances which can be important to the economy of the procedure. Thus in the industrial manufacturing of potassium sintered phosphate, high potassium losses occur through the volatilization of certain potassium compounds in the range of the reaction temperature. If an activated quartz sand is used for the reaction instead of a mechanically non-activated product then potassium silicates come intermediately into being at lower temperatures (< 573 K) which continue to react with the calcium phosphate to potassium sintered phosphates at higher temperatures. Since potassium silicates are little volatile the potassium losses may be kept small with this process variant, and thus an economically working procedure set up [99].

To date solid-solid reactions have been investigated and used in industry to a lesser extent in comparison to tribochemical gas-solid and liquid-solid reactions. The chief reason may be that because of the higher temperatures required for the run of solid reactions the effect of the mechanical treatment is of less importance than, for instance, in the dissolution processes which in most cases are conducted at very low temperatures. While also taking into account the tribochemical processes discussed in 7.2.3.3. and 7.2.3.4. nevertheless it is estimated that the possibilities for technological application have still only been exhausted to a minor extent and promise to be economically advantageous for a series of procedures.

7.2.2.4. *Tribochemical reactions of inorganic solids with organic reactants*

7.2.2.4.1. *The importance of modification reactions for the manufacturing of composites*

The use of materials consisting of mixed inorganic and organic components is assuming increasing importance in engineering. These material combinations described as composites are used in various branches of engineering, e.g. as filled and reinforced polymers, plastic-coated metals, printed circuit boards, as well as colours, pastes and latices. Composites with a wide variety of properties can be produced and adapted to the specific purpose by the selelction of appropriate components, by varying the mixture ratio, and by different pretreatment of the organic and inorganic raw materials.

The main problem of the production of composites consists above all in the formation of a phase boundary between the inorganic and organic constituents of the composite, which is sufficiently stable in respect of outside influences. As is known inorganic substances such as silicates, carbonates, and sulphates used for the production of composites are hydrophilic, so that on contact with mainly hydrophobic organic substances only a weak adhesion occurs and the strength of such structures is small [100]. The reason for this may be found in the different bond conditions of the two classes of substances. Whereas the ionic bond dominates in the structure of most inorganic materials, organic solids are mainly formed on the basis of covalent bonds. Consequently the manufacturing of high-grade materials first of all requires the creation of a sufficiently high bonding strength between the inorganic and organic phases of a composite. In most cases an increase in the bonding strength at the interface is obtained by the modification of surface of the inorganic solid. In this usually the inert

Table 7.9. Methods for the modification of surfaces of inorganic solids in order to manufacture inorganic-organic composites.

Method	Reaction example	Ref.
Use of adhesive agents	Modification of glass fibres with silanes, adsorption of octodecylamine on rutile	[101, 102]
Ion exchange process	Exchange of metal cations on silicates surfaces into organic quaternary ammonium bases	[103]
Chemical grafting	Reaction of organosilicone compounds with silicate surfaces containing OH groups $\rangle Si{-}OH + Cl{-}SiR_3 \rightarrow \rangle Si{-}O{-}SiR_3 + HCl$	[104]
Grafting by thermal shock	Hydrophobing of glass fibres by the action of silicon solutions	[105]
Precipitation copolymerization	Polymerization of an acrylic acid-methylmethacrylate mixture in the presence of the inorganic component	[100]
Plasma modification	Plasma treatment of quartz meal in the presence of a monomer	[106, 107]
Modification by means of particles rich in energy	Development of adhesive points by γ-irradiation causing polymerization in the presence of monomers	[108]
Tribochemical modification		
A) by ultrasonics	Establishment of OH groups capable of grafting on laminated silicates and grafting of organic radicals	[109]
B) by activation in mills	Vibration milling of slate meal in the presence of a silane	[21]

surface is made hydrophobic and provided with certain reactive groups or centres enabling a better interaction with the organic matrix. A series of methods suitable for the modification of inorganic surfaces are summarized in Tab. 7.9. The use of adhesive agents is the method most frequently used in practice and best investigated. The method of tribochemical modification is still relatively new and will be discussed in summary in the following [109].

7.2.2.4.2. *Tribochemical modification*

With the treatment of inorganic solids in activation sets both the dispersity is changed (see 5.5.) and diverse electronic (see 3.3. and 5.4.) and structural (see 5.9.) defects are produced. Generally an enlargement of the surface area and an increase in reactivity takes place this. Numerous solids which like quartz are little reactive in an undisturbed state relative to organic substances, are capable of entering into the most varied modification reactions with these latter after mechanical activation. For instance, in the mechanical activation of quartz, reactive groups having unsaturated valencies (see equ. (5.65.)) come into being due to the breaking up of Si—O bonds, which easily react with water [110] and organic substances such as methyl chloride, aliphatic and aromatic alcohols [111] by forming surface groups. A grafting followed by a polymerization takes place with mechanical treatment in the presence of monomers such as ethylene, vinyl chloride, and styrene (see 5.4.), (Chapt. 5., [50, 57], [112—115]).

Polymers are also capable of grafting onto surfaces of inorganic solids under mechanical treatment, since according to 6.3. reactive radical groups develop through the mechanical degradation, which enter into chemical interaction with the surface of the inorganic solid, e.g. after the following equation [116, 117]:

$$\rangle\!\!\!-Si{-}O{-}Si\!\!\!-\!\!\langle + R{-}R' \xrightarrow[\text{energy}]{\text{mechanical}} \rangle\!\!\!-Si^{\cdot} + \rangle\!\!\!-Si^{\cdot}{-}O^{\cdot} + R^{\cdot} + R'^{\cdot}$$

$$\rightarrow \rangle\!\!\!-Si{-}R + \rangle\!\!\!-Si{-}O{-}R' \qquad (1)$$

As has been shown by the example of the mechanical degradation of polyvinyl chloride this process is substantially accelerated by simultaneous milling with quartz. It is assumed that the SiO_2 surface newly produced during the treatment has a catalytic effect on the degradation of the polymer. In this way the tribochemical grafting is also accelerated [118].

As appears from Tab. 7.9. a tribochemical grafting also appears by the action of ultrasonics upon the inorganic solid surface. In this method, no comminution for instance of the silicate particles takes place, but only a splitting of very fine particles from the silicate surface occurs. The grafting is caused by the disturbance of the solid surface on account of a cavitation stress. The carbon content of the ultrasonically grafted solid reaches values up to 4 per cent [109].

7.2.2.4.3. *Technological application of tribochemical modification*

Tribochemically modified inorganic solids have great importance as fillers for the manufacturing of inorganic/organic composites. The objectives of the use of fillers are generally:

— dilution of polymers with cheaper mineralic raw materials and consequently a reduction of manufacturing costs,

— improvement of certain practical properties, such as the increase in stiffness and strength, the reduction of the cold flow, the increase in the thermal dimensional stability, and the thermal stability as well as the prevention of embrittlement.

If unmodified fillers in the form of a milled powder are used, apart from the stiffness important properties of the initial polymer such as the flexural and tensile strength are deteriorated [119]. If a reduction of these parameters is acceptable in terms of use, then the sole economic advantage consists in the dilution of the composite. Nevertheless it is generally the objective to make both the above advantages simultaneously effective.

The special advantage of tribochemical modification as compared to other methods (see Tab. 7.9.) consists in the fact that the comminution to the production of a fine powder necessary for the manufacturing of the filler, may be directly combined with the modification. In this the following posibilitis have proved their worth:

— The milling of the filler occurs with the continuous addition of an adhesive agent. This way, for instance, slate chips may be processed into a high-grade filler in the presence of silanes, which not only has a diluting but also a reinforcing effect, as can be seen in Tab. 7.10. Simultaneously the iron contained in the slate is masked by the formation of the adhesive layer [21]. If surface-active substances such as calcium stearate are also added, hydrophobic fillers modified by tribochemical interaction come into existence due to the milling. Thus in this way modified calcium sulphite has proved its worth as a filler for polyvinyl chlorid. Such modified fillers may be worked into a polymer in larger quantities than an unmodified filler, without the strength properties being substantially deteriorated [120]. Moreover the milling effect is also improved by the addition of adhesive agents, since the agglomeration is repressed when appearing during the milling [116].

Table 7.10. The improvement of the properties of a PVC composite by tribochemical modification of the filler [21].

Property	Use of an untreated slate meal	Use of a modified slate meal
Notch impact bending strength, MPa · cm	0.7	1.1
Impact bending strength, MPa · cm (at 253 K)	5.1	6.8
Limiting bending stress MPa	47.1	58.8
Tensile strength MPa	33.0	38.9

The reinforcement effect of such additives has also been shown to depend on their concentration in the filler and it is only then observed if the adhesive agent enters into a chemical interaction with the filler surface. As Fig. 7.17. shows no reinforcing effect appears with purely physical adsorption of the modificator [102].

— The mechanical stress of the filler takes place in the presence of a monomer, e.g. ethylene, butadiene, vinyl chloride, or styrene. As results from Fig. 4.42. ethylene for instance is thus strongly tribosorbed. Grafted polymerisates could be detected at the surface of the filler after the milling of slate meal in the presence of styrene (see Tab. 5.11.). If slate meal is milled in a butadiene atmosphere for longer time then the polymer content in the filler increases to over 30 per cent (see Fig. 5.21.). Consequently a tribochemical modification is possible by using the most different initial components and may be adapted for the purpose by varying the milling conditions such as the duration of milling.

As Tab. 7.11. shows, such modified fillers improve both the properties of the composites produced by them and the processibility in the manufacturing process [112].
— The milling of the filler takes place in the presence of the polymer, e.g. in the form of a polymeric solution. For instance, effective fillers are obtained, if asbestos is activated in caoutchouc solution. The hydrophobic product contains up to 10 per cent

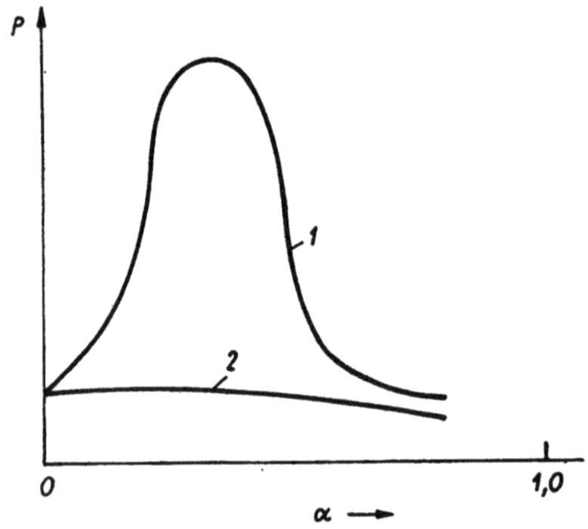

Fig. 7.17. Changes in the strength *P* of a filled polymer in dependence on the degree of saturation α of the filler surface by a modificator [102].

1 — at chemisorption of the modificator, 2 — at physical adsorption of the modificator.

of bonded carbon. On the other hand a too strong destruction of the asbestos is prevented by the presence of the caoutchouc solution [117]. Asbestos is also modifiable by ultrasonic influences without any noticeable destruction of the fibrous structure

Table 7.11. The effect of a filler produced by milling of calcium sulphite with the simultaneous action of vinylchloride on the composite formation as compared to unmodified filler [112].

Filler	Average particle diameter, μm	Kneading time, min	Tensile strength, MPa	Failure strain, per cent
Modified by vinylchloride	1	9	17	250
Unmodified	1	13	15	180

[121]. By using tribochemically modified fillers the thermal stability of the composites is also essentially increased. As an example in Fig. 7.18. the mass loss of filled and unfilled polyethylene is given, in dependence on temperature. Thermal stability is increased by mechanical activation of the raw materials [122].

Tribochemical modification is not solely restricted to the manufacturing of fillers for the plastics industry but has a general importance for the manufacturing of disperse products with changed interfacial properties. For instance, tribochemically modified amorphous silicas have proved to be suitable as additives for lubricants and have led to a reduction of wear [123].

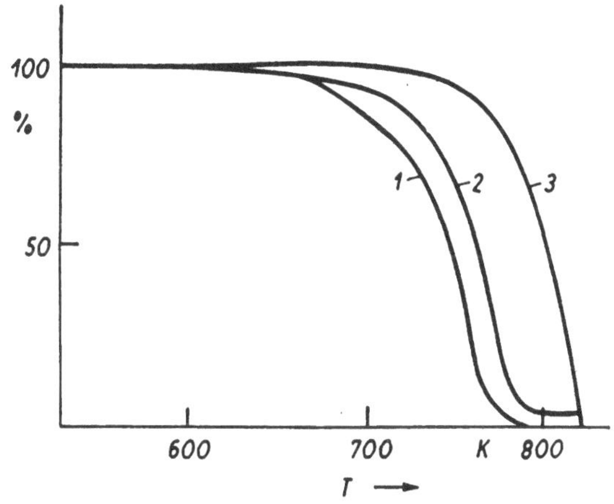

Fig. 7.18. Thermostability of filled and unfilled polyethylene (PE) [122].

ordin. — weight of the sample (per cent of the initial weight), T — temperature, 1 — pure PE, 2, 3 — PE mixed with quartz, 3 — with mechanical treatment.

7.2.2.5. *Summary*

Tribochemical procedures may be counted as impulse procedures. These latter are distinguished by the fact that states highly excited for a short period by impacts are built up in the system in hand and the macroscopic process thereby initiated takes place by summing up the microprocesses occurring independently of each other.

From this result the numerous technological possibilities mentioned above which a purely thermal procedure is not able to provide.

7.2.3. *Tribochemistry of building materials*

7.2.3.1. *Tribochemical activation of SiO_2 for the manufacture of silicate concretes*

7.2.3.1.1. *Introduction*

Lime and sand are among the building materials which have been used by man for thousands of years. As archaeological investigations have shown, a lime-sand mixture having the ratio 1 : 3 to 1 : 5 was already in ancient times used as mortar for the bond of wall stones. Noteworthy is the great strength which these binders have obtained in the course of thousands of years. However, in spite of many attempts it was only at a very late date that an artificial stone could be produced from lime and sand because of the extremely slow solidification of these building material mixtures. It was only when Michaelis found in 1880 [124] that the curing already occurs in the course of few hours under hydrothermal conditions and at temperatures of 450 K, that it became possible to produce bricks from lime and sand having a compressive strength of $1.5 \cdot 10^7$ Pa.

However this compressive strength is not enough for the manufacture of large size or self-supporting stuctural units. The compressive strength of $3 \cdot 10^7 \ldots 4 \cdot 10^7$ Pa required for this could so far only be reached for cement concrete. The main reason for this is that the chemical reactions between lime, SiO_2, and water leading to solidification run too slowly or in too small an extent if SiO_2-sand is used.

Therefore great efforts have been made to clarify the chemical reaction of the curing and the external factors acting upon this process and to come to conclusions with respect to the manufacture of high-strength silicate concrete. The compressive strength of the hardened building stone has been shown to depend on a lot of factors mostly overlapping a complicated way, e.g. on the density of the building stone, on the relative quantity of CaO in the mixture, on the contamination of the raw materials, on the structural strength of the sand used, on the grain shape of the milled particles, on the homogenization of the mixture, etc.

The decisive factors in obtaining a high compressive strength are however the chemical processes running during the hardening (Chapt. 3., [42], [125, 126]).

7.2.3.1.2. *Chemical processes in the hydrothermal hardening*

When silicate concretes are hydrothermally hardened the following processes run in the $Ca(OH)_2-SiO_2-H_2O$ system (Chapt. 3., [42], [126, 127]):

— Dissolution of $Ca(OH)_2$ in water and formation of Ca^{2+} and OH^- ions,
— dissolution of SiO_2 in water in the form of $(H_3SiO_4)^-$ or $(H_2SiO_4)^{2-}$ complexes,
— chemical reaction of the dissolved ions to a dissolved calcium silicate-hydrate rich in lime (CSH phase),
— formation of calcium silicatehydrate nuclei,
— growth of the nuclei,
— precipitation of silicate crystals of different compositions,
— further growth of the newly formed CSH,
— rearrangement to products poorer in lime.

This mechanism holds for the temperature range of 300···600 K [126, 128].

The dissolution of $Ca(OH)_2$ is a relatively rapid process so that the Ca^{2+} ions are already evenly distributed in the solution after a short time. The dissolution of SiO_2 in water, however, is the slowest partial process and consequently the velocity-determining process of the whole hardening. The crystal nuclei of the CSH phases appear on account of the low solubility in water at the interphase between the solid SiO_2 surface and the aqueous solution. The developing calcium silicate-hydrates are still significantly less soluble than the raw materials so that the newly formed phases precipitate from the solution at a time when the saturation concentration of the dissolved silicic-acid ions has not yet been reached.

The diffusion of the silicate ions is increasingly inhibited by the newly formed layer due to the growth of the latter at the SiO_2 surface so that at a later stage the process becomes determining for the velocity.

Approximately 20 different $C_xS_yH_z$ phases are known to appear in crystalline, semicrystalline or nearly amorphous forms. Among the variety of compounds the tobermorite plays the most important role in the new formation of CSH phases. In order to accelerate the formation of CSH phases the dissolution velocity of SiO_2 has to be fundamentally increased.

— The increase in temperature is an important factor for the acceleration of this partial process. At the same time the greater dissociation of the water as well as the lower viscosity and the reduced surface tension of the aqueous solution have a promoting effect on the reaction behaviour [129].

— With growing pH the SiO_2 dissolution is accelerated, e.g. by the addition of NaOH to the building material mixture. However, the presence of NaOH also leads to a changed precipitation of the CSH phases and of their composition and in this way can also fundamentally influence the properties of the finished building materials [130].

— The structure of the SiO_2 applied has a significant influence on the SiO_2 dissolution. Thus the dissolution of SiO_2 in water is assumed to run via defects of the crystal lattice [131].

At distorted Si—O—Si bonds of the surface there is a higher probability of hydrolytic decomposition accompanied by the formation of HO—Si groups and consequently also of the detachment of a lattice component from the crystal. Therefore SiO_2 having a high concentration of defects promotes the dissolution process. Accordingly amorphous SiO_2 has a strongly increased solubility over crystalline SiO_2, as Fig. 7.19. shows.

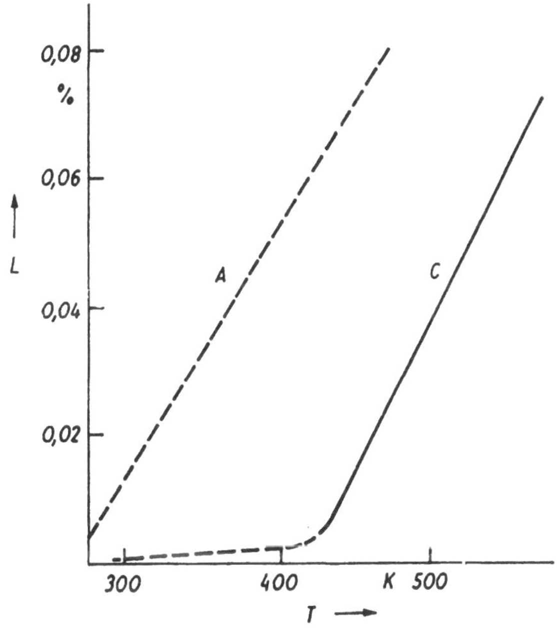

Fig. 7.19. Solubility L of amorphous SiO_2 (A) and crystalline (C) quartz in dependence on the temperature T [132].

7.2.3.1.3. *The mechanical activation of SiO_2*

A significant step forward in the production of high-strength lime-sand building materials was made when the SiO_2 component began to be activated mechanically.

As has been explained in detail in 6.1.2. through this the dissolution velocity of SiO_2 can be considerably increased. This leads to an increased formation of CSH phases and consequently to an increased solidification of the building material structure. As an example the compressive strength of lime-sand building materials in dependence on the treatment time using a vibration mill, and on the hardening time at a temperature of 373 K, is represented in Fig. 7.20. Whereas the compressive strength of non-activated samples does not exceed the value of $2 \cdot 10^7$ Pa even after a longer hardening time (10 h), a short-term treatment of a few minutes is already sufficient to obtain a concrete strength of $3 \cdot 10^7$ Pa at hardening times of 0.5 days.

If a disintegrator is used as the activation device then according to Fig. 7.21. an almost linear correlation exists between the speed of revolution of the rotors (for opera-

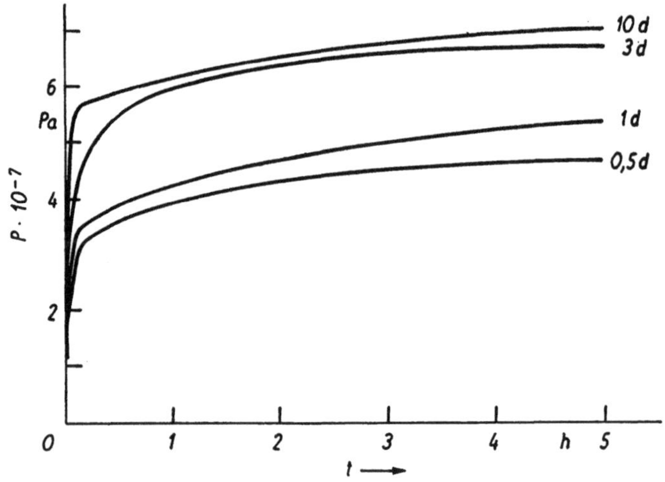

Fig. 7.20. Compressive strength P at silicate concrete samples hardened for differently long times at 373 K in dependence on the time t of the treatment of the SiO_2 component in a vibration mill [133].

d — time of hardening in days.

tion see 3.1.2.2.4.) and the compressive strength of the building-material bodies produced by using the activated SiO_2.

The parameters acting upon the increase in the activation of SiO_2 during the disintegrator treatment are represented in detail in Fig. 7.22. The surface of the stressed SiO_2, the quantity of SiO_2 converted by hardening to CSH phases and the compressive strength of the hardened building material samples increase in almost identical way with increasing speed of revolution of the rotors. Comprehensive investigations of the

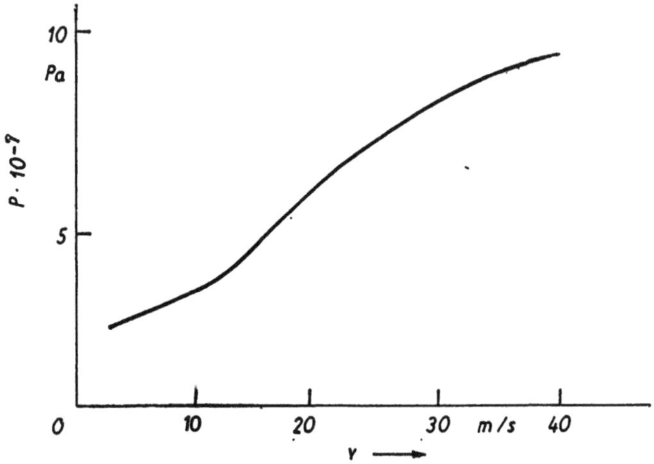

Fig. 7.21. Dependence of the compressive strength P of silicalcite building materials on the peripheral speed v of the rotors in the disintegrator [134].

CSH phases produced from the reaction of the SiO_2 have led to the result that there is a particularly close correlation between the increase in compressive strength and the increase in the development of highmolecular silicates rich in SiO_2 [136]. In other investigations, a linear correlation within certain limits has also been observed between the SiO_2 surface developing due to the mechanical treatment and the compressive strength of the building-materials body. In this way the specification of the dispersity of the sand applied has become an important activity parameter in the building material industry (Chapt. 3., [42], [125, 134]).

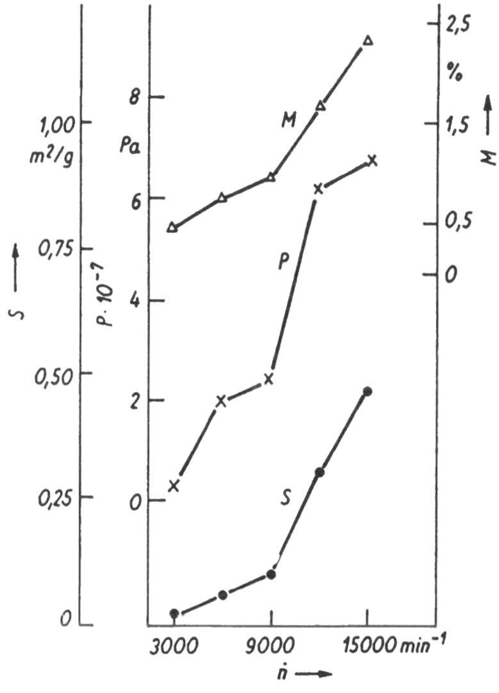

Fig. 7.22. Specific SiO_2 surface area S (after Blaine), SiO_2 M converted by the formation of CSH phases, and compressive strength P of the hardened building material bodies in dependence on the rotational speed $\dot{n}$ of the rotors of a disintegrator used for the activation of SiO_2 [135].

However as has been described in detail in 6.1.2., the reactivity of quartz, is in the first instance not determined by its dispersity but decisively by the degree of amorphization (see Fig. 6.22.). This result seems at first sight to be contradictory to experience. But comparative investigations of the surface sizes occurring in the building material industry have demonstrated that a linear correlation between the specific surface area and the degree of amorphization of quartz similarly exists [3, 50, 137]. This correlation is illustrated for samples activated in disintegrators and vibration mills in Fig. 7.23.

The way in which the amorphization decisively determines the dissolution of SiO_2 and consequently the entire hardening process can also be derived from the reaction-time curves given in Fig. 7.24.

A detailed interpretation of the course of the curves given in Fig. 7.24. is possible if account in taken of the conception developed in 3.1.2.3. and 6.1.2. that the SiO_2 particles of samples activated through the disintegrator do not have such greatly disturbed surface layers, but on the other hand more strongly disturbed volume regions while

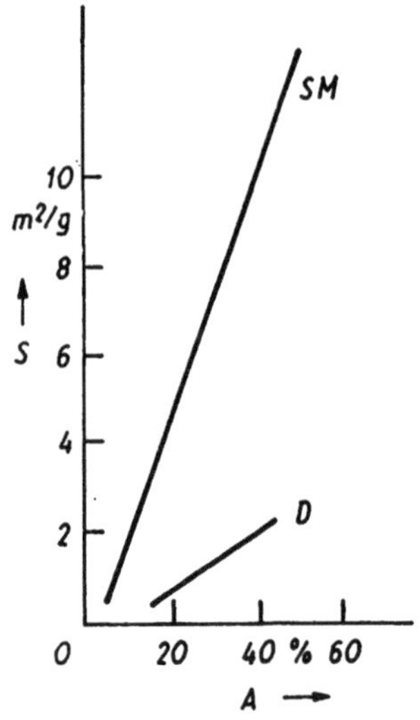

Fig. 7.23. Correlation between the specific surface area *S* of mechanically activated quartz and the amorphous part *A* using a disintegrator (*D*) and a vibration mill (*SM*) [137].

the SiO_2 particles of vibration-milled samples have marked amorphous surface layers and a relatively undisturbed nucleus. At the beginning the reaction is considerably smaller for SiO_2 activated in the disintegrator than for SiO_2 activated in the vibration mill. But after a relatively short period of time the curves *A* and *B* change into straight lines, indicating that the reaction crystal regions have a nearly constant degree of amorphization. The change of the curves *C* and *D* into straight lines appears only

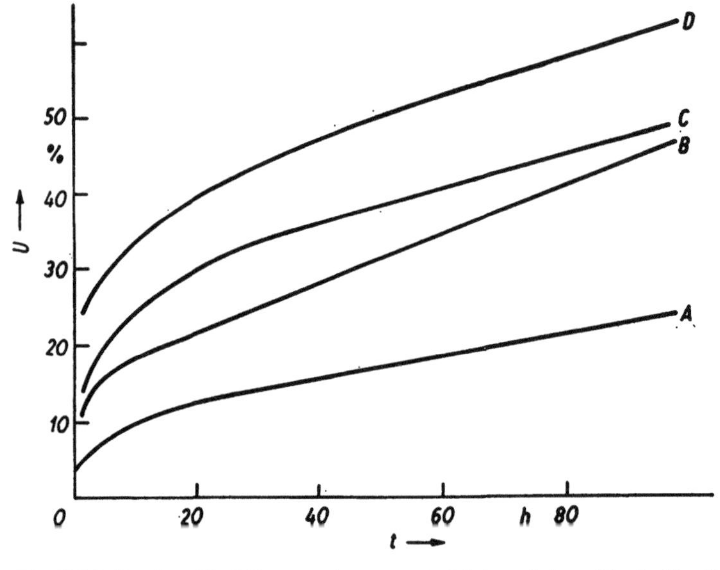

Fig. 7.24. Portion *U* of the quartz converted into calciumsilicated hydrates in dependence on the hardening time *t* and different activation conditions [137].

A – disintegrator, 9,000 rpm, *B* – disintegrator, 18,000 rpm, *C* — vibration mill, treatment time: 45 min, *D* – vibration mill, treatment time: 4 h, reaction temperature – 373 K.

after a relatively long period of time, with the slope of the curve *C* being even smaller than that of the curve *B* after having changed into a straight line. This result also corresponds with the model established in 6.1.2. according to which the inner volume regions of SiO_2 particles activated by the disintegrator may be disturbed much more strongly than those of SiO_2 particles activated by vibration milling.

Table 7.12. Compressive strength *R* of silicalcite in dependence on the specific surface area *S* and density ϱ of the raw material mixture and on the type of the activation set used (Chapt. 3,[42], [138]).

Activation set	S, m^2/g	ϱ, g/cm^3	$R \cdot 10^{-5}$, Pa
Disintegrator	0.16	1.7	909
Ball mill	0.16	1.7	795
Vibration mill	0.16	1.7	547
Disintegrator	0.32	1.7	943
Ball mill	0.32	1.7	559
Vibration mill	0.32	1.7	579
Disintegrator	0.32	1.9	1,091
Ball mill	0.32	1.9	966
Vibration mill	0.32	1.9	718
Disintegrator		1.9–2.0	up to 3,500*)

*) In laboratory experiments.

In Fig. 7.23. the amorphization of SiO_2 activated by the disintegrator related to the surface unit is shown to be always higher than that of SiO_2 activated by vibration milling. If the amorphization determines the hardening process then the SiO_2 samples activated in a disintegrator should lead to a higher compressive strength as compared to vibration-milled or ball-milled samples, if products are compared having the same specific surface area. As Tab. 7.12. shows this precondition is also always met with different initial parameters of the specific surface area *S* and of the density ϱ. The

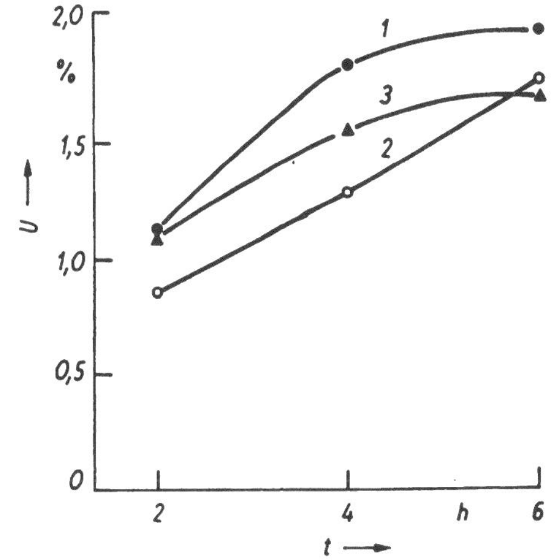

Fig. 7.25. Portion *U* of soluble SiO_2 of quartz having the same surface milled in various devices in dependence on the time *t* of the action of steam at $8 \cdot 10^5$ Pa (Chapt. 3., [42]).
1 – disintegrator, 2 – ball mill, 3 – vibration mill.

increased solubility of these samples speaks for a higher degree of amorphization of the SiO_2 samples activated by the disintegrator compared to those samples activated by ball and vibration mills which are milled onto the same surface area (see Fig. 7.25.).

Altogether the investigations show that the dissolution velocity of the SiO_2 component in the formation of silicate concrete is above all determined by both the degree of amorphization and the distribution of the amorphous regions in the SiO_2 particle, with the dissolution of SiO_2 decisively influencing the course of the hardening. The often measured linear dependence of compressive strength on the specific surface area may be explained by the fact that there is also a linear correlation between the degree of amorphization and specific surface area.

7.2.3.1.4. *The importance of the tribomechanical activation for the manufacture of silicate concretes*

The various technological aspects of the application of tribomechanical activation will be discussed in the following:

— First of all the manufacture of high-strength silicate concretes is facilitated by this principle. According to Tab. 7.12, silicate concretes with strengths above $3 \cdot 10^8$ Pa can already be produced today. In particular the possibility exists to manufacture large size structural units used in industrial building from silicate concrete (Chapt. 3., [42], [130]). In the Soviet Union in particular, but also in other countries a great number of building material factories have been constructed in the last 2···3 decades based on activation by disintegrators. They produce the new building material under the name "SILICALCITE". As an example a SILICALCITE-wall section is shown in Fig. 7.26.

Fig. 7.26. Manufacture of a SILICALCITE exterior wall panel [139].

— Porous lime-sand building materials having densities below 1 and sufficient strengths can be manufactured by the tribomechanical activation of sand. This builing material is being increasingly used in many countries because of its excellent heat dinsulation [140].

— Numerous raw and waste materials which because of their small reactivity it has not been possible to use until now as initial products in the manufacture of high-grade building materials, will be utilized through an appropriate mechanical activation in future [141—143].

— It is possible to rationalize the production process by the selective use of the tribomechanical activation of sand. As is shown as example in Fig. 7.27. with the use of a vibration mill, higher degrees of activation lead for instance to shorter hardening times if only a certain compressive strength is to be attained in production.

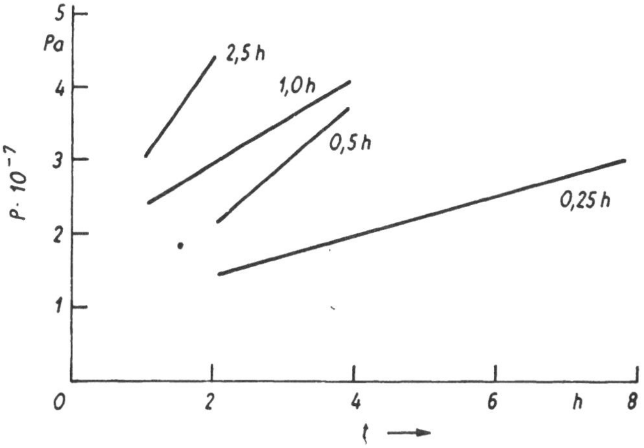

Fig. 7.27. Compressive strength P of silicate concrete in dependence on the isothermal hardening time in hours at 464 K and on the treatment time t of the binder in a vibration mill [133].

— In the knowledge that the degree of amorphization of SiO_2 plays a leading part in the hardening of lime-sand building materials, amorphous SiO_2 obtained from other branches of industry may be successfully used as an additive to the silicate building materials. As Fig. 7.28. shows, small quantities of an amorphous SiO_2 obtained from

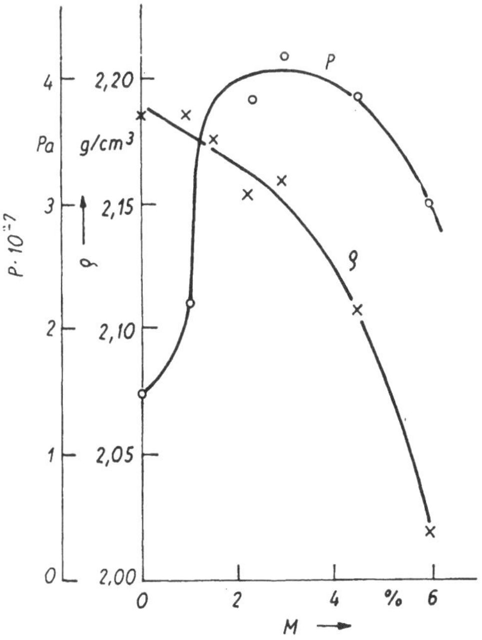

Fig. 7.28. Compressive strength P and density ϱ of silicate concrete in dependence on the addition of amorphous SiO_2 M at pressure-free hardening [144].

the preparation industry are already sufficient to fundamentally increase the strength of a building material body manufactured from lime and sand. However, higher additions lead again to a decrease in strength, since the density which decreases with growing quantities of additions causes a reduction in strength. The effect of amorphous SiO_2 additions is particularly marked if the hardening is conducted at relatively low temperatures, e.g. in the range of the boiling point of water [145].

— If the tribomechanical activation is combined with a chemical activation, e.g. by the addition of NaOH promoting the dissolution of SiO_2 and simultaneously elevating the boiling point (see 7.2.3.1.2.), then the possibility exists to manufacture large size structural units from silicate concrete without using autoclaves [138].

7.2.3.2. *Tribomechanical activation of cement and cement aggregates*

A considerable number of works deal with the tribomechanical activation of cements and its individual phases (Chapt. 4., [49], [146—158]. Different practical properties of cements may be varied by intensive mechanical treatment within wide limits, e.g. setting behaviour, storage stability as well as the compressive and tensile bending

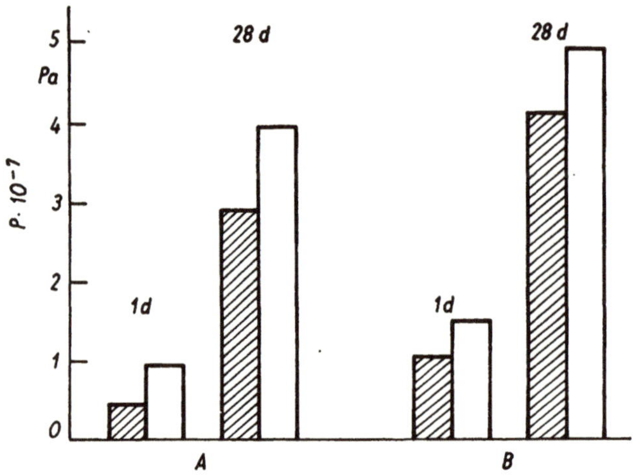

Fig. 7.29. Changes in the compressive strength *P* for two different cements (*A*, *B*) after different hardening times [149].
d — hardening time in days, □ — activated, ▨ — non-activated.

strength of the concretes produced from activated cements. By way of illustration the compressive strength for two different cements after different times of hardening is represented in Fig. 7.29. Particularly at the beginning but also after the end of the hardening process, the mechanical activation of cements leads to an important increase in the compressive strength of the cement concrete. The reasons for this are mechanically released changes in surface and structure which for their part accelerate the hydration velocity of the cement during hardening. As Fig. 7.30. shows, the change in the compressive strength in dependence on milling duration may be correlated above all with the lattice imperfections measured by X-rays, whereas the changes in the surface have a smaller effect on the increase in the compressive strength.

The changes in the hydration velocity of the cement clinker and of two of its individual phases for short treatment times are given in Fig. 7.31. It is seen that the individual phases of the cement are activated with different velocities in the mechanical treatment. C_3S (tricalcium silicate), which is held responsible for the high-early-strength on account of its high hydration velocity, may already be optimally activated after a short treatment time. β-C_2S (dicalcium silicate) requires a longer treatment time for an optimum activation and on the other hand influences to a larger extent the late-strength. In this way the parameters important to the use of cement may be influenced to a certain extent by the activation energy (Chapt. 4., [49]).

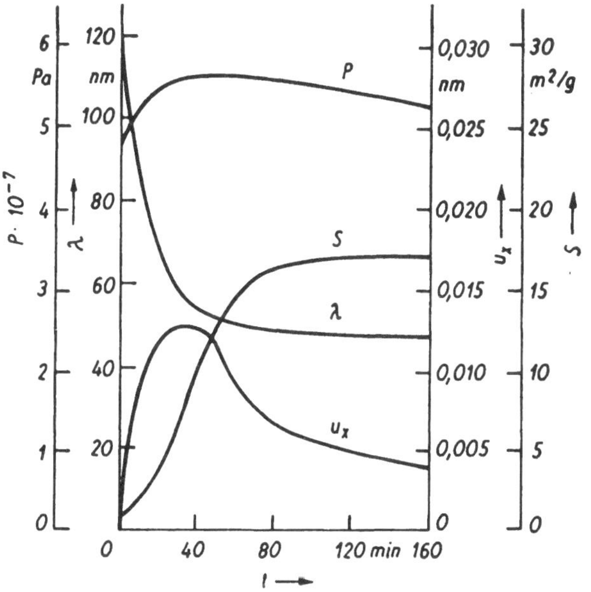

Fig. 7.30. Changes in he lattice imperfections u_x, in the specific surface area S, in the primary crystallite size λ of mechanically activated cement, in the compressive strength P of the hardened concrete in dependence on the treatment time t of cement [159].

Certain practical properties of the cement may also be optimized selectively by the choice of the activation set. As Fig. 7.32. shows, the vibration mill is particularly suitable for the manufacture of high-early-strength cements and has been used to this and under industrial conditions [148]. If cements activated by vibration milling are added the concretes harden so rapidly that they can already be demoulded after less than 24 h [158]. If milling auxiliaries such as sulphite waste liquors are used in the vibration-milling activation of cements, a further increase takes place in the compressive strength up to 30 per cent under otherwise similar stress conditions [148, 160, 161].

If highly activated cements are stored under atmospheric conditions an only gradual annealing of the lattice imperfections takes place. The annealing is even smaller in closed vessels (Chapt. 4., [49]). However the gain in strength achieved by milling is lost again very rapidly with certain less activated cements, in the period of storage [162—164]. Evidently the ageing process is very strongly determined by the specific properties of the individual kinds of cement and by the way the ambient atmosphere acts.

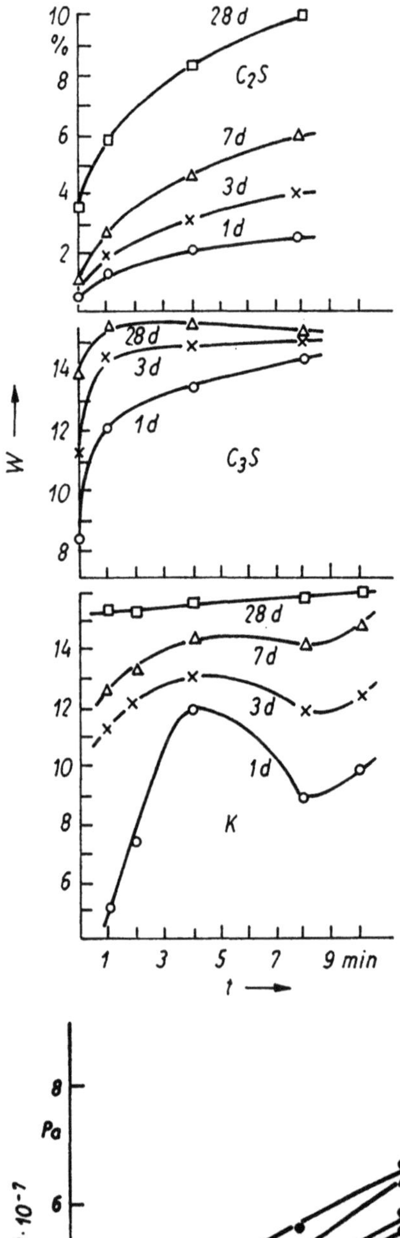

Fig. 7.31. The influence of the milling on the velocity of the hydration of dicalcium silicate (C_2S), tricalcium silicate (C_3S), and cement clinker (K) at different hardening times [153].

W – degree of hydration, t – treatment time, d – hardening time in days;

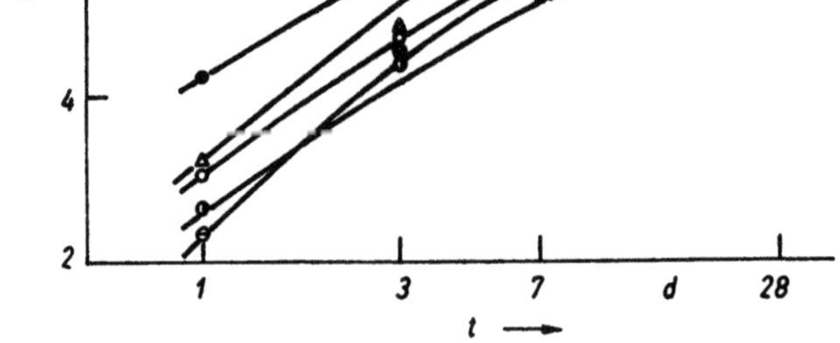

Fig. 7.32. Compressive strength P in dependence on the hardening time t and on the activation of the cements in various activation sets [153].

1 – vibration mill, 2 – eccentric mill (amplitude – 3 mm), 3 – eccentric mill (amplitude – 5 mm), 4 – ball mill, 5 – jet mill.

In certain cases cements are mixed with substances having latently hydraulic properties, such as slags and power station ash, in order to dilute the cements. However, the addition mostly amounts to a few per cent only. If the substances, e.g. lignite filter ash, are mechanically activated, on account of their growing hydraulic properties they can be added to the cement up to 30 per cent without falling below the standard values of pure cement [165, 166]. Since it is not always possible to provide filter ash of constant quality due to the varying composition of the combustion products, the technological utilization of this effect is rendered more difficult.

In summary the investigations have indicated that the hydration of cement associated with the structure formation may be substantially accelerated by mechanical activation in analogy to the formation of CSH phases in silicate concrete. The increased reaction velocity is brought about by the reduction of the activation energy of the hydration process in the order of 10···25 kJ/mole [153]. As reasons for the acceleration of hydration, lattice imperfections (Chapt. 4., [49]) and growing amorphization particularly in the surface layer, are discussed.

The following fields of practical application result from this:

- Manufacture of cements with higher high-early-strength,
- manufacture of cements with higher final strength,
- maintenance of the same concrete quality, but reduction of the cement portion in the concrete mixture,
- dilution of the cements with mechanically activated aggregates while maintaining the quality parameters.

7.2.3.3. *The formation of binders from tribomechanically activated raw materials*

The effect of tribomechanical treatment on the formation of binder phases has been extensively investigated. The clarification of relationships has attained a different level for the individual substance systems. In individual cases only the qualitative proof of a tribomechanical effect has been furnished. In some cases this has already been used in the development of technological procedures.

In the following the most important systems of substances will be discussed:

— As has already been shown by numerous examples in 5.6. and 7.2.2., the temperature required for the appearance of a solid reaction may be decreased in mechanical treatment by some 100 K and sometimes to room temperature. This effect could also be detected by several examples for the tribomechanical activation of raw materials needed for the manufacture of cement binders. After having pretreated a mixture of CaO and kaolinite in a disc-type vibration mill, the solid reaction took place accompanied by the new formation of phases in a temperature range of 700···1,100 K [167]. The decrease in the reaction temperature is particularly strong if the reaction is released by a combined thermal and mechanical activation of the initial components. Thus if CaO is milled together with pumic stone, fly ash, clay, sand, alumina, shale, and infusorial earth in a heated mill, chemical reactions binders already appear at 700 K with which building material bodies of considerable strength can be produced [168].

Clinker phases having hydraulic properties may be manufactured with an output of 64 per cent through the treatment of cement raw meal at a temperature of 1,123 K in a vibration mill [169].

However, further scientific and technological investigations are still required for an industrial utilization of the tribochemical manufacture of cement binders.

— A tribochemical slaking method has been developed for the manufacture of building limes up to technological maturation, which is particularly suitable for the processing of dolomite limestones. With milling a higher reaction velocity through the removal of covering films (see 5.6.) and a high adjustable final fineness are simultaneously obtained [170].

— In the preparation of synthesis-raw gypsum, too, the utilization of the tribochemical effect leads to a simplified technology and to an increased mechanical strength of the building material body made from activated raw gypsum. In order to produce the α-hemihydrates, water-soluble compounds are needed such as phosphates which if activated products are applied, apart from a small residual content, may be dispensed with. Simultaneously the commencement of stiffening of anhydrite produced from the synthesis-raw gypsum may be controlled by the gauged delivery of milled additives, e.g. calcium hydroxide [171].

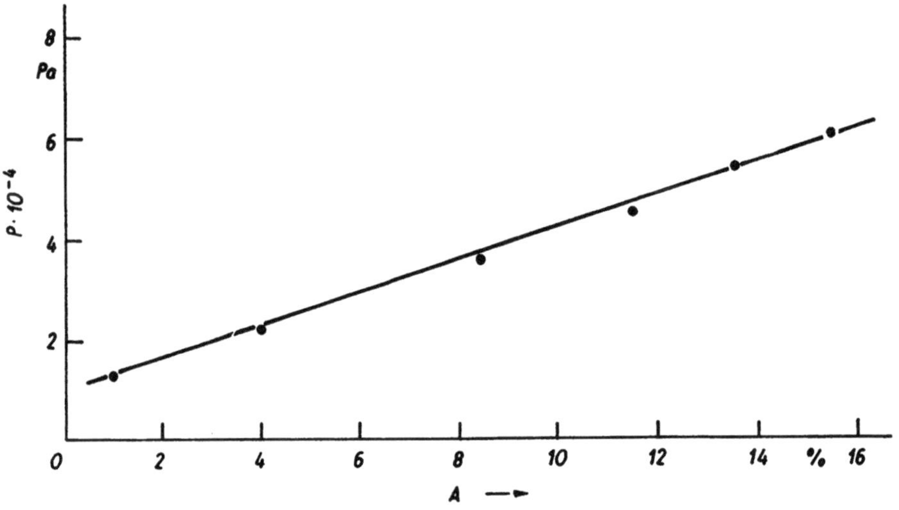

Fig. 7.33. Green compressive strength P of moulds used in foundaries in dependence on the amorphous part A of SiO_2 in the binder [172].

— Binders are also needed to a larger extent for the manufacture of moulding material mixtures in foundries. Especially suitable as binders are specific bentonites which are made into moulding material mixtures by milling with quartz sand.

Clays not suitable for this purpose may also be made into highgrade binders by mechanical activation, with in addition the hardening time being considerably shortened. Furthermore mechanically activated quartz may also be used as a binder.

As Fig. 7.33. shows, in analogy to silicate concrete, the compressive strength increases almost proportionally with the amorphous part of the tribomechanically treated SiO_2 phase in this system as well. It may be concluded that, as with silicate concrete and cement concrete, the amorphous phase plays a decisive part in the dissolution and solidification processes [172].

Even though no system on the tribomechanically activated formation of binders exists at present, the examples mentioned above, which differ widely in terms of substance system and application, show that mechanical treatment often very substantially influences the formation and the quality parameters of binders and through this in some cases the design of new rational technologies for the production of binders is possible.

The different effects may be attributed uniformly to the tribomechanically determined acceleration of those solid reactions which cause the formation of the binder phase and the hardening of the building material, respectively. The formation of amorphous phases has a crucial importance in the activation of silicate binders.

7.2.3.4. The sintering of mechanically activated solids

Sintering plays an important part in the production of building and other materials in technology. Thus the high-refractory stones used for lining of kilns are manufactured from MgO and CaO by sintering processes. Sintering is also the most important process step in the manufacture of ceramic products.

Sintering is defined as the thermal treatment of a porous body or powder product with the object of compaction or solidification without the action of external forces upon the body. Usually high temperatures are used in the sintering, amounting to 50···95 per cent of the melting temperature for oxides.

Under the influence of the high temperatures a number of processes run in the solid, such as annealing of the crystal, recrystallization, grain growth and chemical reactions if reaction possibilities are present. The processes are based on the diffusion, or self-diffusion in one-component systems, of the lattice components as the prevailing transport mechanisms.

As has already been presented in 5.7. the diffusion in solids is a structure-sensitive quantity and can be particularly influenced by mechanical treatment of the solid phase. Usually a loosening of the lattice occurs under the action of mechanical energy, accompanied by the formation of zero-dimensional or multidimensional defects (see 3.2.). Since the sintering mainly consists of the diffusion of lattice components, both grain boundaries and lattice vacancies are of particular importance to the transport of materials and pores. Generally the mechanical treatment of raw material to be sintered results in a strong increase of the inner and outer surfaces (see 5.5.) leading to an easier diffusion over grain boundaries and surfaces. But through mechanical treatment, the number of vacancies in the crystal lattice is also increased beyond the equilibrium defect concentration, which leads to an acceleration of the lattice diffusion and to an increase in the diffusion coefficient, respectively. For instance, as Fig. 7.34. shows, the values of the self-diffusion coefficient of Ca^{2+} ions are considerably higher for active CaO than the values measured for single crystals or polycrystals in the whole temperature range. Surface diffusion processes dominate up to temperatures of 1,400 K, and volume diffusion processes beyond that.

An increase in both the bending and tensile strengths as well as in the bulk density of the sample occurs during the sintering. The following simple relationship is frequently

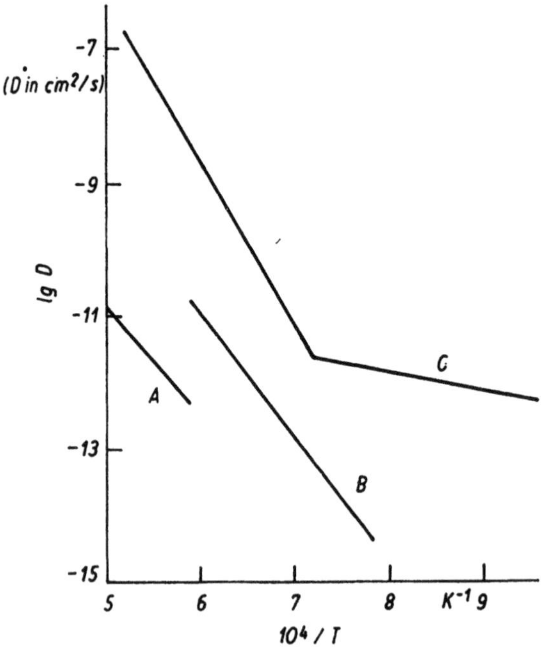

Fig. 7.34. Self-diffusion coefficients D of Ca^{2+} ions in CaO in dependence on the temperature T for monocrystals (A) and polycrystals (B) as well as for active CaO (C) [133].

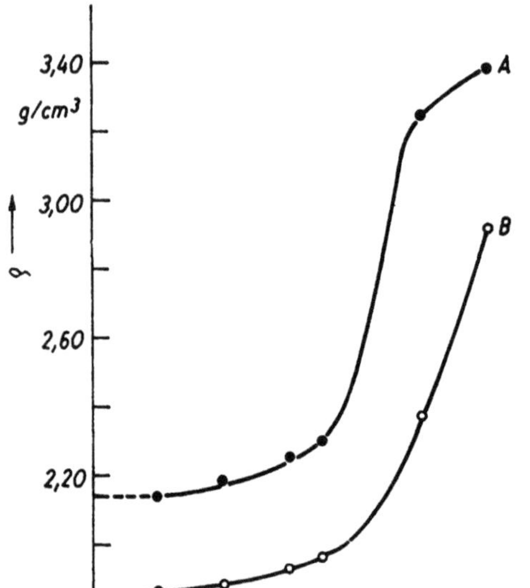

Fig. 7.35. Dependence of the bulk density ϱ on the sintering temperature T for activated MgO (A) and for MgO (B) not stressed mechanically [177].

observed [173]:

$$P = K \cdot \varrho^m \tag{2}$$

(P — tensile strength, ϱ — bulk density, K, m — constants.)

With the sintering of CaO the strength increases proportional to the density ($m = 1$), [174]. Consequently in order to obtain a high strength it is necessary to so design the

Table 7.13. Changes in the properties of sintered materials by tribomechanical activation of the initial products.

No.	Material	Activation set	Effect	Comments	Ref.
1	Dolomite	Vibration mill	Increase in compressive strength	Appearance of a maximum in dependence on the milling time	[175]
2	Dolomite	Press or roll	Reduction of the sintering temperature		Chapt. 4., [141]
3	MgO	Disintegrator	Increase in density		[176]
4	MgO	Vibration mill	Increase in density		Chapt. 5., [209], [177, 178]
5	MgO	Vibration mill	Decrease in porosity at reduced sintering temperature		[179]
6	MgO	Jet mill, attritor	Increase in density		[180]
7	Aluminium silicate	Vibration mill	Increase in compressive load	Appearance of minima and maxima	[175]
8	CaO	Press	Changes in strength		[181]
9	CaO	Vibration mill	Reduction of the sintering temperature		[133]
10	α-Al_2O_3	Vibration mill	Reduction of the sintering temperature		[182]
11	$BaTiO_3$	Vibration mill	Reduction of the sintering temperature Increase in the sintering interval		[183]
12	SiO_2	Vibration mill	Decrease in porosity		[133]
13	SiO_2	Vibration mill	Sintering of quartz runs similarly as in a glass		[184—186]
14	SiO_2	Vibration mill, attritor	Increase of the flexural tensile strength		[187]

sintering process that a body as free of pores as possible develops. As Tab. 7.13. shows, in a larger number of substances the sintering process is stimulated by mechanical treatment.

The increase in the bulk density in dependence on the sintering temperature is represented for MgO in Fig. 7.35. The activated sample already shows an increased density upon press-working. This effect may be attributed to the fact that porous agglomerates are destroyed, voids are removed, and the particles are comminuted, particularly at places with high self-porosity, and in this way the particle arrangement approximates to the densest packing of spheres [177].

The density differences between activated and non-activated samples continues to grow in a temperature range in which the surface diffusion becomes the prevailing mechanism. This is particularly marked in the first stage of the sintering process.

Usually the sintering is conducted at temperatures at which the defects responsible for the acceleration of diffusion processes are annealed. However, it can be seen that the diffusion coefficients of activated powders only slowly approach the standard value during the sintering [188]. It is assumed that the many vacancies coagulate into micropores and leave a network of dislocations in the crystal. These defects are only slowly eliminated at high temperatures [189]. However, if the activated substances to be sintered undergo a preliminary annealing the specific effects of activated substances decrease more and more with rising temperature, since an increasing annealing of the defects already appears in the preliminary annealing phase (Fig. 7.36.).

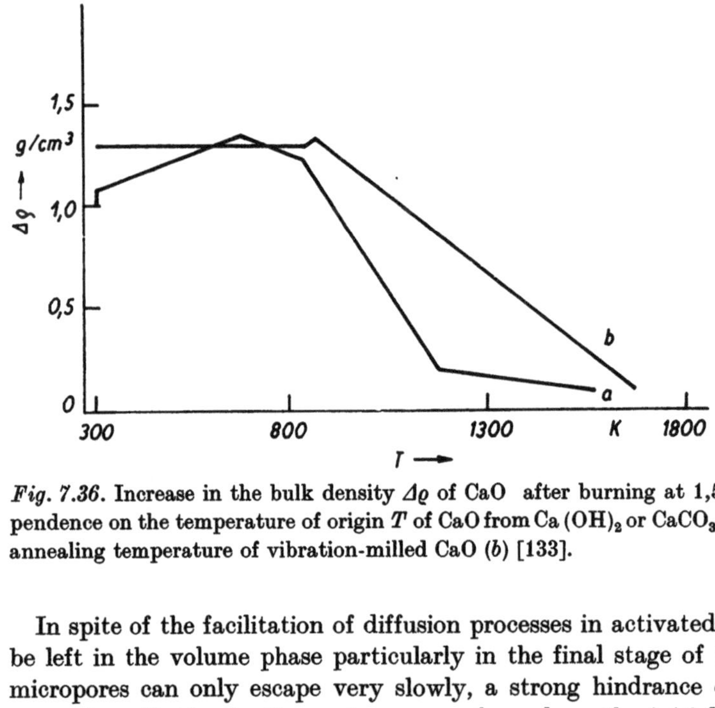

Fig. 7.36. Increase in the bulk density $\Delta\varrho$ of CaO after burning at 1,573 K for two hours in dependence on the temperature of origin T of CaO from $Ca(OH)_2$ or $CaCO_3$ (*a*) and on the preliminary annealing temperature of vibration-milled CaO (*b*) [133].

In spite of the facilitation of diffusion processes in activated solids, micropores may be left in the volume phase particularly in the final stage of the sintering. Since the micropores can only escape very slowly, a strong hindrance of the compaction process arises. Evidently these phenomena depend on the initial structure, for maxima and minima of the compressive strength occur in certain cases depending on the duration of treatment (see Tab. 7.13.). How in detail the treatment intensity influences the change in structural disorders and the latter in turn affect the sintering ability of high-refractory products will be demonstrated by the example of a magnesium oxide activated by a disintegrator. Some structural quantities and the bulk density of MgO are given in dependence on the peripheral speed of a disintegrator in Fig. 7.37. and Fig. 7.38., respectively.

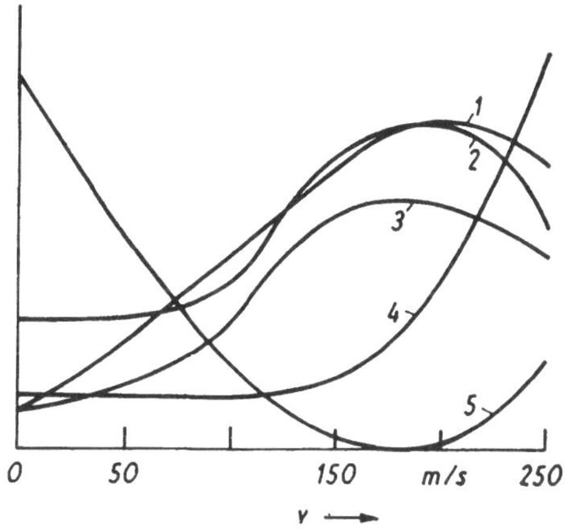

Fig. 7.37. Changes in the solid properties of MgO by disintegrator treatment in dependence on the peripheral speed v of the rotors [176].

1 – relative lattice distortion, 2 – specific surface area, 3 – particle size frequency H at 1 μm, 4 – X-ray amorphous portion, 5 – primary crystallite size.

Three ranges may be distinguished with respect to the increase in bulk density: Range I: 0···85 m/s; Range II: 85···170 m/s; Range III: 170···250 m/s.

The reason for the relatively small increase in bulk density in range I is in the main the absence of enlargement of the specific surface area and the little changed particle

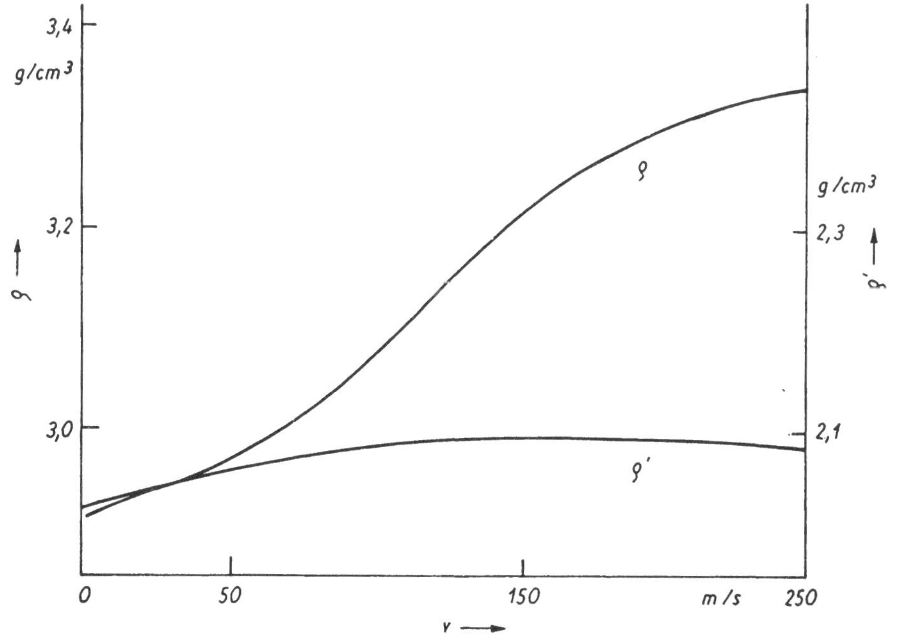

Fig. 7.38. Bulk density of sintering ϱ and unsintering ϱ' test bodies in dependence on the peripheral speed v of the rotor using a disintergrator [176].

size of MgO powder. Range II which is characterized by a strong increase of bulk density is distinguished by the fact that important solid properties such as lattice distortions, specific surface area, particle size, primary crystallite size — with the exception of the X-ray amorphous part — are influenced towards the promotion of the sintering process. In range III most structural quantities no longer reach the maximum values of range II in spite of a further increase in the peripheral velocity of the rotors. Only the X-ray amorphous part increases, and this is considered to be the reason for a further but smaller increase in the bulk density of the sintered sample.

As the last example shows there is no single quantity which changes the sintering process alone. The structural parameters act in a complex manner upon the sintering and are able to mutually compensate each other as has been shown above. From this arises the difficulty of arriving as a theory of the sintering of activated solids valid for all substances. However, the qualitative statement that generally an acceleration of the sintering process can be obtained by tribomechanical activation, may be regarded as a significant result.

The technological advantages of using tribomechanically activated materials to be sintered consists in reaching higher bulk densities [2] and in decreasing the sintering temperature [133].

In summary, the results discussed in 7.2.3. show that the most different building material properties may be varied through tribomechanical structural changes, and therefore a modification of these properties to suit the desired purpose is possible. In general by mechanical activation building materials are obtained which are characterized by their higher quality parameters. Whilst adhering to the usual standard values, the preconditions are given to manufacture the respective building materials by a more rational technology, e.g. by the reduction of the firing temperature.

Possibilities also arise for the development of completely new technologies. Finally the range of raw materials which can be utilized in the manufacture of building materials is extended by using the principle of mechanical activation.

7.3. The ignition of metal fires by tribochemical reactions

7.3.1. Metal fires in engineering

In recent decades there has been a considerable increase in metal fires, leading to serious accidents and the destruction of industrial plant (Chapt. 4., [130], [190, 191]). The increased appearance of metal fires may basically be attributed to the following reasons:

— While most utility metals are not able to catch fire in air if they exist in a compact form, in pure oxygen or an oxygen enriched air atmosphere, they are able to burn with the appearance of flames, producing partly very hot flames with temperatures of about 3,000 K [192]. The dangers connected with the production, delivery, and utilization of O_2, particularly in industrial plants, have constantly increased with the development and continuous growth of oxygen production in the chemical and metallurgical industries since the beginning of this century.

— The ability to burn in an oxygen atmosphere is manifest to greatly varying degrees in the individual utility metals. While copper in the form of fine wires can only be set on fire with great difficulty, and in the form of larger work pieces not at all [193], iron is one of the metals which ignite relatively easily and burn fiercely within seconds (Chapt. 4., [130]). Therefore parts of machines particularly endangered by ignition are made of more precious metals such as copper and copper alloys. However, cheap iron materials are extensively used for the construction of industrial plant with substantial piping systems thus considerably increasing the ignition hazard for such plants.

— Moreover oxygen is increasingly used in a compressed form. Since the combustion velocity increases proportionally to the square root of the oxygen pressure, not only is the danger of a fire ignition increased but also an increased combustion intensity occurs [194].

— For many technological purposes metals are produced in a finely distributed form, e.g. by milling or the reduction of metallic compounds such as nitrates, oxalates, carbonates. Depending on the production conditions these are sometimes already self-igniting in air and often burn with great intensity [195].

As a result in recent decades have been made to clarify the mechanism of metal fires and the causes of ignition leading to fire [196, 197].

7.3.2. The mechanism of metal fires

Metallic oxides appear during the oxidation of metals which lead to the formation of solid reaction layers in the lower temperature range and with growing thickness increasingly hinder the course of the reaction. Frequently, complicated diffusion processes of the reactants through the covering layer are needed at this stage to enable the reaction to continue. These processes, described as starting or scaling processes, determine the reaction output of the heterogeneous reactions between the gaseous and solid phases in a large temperature range (Chapt. 4., [166]).

If the reaction heat appearing during the oxidation leads to the melting of the oxide layer a precipitous increase in the oxidation velocity occurs. This reaction stage is described as metal fire because the reaction runs accompanied by flames. The combustion takes place at the surface of the melted oxide layer continuously dissolving the metal with the formation of a lower oxide. The following mechanism is assumed to act in the case of iron [192]:

$$3FeO + 1/2O_2 \rightarrow Fe_3O_4 \quad (3)$$

$$Fe_3O_4 + Fe \rightarrow 4FeO \quad (4)$$

For some metals having a low boiling point, such as magnesium, the metal combustion runs in a homogeneous gaseous phase [198]. In order to maintain the combustion process the reaction enthalpy has to be so high that the oxide developing on the metal does not solidify. When the oxide enters the solid phase the combustion is immediately

Table 7.14. Order of reaction and mechanism of different oxidation reactions at iron.

Kind of reaction	Reaction order	Mechanism	Ref.
Tribochemical reaction	0	Reaction at disturbed fresh surfaces	Chapt. 4., [130]
Thermal reaction with solid covering film	0.28	Diffusion through solid covering films	Chapt. 4., [166]
Fire reaction	0.5	Diffusion in liquid covering films and homogeneous gaseous phase reaction, respectively	[192, 198]

finished. Temperatures of over 1,500 K are required to maintain the burning of utility metals.

The order and the mechanism of the tribochemical reaction, of the thermal reaction on solids with solid covering films, and of the metal fire of iron are summarized in Tab. 7.14. All 3 types of reaction are different in their course. As will be shown in 7.3.3. and 7.3.4. the first two processes named may play the role of initial igniter in initiating a fire.

7.3.3. *Reasons for the ignition of metal fires*

It is relatively difficult even in oxygen atmospheres to ignite compact metals and alloys, since on the one hand high local temperatures are required, and on the other hand the heat provided locally is rapidly carried off and distributed in the metal volume on account of the high heat conductivity of the metals. The high activation threshold for the initiation of a fire permits the use of iron and aluminium materials in oxygen plants with an acceptable risk.

Nevertheless certain physical and chemical processes may lead to the accumulation of energy needed for the initiation of a fire.

Among these are the following:

— Ignition by compression heat:

If the compression takes place very rapidly the adiabatic release of compression energy leads to locally high temperature peaks which especially in fast running pressure impacts have led to fire in oxygen valves [197].

— Ignition by eddy formation:

The flow velocity of gases often reaches high values of > 100 m/s in oxygen plants, particularly in turbocompressors and in valves.

An intensive eddy formation and by that a local heating up of the system and the initiation of fire may occur with an incorrect guidance of the flow (example in [199, 200]).

— Ignition by friction:

Friction processes may result in high short-term temperature peaks sometimes exceeding 1,000 K (see 5.3.). Mobile machine parts in oxygen plants are particularly endangered. Thus shafts and bearings becoming not as well as frictional processes on rotors inadequately fitted in turbocompressors have often led to a local overheating and to the initiation of fires.

— Ignition by chemical reaction:

Chemical reactions play a decisive part in the fire ignition of metals.

Sometimes such high quantities of heat are released during the course of chemical reactions that temperatures over 2,000 K may frequently appear. Therefore thermal sources of ignition are particularly dangerous, because the energy released by the reaction is often concentrated in statu nascendi in the smallest space. In many cases the high energy density thus developing provides the activation energy for the fire ignition. For instance, steel sparks may have temperatures higher than 2,500 K [201].

7.3.4. *Tribochemical fire ignition*

7.3.4.1. *Fire ignition during the treatment of metals in vibration mills*

The occurence of a tribochemical fire ignition has been thoroughly investigated in the mechanical treatment of carbonyl iron powder. Immediately after the beginning of the treatment the tribochemical reaction runs with a constant velocity dependent on the treatment intensity (see 4.4.). The temperature in the standard-condition reaction vessel measured here mostly only very slightly exceeds the ambient temperature. Hence the heat developing due to the reaction flows off sufficiently fast and has practically no effect on the velocity of the oxidation reaction. The reaction is of a zero order above 1 Pa and runs with the same velocity in pure oxygen as in air (see 4.6.). However, the treatment processes in a vibration mill are of a statistic nature, so that at individual points in the reaction vessel the tribochemical reaction is more accelerated relative to the measured average and consequently a higher quantity of heat is released. In this way the tribochemical reaction may change into a thermally activated reaction and with a sufficiently high storage heat also release the combustion reaction. Such a transition from the tribochemical into the combustion reaction is shown in Fig. 5.8.

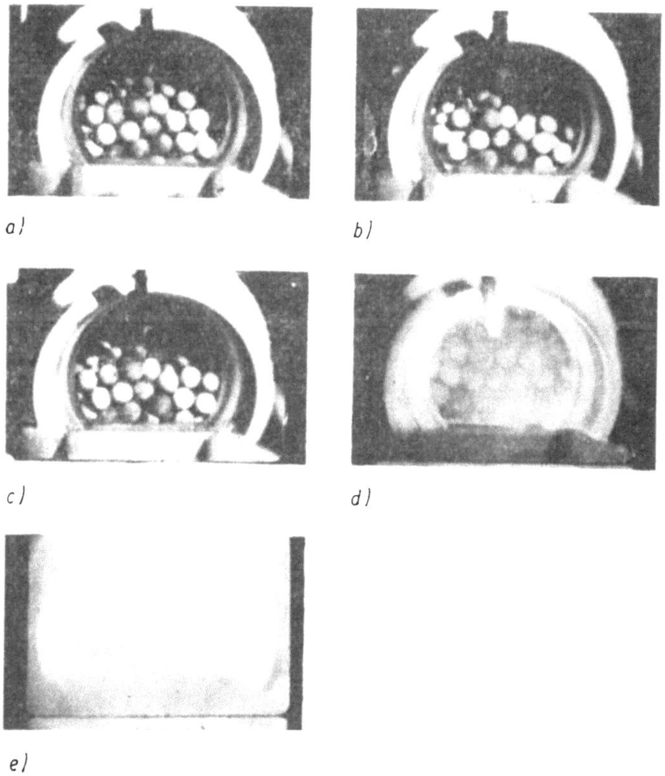

Fig. 7.39. The ignition of manganese in a pure oxygen atmosphere immediately at the beginning of the treatment in a vibration mill (Chapt. 4., [130]).

Temporal interval of the photos — 1/24 s; a — beginning of the treatment, d — beginning of the fire at one point, e — fire in the whole vessel.

The ignition of the metal powder only takes place after a milling period of several minutes. Due to the oxygen deficiency in the reactor the fire is extinguished after a short period of time. If oxygen is once again applied then in turn a certain period of milling will be necessary to initiate the fire for the second and third times. The statistical nature of the fire ignition is expressed by the fact that the time of the mechanical pretreatment up to the appearance of the fire is subject to strong variations.

If iron is pre-milled in the vacuum for varying lengths of time and treated with oxygen after interruption of the stress, no burning will be observed. Hence the ignition of a fire is directly connected to the treatment process. However, as will be shown in Fig. 7.40. below, the probability of ignition may be strongly increased by pretreatment in the vacuum and subsequent mechanical stress in oxygen.

The probability of ignition is different for individual metals and metallic alloys. Even if for example manganese is used in the form of relatively coarse-disperse granules, the tribochemical reaction immediately turns into metal fire at the beginning of the treatment. The metal fire consists of white-yellow flames and ususally leads to the bursting of the reaction vessel.

As Fig. 7.39. shows, the fire emanate from one point of the reaction vessel and very quickly overcomes the entire milling volume. The transition from the tribochemical to the burning reaction is bound up with the increase in the oxidation velocity by more than 4 orders of magnitude (see Tab. 7.15.).

Table 7.15. Velocity v of the tribochemical reaction and of the metal fire at the treatment of manganese in a vibration mill (Chapt. 4., [130]).

$v_{tribochem.}$, mole/h	v_{fire}, mole/h
$2 \cdot 10^{-3}$	> 40

Fig. 7.40. shows the probability of ignition of a series of iron and aluminium alloys used in industry, in the tribomechanical stress of the materials pre-comminuted to cuts using a vibration mill. The alloy GS-45 has proved to be easily inflammable, with an ignition occurring immediately during the addition of pure oxygen after pretreatment in a nitrogen atmosphere. Chrome steel GS 15 Cr 13 and the aluminium alloy GAlSi5Cu1 could not be ignited under the conditions used; the reason for the latter evidently being that a particularly high activation threshold has to be provided for the ignition process of aluminium.

The probability of ignition depends on the pressure. Thus below a certain pressure a transition into the burning reaction is no longer observed. Accordingly no ignition is realized in air (Chapt. 4., [130]). The reason for this may be found in the fact that the reaction velocity and consequently quantity of energy released per unit of time, decreases proportionally with $\sqrt{p_{O_2}}$ with dropping O_2 partial pressure, and is no longer sufficient for maintaining the fire (see 7.3.1.).

The probability of ignition depends additionally on the intensity of the treatment. By reducing the intensity in the milling of manganese, e.g. by the use of smaller, less intensive mills, the usually spontaneously appearing ignition process can be delayed (Chapt. 4., [130]).

When aluminium is ignited, the oxygen required for the oxidation of this metal may also exist in a bounded form, since in the bombardment of aluminium with fine oxide particles, e.g. Fe_2O_3 or Fe_3O_4 particles, the thermite reaction runs. Even in a pure nitrogen atmosphere the initiation of the thermite reaction may be detected by the appearance of flashes of light when Fe_2O_3 is used [202].

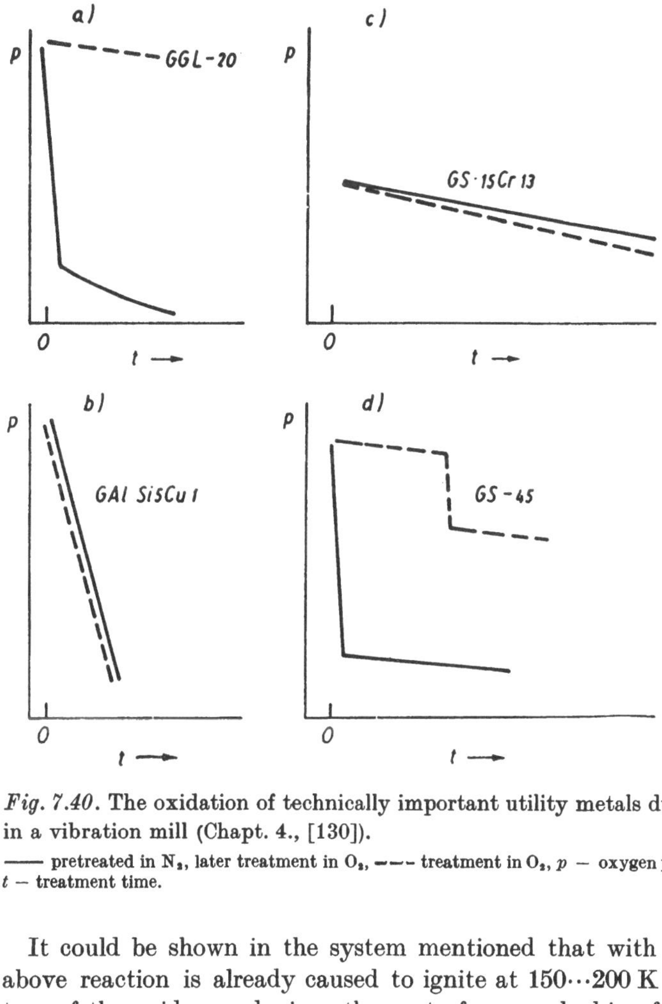

Fig. 7.40. The oxidation of technically important utility metals during the tribomechanical stress in a vibration mill (Chapt. 4., [130]).
—— pretreated in N_2, later treatment in O_2, ---- treatment in O_2, p – oxygen pressure in the reaction vessel, t – treatment time.

It could be shown in the system mentioned that with continuous heating up, the above reaction is already caused to ignite at 150···200 K below the melting temperature of the oxides used, since the part of energy lacking for the melting of the oxides is evidently locally released through the commencement of the solid phase reaction between aluminium and iron oxide [203].

Another source of ignition for the release of fires consists in the spontaneous demolishing of tribomechanically produced lattice imperfections which leads to a high short-term heating up of the stressed solid. For instance, glowing powder particles could be shown to leave the rolling mill when various oxides are rolled, such as Fe_2O_3 [204] or WO_3 (Chapt. 4., [141]), (see Fig. 7.41.B). If the rolling process is repeated such violent explosions occur, accompanied by sparks, that for instance a steel roller having a diameter of 20 cm is smashed into several parts, see Fig. 7.41.C.

Fig. 7.41. The appearance of explosions during the rolling of Fe_2O_3 [204].

A – rolling of Fe_2O_3 for 1–3 times without explosion, B – rolling for 3–5 times with partially vehement explosions, C – a shaft destroyed by explosion (⌀ ≈ 200 mm).

7.3.4.2. *Tribochemical fire ignition in technical plants*

The dramatic increase in the need for oxygen in the most varied technical procedures requires the construction of high-performance large-scale plants in which oxygen is conveyed with pressures up to $1.5 \cdot 10^7$ Pa, with velocities of more than 100 m/s, and with a throughput of some 10,000 Nm^3/h. For various reasons it has been necessary in the construction of these large-scale plants to make the pipelines of unalloyed steel instead of copper [205]. However, this greatly increases the risk of a fire ignition. In order to investigate the causes of pipeline fires various powders were blown at different velocities and oxygen pressures through bent pipelines [206]. From this the results were obtained listed in Tab. 7.16. They may be divided into 3 categories:

1. No light effects during the blowing through of the powder mixture,
2. sparking, and
3. burning.

Using quartz sand neither sparking nor burning is observed even with high velocities. The addition of 20% iron powder to the quartz sand is sufficient to cause mainly sparking, and at high velocities, that is at high stress intensities, burning. Similar conditions are found when mill scale is used [207].

These investigations show that on their own, the frictional processes of the powder particles used are not sufficient to cause sparking or even fire ignition. Powder particles act as fire igniters only in association with the mechanical activation of chemical reactions.

Moreover, in bent pipe segments a substantially higher spark concentration occurs already at smaller velocities than in a straight pipe segment.

Fires in pipe networks were always observed immediately after the bends, i.e. at a point where the particles are especially strongly activated mechanically and consequently chemically, too [207], (see 4.4.). Moreover in a bent section there exists the greatest probability of a complex effect in terms of energy accumulation occurring, through the amassing of several activated particles.

Table 7.16. Release of metal fires in the oxygen flow in dependence on the experimental conditions in a pipe system fitted with bends [207].

Solid added	Velocity range Δv, m/s	Pressure range $\Delta P \cdot 10^{-5}$, Pa	Number of experiments + sparks observed − sparks not observed
Quartz sand	30···36	25···29	4 −
	43···57	21···28	3 −
	52···82	18···29	12 −
Mixture of 80% quartz	13	29	2 +
sand and 20% iron powder	28···31	26···29	1 −, 2 +
	28···31	26···29	fire
	42···56	22···29	fire
Mill scale	28···36	26···29	11 −, 2 +
	42···57	22···27	5 −, 9 +
	52···82	18···29	3 −, 4 +
	52···82	18···29	fire

At 200···300 m/s, gas velocities are even higher in oxygen turbocompressors than in pipelines, and today the former are being constructed with a throughput of up some 10,000 Nm³/h. Hence it is not surprising that during the last years repeated accidents

Fig. 7.42. Element of an oxygen compressor destroyed by fire [208].

have especially occurred in these units, which sometimes have been completely destroyed by fire [208]. A section of an oxygen compressor destroyed by fire is shown in Fig. 7.42.

7.3.5. *Conclusions on the prevention of metal fires*

Here only those conclusions will be drawn which especially derive from processes of tribochemical fire ignition.

— Since the oxidizable powder particles existing as sediment inside an oxygen plant or brought by the oxygen flow into the plant may initiate a burning during their movement through the pipelines or with intensive mechanical stress in the turbo-compressors, in order to prevent a fire, the presence of not only coarse pieces of metal such as cuts but also of finest powders and dusts has to be excluded.

— Since the mechanical treatment of solid particles, particularly of iron alloys in an inert gas atmosphere results in an increase of reactivity which in turn increases the ignitability of the activated particles, greater dangers arise during the starting up of oxygen plants in a nitrogen atmosphere when they are switched over to pure oxygen atmosphere. Hence the flushing of oxygen plants is better carried out with air or an oxygen-nitrogen mixture.

— When an oxygen plant is constructed, only materials and metals should be used for particularly endangered parts of machines and plants, which have an as low as possible ignition probability.

— Corrosion processes have to be excluded inside oxygen plants, which lead to the formation of freely mobile rust particles and create a risk of tribochemical fire ignition.

In practice experience has shown that the ignition probability may be significantly reduced by taking into consideration these factors and that it is possible to control technically the systems of an oxygen plant which are chemically very metastable (Chapt. 4., [130]).

7.4. The importance of tribochemical reactions in friction, lubrication, and wearing processes

7.4.1. *The technical importance of friction, lubrication, and wear*

A large part of products and basic units is exposed to tribochemical stress when used. The mechanical effect is particularly intensive between moved basic units of machines and equipment acting upon each other, especially in bearings, drive units, and guideways. Mechanical energy is lost at the power train due to the occurring friction. Great losses of materials occur by virtue of the fact that tribomechanically stressed basic units undergo a wear limiting their life time.

With intensification of production, due to increasing stress on technical equipment there is a tendency to ever greater losses of energy and to a reduction in functionality and lifetime of the basic units. Therefore, great efforts have been made to minimize the losses arising because of this. However, about 90 per cent of the basic unit losses today are estimated to be still attributed to the non-control of friction and wear [209]. In view of the immense economic losses resulting from this, amounting worldwide to some 10 milliard Dollars annually [210], advance in this field is of major importance.

7.4.2. Tribochemistry as a subsection of tribology

In order to understand the very complex phenomena and processes occurring with friction and wear, and to make use of them for increasing the reliability of technical systems, a greater integration of various technological and scientific branches and the development of an independent branch "Tribology" has taken place, particularly over the last two decades. The term tribology is based on the following definition [211, 212]:

> *"Tribology is the science and technology of surfaces interacting in relative motion and of related subjects and practices."*

The field of tribology is subdivided into two main branches, namely tribotechnology and triboscience [213]. While tribotechnology covers the technical behaviour of certain basic units and sets, triboscience deals with clarifying the fundamental processes occuring during friction and wearing processes and with the establishment of universally valid laws. This task requires an ever greater involvement physical and chemical approaches and the investigation of the interaction between "tribophysical" and "tribochemical" processes.

Triboprocesses are bound up with the occurrence of friction. Since friction is described by definition [214] as the loss of mechanical energy during the course or at the beginning or end of a relative motion of contacting material regions, the essence of triboprocesses may also be seen in the fact that they run under the influence of mechanical energy or are released by the effect of mechanical energy, and lead to a dissipation of mechanical energy. Consequently tribophysical and tribochemical processes are always connected with the irreversible transformation of mechanical energy accompanied by an increase in the entropy of the total system [215—218]. As the statements given in 4.8.2. and 6.6. show, tribochemical processes may also therefore be described by the relations of irreversible thermodynamics. The designation of a conversion as a "tribochemical reaction" is independent of the way the mechanical energy is applied to the solid, e.g. by slipping, rolling, beating, drilling, impacting, oscillating, etc., since each of these processes is connected with the dissipation of mechanical energy. Therefore chemical reactions initiated mechanically by milling also have to be described as tribochemical reactions.

The difficulty of clarifying tribophysical and tribochemical processes consists in the fact that there are a whole number of elementary processes via which the dissipation of the mechanical energy may run. The most important physical processes occurring upon tribomechanical treatment are given in Tab. 3.3. The part which the individual processes contribute to the total event depends to a major degree on the nature and intensity of the action of mechanical energy. The same applies to tribochemical reactions which for their part are closely linked to the tribophysical processes and depend to a high degree on the chemical properties of the substances acting upon each other.

Tab. 7.17. shows schematically how tribochemistry is integrated into the field of tribology [219, 220]. According to this scheme tribochemistry is subdivided into a number of subsections in which specific processes such as sorption, diffusion, reaction are separately investigated. The respective mechanisms of these processes closely interact with

Table 7.17. Scheme of the position of tribochemistry in the field of tribology [219, 220].

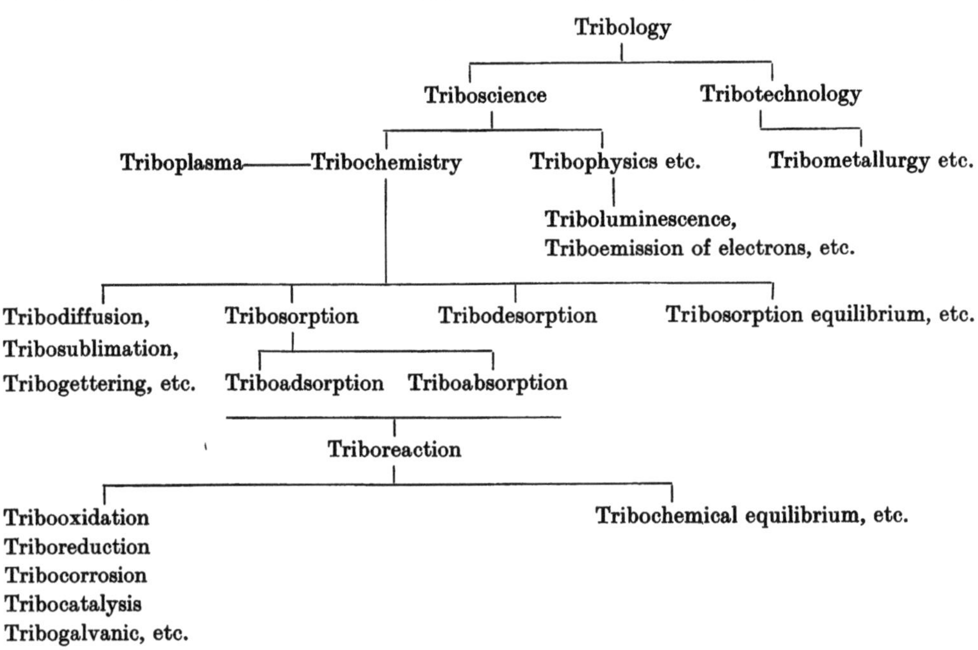

the mechanically initiated tribophysical processes and determine direction and velocity of the chemical reaction. Some aspects of the tribochemical initiation of reactions important to tribological research will be presented in the following section.

7.4.3. *The tribochemical dissipation model*

The reasons for the tribochemical initiation of reactions are discussed in detail in Chapt. 5. and 6. They are comprehensively listed together with accompanying reaction examples in Tab. 7.18. According to 4.1. the individual reasons do not occur independently of each other but fit in with the stepped scheme of energy dissipation (see Fig. 4.1.). Triboplasma is at the top of the hierarchy of excited states.

The stochastic reactions caused by the plasma run extremely fast, but they lead to a very small conversion only. Ultimately "hot spots" will appear via excited vibrational states in the nonequilibrium. Tribochemical reactions run in the relatively slowest way on the mechanically disturbed solid. They are capable of leading to relatively large conversions when the energy required for releasing the reaction is provided by the excess energy of the disturbed solid. Although in mechanically activated reactions there is no direct coupling with the dissipating mechanical energy the former are still ranked among the group of tribochemical reactions. On the one hand the production of the defect structure frozen-in in the solid is an essential part of the dissipation process and on the other hand the chemical effects based on it also influence the total reaction.

As an example, the reaction of iron with sulphur and different sulphureous compounds may be given which is important for the lubrication of bearings. As results from Fig. 7.43. if thermally

Table 7.18. Summary of the reasons for tribochemical reactions.

Reason for reaction	Reaction example	Chapter	Ref.
Triboplasma	$Au + \frac{3}{4}CO_2 \rightarrow \frac{1}{2}Au_2O_3 + \frac{3}{4}C$	5.1.	Chapt. 5., [2]
Elastic stress	Degradation of polymers	5.2.	Chapt. 4., [103]
Vibration states in thermal non-equilibrium	Reactions in crack tips	5.3.	Chapt. 5., [23, 24], Chapt. 4., [33]
Vibration states in thermal equilibrium (hot spots)	Release of explosion reactions	4.8.2.	Chapt. 4., [12]
Triboemission of electrons	$Ni + 4CO \rightarrow Ni(CO)_4$	3.3.5.	Chapt. 4., [33]
Triboelectrical excitation	Grafting of monomers on solid surfaces	5.4.	Chapt. 5., [50]
Surface alterations	$Ni + 4CO \rightarrow Ni(CO)_4$	5.5.	Chapt. 5., [106]
Production of fresh surfaces	$Ni + 4CO \rightarrow Ni(CO)_4$	5.6.	Chapt. 5., [106]
Transport effect	Reactions between several solids: $Ni + J_2 \rightarrow NiJ_2$	5.6.	Chapt. 4., [47]
Plastic deformation, Migration of lattice defects	Tribodiffusion of hydrogen in solids	5.7.	Chapt. 4., [46]
Structural effects	$Ni + 4CO \rightarrow Ni(CO)_4$	5.9.	Chapt. 5., [106]
Tribocatalysis	$Ni + 4CO \xrightarrow{H_2S} Ni(CO)_4$	6.2.	Chapt. 6., [150]

initiated, the reaction of the individual compounds commences at different temperatures dependent on their reactivity.

At the same time Fig. 7.43. shows that these temperatures are closely related to the load-carrying capacity of the sulphides used as lubricants. If, however, mechanically activated iron powder is reacted with the same sulphuric compounds a considerable reduction in the temperature at the beginning of the reaction occurs through the facilitation of the the material transfer. The differences of temperature at the beginning of the reaction in activated and non-activated metals have almost the same quantity as the differences between the most reactive and most inactive sulphuric compounds studied [221].

The reaction of copper(II) oxide with diphenyldisulphide is also very greatly accelerated by the mechanical activation of the oxide (comp. curves 1 and 2 in Fig. 7.44.).

The "hot spots" appearing in the dissipation process have been especially frequently investigated (Chapt. 4., [1], [223]). They represent the thermal portion of energy in the dissipation of mechanical energy. However, as has already been stressed in 4.1. "hot spots" are only one of the possible reasons for the tribochemical initiation of reactions. They may be overlapped in their effect by other causes of reactions as the numerous examples given in 4.1. show [224].

Tribochemical and thermally initiated chemical reactions are often compared in the analysis of chemical reactions on mechanically treated solids (e.g. [225]).

Belonging to the latter are those reactions whose course is determined by the ambient temperature or by the temperature developing stationarily in the contact zone of two solids acting upon each other e.g. in the contact area of a slide bearing, or in a milling vessel. These static or quasi-static temperature conditions make the use of reversible thermodynamics possible by delineating certain spaces as a closed system. The thermal states occurring briefly due to the dissipation of mechanical energy re-

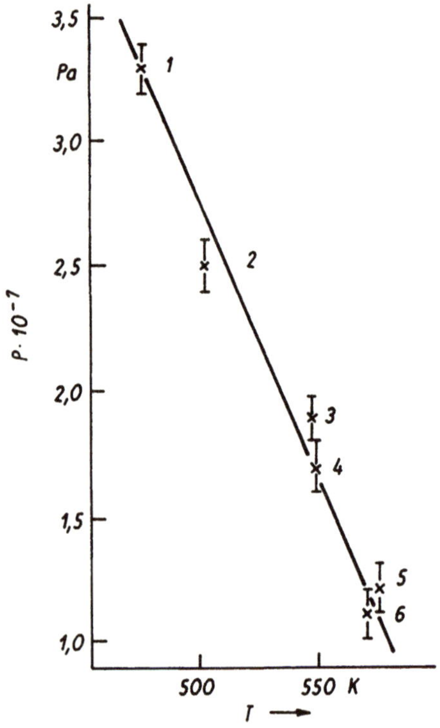

Fig. 7.43. Load-carrying capacity P in dependence on temperature T at the beginning of the sulphidizing [221].

1 — sulphur, 2 — dibenzyldisulphide, 3 — m, m'-dichlorodiphenylsulphide, 4 — diphenyldisulphide, 5 — dihexylsulphide, 6 — diphenylsulphide.

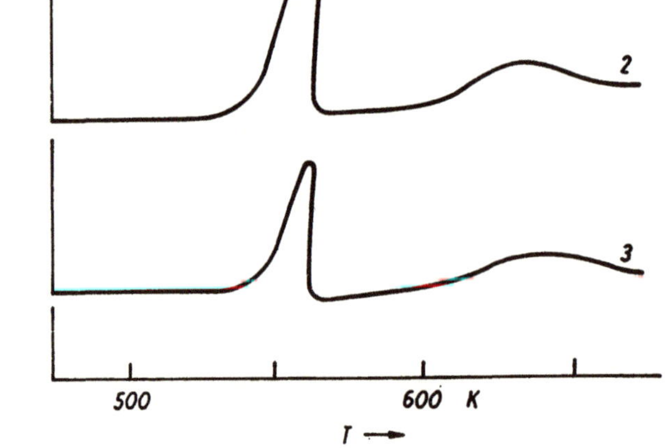

Fig. 7.44. Differential thermograms of the diphenyldisulphide reaction with copper(II) oxides before and after mechanical treatment [222].

1 — untreated, 2 — treated, 3 — treated and annealed, T — temperature.

quire the application of irreversible thermodynamics. In comparison to the reactions running at constant temperature they often also change the kinetics of the chemical processes [226]. In order clarify the mechanism of tribochemical reactions in a given tribological system, their causes have to be analysed and the parts of the reaction which are based on individual mechanisms have to be studied separately. The characterization of the reactivity of pretreated solids is relatively easy, whereby the milling of metal powders graded with respect to time and intensity having proved to be especially suitable for the manufacture of samples with different activity (Chapt. 5., [106], [222]).

In order to simulate the effect of "hot spots" in tribochemical systems the method of pulse heating by current impulse in the range of 10^{-3} s has been developed [221, 227, 228]. It should be emphasized that the highly excited states existing in thermal non-equilibrium and induced by impact do not appear by this method. As Tab. 7.19.

Table 7.19. Emission of electrons with separate and simultaneous action of strain and great current impulse upon lead and zinc single crystals [229].
Current density $I = 1{,}000$ A/mm², Impuls length $t = 500$ μs, $\Delta T \approx 450$ K, Deformation velocity $v = 0.6$ mm/min.

Metals	Electron emission N, Impulse/s		
	Current impulse action without strain	Strain without current impulse action	Simultaneous action strain and current impulse
Lead	—	300	1,050
Zinc	—	350	940

shows, an emission of electrons will not occur either when this method is used, since the temperature peaks existing in the range of some 100 K are not capable of providing the work function required for the emission of electrons. On the other hand a metal surface emits electrons during the deformation process (see 3.3.5.). The synergistic intensification of the emission effect with the simultaneous action of strain and current impulses is of importance to the initiation of tribochemical reactions (see Tab. 7.19.), [229].

According to Fig. 7.45. the tribooxidation of mechanically stressed metal surfaces runs faster by 2···3 orders of magnitude than that of mechanically untreated surfaces.

According to the statements in Chapt. 5. at least 3 tribochemical effects are involved in the mechanically accelerated formation of an oxide layer, namely the surface effect (5.5.5.), the fresh surface effect (5.6.1.), and tribodiffusion (5.7.). According to Tab. 7.18. at least 6 different tribochemical effects determine the formation of nickel carbonyl measured integrally. It is understandable from this complexity of the reactions that the mechanism of a tribochemical reaction may already change due to the alteration of individual stress conditions that in individual cases relations are even reversed as a result. As an example the dependence on relative humidity of the sliding wear of soft iron is given, which runs in contrary manner with different loads (see Fig. 7.46.), [231]. (For more details of the effect of water on friction and wearing processes see 7.4.4.4.).

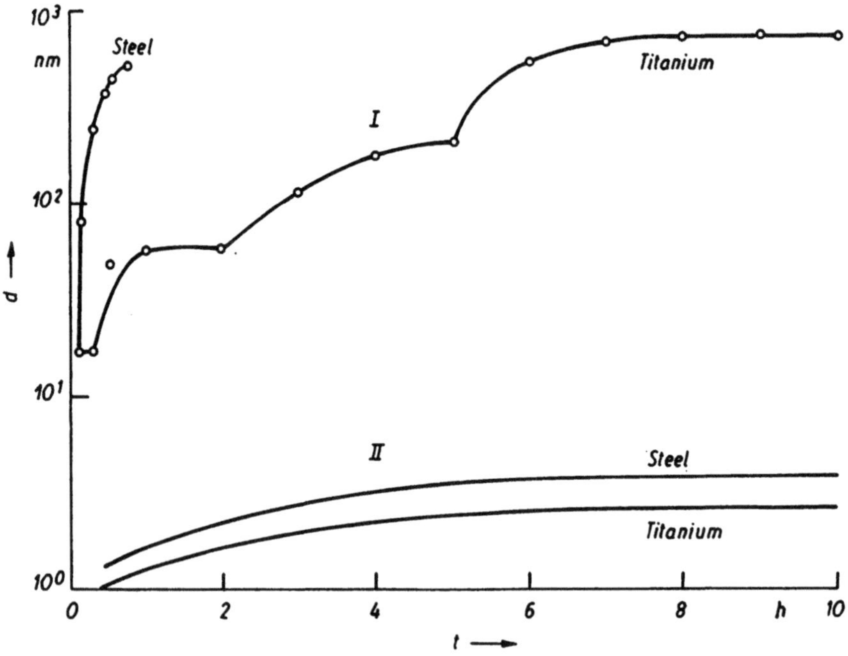

Fig. 7.45. The formation of oxide layers on steel and titanium with (I) and without (II) mechanical treatment [230].

d — thickness of the oxide layer, t — time, I — tribooxidation, II — oxidation by thermal excitation.

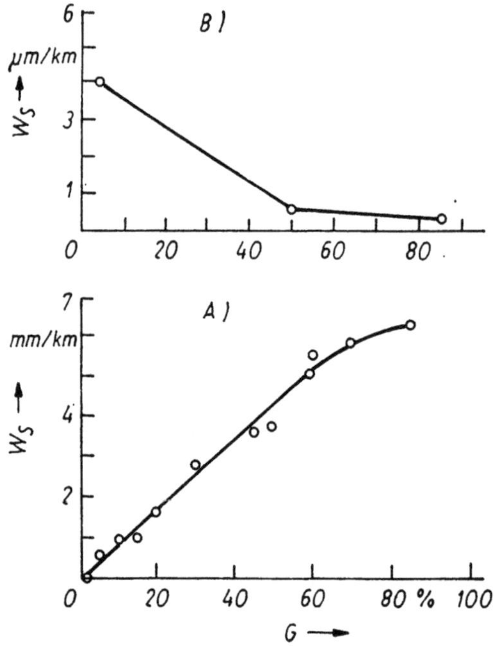

Fig. 7.46. Dependence of the specific wear quantity W_S in the soft iron — soft iron system on the relative humidity G at different load [231].

A — $p = 10^6$ Pa, B — $p = 5 \cdot 10^4$ Pa.

In all it is seen that tribochemical reactions are very closely connected with the excitation processes which run with the dissipation of mechanical energy at the solid [232]. The integral reaction measured at the point of friction mostly consists of several reaction portions based on different mechanisms. The respective dominating mechanism is determined by the specific dissipation process dependent on the kind of mechanical stress, by the pairing of substances, and by external conditions such as pressure and temperature.

7.4.4. *The effect of tribochemical reactions on friction and wearing processes*

7.4.4.1. *The formation of friction- and wear-reducing protective layers on solid surfaces by tribochemical processes*

7.4.4.1.1. *Triboadhesion*

When surfaces of metals are cleaned by reduction with hydrogen or by heating up in the high vacuum, they have a great tendency to adhesion and lead to "fusion welding" upon direct metallic contact. The friction coefficient between metallic fresh surfaces is very high and can exceed the value 50 [233—235].

The strength of such adhesive bonds depends on the metal pairings. The contact of the fresh surfaces leads to a strong triboadhesion when the metal pairings are mutually mixable and capable of forming alloys. The formation of alloy phases with tribomechanical treatment could be detected in the Zn—Cu system (see 6.6.).

If, however, the pairings are not mixable or they only form quite specific compounds the tendency for the materials to weld each other is less strongly developed [236]. The strength of the adhesive bonds depends to a great extent on the position of the metals in the periodic system of elements and consequently on their chemical nature [237].

A tribodiffusion also takes place parallel to the triboadhesion in the mechanically excited phase boundary zone. In respect of crystallites this material transfer mainly leads to the incorporation of other atoms or molecules and by this to the alteration of the chemical composition of the crystalline phase and in individual cases to the new formation of phases as well [238].

7.4.4.1.2. *The formation of protective layers by tribosorption*

If the contact of clean surfaces does not take place in high vacuum but with a pressure of over 100 Pa in a noble gas atmosphere, then with mild friction and impact stress, it is possible to prevent the welding of the surfaces stressing themselves on account of the physical adsorption and triboabsorption of the gas inside the phase boundary layer [239]. As Fig. 7.47. shows, this results however in micro-weldings and continuous material transfer. Surfaces labelled and non-labelled by radioactive isotopes were used. With appropriately long treatment, the same radioactivity appeared on all the surfaces because of a material transfer occurring in both directions [239].

Since noble gases only tribosorb and do not react with the matrix this state is reversible. When the pressure is lowered to 0.1 Pa, a desorption of the triboadsorbed and absorbed gas occurs, and following this, a complete welding of the clean surface already after a short stress time [239].

7.4.4.1.3. *Oxidic protective layers*

In order to effectively repress the welding between frictional surfaces, adherent covering films have to form on the metal surfaces which for their part only have a small tendency to cold welding. This task is fulfilled by reaction layers on metals, particularly oxide layers [241]. Therefore the influence of the ambient atmosphere on fric-

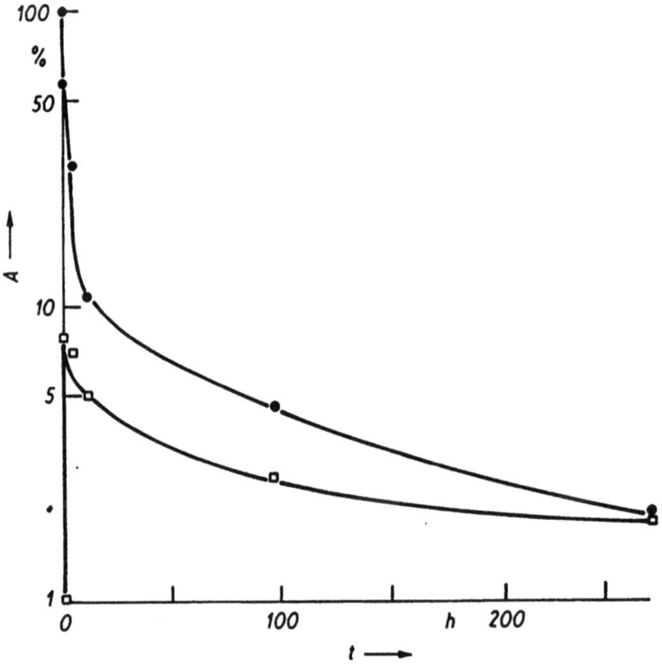

Fig. 7.47. Material transfer by mutual mechanical stress of copper samples labelled and non-labelled by ^{85}Kr [239].

• — radioactivity A of the labelled samples, □ — radioactivity of the initially inactive samples, t — treatment time.

tion and wear is very great. For instance, when an inert gas atmosphere was replaced by an air atmosphere the wear between steel samples rubbing upon each other strongly decreased and could be reduced further by the factor 180 with the change-over from an air atmosphere to a pure oxygen atmosphere [240]. Due to their great hardness oxidic layers are relatively well able to protect the underlying metal layers and to significantly reduce the adhesive interaction. Furthermore oxides represent a barrier to the liberation of dislocations on the frictional surface [243]. The degree of protection provided by the oxidic layers during dry friction depends on surface roughness, thickness of the oxidic film, and in particular on the relative hardness of the oxide and of the metal [243–246].

The effect of air atmosphere on carbon brush wear in the copper-graphite system is particularly marked with small gas pressures. According to Fig. 7.48. this increases considerably at pressures below 10^4 Pa, apparently caused by an insufficient protective layer formation. When the gas pressure is reduced to $0.5 \cdot 10^{-2}$ Pa it will no longer be sufficient for the formation of an anti-wear protective layer and the wear will increase by the factor 10^6.

If following frictional stress at higher pressures sufficient quantities of gas are stored by tribosorption in the phase boundary layers of copper and graphite, then for a certain time (about 2 h) they may prevent a precipitous increase of the wear at the critical pressure of $0.5 \cdot 10^{-2}$ Pa. However, if the system is degassed in the high vacuum for a longer period of time an extremely high wear will occur immediately upon commencement of the frictional stress [247].

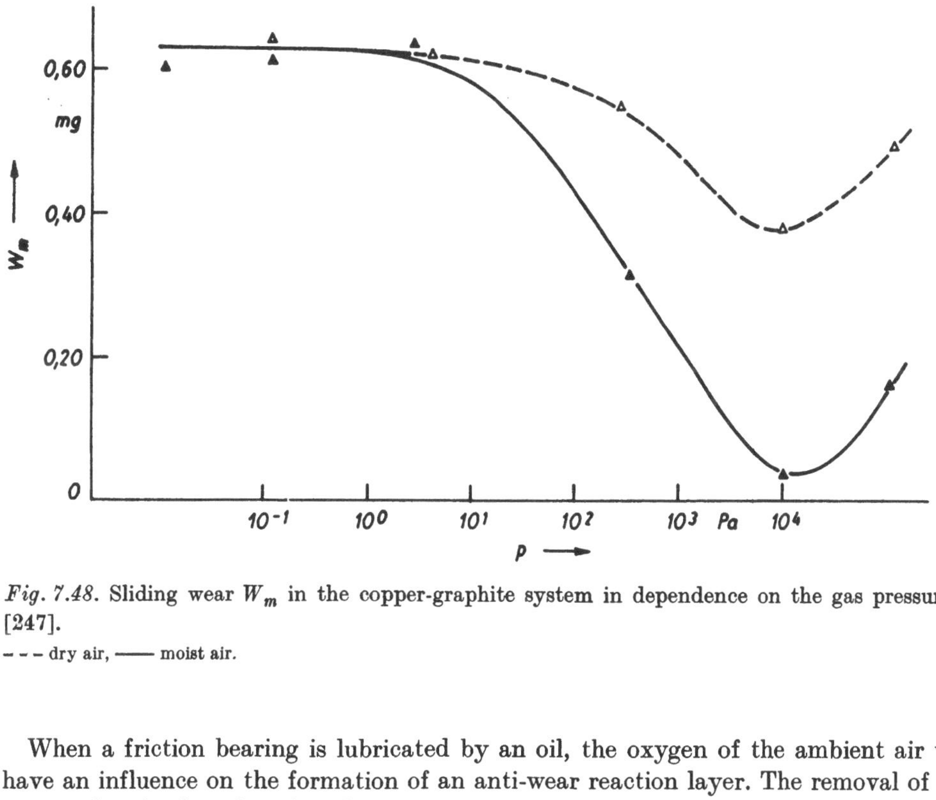

Fig. 7.48. Sliding wear W_m in the copper-graphite system in dependence on the gas pressure p [247].
- - - dry air, —— moist air.

When a friction bearing is lubricated by an oil, the oxygen of the ambient air will have an influence on the formation of an anti-wear reaction layer. The removal of the oxygen dissolved in the oil leads to a precipitous growth of the abrasion and to a lower welding limit, as a selected example showed [248]. The presence of oxygen is also an indispensible precondition for the reaction of EP additives [249]. The load-carrying capacity of the bearing increases with growing oxygen concentration [225].

7.4.4.1.4. *The formation of protective layers through chemisorption and reaction of surface-active substances*

Fatty amines, fatty alcohols, and fatty acids have also proved to be specially suitable for the building-up of anti-wear protective layers. Already as a one-molecular layer they act as a good boundary lubrication and greatly reduce the friction coefficient. Thus a μ-value of 0.09 has been found for highly fluorinated paraffin derivatives during the friction of steel on glass as compared to 1.1. in dry friction [234].

Such layers are capable of reducing the metal transfer by the factor 20,000 in comparison to dry friction [250]. These layers too, are combined with the metallic surface by chemisorption or by chemical reaction via the metallic oxide film and relatively resistant to shearing [251]. By contrast, in stainless high-alloy steels extensive wear already occurs at low stress because the ability of the surface to adsorb lubricants is reduced [252]. On gold or platinum too, there is either a very low adsorption of the additive or none at all [253].

If adsorption heat is used as a measure of the bond strength then in comparing very similar additives, with growing adsorption heat a decrease in the wear can be detected. For instance in an aluminium-steel pairing, C_{12} amines are more strongly bound to the frictional surfaces than alcohols and these in turn more strongly than acids of the same chain length [254]. As Tab. 7.20. shows, amines also reduce the metal-metal contact the most strongly. This finds expression in the portion of aluminium and iron contained in the lubricant [255]. Surface-active substances are also capable of lowering the wear of ceramic products such as steatite by the tribochemical formation of a protective layer. The reaction layer developed tribochemically and bound to the base by strong adhesion leads to a reduction of the μ-value from 0.16 to 0.038 and thus to a considerably improved lubrication. The dislocation migration, the exoelectron emission, and discharge actions by field ionization are discussed as mechanisms initiating the reaction [256].

Table 7.20. Dissolved aluminium and iron in oil during wear [249]. $T = 423$ K.

	C_{12}-acid in paraffinic oil		C_{12}-alkohol in paraffinic oil		C_{12}-amine in paraffinic oil	
	Al	Fe	Al	Fe	Al	Fe
Dissolved Al and Fe in oil, ppm	40	18	21	8	14	4

7.4.4.1.5. *Protective layers by EP additives*

Another group of additives leading to the formation of protective layers upon high loading of the friction bearing, are the EP additives. They consist for the most part of chlorine, sulphuric, and phosphoric compounds reacting tribochemically with the exposed metal during mechanical treatment and developing a well adhering and easy to shear protective layer [257, 258]. The measured contact resistance serves as evidence for the destruction of the protective layer by frictional stress and for the occurrence of metal-metal contacts [259—261].

In order to prevent the appearance of larger metal-metal contact regions and thereby the feared fretting wear, the new formation of the layer by reaction of the EP additives with the friction area has to follow the layer destruction before a further mechanical stress takes place. Hence in order to meet this conditions the additive has to react with the metal with a sufficiently high velocity. As Fig. 7.49. shows, as expected the load-carrying capacity of the friction point investigated also increases linearly with the growing layer formation velocity [221]. Moreover a correlation was detected between the concentration of sulphur on the gear-tooth flanks of a reference load and the specific fretting load attained [262]. Using phosphorus-bearing additives the differences in reactivity are also reflected by their antiwear effect [263]. The load-carrying capacity of organic additives containing chlorine, bromine, and iodine measured as size of the welding load is directly proportional to the velocity of the formation of protective layers.

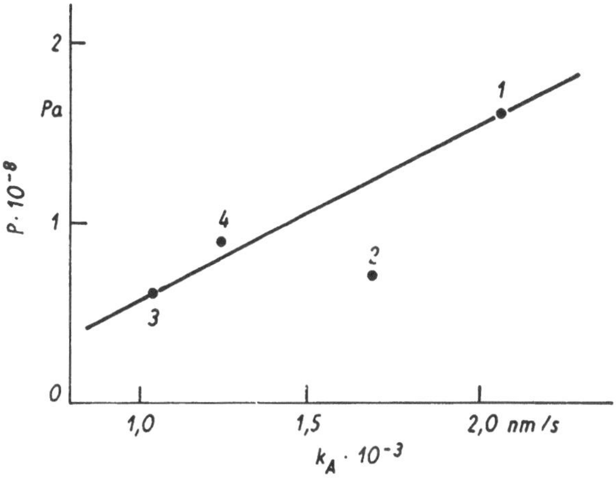

Fig. 7.49. Specific load-carrying capacity P of some organosulphuric compounds in dependence on the initial layer formation constant k_A [221].

1 — dibenzyldisulphide, 2 — diphenyldisulphide, 3 — dibenzylsulphide, 4 — diphenylsulphide.

As Fig. 7.50. shows, a direct proportionality also exists between the load-carrying capacity and the lability of the C-X bond measured as the reciprocal value of the bond energy. Consequently additives having the relatively most labile C-X bonds are best suited for the fast formation of an antiwear protective layer [264]:

The following fundamental ways for the formation of layers are possible [263]:

I. Spontaneous formation of layers from the reactants:

$$RX + Fe \xrightarrow{k_1} FeX + R^{\cdot} \tag{5}$$

II. Formation of layers via the formation of sorption stages:

$$RX + Fe \xrightarrow{k_1'} R{-}X \ldots Fe \tag{6}$$

$$RX \ldots Fe \xrightarrow{k_2'} FeX + R^{\cdot} \tag{7}$$

III. Formation of layers following the preceding decomposition or reformation of the additives:

$$RX \xrightarrow{k_1''} R^{\cdot} + X, \quad RX \xrightarrow{k_2''} R'X + R'' \tag{8a, b}$$

$$X + Fe \xrightarrow{k_3'''} FeX, \quad R'X + Fe \xrightarrow{k_4''} FeX + R''^{\cdot} \tag{9a, b}$$

In the case of multistage reactions the smallest velocity constant determines the formation of the protective layer. For instance for the halogen-containing additives, a spontaneous formation of layers was detected upon impact onto the metal surface, accompanied by the formation of iron halides and residual products [264]. Tribo-induced electrons are said to release the reaction in this [265]. Iron phosphate [266, 267] and iron sulphides [221] are important reaction products in the reaction with

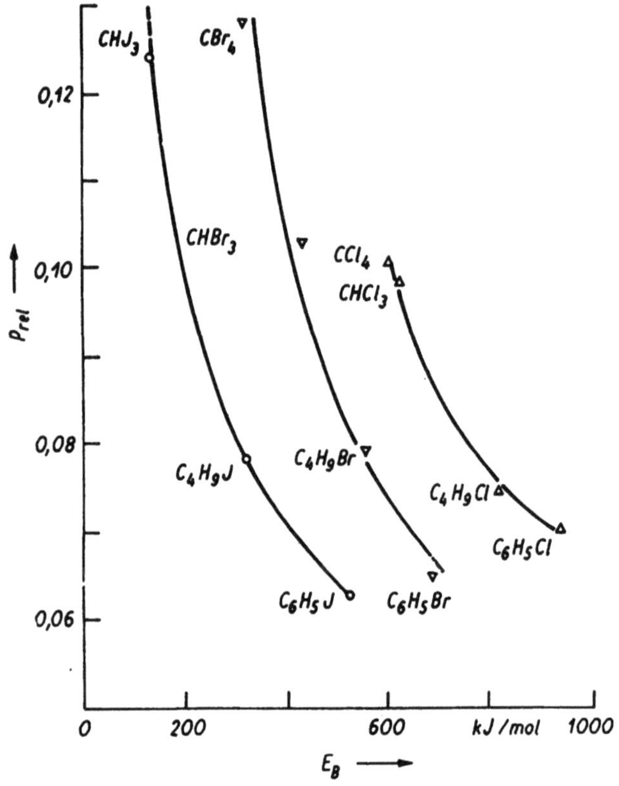

Fig. 7.50. Load-carrying capacity P_{rel} of an anti-wear layer on steel in dependence on the C-X bond energy E_B of the used additive taking into consideration the effect of iron halides [264].

phosphorus-bearing and sulphur-containing additives, respectively. Polar substances which enter into competition with the metal surfaces in the interaction [268] as well as oxygen which passivates the surface by the formation of an iron-oxide layer, decrease the layer formation velocity of the additives [269, 270].

7.4.4.1.6. *High-temperature protective layers*

In engineering lubrication problems at temperatures above 700 K also have to be solved, especially in the heat working of steel carried out about 1,100 K. The metal surfaces are so reactive at this temperature that, for instance, in the application of graphite-oil mixtures a rapid carburization of the peripheral layers occurs. This has disastrous consequences particularly for alloyed material. Apart from materials which mainly act as a parting agent such as slags and glasses [271] common salts and phosphates have proved to be successfull as chemically acting lubricants. Thus common salt very quickly converts into ferrous chloride which together with the iron oxides of the metal surface forms an eutectic non-metallic lubricating interlayer [272]. When phosphates are added eutectic phosphides come into existence for the most part, which as non-metallic interlayers enable a sufficient sliding even at very high temperatures [273, 274]. As in all the lubrication processes connected with a chemical reaction of the lubricant or additive, with the metallic surface of the friction body, a rapid corrosion takes place especially at high temperatures. An advance in this field resulted in particular from the use of relatively small quantities of phosphate making a pure boundary phase lubrication possible with very low wear [275].

At low temperatures too, with chemically reactive additives, the reaction of additive and metal may lead to a strong loss of metal and thus to corrosion. This effect is desired in running-in processes since it leads to a levelling of the roughness with only minor burr phenomena [276]. However in general the use of those additives is sought which are sufficiently reactive to ensure an adequate lubrication even with severe loads on the one hand, but which are not so reactive that they reduce the lifetime of the friction bearing because of their corrosive effect on the other. The selection of chemical substances having reactivities suited to their intended purposes represents the heart of the search for appropriate additives.

7.4.4.1.7. The co-operation of tribochemical reactions in the formation of protective layers by means of solid lubricants

Solid lubricants such as MoS_2 also have a great importance for the lowering of adhesive wear. They are for instance applied in the form of dry films or as a paste, to the sliding material. The protective layers thus coming into existence have a high-anti-wear and anti-friction effect in numerous cases in practice.

On the one hand good lubricating effect of a number of solid lubricants can be attributed to their special crystal structure, e.g. the laminated structure in the case of MoS_2. On the other hand such lubricants must nevertheless have a sufficient adhesion strength in analogy to the above protective layers and resist the shear stresses occurring in the sliding bearing. If the last mentioned condition is not met, or only partially so as is the case with boron nitride, the products do not show any lubrication effect [277]. Thus the lubricating behaviour of MoS_2 also depends very substantially on the pretreatment of the surface to be lubricated and on the method of application of the MoS_2-lubrication film.

In this tribochemical processes play a leading part as solid reactions between MoS_2 and the metallic base [278—281]. As discussed in 5.8., MoS_2 reacts with the metallic base, forming metal sulphide and molybdenum (Chapt. 5., [154]):

$$MoS_2 + 2Me = 2MeS + Mo \tag{10}$$

This reaction can be detected for nickel directly during the mechanical treatment, whereby the developing nickel sulphide catalyzes tribochemically the formation of nickel carbonyl (see 6.2.). As is seen from Fig. 7.51. the chemical reaction of MoS_2 is indicated by an increased chemical reaction [281].

In the reaction of MoS_2 with Fe, the elimination of sulphur from MoS_2 with an activation energy of 100 kJ/mole is the velocity-determining process [279]. At the same time in addition to the release of metallic molybdenum there occurs a tribochemical formation of a relatively acid-resistant iron-molybdenum mixed crystal.

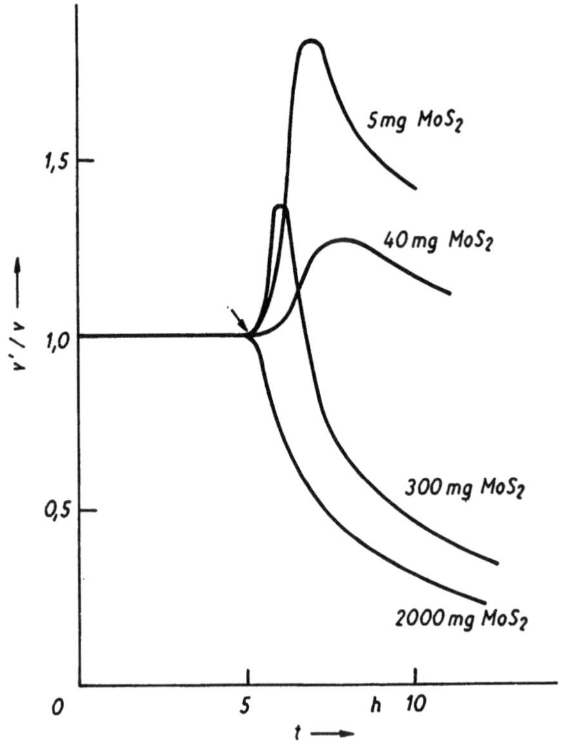

Fig. 7.51. Course of the mechanically initiated formation of nickel carbonyl with the addition of different amounts of MoS_2 [281].

v — steady-state reaction velocity before the addition of MoS_2, v' — reaction velocity, t — reaction time, ↓ — addition of MoS_2.

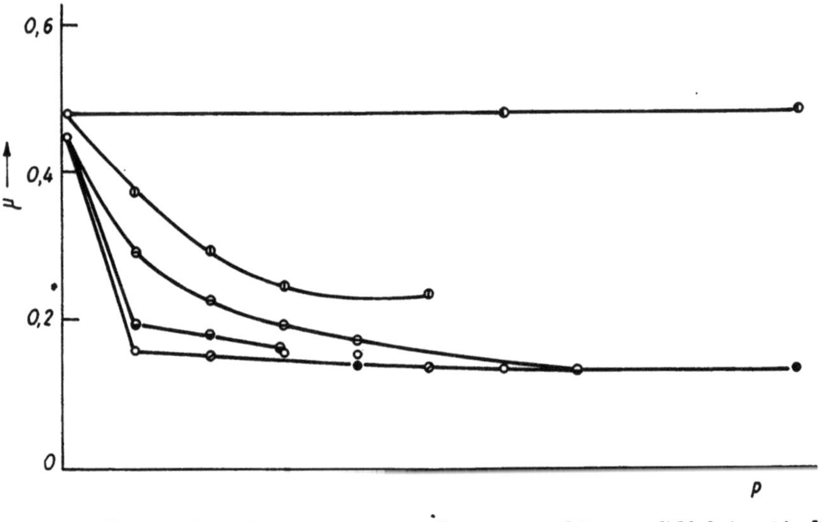

Fig. 7.52. Frictional number μ using spectrally pure graphite as solid lubricant in different gaseous atmospheres [282].

◐ — nitrogen, ⌽ — hydrogen, ⌽ — ethyl alcohol, • — oxygen, ⊖ — water vapour, ◒ — dry air, ○ — heptane, p — gas pressure.

Graphite has a only small lubrication effect in an inert atmosphere, too. By contrast, the lubrication effect of graphite is considerably improved in a gaseous atmosphere with reactive gases such as H_2O, O_2, H_2, etc. (see Fig. 7.52.). In mechanical treatment, graphite and the substantially less reactive silicon carbide could be shown to enter into a larger number of tribochemical reactions with the gases mentioned (see 6.1.1.).

7.4.4.1.8. Polymeric protective layers

Organic lubricants are also involved in the formation of a protective layer and are transformed tribochemically. Thus metallo-organic compounds such as acetylacetonates decompose under frictional stress and precipitate metal oxide [283]. Organic decomposition products are radicals for the most part and have a tendency towards polymerization [284].

Thus in the mechanical stress of aluminium surfaces, e.g. by the rolling of aluminium in the presence of various basic oils, the formation of a polymeric layer could be detected on the freshly-formed aluminium surface. This layer substantially contributes to the wear of the aluminium surface and to the friction [255]. According to 6.1.1. paraffins, being the main components of the basic oil, are tribochemically decomposed up to the separation of carbon and reacted to higher molecular compounds as well. In the presence of precious metals benzene also reacts tribochemically into a high-molecular white substance having an area-like structure [285]. When aluminium is used as a bearing metal the formation of polymers takes place more rapidly and more intensively by using an aromatic basic oil than by using a paraffin-based basic oil, and causes a prolongation of the lifetime of the bearing [286]. Because of organo-chemical reactions in the lubrication gap, the protective layer is often composed in a very complex way and is rich in organic substances [270, 287].

7.4.4.1.9. Protective layers through selective material transfer (GARKUNOV *effect*)

Under special conditions ($T < 353$ K, sliding velocity < 6 m/s) thin metal layers have a strong anti-wear effect, too. These metal layers develop, for instance, during the sliding of certain bronzes on steel in the presence of glycerol-containing lubricants and reduce the wear to a very small value (zero wear) [288, 289]. A thin copper layer develops on the frictional surface of both materials due to the friction of the two materials mentioned. Since this process runs in the presence of an electrochemically active and reducing medium (see 6.6.) no oxides are developed on the surface of the copper. The triboelectrochemical reaction leads to the formation and adsorption of atomic hydrogen on the copper surface and consequently to the blocking of the oxidation process. Simultaneously glycerates come into existence as chemisorbed adsorption layers on the copper whereby the surface-active substance penetrates up to 1 μm into the boundary layer through selective transition and tribodiffusion. Because of the absence of oxide layers good preconditions also exist for the migration of dislocations so that the layer does not consolidate and retains its suitability for a multiple plastic deformation. On the other hand the migration of dislocations is also facilitated by the effect of surface-active agents (REHBINDER effect). The great deformability of copper causes a strong decrease of the friction coefficient. The tribocatalytic formation of polymerization products (see 7.4.4.1.8.) contributes to the protection against a direct contact of the active copper surfaces. Mechanically induced local-anode processes on roughness tips and in stress zones in the copper layers are the basis for triboelectrochemical processes.

The utilization of the GARKUNOV effect has a great economic importance. The potential application of the effect extends beyond the bronze-steel frictional pairing to frictional pairings such as plastic-steel and steel-steel [289].

7.4.4.1.10. *Summary*

It is characteristic of the tribochemical formation of anti-frictional and anti-wear protective layers that this process is self-regulating within certain limits. This behaviour is represented graphically in Fig. 7.51. taking into consideration 5.8. If small quantities of MoS_2 are applied to the metallic frictional surface the protective layer is still incompletely developed and the chemical reaction of MoS_2 takes place on the tribochemically produced fresh surfaces according to Equ. (10). Through this the formation of an increasingly better acting protective layer takes place, which in turn leads to the decline of the tribochemical reaction. This process can be discerned by the course of the reaction curves describing a maximum followed by a dropping as in Fig. 7.51. If the metallic fresh surface are exposed then the tribochemical reaction starts again at the interface.

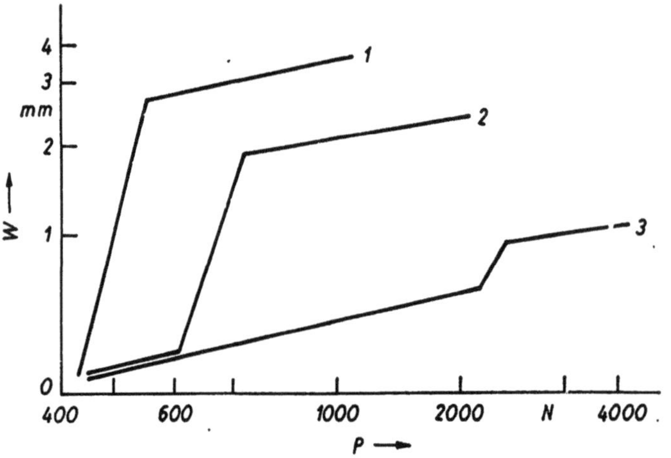

Fig. 7.53. The effect of additives on the wear reduction by the formation of different protective layers [291].

1 — oil without additive, 2 — oil with 1% chlorinated paraffin, 3 — oil with 1.8% chlorinated and sulphurized oleic acid, *W* — wear, *P* — load.

Generally several processes contribute to the formation of a protective layer, e.g. the development of oxidic, polymeric, and other reaction layer. In this way protective layers may be produced with an appropriate combination which remain constant and do not loose their lubricating effect even with deformation of the basic metal [290, 292].

As an example for the combined effect of different additives the degree of wear is given for different applications of additives to a basic oil in dependence on the load used in Fig. 7.53. The relatively smallest wear appears with the combined application [291].

In summary it may concluded that tribochemical reactions play a decisive role in the formation of anti-frictional and anti-wear protective layers, and by the formation of this new phase protect the backing material above all against adhesion wear but beyond this also against surface fatigue and from an external chemical attack [293].

7.4.4.2. *Tribocorrosion*

7.4.4.2.1. *Forms of manifestation of tribocorrosion*

As has been shown in 7.4.4.1.6. tribochemical reactions can also lead to increased wear if the production of anti-wear protective layers caused by them and their wear takes place with a velocity which is too high in relation to the lifetime of the friction bearing. These processes, which are connected with a severe loss of metal are described as "TRIBOCORROSION" [294].

In engineering tribocorrosive wear always appears when the frictional surfaces acting upon each other are in tangential or normal relative vibrational motions with small amplitude. Such processes were first observed at non-positive and positive joints of machine elements such as rolling bearing fits or other press, adhesion, or shrink fits. If there is no relative macroscopic displacement of the contacting interfaces to each other so that the elements remain outwardly in rest to the observer, then this is qualified as "static tribocorrosion". By this an elastic displacement of the interfaces comes into existence in places but also a partial microslipping in individual regions of

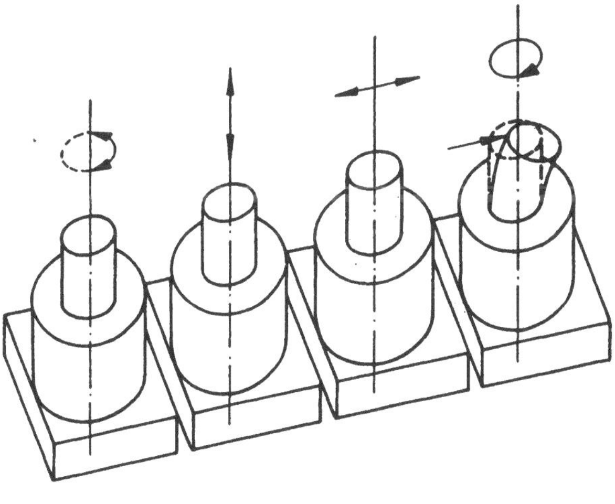

Fig. 7.54. Frequently used shaft-hub arrangements leading to tribocorrosion [296].

the surface. In "kinetic tribocorrosion" however a free oscillating sliding on the whole friction surface occurs [295]. As an example a frequently occurring shaft-hub arrangement is represented in Fig. 7.54. which in the sequences of motion depicted often leads to tribocorrosion [296]. The forms in which such tribocorrosive damage manifests itself are extremely diverse [297]. The formation of corrugation may be presented by ways of example caused by tribocorrosion in the running surfaces of rails and rolling-bearings [298]. Particularly exposed to the danger of tribocorrosion are sliding seats on rotating or vibrating shafts, e.g. in adjustable drives or on the wheel axles of rolling stock [299]. Often oxidation products are obtained in such quantities in vibration-stressed joints and fits that the mobility of the machine elements is lost and the reaction products come out of the mating joints. Since tribocorrosion especially occurs in oscillating relative movements the designation "vibrational frictional wear" has been proposed [300]. However, vibratory movements are not a necessary precondition for tribocorrosion. The latter already appears with the solitary passing of a defined displacement distance in one direction [301] or with a vertical impact [302], (see 3.1.). Tribocorrosion can be expected for every kind of mechanical stress which with sufficient velocity leads to an initiation of the tribochemical reaction.

7.4.4.2.2. Tribocorrosion as a fundamental wear mechanism

Among the wear mechanisms tribocorrosion is directly connected with the course of tribochemical reactions [294]. The following definition is valid for this term [294, 295]: "*Tribocorrosion is the damage to components caused by an increased chemical attack as a result of tribomechanical treatment.*"

In the past the most varied terms were used for tribocorrosion in individual countries. As Tab. 7.21. shows the new term has substantially contributed to an internationally uniform definition.

Table 7.21. Comparison of terms used until now and new terms for the wear caused by tribochemical reactions [294, 295].

	Terms used until now	New terms
German	Reiboxydation, Reibkorrosion, Reibrost, Paßflächenkorrosion, Passungsrost, Druckrost, Edelrost, Preßrost, Transportrost, Spaltrost, Lagerbluten, Schwitzen, Scheuern, Prellen, Schwingungsreibverschleiß	Tribokorrosion
English	Fretting corrosion, fretting damage, contact erosion, false brinelling, seizing, formation of cocoa, wear oxidation, friction oxidation, frettage corrosion	Tribocorrosion
Russian	Коррозия при трении	Трибокоррозия
French	Corrosion par frottement	Tribocorrosion

In the majority of cases tribocorrosion is based on an oxidation process, namely "tribooxidation". This includes in addition to oxidation by oxygen, carbon dioxide, carbon oxide and water, the halogen, sulphide, and nitride forming reactions. Therefore the wearing process discussed here is partially designated as tribooxidation [303]. This term is however too narrowly restricted for a fundamental wearing process, since tribocorrosion is not necessarily to be attributed to a tribooxidation process but can also be based on hydrogenation, carbonylization, and decarburization reactions, too. For instance, tribocorrosion damage appearing in CO pipelines through the formation of carbonyl is not caused by an oxidation process [304]. And the tribooxidation reactions performed in the chemical industry, e.g. lead oxidation in yet mills, also have nothing to do with a wearing process (Chapt. 3., [48]). Finally tribooxidation reactions also participate in the lowering of friction and wear by the formation of covering films.

Consequently "tribooxidation" denotes the mechanically initiated chemical process without taking its effect on friction and wear into consideration, whereas "tribocorrosion" represents the wearing process running on the basis of tribochemical reactions, e.g. of tribooxidation reactions.

7.4.4.2.3. Causes of tribocorrosion

Tribocorrosion may occur both in dry atmosphere and in the presence of an electrolytical conducting medium. As Fig. 7.55. shows, the velocity of the tribochemical reaction may be essentially influenced by this. The triboelectrochemical reactions run with high reaction velocity (see 6.6.). They have to be precluded in friction bearings in order to prevent a rapid tribocorrosion. Only the causes of tribocorrosion occurring in the course of tribooxidation reactions will be discussed in the following.

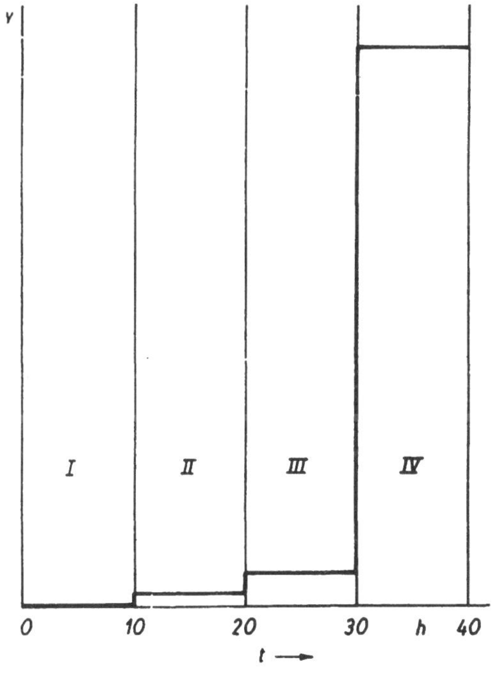

Fig. 7.55. The effect of water on the oxidation velocity v of iron with and without treatment (Chapt. 5., [141]).

I – in O_2 without treatment (tarnishing reaction) II – in O_2 with treatment (tribooxidation), III – in O_2/H_2O without treatment (electrochemical corrosion), IV – in O_2/H_2O with treatment (triboelectrochemical corrosion), t – treatment time.

As already explained in 7.4.3. tribooxidation is a very complex process and can be attributed to several fundamental processes. The tribomechanically determined formation of fresh surfaces leading to a spontaneous reaction with the oxygen of the ambient air is one of the most important initiating factors. The plastic deformation [305] which can always be observed in tribocorrosion processes and leads to the breaking up of the oxidic layer, and the surface fatigue connected with the development of wear particles [306] come into consideration as causes for the destruction of the oxidic covering films (see 7.4.4.3.).

Table 7.22. Tribooxidation of metals in different gas atmospheres [219].

Material	Reaction velocity		
	$mmole_{H_2O}/h$	$mmole_{O_2}/h$	$mmole_{CO_2}/h$
Aluminium	—	3.7	4.2
Iron	0.071	0.076	—

As emerges from Tab. 7.22. the conversion of the fresh surface reaction is almost independent from whether the oxygen exists in free or in bounded form, e.g. as H_2O or as CO_2 (see Fig. 5.43.). According to 5.6.1. the reaction rate is first of all determined by the velocity of the fresh surface formation.

Another process is the growth of an oxide layer on a disturbed fresh surface. In the case of nickel surfaces, at room temperature this is already completed for the most part some seconds after the formation of 4···5 oxygen layers [307]. For iron and copper

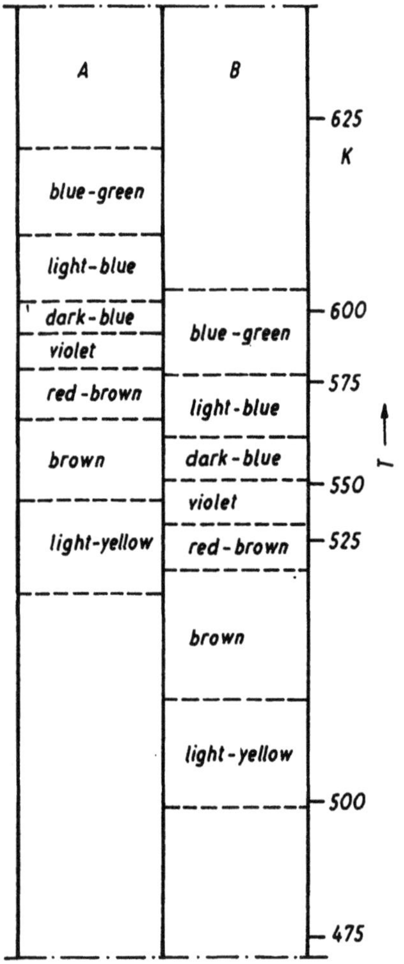

Fig. 7.56. Temper colours on a sheet metal strip made of soft steel and activated mechanically to the half [310].

A — preliminarily annealed, *B* — mechanically activated, *T* — temperature.

the number of the oxygen layers formed is greater [308]. This process may be investigated independently from the treatment by exchanging the inert gas atmosphere for a reactive one after interruption of the stress. In this way, for instance, an increase in weight can still be found in stressed copper test rolls 1,000 min after the mechanical treatment by the action of oxygen [309]. Since the layer grows on a highly disturbed, very finely crystalline base, the conditions are provided for an accelerated diffusion during the growth of layers. As Fig. 7.56. shows this effect is observed at increased temperature up to the appearance of coloured tarnished layers. If the formation of the layer takes place under constant stress the above processes are additionally accelerated by the dissipation of the mechanical energy. It could be seen in fact that considerably more oxygen was converted during the treatment than following it [311].

Another process is the penetration of the gas inside the solid by tribosorption (see 5.7.). As can be seen from Fig. 4.48. in the Cu/CO_2 system the gas still exists in a mainly molecular form in the disturbed boundary layer at the beginning of the treatment. With continuous treatment it reacts with the metal matrix while forming an oxide to establish the tribochemical equilibrium so that an oxidation occurs inside the metal, too.

All the effects finally lead to the manifestations of tribocorrosion known in practice. The way in which the individual effects overlap mainly depends on the tribomechanical conditions and on the type and intensity of the dissipation processes, respectively. The physical state of the corrosion products such as colour, particles size, etc. as well as the chemical composition [312] are also influenced by this.

7.4.4.3. *The effect of tribosorption and triboreaction on fatigue wear*

Surface fatigue is a fundamental wear mechanism occurring for instance in sliding, rolling, or impact stresses [303]. It is characterized by an embrittlement of the layer close to the boundary of the stressed body under the effect of a constant change of multiaxis tensile and compressive stresses, which lead to the formation of cracks and finally to the breaking out or flaking off of coarser particles.

As already discussed in 7.4.4.1.9. thin metallic layers may retain their capacity for plastic deformation for a very long time with mechanical stress under appropriate external conditions, e.g. in the absence of oxide layers. Generally however, in the presence of an ambient atmosphere the mechanical stress leads to tribosorption and triboreaction of gases with the activated solid surface and thereby to a solidification of the boundary layer. As emerges from Fig. 4.44., for instance, CO_2 molecules penetrate the boundary layers up to a depth of some 10 μm and react with the metallic matrix into metal oxide and carbon [313]. The increase in microhardness running parallel to this is shown in Fig. 7.57. It further results from Fig. 7.57. that tribosorbed noble gases also lead to an increase in hardness. According to 5.7. the noble gases are for the most part absorbed by vacancies. Since the latter have a great influence on the mobility of dislocations, the dislocation migration is also influenced by tribosorption [314—316]. However, the tribosorption of noble gases leads relatively quickly to a tribosorption equilibrium and by this to the establishment of a microhardness value which does not change with further treatment. In the case of carbon dioxide however the gas is removed from the equilibrium by tribochemical reaction, and non-metallic impurities are increasingly accumulated in the copper matrix. An embrittlement of the boundary layer results followed by a macroscopic rupture via the formation of submicroscopic cracks.

In general oxygen and nitrogen of the air as well as water and carbon dioxide come into consideration as gases for the tribosorption. The occurrence of oxidation products in the phase boundary layer is one of the most frequent reasons for surface fatigue. But the penetration of nitrogen into the boundary layer has often been observed, too [317, 318].

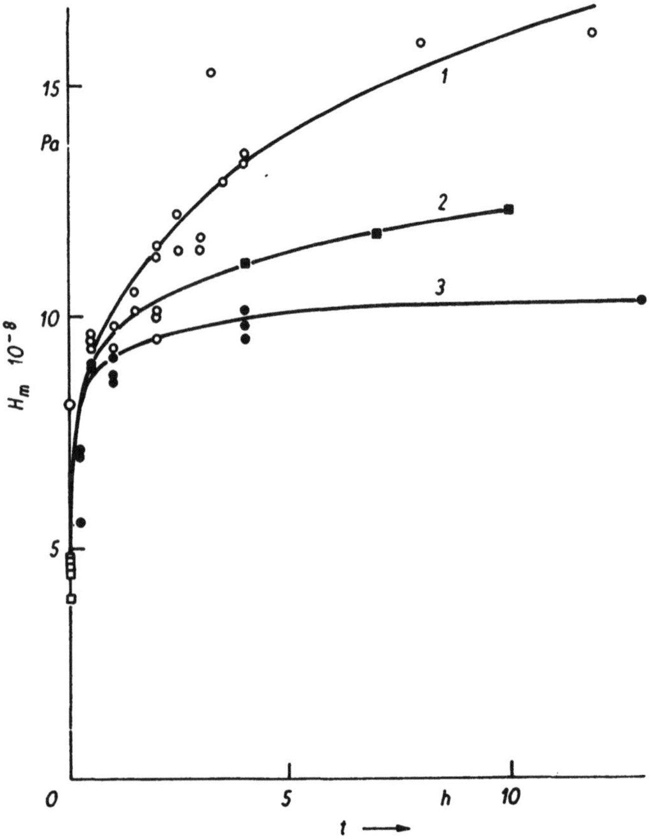

Fig. 7.57. The microhardness H_m of copper in dependence on the treatment time t and on the gas atmosphere [219].

1 — CO_2 (> 10^{16} molecules tribosorbed), 2 — Kr (10^{15} – 10^{16} atoms tribosorbed), 3 — Kr (10^{10} atoms tribosorbed).

Layers consisting of solid lubricants may similarly undergo fatigue wear by tribochemical reactions. As an example the chemical reaction of MoS_2 with an oxidizing atmosphere may be mentioned [319]. The lifetime of the MoS_2 film is so high in a noble gas atmosphere that it meets the current conditions of space travel [320]. As Fig. 7.58. shows, small quantities of oxygen already reduce the lifetime of a MoS_2 film from more than one year to less than one day. The lifetime continues to diminish with increasing oxygen concentration. The destruction process is initiated by blistering, as is seen in

Fig. 7.59. First of all corrosion products are built up by oxidation at individual points leading to an embrittlement of the lubricating film. When the embrittlement has sufficiently spread a flaking off of the MoS_2 film occurs [321].

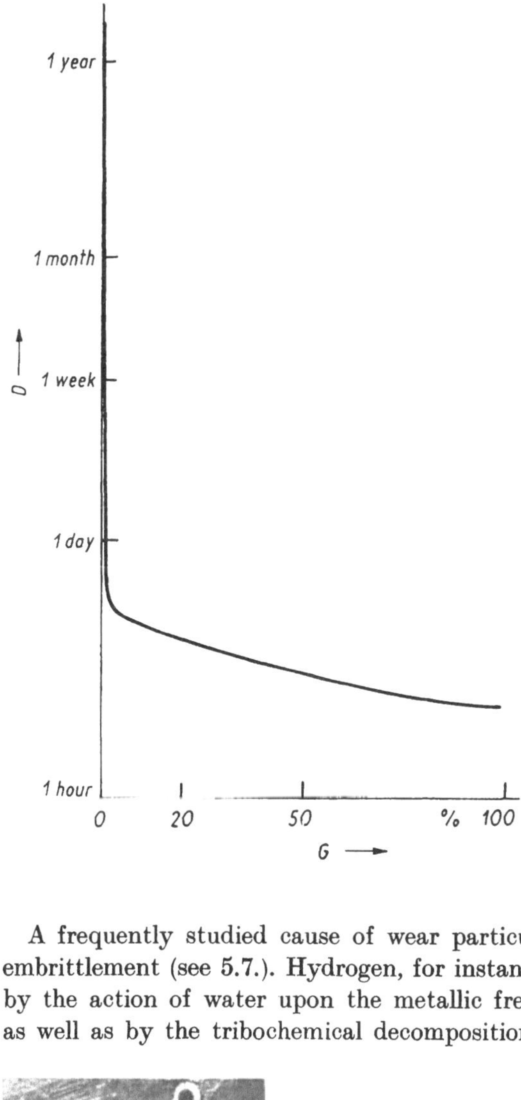

Fig. 7.58. Life time D of a mechanically stressed MoS_2 film in dependence on the oxygen content G of the atmosphere [321].

A frequently studied cause of wear particularly in high-alloyed steels is hydrogen embrittlement (see 5.7.). Hydrogen, for instance, comes into existence tribochemically by the action of water upon the metallic fresh surface (Chapt. 5., [141], [322, 323]) as well as by the tribochemical decomposition of lubricating oil and additives [324].

Fig. 7.59. Blistering on a MoS_2 film by tribo-oxidation reactions [321].

Hydrogen is also capable of reacting tribochemically inside the phase boundary layer. As Fig. 7.60. shows, during treatment hydrogen reacts with the residual carbon of Armco iron up to extensive edge decarburization with the formation of hydrocarbons (see 6.1.1.). This way the methane developed in the bulk phase of the boundary layer plays an important part in the embrittlement of this layer (Chapt. 5., [199]).

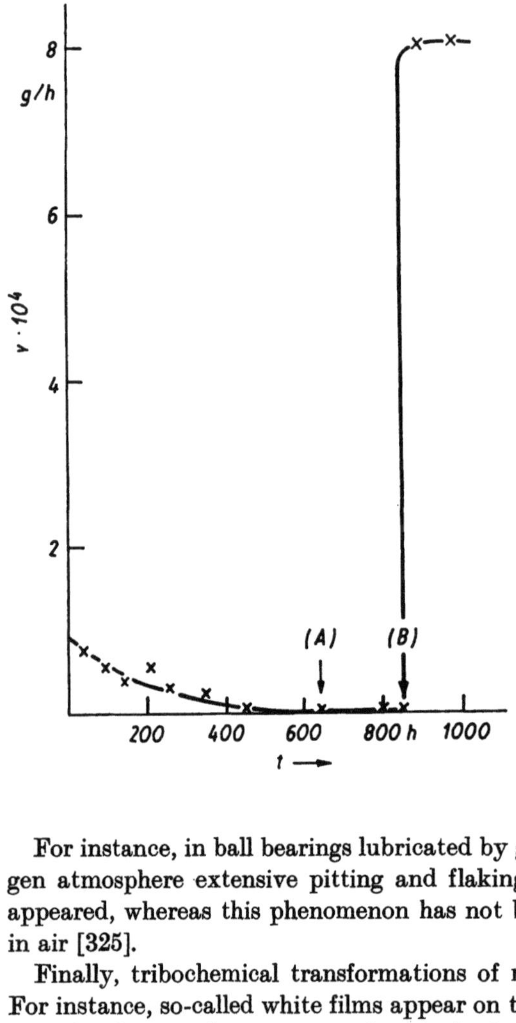

Fig. 7.60. Edge decarburization during the mechanical treatment of Armcoiron with corundum in hydrogen atmosphere (Chapt. 4., [46]).
v – velocity of the methane formation, *t* – treatment time, *A* – corundum changed into fresh corundum, *B* – corundum changed into SiC.

For instance, in ball bearings lubricated by grease, with mechanical stress in a hydrogen atmosphere extensive pitting and flaking off of the material layers on the balls appeared, whereas this phenomenon has not been observed under the same conditions in air [325].

Finally, tribochemical transformations of modifications are also possible (see 4.9.). For instance, so-called white films appear on the treads of railway wheels, which have a great hardness and are connected with the formation of martensite [326].

Altogether the tribosorption processes and tribochemical reactions discussed here contribute to the fact that the boundary layer shows an increasingly marked heterogeneousness, which leads to the embrittlement of the layer and finally to cracks, to pitting, and to the flaking off of the layers.

7.4.4.4. *Tribochemical effects of water on friction and wear processes*

It will be shown by means of the tribochemical interaction of water with mechanically stressed metal surfaces as a model reaction, how tribochemical reactions influence friction and wear in a complex way and how, in accordance with the respective external conditions, such as stress, gas pressure, and given frictional pairing, very different mechanisms underlie these reactions.

It has been found under the most varied stress conditions that both the μ-value and the wear rate depend very sensitively on the whether the investigations have been conducted in the presence or absence of water vapour [327—329] (see Tab. 7.23.). Under certain conditions the wear increases with growing partial pressure of water in the ambient gaseous atmosphere [231, 329, 330]. Under other conditions an increasing partial pressure of water leads to a decrease in the wear rate [231, 330—335], (see Fig. 7.46.).

Table 7.23. Tribocorrosion wear in dry and moist air at tribomechanical stress of different metals following 100,000 frictional vibrations [330].

Material	Atmosphere		Weight loss ΔG, mg
Copper	Air	moist	58.5
		dry	59.1
Steel	Air	moist	6.0
		dry	19.5
Aluminium	Air	moist	5.7
		dry	3.6

In the following, the various mechanisms which form the basis of the actions of water upon friction and wear processes will be discussed:

— Rehbinder effect:

Water is one of the surface-active substances which in certain groups of substances reduce the strength of the mechanically stressed solid on account of the Rehbinder effect thus also changing the wear (Chapt. 5., [121], [336]).

— Sorption of water:

As results from Fig. 7.52. the coefficient of friction is reduced by using spectrally pure graphite as a solid lubricant, if the frictional process is conducted in the presence of water vapour. Its lubricating effect is improved by chemisorption of water molecules to the surface of the graphite. In general strongly adherent adsorbate layers of water change the frictional and wearing processes in a friction bearing and have a more or less marked protective-layer effect.

— Competitive sorption of water:

If oxygen and water vapour interact simultaneously with a mechanically produced fresh surface, then a competitive sorption will occur, leading to a partial blocking of the surface against oxygen adsorption (see 5.6.2.).

The reduction of wear occurring in the tribocorrosion of iron and copper through the effect of moisture is for instance attributed to this effect (see Tab. 7.23.), [330].

— Reaction of water with the metallic base and the oxide layer:

The oxidation of the metal with the formation of metal oxide and hydrogen (Chapt. 5., [141]) and hydrolysis with the formation of hydroxides [335] belong to the most important reactions of water in frictional and wearing processes. Since even gold may be oxidized by tribomechanical treatment [49] the reduction of the coefficient of friction by about one order of magnitude with a water vapour pressure of $3 \cdot 10^3$ Pa is attributed to the adsorption of water and its reaction with the surface of the gold [334].

In moist air, very stable covering films come into existence through the simultaneous reaction of water and oxygen with the metal surface. These films strongly reduce the wear and lower the coefficient of friction as well, as compared to dry friction in air (see Fig. 7.46.). The adherence of the developing protective layer is so good that it can transmit increasing torques (see Fig. 7.46. and Fig. 7.48.), [335]. Furthermore this layer develops more rapidly than a pure oxide layer and counteracts more severely the adhesion wear in this way [219].

— Hydrogen embrittlement:

The hydrogen appearing due to the chemical reaction of water with the metal causes an accelerated surface fatigue via the hydrogen embrittlement (see 7.4.4.3.). Thus, for instance, the lifetime of ball bearings is reduced by 39 per cent through the addition of 0.5 per cent of water in the lubricating oil [337].

In the "severe wear" stage the wear increases with growing relative moisture (see Fig. 46a). Since the abrasion mainly consists of a gray metal powder, fatigue phenomena are apparently involved in the wear process here too [231].

— Hindrance of diffusion by the presence of hydrogen:

The time between the treatment phases in a certain metallic micro-region is decisive for the course of the reaction on surfaces stressing themselves by vibrational motion (frequency effect [338]). Since a certain time is required for the delivery of the gaseous reactant, with a high treatment frequency an incomplete adsorption results [339]. When the tribocorrosion takes place in the presence of water then hydrogen will result from the chemical reaction, diffusing out of the mating groove and thereby hindering the contrary running oxygen delivery and in this way influencing the tribocorrosion velocity (Chapt. 5., [141]).

— Triboelectrochemical processes:

When in the presence of a partial water vapour pressure the dew point is fallen short of, resulting in the development of electrolytically conducting fluid films, then the preconditions are given for the cycle of triboelectrochemical processes (see 6.6.). As Fig. 7.55. shows these reactions run with relatively high velocities.

Generally an overlapping of several mechanisms takes place. Which mechanism is the decisive one for the friction and wear processes is above all determined by the mode and intensity of the teratment and by the material pairing.

It will be the objective of future research to establish quantitative relations for the topics discussed here.

7.4.5. The effect of tribochemical reactions on energy exchange processes

Since chemical reactions are combined with the input or output of energy, they also change the energy balance at the friction point. Thus reactions running exothermally may still increase the concentration of energy at the friction point, whereas desorption processes redue it because they remove energy from the volume of deformation [340]. The parts of energy coming from the chemical processes and required for the initiation of certain processes should not be neglected for an analysis of the energetic conditions at the friction point. Such a process is for instance, chemoemission, discussed in 3.3.5.3.

In summary, tribochemical reactions may be assessed to have the most diverse effects on friction, lubrication, and wear processes. They play a central role in the formation of the surface structure with the formation of the so-called BEILBY layer. The strongly disturbed structure is stabilized by the tribosorption of the reaction products and in this way a high stability of the "run pattern" is effected [341].

Moreover they are decisively involved in the formation of anti-friction and anti-wear covering films.

The locating of solid lubricants on the bearing surface and the reduction of friction and wear thereby effected, also take place with the decisive participation of tribochemical reactions [342]. A deepened knowledge with respect to these processes will contribute to the reduction of friction and wear in machines and plants.

7.5. Final remarks

According to the comments in 7.1.—7.4., tribochemical reactions have already assumed a technological importance in the most varied fields today. Tribochemical reactions will play a growing role with the increasing intensification of industrial production.

In spite of the great breadth of potential applications and the heterogeneous outward appearances of the conduct of tribochemical reactions, the processes discussed in Chapt. 7. may be summarized from uniform points of view according to their nature. They are initiated by the dissipation processes running during the transfer of mechanical energy to solids, which are irreversible and in a series of cases can be described by the application of irreversible thermodynamics. Coupled tribochemical reactions run as pulse reactions in which in particular short-term excitation states determine the reaction events.

Since the efficiency of tribochemical reactions is relatively small the introduction of numerous tribochemical procedures is still balanced against an excessive specific energy consumption.

Advances in this field will above all result from the further development of economically working triboreactors and activation sets. On the other hand it is estimated that a great number of technically realisable possibilities are not yet being used. As could be shown by several examples the utilization of tribochemical reactions not only leads to a further development of known procedures but to completely new procedures which through the application of thermal energy cannot be realized or only with a very great expenditure.

References

[1] Rumpf, H.: „Viertes Europäisches Symposium Zerkleinern", p. 36, Karlsruhe 1975
[2] Brendel, H.: „Wissensspeicher Tribotechnik", p. 15, Leipzig 1978
[3] Bernhardt, C.: „Untersuchungsmethoden zur Charakterisierung mechanisch aktivierter Festkörper", p. 26, Budapest 1978
[4] Bej, V. P.: Chapt. 1., [10], p. R 1
[5] Chapt. 3., [48], p. 240
[6] Ginsberg, W.: Aufber.-Techn. **1** (1960), 86
[7] Ref. [3], p. 28
[8] Bernhardt, C., Gottschalk, J., Heegn, H., Husemann, K.: Silikattechnik **25** (1974), 183
[9] Husemann, K., Bernhardt, C., Gottschalk, J., Heegn, H.: Silikattechnik **25** (1974), 238
[10] Gottschalk, J., Bernhardt, C., Ulbrich, P., Hoffmann, B.: Banicke listy, p. 242, Košice 1976
[11] Chapt. 4., [6], p. 138
[12] Chapt. 4., [141], p. 4
[13] Chapt. 3., [48], p. 249
[14] Köster, A.: Angew. Chem. **69** (1957), 563
[15] Köster, A.: BRD-Patent 1045985, 1067784
[16] Chapt. 5., [110], p. 120
[17] Chapt. 6., [100], p. 648
[18] Beyer, S., Heinicke, G., Kirmes, D., Michel, W., Paudert, R., Peper, F. K., Schuster, G., Wolf, K.: DDR-Patent 72852
[19] Schrader, R.: „Ein neues Verfahren zur Herstellung von Gasreinigungsmasse aus Eisenoxiden durch mechanische Aktivierung", Freiberger Forschungsh. **A 292**, Leipzig 1964
[20] Chapt. 4., [6], p. 93
[21] Paudert, R., Harenz, H., Auerbach, J., Stollberg, B., Richter, H., Stendel, R., Gähde, J., Müller, U., Heinicke, G., Hennig, H.-P., Bräutigam, L., Neumann, E., Berger, K., Aust, R.: DDR-Patent 136848
[22] Chodakov, G. S., Plucis, Z. P.: Dokl. AN SSSR **123** (1958), 725
[23] Šakov, P. I.: „Kačestvennyj analiz rud i mineralov metodom rastiranija poroškov", Moskva 1955
[24] Oengo, H.: Silikattechnik **23** (1972), 246
[25] Heinicke, G., Bock, N., Eckstein, H.: Wissenschaft und Fortschritt **25** (1975), 10
[26] Heinicke, G., Bock, N.: Chapt. 1., [9], II, p. 45
[27] Schrader, R., Petzold, D.: GWF **110** (1969), 1146, 1199
[28] Matthes, F., Möbius, R., Steinbrecher, B., Zingler, F.: Chem. Techn. **23** (1971), 270
[29] Schrader, R., Werner, K.: Chem. Techn. **27** (1975), 156
[30] Schrader, R., Dreier, K., Werner, K.: Chem. Techn. **20** (1968), 475
[31] Schrader, R.: Freiberger Forschungsh. **A 392** (1966), 86
[32] Kubo, T., Miyazaki, T.: Kogyo Kagaku Zasshi **71** (1968), 1307
[33] Molčanov, V. I.: Dokl. AN SSSR **182** (1968), 445
[34] Naeser, G., Scholz, W.: Ber. dtsch. keram. Gesell. **39** (1962), 109
[35] Naeser, G., Fiedler, A.: Arch. Eisenhüttenwesen **38** (1967), 347
[36] Naeser, G., Dautzenburg, N.: Arch. Eisenhüttenwesen **40** (1969), 297
[37] Hint, J. A.: Chapt. 1., [9], I, p. 18
[38] Schrader, R., Westphal, H.: Chapt. 1., [7], p. 19
[39] Heinicke, G., Bock, N.: DDR-Patent 73516
[40] Chapt. 4., [6], p. 53
[41] Chapt. 4., [6], p. 102
[42] Merson, G. A., Michajlova, V. G.: Svetnye metally **40** (1967), 69

[43] MERSON, G. A., CHAVSKIJ, N. N.: Metallurg. i Toplivo (1961), 36
[44] TARANCOVA, M. I., ČAJKINA, M. V., KOLOSOV, A. S., BOLDYREV, V. V.: Chapt. 1., [11], p. 143
[45] SCHULZ, M.: unpublished
[46] LIEBSCHER, S., TRINKS, W., KÖRLIN, P., PAUDERT, R.: Neue Hütte **13** (1968), 27
[47] MÖBIUS, R.: Chapt. 1., [10], p. 50
[48] SCHRADER, R., MARCY, J., KOSTMANN, D.: Z. Chem. **7** (1967), 437
[49] Ref. [3], p. 36
[50] PETZOLD, A.: Chem. Techn. **29** (1977), 391
[51] PETZOLD, A., UNG, L., HENNEK, H.: Silikattechnik **28** (1977), 4
[52] BERNHARDT, C., HEEGN, H.: Silikattechnik **29** (1978), 373
[53] BERNHARDT, C., HEEGN, H.: Banicke listy, p. 296, Košice 1976
[54] ŠUMSKAJA, L. G., LEMINA, N. M., GUSEV, G. M.: Chapt. 1., [9], III, p. 79
[55] SZANTHO, E. v., SCHULTESS, W.: Erzmetall **22** (1969), 465
[56] SCHRADER, R., RUMP, H., KRESSNER, R.: Chem. Ing. Techn. **39** (1967), 14
[57] JUSUPOV, T. S., MOLČANOV, V. M., KIRILLOVA, E. A., KORNEVA, T. A., LAPUCHOVA, E. S.: „Fiziko-chimičeskie issledovanija mechaničeski aktivirovannych mineral'nych veščestv", p. 74, Novosibirsk, 1975
[58] GREGG, S. J., PARKER, T. W., STEPHENS, M. I.: Appl. Chem. **4** (1954), 666
[59] TAKASAKA, A., DAIMON, Y.: J. Soc. Mat. Sci., Japan **21** (1972), 225
[60] PAUDERT, R., LIEBSCHER, S., TRINKS, W.: Chem. Techn. **19** (1967), 356
[61] KIRILLOVA, E. A., KRIVOPUCKAJA, L. M., JUSUPOV, T. S.: Chapt. 1., [8], p. 236
[62] PAUDERT, R., LIEBSCHER, S., TRINKS, W.: Mber. Dt. Akad. Wiss. **9** (1967), 66
[63] JUSUPOV, T. S., KRIVOPUCKAJA, L. M., KIRILLOVA, E. A.: Ref. [57], p. 37
[64] CILOGIN, C.: Dissertation, Berlin-West 1974
[65] Chapt. 2., [29], p. 672
[66] GUSEV, G. M., NOVGORODOVA, S. V.: Ref. [57], p. 46
[67] BIAGANDI, S., PIETSCH, H.: Erzmetall **29** (1976), 73
[68] GOSH, S. K.: Dissertation, Berlin-West 1973
[69] VEZINA, J. A.: Can. mining metallurg. bull. **66** (1973), 733
[70] PAVLIK, F.: CS-Patent 2451565
[71] Ref. [26], p. 89
[72] BEJ, V. P.: Chapt. 1., [9], III, p. 75
[73] SZANTHO, E. v., GERLACH, J., GOCK, E.: Chem. Ing. Techn. **41** (1969), 849
[74] GOCK, E., SZANTHO, E. v.: Erzmetall **23** (1970), 165
[75] BARTHOLOMÄ, H. D., SCHWIETE, H. E.: Ziegelindustrie **13** (1960), 97
[76] Chapt. 5., [92], p. 410
[77] UNG, L.: Diplomarbeit, Freiberg 1975
[78] PAUDERT, R., LIEBSCHER, S., KÖRLIN, P., TRINKS, W.: Chem. Techn. **20** (1968), 276
[79] ESNA-ASHARI, M., KAUSEL, E., NISSEN, R., PASCHEN, P.: Erzmetall **30** (1977), 262
[80] PAUDERT, R., HARENZ, H., HEINICKE, G., SCHUMANN, H.: Chem. Techn. **31** (1979), 27
[81] VOLFKOVIČ, S. I.: Chimija v sel'sk. chozjajstve **14** (1976), 23
[82] MÜLLER, S., HAGEMANN, O., ANSORGE, H., JANERT, R.: Arch. Acker- und Pflanzenbau u. Bodenkunde **18** (1974), 10, 721
[83] BOLDYREV, V. V.: Wissenschaft u. Fortschritt **28** (1978), 17
[84] HARENZ, H., PAUDERT, R., KLUDSSUWEIT, H., LINKE, E.: Arch. Acker- und Pflanzenbau u. Bodenkunde **23** (1979), 7
[85] PAUDERT, R., HARENZ, H., PÖTHIG, R., HEINICKE, G., DÜNKEL, L., SCHUMANN, H.: Chem. Techn. **30** (1978), 470
[86] SCHEFFER, F., ULRICH, B., FASSBENDER, H. W.: Z. Pflanzenernährung, Düng. Bodenkunde **113** (1966), 1
[87] HARENZ, H., PAUDERT, R., MÜLLER, S., LINKE, E., MÜLLER, H.: Arch. Acker- und Pflanzenbau u. Bodenkunde **23** (1979), 707

[88] Heinicke, G., Paudert, R., Harenz, H., Steinike, U., Pöthig, R.: Ž. prikl. chim. **50** (1977), 969
[89] Schrader, R., Kölling, W.: DDR-Patent 50043
[90] Lasiewicz, K.: Przemysl chemiczny **47** (1968), 76
[91] Paudert, R., Harenz, H., Heinicke, G., Steinike, U., Pöthig, R., Richter, H., Schumann, H.: DDR-Patent 107195
[92] Paudert, R., Harenz, H., Schumann, H., Richter, H., Heinicke, G., Steinike, U.: DDR-Patent 119569
[93] Paudert, R., Harenz, H., Heinicke, G.: DDR-Patent 119406
[94] Paudert, R., Harenz, H., Heinicke, G., Schumann, H., Bornemann, H., Müller, S., Kludssuweit, H.: DDR-Patent 138198
[95] Paudert, R., Harenz, H., Heinicke, G., Dünkel, L., Pöthig, R., Kolosov, A. S., Boldyrev, V. V., Tschaikina, M. V., Pamostschikov, S. E., Gordejeva, G. J.: DDR-Patent C05B/205660
[96] Hennek, H., Hofmann, H., Schönborn, N., Schumann, H.: DDR-Patent 125513
[97] Schmalzried, H.: Ber. dtsch. keram. Ges. **42** (1965), 11
[98] Albrecht, R., Blechschmidt, U., Möbius, R.: Chem. Techn. **23** (1971), 473
[99] Schönborn, N., Hofmann, H.: Chem. Techn. **29** (1977), 325
[100] Thinius, K., Hösselbarth, B.: Plaste u. Kautschuk **17** (1970), 474, 567, 653
[101] Selden, P. H.: „Glasfaserverstärkte Kunststoffe“, p. 210, Berlin/Heidelberg/New York 1967
[102] Tolstaja, S. N.: „V. Kongreß über oberflächenaktive Substanzen“, p. 605, Barcelona 1968
[103] Kukowski, E. G.: „Besonderheit des Baus und physikalisch-chemische Eigenschaften von Tonmineralien“, Kiew 1966
[104] Deuel, H., Huber, G.: Helv. **34** (1951), 1967
[105] Konzina, M. V., Kucharskaja, Ė. V., Skorik, Ju. I.: Ž. prikl. chim. **43** (1970), 2058
[106] Frommelt, H.: Plaste u. Kautschuk **20** (1973), 721
[107] Gähde, J., Wittrich, H., Schlosser, E., Friedrich, J., Kaiser, G.: DDR-Patent 106052
[108] Posselt, K.: „1. Intern. Mikrosymposium über Polymerkompositionen“, Berlin 1973
[109] Skorik, J., Heinicke, G.: Silikattechnik **23** (1972), 367
[110] Weyl, W. A., Forland, T.: Ind. Eng. Chem. **42** (1950), 257
[111] Grohn, H., Paudert, R.: J. prakt. Chem. **11** (1960), 63
[112] Takahashi, H., Yui, H., Takayuki, I., Yokkaidi, M.: BRD-Patent 2204707
[113] Momose, Y., Yamada, K., Nagayama, K.: Polymer letters **12** (1974), 623
[114] Schönherr, S., Schrader, R., Steinert, K. H.: Z. anorg. allg. Chem. **342** (1977), 215
[115] Takahashi, H.: Seisankenkyuu **26** (1974), 90
[116] Yasue, T., Arai, Y.: Sekkoo to sekkai **139** (1975), 7
[117] Kucharskaja, Ė. V., Skorik, Ju. I., Gileva, K. G.: Ž. prikl. chim. **40** (1967), 2370
[118] Yasue, T., Aizawa, T., Arai, Y.: Nippon kakuka kaiski (1976), 415
[119] Gähde, J.: Faserforsch. u. Textiltechnik, Z. Polymerforsch. **29** (1978), 5
[120] Yasue, T., Miyamoto, K., Arai, Y.: Sekkoo to sekkai **151** (1977), 261
[121] Skorik, Ju. I., Kucharskaja, Ė. V., Fedoseev, A. D., Klimova, K. P.: Ž. prikl. chim. **38** (1965), 510
[122] Skorik, Ju. I., Heinicke, G., Kučaeva, S. K., Hennig, H.-P.: Ž. prikl. chim. **48** (1975), 1168
[123] Kucharskaja, Ė. V., Vološuk, L. L., Visnevskaja, M. M.: Chapt. 1., [9], p. 124
[124] Michaelis, W.: DRP 14195
[125] Gundlach, H.: „Dampfgehärtete Baustoffe“, Wiesbaden, Berlin 1973
[126] Merkle, G., Ohnemüller, W.: Tonind.-Ztg. **91** (1967), 5, 161
[127] Greenberg, S. A.: J. phys. Chem. **65** (1961), 1, 12
[128] Moorehead, D. R., Mc Cartney, E. R.: "Symposium Autoclaved Silicate Buildings Products", London 1965

[129] RÖBERT, S.: „Silikatbeton", Berlin 1970
[130] KLINGER, S., WOLF, R., KOHL, S., WIEKER, W., WINKLER, A., HEINICKE, G., SIGRIST, K., JOST, H., JEDAMZIK, J., HEMPEL, G., SCHUSTER, D., KÜHN, H. J., WEISSBACH, K.: DDR-Patent 118852
[131] WEYL, W. A., MARBOE, E. C.: Sprechsaal **107** (1974), 23, 1030
[132] MČEDLOV-PETROSIAN, O. P.: „Thermodynamik der Silikate", Berlin 1966
[133] SCHLEGEL, E.: Dissertation B, Freiberg 1976
[134] Chapt. 3., [44], p. 343
[135] HEINICKE, G., JOST, H., KRETZSCHMAR, U.: unpublished
[136] WINKLER, A., WIEKER, W., SIGRIST, K., JOST, H.: „5. Ibausil", Weimar 1973
[137] JEDAMZIK, J.: Dissertation, Berlin 1982
[138] OENGO, H.: Silikattechnik **23** (1972), 246
[139] HEINICKE, G.: Wissenschaft u. Fortschritt **22** (1972), 11; **23** (1973), 32
[140] KÜHN, H. J., KÄHLER, G., WEISE, W., ALTMANN, W., FRACKMANN, W.: „Silikatbeton", Schriftenreihe der Bauforschung 9, Berlin 1970
[141] SAVINKINA, M. A., LOGVINENKO, A. T.: Izv. Sib. otd. AN SSSR, Ser. chim. nauk, **16**, 6 (1974), 141
[142] LOGVINENKO, A. T., SAVINKINA, M. A.: „Mechanochimičeskie javlenija pri sverchtonkom izmelčenii", p. 79, Novosibirsk 1971
[143] BEJ, V. P.: Chapt. 1., [9], III, p. 71, 75
[144] HEINICKE, G., SIGRIST, K., JOST, H., JEDAMZIK, J.: unpublished
[145] HEINICKE, G., SIGRIST, K., JOST, H., JEDAMZIK, J.: DDR-Patent 117437
[146] REBINDER, P. A.: Stroit. mat. **2** (1956), 8
[147] OPOCZKY, L. BEKE, B.: Zement—Kalk—Gips **20** (1967)
[148] KNAPE, G., KÖLLING, W., SCHUMANN, H.: Silikattechnik **18** (1967), 388
[149] SCHEIBE, W.: Silikattechnik **22** (1971), 339
[150] RIMPLER, W., SCHEIBE, W.: DDR-Patent 103227
[151] KUDRJAVCEVA, N. L., CHODAKOV, G. S.: Kolloidn. ž. **33** (1971), 234
[152] CHODAKOV, G. S., KUDRJAVCEVA, N. L.: Ž. prikl. chim. **43** (1970), 1453
[153] KUDRJAVCEVA, N. L., CHODAKOV, G. S.: Silikattechnik **24** (1973), 63
[154] BURTON, T. G.: Trans. Instn. Chem. Engrs. **44** (1960), T 37
[155] BRUTHANS, Z.: Sprechsaal **101** (1968), 106
[156] BRUTHANS, Z.: Technical Digest 8 (1966), 34
[157] BEKE, B., OPOCZKY, L.: „Zweites Europäisches Symposium Zerkleinern", Dechema-Monogr. **57** (1967), 459
[158] BRUTHANS, Z.: „9. Conf. Silicate Ind.", p. 207, Budapest 1968
[159] SCHRADER, R.: Chapt. 1., [4], p. 539
[160] REINSDORF, S.: Schriftenreihe Baustoffe 8 (1956), 1
[161] RIMPLER, W., SCHEIBE, W.: DDR-Patent 103227
[162] GRIMME, H.: Silikattechnik **7** (1956), 303
[163] CHOLIN, J. H.: Silikattechnik **7** (1956), 431
[164] WEIGERT, K.: Bauzeitung **7** (1957), 197
[165] SCHRADER, R.: DDR-Patent 57542
[166] VOIGT, U.: Wiss. Z. der Hochschule f. Architektur u. Bauwesen, Weimar 8 (1969 315
[167] SCHWIETE, H. E., KRÖNERT, W., RENNEN, H.: Ber. dtsch. keram. Ges. **43** (1966), 91
[168] CONEY, J. J.: Belg. Patent 623417
[169] SCHRADER, R., KÖHNKE, K.: Silikattechnik **21** (1970), 47
[170] BARTEL, A.: Baustoffindustrie **16** (1973), 11
[171] KLAUS, G.: BRD-Patent 2136858
[172] KOCH, E., SCHRADER, R.: Gießereitechnik **16** (1970), 406
[173] BALŠIN, M. Ju.: Dokl. AN SSSR **67** (1949), 831
[174] Ref. [13], p. 43

[175] Ref. [34], p. 106
[176] Kneschke, G., Zachmann, L., Heinicke, G., Steinike, U., Hennig, H.-P.: Silikattechnik **24** (1973), 118
[177] Schrader, R., Kneschke, G.: Silikattechnik **18** (1967), 239
[178] Kneschke, G., Schrader, R.: Ber. dtsch. keram. Ges. **46** (1969), 8
[179] Tresvjatskij, S. G.: Ogneupory **23** (1958), 229
[180] Reinhardt, H., Kneschke, G., Harjes, W.: Silikattechnik **24** (1973), 117
[181] Naeser, G., Scholz, W.: Verfahrenstechnik **6** (1972), 299
[182] Kieffer, R., Heimke, G., Wruss, W., Vendl, A.: Tonind.-Ztg. **96** (1972), 313
[183] Paudert, R., Stellenberger, K.: Hermsdorfer techn. Mitt. **11** (1971), 1001
[184] Widmark, H.: Jemshontorets Annaler **143** (1959), 426
[185] Chakladen, A. C. D., Roberts, A. L.: Trans. Brit. Ceram. Soc. **59** (1960), 323
[186] Moriya, Y.: Vacuum **11** (1961), 158
[187] Steier, K., Fuerstenau, D. W.: Chem. Ing. Techn. **48** (1976), 569
[188] Thümmler, G.: „Fortschritte der Pulvermetallurgie", I, p. 329, Berlin 1963
[189] Geguzin, Ja. E.: „Physik des Sinterns", Leipzig 1973
[190] Erenz, A.: Moderne Unfallverhütung **5** (1961), 100
[191] Erenz, A.: Die Berufsgenossenschaft (1957), 448, 500
[192] Kirschfeld, L.: Moderne Unfallverhütung (1961), 23
[193] Kirschfeld, L.: Metall **14** (1960), 792
[194] Kirschfeld, L.: Metall **15** (1961), 873
[195] Freytag, H. H.: „Handbuch der Raumexplosionen", Weinheim/Bergstr. 1965
[196] Selle, H.: Arbeitsschutz (1960), 160
[197] Erenz, A.: Die Berufsgenossenschaft (1963), 1
[198] Kirschfeld, L.: Angew. Chem. **21** (1959), 663
[199] Ref. [197], p. 6
[200] Bammert, K., Milch, R.: Arch. Eisenhüttenwesen **42** (1971), 743
[201] Ref. [195], p. 371
[202] Thiessen, P. A., Zabel, E., Siegling, W.: Abh. dtsch. Akad. Wiss., Kl. Math. Physik, Technik **1**(1967), 179
[203] Thiessen, P. A., Franke, E.: Mber. Dt. Akad. Wiss. **7** (1965), 635
[204] Heinicke, G.: Chapt. 1., [1], p. 140
[205] Wegener, W.: Moderne Unfallverhütung **6** (1962), 57
[206] Wegener, W.: Arbeitsschutz (1962), 164
[207] Wegener, W.: Stahl u. Eisen **84** (1964), 469
[208] Heinicke, G., Selge, G.: Wissenschaft u. Fortschritt **25** (1975), 7
[209] Reschetow, D.: Fertigungstechnik u. Betrieb **22** (1972), 6
[210] Jost, P.: Schmiertechnik u. Tribologie **21** (1974), 2
[211] Committee on Tribology: "The introduction of a new technology", London 1973
[212] Göttner, G. H.: Schmiertechnik u. Tribologie **17** (1970), 285
[213] Czichos, H.: Schmiertechnik u. Tribologie **22** (1975), 126
[214] Ref. [2], p. 13
[215] Knappwost, A.: Schmiertechnik u. Tribologie **20** (1973), 43
[216] Kosteckij, B. I.: Chapt. 1., [11], p. 36
[217] Karaulov, A. K., Kosteckij, B. I., Romanov, V. S.: Chapt. 1., [11], p. 123
[218] Fleischer, G.: Wiss. Zeitschr. TH, Magdeburg **14** (1970), 415
[219] Heinicke, G., Fleischer, G.: Technik **31** (1976), 458
[220] Fleischer, G.: Wiss. Zeitschr. TH, Magdeburg **12** (1968), 493
[221] Meyer, K., Keil, G., Berndt, H., Kloss, H., Essiger, B., Krüger, J., Wagner, K., Homann, K.: Chem. Techn. **31** (1979), 411
[222] Berndt, H., Essiger, B., Meyer, K.: Schmierungstechnik **9** (1978), 144
[223] Krause, H., Christ, E.: VDI-Z. **118** (1976), 517

[224] HEIDEMEYER, J.: Schmiertechnik u. Tribologie 22 (1975), 84
[225] SALOMON, T.: Schmiertechnik u. Tribologie 24 (1977), 114
[226] BOLDYREV, V. V.: Izv. Sib. otd. AN SSSR, Ser. chim. nauk 14 (1978), 3
[227] BARCROFT, F. T.: Wear 3 (1960), 440
[228] SAKURAI, T., SATO, K.: ASLE Trans. 13 (1970), 252
[229] TROITZKIJ, O. A., LINKE, E.: „Symposium für Exoelektronenemission", Rostock 1979
[230] KRAUSE, H., SCHOLTEN, J.: J. Lubr. techn. 100 (1978), 199
[231] UETZ, H.: Werkstoffe u. Korrosion 19 (1968), 665
[232] UETZ, H., FÖHL, J.: Wear 49 (1978), 253
[233] BOWDEN, F. P., YOUNG, J. E.: Proc. Roy. Soc., London A 208 (1951), 311
[234] STUDT, P.: Schmiertechnik 11 (1964), 211
[235] GWATHMEY, A. T., LEIDHEISER, H., SMITH, P.: Proc. Roy. Soc., London A 212 (1952)
[236] ROACH, A. E., GOODZEIT, C. L., HUNNICUTT, R. P.: ASME Transactions 78 (1956), 1659
[237] PURSCHE, G.: Technik 27 (1972), 457
[238] THIESSEN, K.: Mber. Dt. Akad. Wiss. 13 (1971), 46
[239] HEINICKE, G., SCHOBER, E.: Schmierungstechnik 1 (1970), 11
[240] SIEBEL, E., KOBITSCH, R.: „Verschleißerscheinungen bei gleitender trockener Reibung", Berlin 1941
[241] ROWE, G. W., SMART, E. F.: British Journ. Appl. Phys. 14 (1963), 924
[242] KRAGEL'SKIJ, I. V., DOBYČIN, M. N., KOMBALOV, V. S.: „Osnovy raščëtov na trenie i iznos", p. 37, Moskva 1977
[243] WHITEHEAD, J. R.: Proc. Roy. Soc., London A 201 (1950), 109
[244] WILSON, R.: Proc. Roy. Soc., London A 212 (1952), 450
[245] WRIGHT, K. H., SCOTT, H. M.: "Proc. of the Conference on friction and wear", p. 286, London 1957
[246] HIRST, W., LANCASTER, J. K.: Proc. Roy. Soc., London A 223 (1959), 324
[247] PFEIL, B., WAGNER, K.: Schmierungstechnik 3 (1972), 299
[248] NASCH, H.: Schmiertechnik u. Tribologie 15 (1968), 32
[249] TOMARU, M., HIRONKA, S., SAKURAI, T.: Wear 41 (1977), 117
[250] RABINOWICZ, E., TABOR, D.: Proc. Roy. Soc., London A 208 (1951), 455
[251] DUNKEN, H.: Freiberger Forschungsh. A 164 (1960), 35
[252] SOMIN, B. K. H., MATKEVICH, S. L.: Friction a. wear in machinery 14 (1960), 167
[253] TAMAI, Y.: Kogyo Kagaku Zasski 71 (1968), 9
[254] TRIPATHI, K. C., GROSZEK, A. J.: "VI. Intern. Congress on Surface Active Substances", Zürich 1972, p. 199, München 1973
[255] TRIPATHI, K. C.: Tribology intern. 8 (1975), 146
[256] FRISCH, B., KÖPER, M.: „Physikertagung 1974", p. 199, Weinheim 1974
[257] BÜCHNER, R.: Dissertation, Merseburg 1972
[258] ALLUM, K., FORBES, E. A.: ASLE Trans. 11 (1968), 162
[259] LECHNER, G.: VDI-Z. 111 (1969), 269
[260] BAIST, W.: Erdöl u. Kohle 18 (1965), 294
[261] BAIST, W.: Schmiertechnik 13 (1966), 209
[262] BÜCHNER, R., WAGNER, J., WEBER, K., DORRN, L.: „2. Intern. Symposium über Wirkungsmechanismen von Additives", Halle 1976
[263] BÜCHNER, R.: Schmierungstechnik 10 (1979), 71
[264] BÜCHNER, R., WAGNER, J., WEBER, K.: Schmierungstechnik 8 (1977), 268
[265] GRUNBERG, L.: Proc. Phys. Soc. B 66 (1953), 153
[266] STUDT, P.: Erdöl u. Kohle 21 (1968), 340
[267] EBERHARDT, E., KEIL, G., WEBER, K.: Schmierungstechnik 1 (1970), 101
[268] STUDT, P.: Erdöl u. Kohle 21 (1968), 784
[269] WEBER, K., BÜCHNER, R., HÖTZEL, D., WAGNER, J.: Schmierungstechnik 10 (1979), 115
[270] ROUNDS, F. G.: ASLE Trans. 15 (1972), 54

[271] Séjournet, J.: BRD-Patent 961611
[272] Graue, G.: Mineralöltechnik 13 (1968), v. 7
[273] Graue, G.: Abh. dtsch. Akad. Wiss., Kl. Chem., Geolog. Biol. 5 (1966), 99
[274] Graue, G., Lückerath, W., Gebauer, G.: Schmiertechnik 9 (1962), 245
[275] Graue, G., Lückerath, W.: Schmiertechnik 12 (1965), 71
[276] Tynnyj, A. N.: Metalloved. term. metalloobrabot. (1961) 15
[277] Fleischer, G.: Freiberger Forschungsh. A 196 (1961), 264
[278] Streckeisen, K.: Techn. Rundschau 20 (1964), 2
[279] Knappwost, A.: „1. Symposium Festschmierstoffe", München 1965
[280] Göttner, G. H., Bartz, W., Schultze, G. R.: Erdöl u. Kohle 19 (1966), 196
[281] Chapt. 4., [11], p. 89
[282] Rowe, G. W.: Wear 3 (1960), 274
[283] Studt, P.: Schmiertechnik 12 (1965), 28
[284] Ming Feng, J., Perilstein, W. L., Adams, M. R.: ASLE Trans. 6 (1963), 60
[285] Chaikin, S. W.: Wear 10 (1967), 49
[286] Ref. [219], p. 116
[287] Rounds, F. G.: ASLE Trans. 9 (1966), 88
[288] Losseva, R.: Ideen des exakten Wissens (1972), 751
[289] Garkunov, D. N.: Schmierungstechnik 5 (1974), 199
[290] Schoon, Th. G. F., Adler, K., Klette, H.: Schmiertechnik u. Tribologie 17 (1970), 119
[291] Vamos, E.: Schmierungstechnik 7 (1976), 233
[292] Suh, N. P., Saka, N., Jakamuir, S.: Wear 44 (1977), 127
[293] Dimitrov, B.: Schmierungstechnik 6 (1975), 151
[294] Heinicke, G., Harenz, H.: Technik 23 (1968), 236
[295] Ref. [2], p. 324
[296] Neukirchner, J.: Schmierungstechnik 4 (1973), 335, 370; 5 (1974), 21, 51, 77, 106
[297] Bartel, A. A.: Der Maschinenschaden 36 (1963), 105, 171; 37 (1964), 48; 44 (1971), 193
[298] Fink, M.: Glasers Annalen 77 (1953), 342
[299] Ref. [296], 5 (1974), 21
[300] Gänsheimer, J.: Mineralöltechnik (1966), v. 1
[301] Hofmann, H., Schmidt, H. J.: Maschinenbautechnik 10 (1961), 251
[302] Wochnowski, H., Fahrentholz, H.: Schmiertechnik u. Tribologie 17 (1970), 280
[303] Czichos, H.: Erdöl u. Kohle 32 (1979), 432
[304] Chapt. 4., [44], p. 337
[305] Krause, H.: Wear 18 (1971), 403
[306] Hurricks, P. L.: Wear 30 (1974), 189
[307] Harenz, H.: Dissertation, Berlin 1966
[308] Weissmantel, Ch.: Werkstoffe u. Korrosion 13 (1962), 682
[309] Fink, M.: Fortschrittsber. VDI-Z. 5, 3 (1967), 18
[310] Naeser, G., Fiedler, A.: Arch. Eisenhüttenwesen 38 (1967), 348
[311] Ref. [294], p. 240
[312] Barwell, F. T.: "Treatise on material science and technology", Vol. 13, London 1979
[313] Wagner, K., Schober, E., Heinicke, G.: Isotoptechnika (Budapest) (1971), 67
[314] Westwood, A. R. C., Goldheim, D. L., Lye, R. G.: Phil. Mag. 17 (1968), 951
[315] Westwood, A. R. C., Goldheim, D. L.: J. appl. Phys. 39 (1968), 3401
[316] Westwood, A. R. C., McMillan, N. H., Kalyoncu, R. S.: J. Amer. ceram. Soc. 56 (1973), 258
[317] Wiester, H. J.: Arch. Eisenhüttenwesen 9 (1935), 525
[318] Schottky, H., Hiltenkamp, H.: Stahl u. Eisen 56 (1936), 444
[319] Salomon, G., De Gee, A. W. J., Zoat, J. H.: Wear 7 (1964), 87
[320] Bartel, A. A.: VDI-Berichte 111 (1966), 65
[321] Salomon, G.: „1. Symposium Festschmierstoffe", München 1965

[322] SCHATZBERG, P.: J. Lubrication Technology **93** (1971), 231
[323] HOLZHOUSER, W. L., BENNETT, T. A.: ASTM Prsc. **62** (1962), 683
[324] SIMAKOV, Ju. S., MATJUŠENKO, V. Ju.: Chapt. 1., [7], p. 121
[325] IRVING, R., SCARLETT, N. A.: Wear **7** (1964), 244
[326] KRAUSE, H., CHRIST, E., BURCHARD, W. G., CHEN, F. S.: „7. Sitzung des Arbeitskreises Rastermikroskopie", 1975
[327] MAILÄNDER, R., DIES, K.: Arch. Eisenhüttenwesen **16** (1943), 385
[328] DANIELS, R. O., WEST, A. C.: Lubr. Engng. **11** (1955), 261
[329] ROWE, G. W.: R. I. C. Reviews (1968), 135
[330] HEINEMANN, R. W., SCHULTZE, G. R.: Schmiertechnik u. Tribologie **16** (1969), 31
[331] UETZ, H., FÖHL, J.: Schmiertechnik u. Tribologie **18** (1971), 183
[332] MING FENG, J., UHLIG, H. H.: J. appl. Mechan. **21** (1954), 395
[333] NEUKIRCHNER, J.: Schmierungstechnik **5** (1974), 56
[334] KRAUSE, H.: Metall **23** (1969), 672
[335] KRAUSE, H.: Fortschr.-Ber. VDI-Z. **5**, 6 (1968), 1
[336] DAVIDGE, R. W., TAPPIN, G.: Proc. Brit. Ceram. Soc. **15** (1970), 47
[337] ARMSTRONG, E., LEONARDI, S., MURPHY, W.: Lubricat. Engng. **34** (1978), 15
[338] HEINEMANN, R. W.: Dissertation, Hannover 1967
[339] DUNKEN, H.: Z. Chem. **11** (1971), 213
[340] FLEISCHER, G.: Schmierungstechnik **4** (1973), 269
[341] THIESSEN, P. A.: „Erfahrungen, Erkenntnisse, Folgerungen", Berlin 1979
[342] CLAUSS, F. J.: "Solid Lubricant and Self-Lubricant Solids", p. 105, New York/London 1972

8. Tribochemical reactions by geomechanical processes

8.1. Geomechanics

Geomechanical processes have led to fundamental changes on the earth in the course of geological history. The geotectonic movements in the terrestrial crust have played an important part, causing folding, overthrust, and faulting processes. Particularly in orogenesis complete earth strata undergo an intensive deformation. The different elastic behaviour of rocks existing in bonds results in the shearing off and the separation of formerly connected rock masses, as well as changes in structure and texture. At present 8,000···10,000 tectonic tremors in the outer earth strata are registered annually.

Microseismic events [1] occur even more frequently. Also of importance are exogenic forces, such as wind and ocean currents, and the compressive and cleavage effects of freezing water which have led and still lead today to the destruction of mountain rocks, to the formation of fineparticle solid substances and to their deposition as sediments [1]. There is a particularly strong mechanical stress of fine mineral particles in river and ocean currents, especially in surf and tidal zones [2, 3]. The transformation of mountain rocks into finely disperse mineral products which amount to about 80 per cent of the sediment layer today, is considered to be one of the most important geological processes in the formation of the outer strata of the terrestrial crust [4].

The effect of the geomechanical processes mentioned on the release and the acceleration of chemical reactions was insufficiently taken into consideration in the past. However, the mechanical energy appearing in many forms in the terrestrial crust may be assumed to be an important source of energy for the excitation of geochemical processes [5].

The geochemical alteration of carbon is of special interest. Carbon existed in the form of methane, carbids, and free carbon on the primeval earth and was subjected to manifold chemical changes, especially for the formation of organic and living matter, in later geological history. The problems of the origin life on earth connected with this will be discussed in 8.2. Moreover it is supposed that natural gas and mineral oil have also developed tribochemically in the terrestrial crust and are continuing to do [6, 7]. These problems connected with the formation of deposits will be discussed in 8.3.

The use of triboreactors together with model systems has proved to be suitable for the simulation of geochemical processes under laboratory conditions. The triboreactor [8] described in 4.2., the vibration mill (3.1.2.2.2.), (Chapt. 4., [10]), and the planetary mill (3.1.2.2.3.), [9] were used to deal with geochemical problems.

8.2. The role of tribochemical reactions in the origin of life on earth

According to the principle of evolution the first forms of life on earth could only come into existence when a so-called primeval ocean had developed with great quantities of complicated organic compounds in course of geological history [10].

Since amino acids belong to the most important basic components these substances must have already developed on earth in the abiologic-chemical evolutionary period. Therefore the formation of amino acids on the earth is one of the important geochemical processes which have to be clarified in the research on the origin of life. The primeval atmosphere still had a reducing nature in the abiologic-chemical evolutionary period. Apart from larger portions of hydrogen, NH_3, CH_4, and H_2O also existed as the most important gases. Oxygen existed only in bound forms as water, metallic oxide, silicate, sulphite, sulphate, phosphite, and phosphate and carbon as graphite and carbide. Nitrides and ammonium salts come into consideration as nitrogen compounds. Water was for the most part bound as water of crystallization, the rest represented the primeval ocean [11, 12]. The change from the reducing atmosphere into the oxidizing one is said to have appeared only in the course of the Sub-Cambrian, caused above all by the photosynthesis of green plants [13—15]. In the view of numerous authors [16—18] the first amino acids probably developed according to the STRECKER synthesis under the conditions of the primeval earth (Equ. (4)—(6)). There are the following reaction stages in the formation of alanine:

$$CH_4 + NH_3 = HCN + 3\,H_2 \tag{1}$$

$$2\,CH_4 = C_2H_2 + 3\,H_2 \tag{2}$$

$$C_2H_2 + H_2O = CH_3CHO \tag{3}$$

$$CH_3CHO + NH_3 = CH_3CH(NH_2)OH \tag{4}$$

$$CH_3CH(NH_2)OH + HCN = CH_3CH(NH_2)CN + H_2O \tag{5}$$

$$CH_3CH(NH_2)CN + 2\,H_2O = CH_3CH(NH_2)COOH + NH_3 \tag{6}$$

The reactions described as Equ. (3)—(6) already run at little increased temperatures so that the most probable assumption is a thermal activation of these partial processes on account of the thermal conditions on the primeval earth. The conditions are not the same for the formation of HCN and C_2H_2 running according to Equ. (1) and (2). Here high temperatures and a rapid cooling are required, particularly in the preparation of C_2H_2 by pyrolysis of methane, in order to prevent the decomposition taking place with slow cooling or the hydrogenation occurring in the presence of hydrogen.

However, such thermal shocks cannot be expected in nature so that the thermal initiation of these prebiotic processes is not very probable. Therefore non-thermal sources of energy come into consideration for the release of the formation of C_2H_2 and HCN, e.g. the electric discharge, the UV-radiation of the sun, the ionizing rays from space and from the decay of radioisotopes. In this way numerous organic compounds such as HCN, aldehydes, hydroxy and amino acids may be detected in addition to CO, CO_2, and N_2 both in the action of UV-radiation and in a spark discharge into a gas mixture corresponding with the primeval atmosphere [19—21].

The above reactions however may also be released by the effect of mechanical energy, e.g. by ultrasonics [22], by shock waves [23], and by the impact stress of carbides in the presence of appropriate gases [1].

The individual steps for the formation of amino acids under the last mentioned conditions have been particularly intensively investigated. Thus reaction 3 already runs with measurable velocity at room temperature accompanied by the formation of aldehydes when it is tribochemically initiated. As discussed in 6.1.1. the formation of acetylene and of hydrogen cyanide may be detected when silicon carbide is treated in a hydrogen atmosphere and in an NH_3/CH_4 atmosphere, respectively. In this it is seen that HCN already appears tribochemically at room temperature in the presence of the elements hydrogen, nitrogen, and carbon, independently of the form of bonding in which the three elements exist. The experimental detection of the products is difficult, since the latter remain at the mechanically disturbed fresh surfaces and may only be tribodesorbed with a sorption displacement (see 5.6.3.). This effect has no influence on the synthesis of amino acids, since according to Equ. (4)—(6) the further reaction also runs in the sorbed state with mechanical treatment. As Tab. 8.1. shows, amino acids are always developed when all the elements required for it are contained in the reaction system. The amino acid yield may be increased by the addition of small amounts of catalysts, e.g. platinum, iron, nickel. However, without tribomechanical stress under otherwise similar conditions no amino acids appear.

Table 8.1. Tribochemical formation of amino acids in different systems [1].

	System	Treatment time, h	Formation of amino acids
1	SiC + Pt	95	(+)
	$CH_4 + NH_3 + H_2O + H_2$	400	+
2	SiC + Fe	0	—
	$CH_4 + NH_3 + H_2O$	365	+
3	SiC + Ni		
	$CH_4 + H_2O$	345	—
4	SiC + Ni		
	$CH_4 + H_2O + NH_3$	355	+

On the basis of these results the hypothesis that prebiotic processes were also determined by tribochemical factors [5], may well have a high degree of probability.

8.3. The importance of tribochemical reactions in the development of deposits

The importance of geomechanical processes to the release of chemical reactions is not limited to the abiologic-chemical evolutionary period. Tribochemical reactions have probably contributed to the formation of the terrestrial crust in all the evolutionary periods, particularly to the development of deposits.

Thus it is manifest that tribochemical reactions were also involved the genesis of natural gas and mineral oil. Numerous studies indicate that the currently known oil fields are of biogenic origin [22–24]. The fact that n-paraffins represent the main components in mineral oils and n-paraffins with uneven or even C-numbers predominate, see Fig. 8.1. a), b), is put forward as an argument for this. On the other hand mineral oils are also known in which n-paraffins with uneven or even C-numbers do not prevail, see Fig. 1. c), [24]. The arguments presented as proof of the biogenesis of mineral oil are also in applicable to many natural gases. Altogehter there are numerous indications of the existence of abiogenic hydrocarbons both on the earth and in the extraterrestrial region [22, 25]. For instance, hydrocarbons have also been detected in meteorites [26, 27].

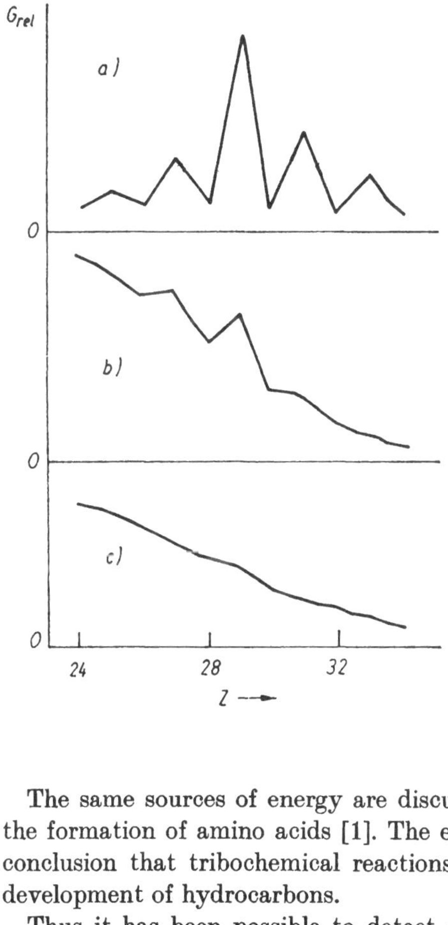

Fig. 8.1. Distribution of n-paraffins in various mineral oils (a, b, c) [34].
G_{rel} – relative content of hydrocarbons, Z – C-number of the hydrocarbon molecules.

The same sources of energy are discussed for the formation of hydrocarbons as for the formation of amino acids [1]. The experimental results shown in 6.1.1. permit the conclusion that tribochemical reactions were and are also involved in the abiogenic development of hydrocarbons.

Thus it has been possible to detect the caming into being of hydrocarbons in the mechanical treatment of carbides (SiC, Fe_3C) and graphite at room temperature and normal pressure in the presence of free hydrogen (see 6.1.1.), [28]. The same result has been obtained in the treatment of carbonaceous substances occurring in nature [2, 7].

The formation of hydrogen from water is possible in a tribochemical manner according to the following equations [29—31]:

$$Fe(Sn, Cd, Al, Cu) + H_2O \rightarrow Fe(Sn, Cd, Al, Cu)\text{-Oxide} + H_2 \quad (7)$$

$$2FeO + H_2O = Fe_2O_3 + H_2 \quad (8)$$

$$3FeS_2 + 12H_2O + 8O_2 \rightarrow Fe_3O_4 + 6H_2SO_4 + 6H_2\uparrow \quad (9)$$

As appears from Tab. 8.2. apart from in the presence of hydrogen hydrocarbons also develop in the presence of oxygen, that is to say under present-day conditions of an oxidizing atmosphere. Tab. 8.2. also shows that hydrocarbons even develop by the action of water upon carbonaceous substances, if there are no other reducing substances in the reaction system.

Table 8.2. Tribodesorption velocity (v_{TDG}) of gases at mechanically stressed SiC (Chapt. 4., [10]).

System	v_{TDG}, μmole/h				
	CO_2	CO	H_2	CH_4	C_2H_6
SiC/O_2	$2 \cdot 10^1$	$< 2 \cdot 10^0$			
SiC/H_2O	$\approx 1 \cdot 10^0$	—	$3 \cdot 10^0$	$6 \cdot 10^{-1}$	$2 \cdot 10^{-2}$
SiC/H_2	—	—	—	$6 \cdot 10^{-1}$	$8 \cdot 10^{-2}$
$SiC/H_2 + O_2$	$1 \cdot 10^1$	$7 \cdot 10^0$		$2 \cdot 10^{-1}$	$< 1 \cdot 10^{-3}$

A results from Figs. 6.7.—6.9. both a straightly dropping curve in analogy to Fig. 8.1. c). and curves with marked maxima according to Figs. 8.1. a) and 8.1. b) are obtained with respect to the distribution of concentration with increasing C-number of the tribochemically developing hydrocarbons. As Fig. 8.2. shows the concentration ratios may well correspond for the low hydrocarbons contained in natural gases and for the hydrocarbons appearing tribochemically under simulated conditions.

The ^{13}C analysis indicates a participation of tribochemical reactions in the development of hydrocarbons on the earth (Chapt. 5., [36]). As Tab. 5.8. shows an isotope effect appears in tribochemical reactions, which leads to a depletion of ^{13}C. $\delta^{13}C$ values between about -25 ‰ and -45 ‰ are obtained for methane of known deposits [32]. Starting out from primeval carbons with a $\delta^{13}C$ value of about -7‰ [33] then a $\delta^{13}C$ value of -16‰ is already reached in one reaction step.

Since the methane is subject to tribochemical decomposition, a $\delta^{13}C$ value of -25‰ can be expected to appear with repeated formation of methane. Consequently tribochemical reactions very probably participated in the ^{13}C depletion in the formation of hydrocarbons of present-day deposits with respect to the isotope ratio occurring in primeval carbon.

The liberation of methane from hard coal is also attributed to mechanical forces, e.g. to a changing orogenic pressure (Chapt. 4., [2]). Thus tectonic rock movements are inter alia held responsible for carbonization processes [34, 35]. The formation of bitumen was also intensively studied in the milling of natural coal products in water and different gases (Chapt. 6., [7]). Solid and fluid bituminous solids extractable by chloroform came into existence, the portion of which accounted for up to 36 per cent of the initial mass of the coal used.

These processes are assumed to run geochemically as a consequence of a mechanical treatment in the finely disperse clay rocks containing carbonaceous substances [6].

As in the hypergenesis of carbon, a reaction release in a tribochemical way is also very probable in many other geochemical processes. Thus it can be shown for instance that the solubility of cassiterite may be markedly increased by a milling process (see Fig. 7.8.). With a corresponding length of milling time, the mineral may already be completely dissolved at room temperature in weakly acid or alkaline media (Chapt. 7.,

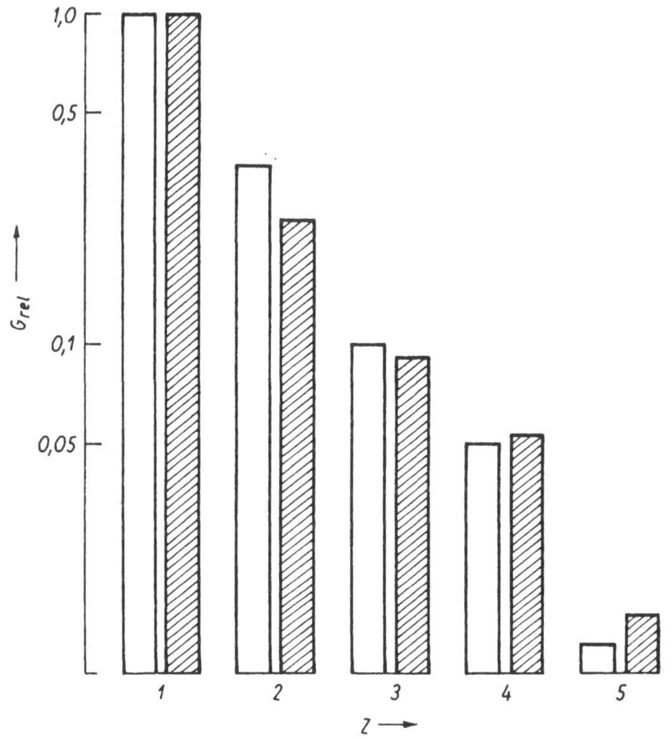

Fig. 8. 2. Example of the composition of a tribochemically produced hydrocarbon mixture (dark columns) and of a natural gas (withe columns) [8].

G_{rel} — relative portion of hydrocarbons, Z — C-number of the hydrocarbon molecules.

[46], [36]). Tin is also assumed to obtain the ability for migration in the terrestrial crust in this way under geological conditions and to react with the formation of tin minerals under hydrothermal and other conditions [37, 28].

The development of elementary gold, silver, copper, mercury, and other metals under exogenic conditions is also attributed to a mechanically released reduction [37].

According to the foregoing tribochemical reactions have apparently generally participated in the geochemical decomposition and new formation of minerals, and contributed to the formation of deposits. The works in the formation of bauxite deposits may be presented as an example [39]. Tribochemical effects may be expected to appear whereover a mutual stress of rocks occurs, or where tribomechanical stress of solids in the presence of gases and liquids occurs. This aspect which has often not been taken into consideration in the past has helped to accelerate the clarification of important geochemical processes, and to amplify and deepen knowledge of the geological evolutionary processes.

References

[1] Thiessen, P. A., Heinicke, G., Hennig, H.-P.: Z. Chem. **11** (1971), 193

[2] Molčanov, V. I., Šugurova, N. A.: „Fiziko-chimiceskie izmenenija mineralov v processe sverchtonkogo izmel'čenija", p. 131, Novosibirsk 1966

[3] Molčanov, V. I.: Dokl. AN SSSR **174** (1967), 1185

[4] Molčanov, V. I.: Chapt. 1., [4], p. 230

[5] Thiessen, P. A.: Z. Chem. **5** (1965), 170

[6] Molčanov, V. I.: Banicke listy, p. 85, Košice 1976

[7] Molčanov, V. I., Goncov, A. A., Andreeva, T. A., Novgorodova, S. V.: „Fiziko-chimičeskie issledovanija mechaničeski aktivirovannych mineral'nych veščestv", p. 125, Novosibirsk 1975

[8] Thiessen, P. A., Hennig, H.-P.: Heinicke, G.: Wissenschaft u. Fortschritt **21** (1971), 205, 247

[9] Molčanov, V. I., Goncov, A. A., Andreeva, T. A., Novgorodova, S. V.: Chapt. 1., [9], III, p. 115

[10] Oparin, A. I.: Die Entstehung des Lebens auf der Erde, Berlin/Leipzig 1949

[11] Böhm, R., Losse, G.: Z. Chem. **7** (1967), 409

[12] Wacker, A.: Angew. Chem. **70** (1958), 519

[13] Urey, H. C.: "The planets", New Naven 1952

[14] Welte, D. H.: Naturwiss. **54** (1967), 325

[15] Vangerow, E. F.: Naturw. Rundschau **20** (1967), 152

[16] Bernal, J. D.: "The Physical Basis of Life", London 1951

[17] Urey, H. C.: Proc. nat. Acad. Sci., Washington **38** (1952), 351

[18] Wald, G.: Sci. Amer. **191** (1954), 44

[19] Miller, S. L.: "Proc. First Intern. Symp. The Origin of Life on the Earth", Moskow 1957, p. 123, London 1959

[20] Bakadur, K.: Nature **173** (1954), 1141

[21] Pavlovskaja, T. E., Passynski, A. G.: Ref. [19], p. 151

[22] Colombo, U.: "Fundamental Aspects of Petroleum Geochemistry", p. 331, Amsterdam 1967

[23] Prinzler, H.: Wiss. Zeitschr. TH Chemie, Leuna—Merseburg **4** (1962), 261

[24] Robinson, Sir G.: Nature **212** (1966), 1291

[25] Klüber, H. v.: „Das Vorkommen chemischer Elemente im Kosmos", Leipzig 1913

[26] Müller, G.: "Sixth World Petroleum Congress", I-29-PD 1, p. 383

[27] Johnson, C. B., Wilson, A. T.: Nature **204** (1964), 181

[28] Thiessen, P. A., Heinicke, G., Senzky, U.: Mber. Dt. Akad. Wiss. **3** (1961), 170

[29] Chapt. 1., [1], p. 122

[30] Molčanov, V. I., Archipenko, D. K.: Ref. [2], p. 86

[31] Molčanov, V. I.: Dokl. AN SSSR **182** (1968), 445

[32] Maass, I., Küstner, E., Hübner, H., Nitzsche, H. M., Schütze, H., Zschiesche, M.: Z. f. angew. Geologie **21** (1975), 536

[33] Galimov, E. M.: „Izotopy ugloroda v neftegazovoj geologii", p. 335, Moskva 1973

[34] Jüntgen, H., Karweil, J.: Freiberger Forschungsh. **A 229** (1961), 27

[35] Gebert, F., Koplin, H., Steiner, H.: Brennstoff-Chem. **45** (1964), 225; **47** (1966), 225, 304

[36] Jusupov, T. S., Molčanov, V. I., Archipenko, D. K., Kirillova, E. A., Krivopyckaja, L. M.: „Mechanochimičeskie javlenija pri sverchtonkom izmel'čenii", p. 135, Novosibirsk 1971.

[37] Molčanov, V. I.: Chapt. 1., [4], p. 231

[38] Molčanov, V. I., Kinė, O. G., Jusupov, T. S., Kirillova, E. A., Krivopyckaja, L. M.: Ref. [36], p. 147

[39] Pavlov, A. L., Molčanov, V. I.: Trudy instituta geologii i geofiziki, AN SSSR, sib. otd., vyp. 249, p. 133, Novosibirsk 1976

Acknowledgements

The author is indebted to P. A. THIESSEN, W. SCHIRMER, and H.-P. HENNIG for their constructive advice when writing this monograph; to E. LINKE, U. STEINIKE and K.-P. THIESSEN for consultation on chapters 4 and 5, to K. SIGRIST for consultation on chapter 4, to M. ROSSBERG for consultation on paragraph 5.3., to J. VÖLTER for consultation on paragraph 6.2., to H. JOST for consultation on paragraph 6.3., to H.-G. JERSCHKEWITZ for consultation on paragraphs 6.4. and 6.5., to A. PINKOWSKI for consultation on paragraph 6.6., to J. JEDAMZIK for consultation on paragraph 7.2.3., and to K. MEYER for consultation on paragraph 7.4.

Theauthor thanks H.-P. HENNIG for his assistance in the scientific editing.

Subject Index

Bei Fragen zur Produktsicherheit wenden Sie sich bitte an:
If you have any questions regarding product safety,
please contact:

Walter de Gruyter GmbH
Genthiner Straße 13
10785 Berlin
productsafety@degruyterbrill.com